D0891897

Human Performance in Planning and Scheduling

Human Performance in Planning and Scheduling

Edited by

Bart MacCarthy
Reader in Operations Management
University of Nottingham

and

John Wilson
Professor of Occupational Ergonomics
University of Nottingham

London and New York

First published 2001
by Taylor & Francis
11 New Fetter Lane, London EC4P 4EE

Simultaneously published in the USA and Canada
by Taylor & Francis Inc.,
29 West 35th Street, New York, NY 10001

Taylor & Francis is an imprint of the Taylor & Francis Group

Printed and bound in Great Britain by
St Edmundsbury Press, Bury St Edmunds, Suffolk

Publisher's Note
This book has been produced from camera-ready copy
supplied by the editors.

British Library Cataloguing in Publication Data
A catalogue record for this book is available from the British Library

Library of Congress Cataloging in Publication Data
Human performance in planning and scheduling / edited by Bart MacCarthy and John Wilson.
p.cm.
Includes bibliographical references and index.
1. Production planning. 2. Production scheduling. 3. Human engineering. 4. Automation-Human factors. I. MacCarthy, Bart (Bart L.) II. Wilson, John (John R.)

TS176. H85 2001 00-051228
658.5-dc21

ISBN 0-7484-0929-7

Contents

Contributors ix
Preface xiii

Part I The Re-emergence of the Domain **1**

1. The Human Contribution to Planning, Scheduling and Control in Manufacturing Industry - Background and Context **3**
Bart MacCarthy and John Wilson, University of Nottingham, UK

2. From Anecdotes to Theory: A Review of Existing Knowledge on Human Factors of Planning and Scheduling **15**
Sarah Crawford, University of Nottingham, UK and Vincent C.S. Wiers, CMG, The Netherlands

3. Lessons from the Factory Floor **45**
Kenneth N. McKay, Memorial University of Newfoundland, Canada

Part II Field Studies of Planners, Schedulers and Industrial Practice **65**

4. A Case Study of Scheduling Practice at a Machine Tool Manufacturer **67**
Scott Webster, Syracuse University, USA

5. Making Sense of Scheduling: The Realities of Scheduling Practice in an Engineering Firm **83**
Sarah Crawford, University of Nottingham, UK

6. Boundaries of the Supervisory Role and their Impact on Planning and Control **105**
Howarth Harvey, University of Sunderland, UK

7. Lingering Amongst the Lingerie: An Observation-based Study into Support for Scheduling at a Garment Manufacturer **135**
Caroline Vernon, University of Nottingham, UK

8. Decision Support for Production Scheduling Tasks in Shops with Much Uncertainty and Little Autonomous Flexibility **165**
Kenneth N. McKay, Memorial University of Newfoundland, Canada and Vincent C.S. Wiers, CMG, The Netherlands

9. Human Factors in the Planning and Scheduling of Flexible Manufacturing Systems 179
Jannes Slomp, University of Groningen, The Netherlands

Part III Plans, Schedules and Computer Systems 199

10. Design of a Knowledge-based Scheduling System for a Sheet Material Manufacturer 201
Vincent C.S.Wiers, CMG, The Netherlands

11. Design and Implementation of an Effective Decision Support System: A Case Study in Steel Hot Rolling Mill Scheduling 217
Peter Cowling, University of Nottingham, UK

12. A Field Test of a Prototype Scheduling System 231
Scott Webster, Syracuse University, USA

13. Architecture and Interface Aspects of Scheduling Decision Support 245
Peter G. Higgins, Swinburne University of Technology, Australia

14. Designing and Using an Interactive MRP-CRP System Based on Human Responsibility 281
Nobuto Nakamura, Higashi-Hiroshima University, Japan

Part IV Context and Environment for Planning, Scheduling and Control 309

15. Assessing the Effectiveness of Manufacturing Information Systems 311
Janet Efstathiou, Anisoara Calinescu, John Schirn, University of Oxford, UK, Lars Fjeldsøe-Nielsen, BizzAdvice, London, UK, Suja Sivadasan, University of Oxford, UK, Julita Bermejo-Alonso, Universidad Alfonso X el Sabio, Spain and Colin J. Neill, Pennsylvania State University, USA

16. Planning and Scheduling in the Batch Chemical Industry 339
Jan C. Fransoo and Wenny H.M. Raaymakers, Eindhoven University of Technology, The Netherlands

17. Engineering a Vehicle for World Class Logistics: From Paradox to Paradigm Shift on the Rover 75 355
Joy Batchelor, Warwick Business School, UK

18. A Socio-technical Approach to the Design of a Production Control System: Towards Controllable Production Units 383
Jannes Slomp and Gwenny C. Ruël, University of Groningen, The Netherlands

19. Planning and Scheduling in Secondary Work Systems **411**
Toni Wäfler, Swiss Federal Institute of Technology, Switzerland

Part V Defining the Future Research Domain **449**

20. Influencing Industrial Practice in Planning, Scheduling and Control **451**
Bart MacCarthy and John Wilson, University of Nottingham, UK

Index 463

Contributors

Dr Joy Batchelor
Warwick Business School, University of Warwick, Coventry CV4 7AL, UK
E:mail: omjb@wbs.warwick.ac.uk

Julita Bermejo-Alonso
Universidad Alfonso X el Sabio, Avda. de la Universidad 1, 28691 Villanueva de la Canada, Madrid, Spain

Dr Anisoara Calinescu
Department of Engineering Science, University of Oxford, Parks Road, Oxford OX1 3PJ, UK

Dr Peter Cowling
School of Computer Sciences, University of Nottingham, Jubilee Campus, Nottingham NG7 2RD, UK
E:mail: Peter.Cowling@nottingham.ac.uk

Dr Sarah Crawford
School of Mechanical, Materials, Manufacturing Engineering and Management, University of Nottingham, University Park, Nottingham NG7 2RD, UK
E:mail: Sarah.Crawford@nottingham.ac.uk

Dr Janet Efstathiou
Department of Engineering Science, University of Oxford, Parks Road, Oxford OX1 3PJ, UK
E:mail: hje@robots.ox.ac.uk

Lars Fjeldsøe-Nielsen
BizzAdvice.com, 17 Abbey Court Road, London NW8 0AU, UK

Professor Jan C. Fransoo
Faculty of Technology Management, Eindhoven University of Technology, PO Box 513, Pav F12, 5600 MB Eindhoven, The Netherlands
E:mail: J.C.Fransoo@tm.tue.nl

Dr Howarth Harvey
School of Computing and Engineering Technology, St Peter's Campus, University of Sunderland, St Peter's Way, Sunderland SR6 0DD, UK
E:mail: howarth.harvey@sunderland.ac.uk

Peter G. Higgins
School of Engineering and Science, Swinburne University of Technology, PO Box 218, Hawthorn 3122, Australia
E:mail: Phiggins@swin.edu.au

Dr Bart MacCarthy
School of Mechanical, Materials, Manufacturing Engineering and Management, University of Nottingham, University Park, Nottingham NG7 2RD, UK
E:mail: Bart.MacCarthy@nottingham.ac.uk

Professor Kenneth N. McKay
Faculty of Business Administration, Memorial University of Newfoundland, St John's, Newfoundland, Canada, AIB 3X5
E:mail: Kenmckay@morgan.ucs.mun.ca

Professor Nobuto Nakamura
Department of Industrial and Systems Engineering, Hiroshima University, 4-1 Kagamiyama 1 chome, Higashi-Hiroshima 739, Japan
E:mail: Nakamura@pel.sys.hiroshima-u.ac.jp

Professor Colin J. Neill
Engineering Department, Great Valley Campus, Pennsylvania State University, Malvern, Pennsylvania, USA
Email: cjn6@psu.edu

Wenny H.M. Raaymakers
Faculty of Technology Management, Eindhoven University of Technology, PO Box 513, Pav F12, 5600 MB, Eindhoven, The Netherlands
Email: w.raaymakers@organon.oss.akzonobel.nl

Dr Gwenny C. Ruël
Faculty of Management and Organization, University of Groningen, PO Box 800, 9700 AV Groningen, The Netherlands
E:mail: g.c.ruel@bdk.rug.nl

John Schirn
Department of Engineering Science, University of Oxford, Parks Road, Oxford OX1 3PJ, UK

Suja Sivadasan
Department of Engineering Science, University of Oxford, Parks Road, Oxford OX1 3PJ, UK

Professor Jannes Slomp
Faculty of Management and Organization, University of Groningen, PO Box 800, 9700 AV Groningen, The Netherlands
E:mail: j.slomp@bdk.rug.nl

Caroline Vernon
School of Mechanical, Materials, Manufacturing Engineering and Management, University of Nottingham, University Park, Nottingham NG7 2RD, UK

Toni Wäfler
Work and Organizational Psychology Unit, Swiss Federal Institute of Technology, CH-8092 Zurich, Switzerland
E:mail: Waefler@ifap.bepr.ethz.ch

Professor Scott Webster
Crouse-Hinds School of Management, Syracuse University, Syracuse, NY 13244-2130, USA
E:mail: Stwebste@syr.edu

Dr Vincent C.S. Wiers
CMG, Luchthavenweg 57, 5657 EA, Postbus JB, Eindhoven, The Netherlands
E:mail: Vincent.Wiers@cmg.nl

Professor John Wilson
School of Mechanical, Materials, Manufacturing Engineering and Management, University of Nottingham, University Park, Nottingham NG7 2RD, UK
E:mail: John.Wilson@nottingham.ac.uk

Preface

Major changes have taken place in organisations across all business and industrial sectors since the early 1980s - downsizing, business process re-engineering, the introduction of new process and information technologies, and globalisation, to name but a few. Change appears ever present. Yet businesses must still deliver their products and services to the marketplace. Our aim in bringing this book together has been to present a fundamental re-examination of the core planning and scheduling processes that make this happen.

The area has traditionally been viewed as primarily technical and largely 'solved'. The reality, as this book testifies, is quite different. Responsive businesses continue to rely on effective planning, scheduling and control processes to compete in tough marketplaces. The underlying theme throughout the book is that improved practice and performance can result only from greater understanding of the nature of these processes. This requires organisational, social and technical perspectives that acknowledge the centrality of people in managing these processes - a human centred perspective.

The domain is complex and presents many challenges. We have endeavoured to select studies that take different perspectives. The international group of contributors span a wide range of disciplines - management science, operations management, ergonomics, human factors and work psychology, industrial engineering and computer science. The emphasis throughout is on the reality of practice. Our aim has been to produce a book that would be valuable to a wide readership - a source for researchers in a range of fields, for systems and solutions developers, for consultants to industry and business and, not least, for industrial practitioners. The studies reported in the book reflect the true complexity of managing operations in a wide spectrum of industrial sectors. Moreover, the issues debated are relevant to operations management across the extended enterprise including supply networks, distribution channels and logistic systems.

A book of this type requires considerable organisation. We are absolutely indebted to Alison Parrett who co-ordinated the project, liaising with contributors, the publishers and ourselves. She has coped valiantly with the vagaries of electronic communication, the inconsistencies in graphics packages and the numerous drafts and revisions that are inevitable for a coherent book of this type. Our sincere thanks must go to Alison for the gargantuan effort.

Thanks also to Tony Moore, our editor at Taylor and Francis, and to Alison Nick, the project manager, who advised on style, consistency and clarity for text and graphics. We believe that this has ensured a high quality book that meets our original intentions. We must of course thank all the contributors. Firstly for accepting the invitation to contribute, secondly for accepting our editorial comments, dealing with our queries, responding to requests for changes and enhancements, and thirdly for their patience. Although the project has had a long gestation period we feel that the additional time has been worthwhile in ensuring high quality contributions. We must also acknowledge the research team at Nottingham for their intellectual stimulation, particularly Sarah Crawford and Caroline Vernon, both contributors to the book. We must salute the practitioners in the field - the studies reported here are a testament to their contributions to successful operations.

Finally, we hope that the book will not only inform the intended readership but that the holistic view advocated here will contribute to a re-evaluation of the domain and its importance in contemporary organisations.

Bart MacCarthy
Reader in Operations Management
University of Nottingham

John Wilson
Professor of Occupational Ergonomics
University of Nottingham
November 2000

PART I

The Re-emergence of The Domain

CHAPTER ONE

The Human Contribution to Planning, Scheduling and Control in Manufacturing Industry – Background and Context

Bart MacCarthy and John Wilson

1.1 INTRODUCTION

This book is concerned with the reality of planning, scheduling and control in contemporary industrial and business organisations. These tend to be critical business processes that occur at the heart of an organisation and are key to what makes the organisation tick. They link customers with the primary manufacturing resources, balance the conflicting constraints on, and competition for these resources, and shoulder overall responsibility for meeting demands. Planning and scheduling comprise technical, organisational and human aspects. At their core are the people who manage and facilitate these processes in dynamic environments and ultimately 'make it happen'. Central to this book is the view that industrial practice and performance cannot be understood, nor can more effective processes be designed, implemented and managed without taking a holistic view of planning, scheduling and control processes.

The book brings together an international group of authors conducting leading-edge research with industry. One of its strengths is that evidence from real industrial practice is emphasised throughout. The sectors covered include automotive, aerospace and a number of engineering industries, PCB assembly, steel, textiles and clothing, pharmaceuticals, printing, furniture manufacture and process industries. Empirical data from over thirty organisations are referred to directly across the contributions and many of the studies are backed up by more extensive fieldwork. Many of the conclusions are based on in-depth longitudinal investigations, a number having been conducted over several years. Studies range from the largest international businesses to small and medium-sized enterprises.

A purely technical view of planning and scheduling has been dominant for many years. Thus it may be surprising to some that a book on the subject should contain so few equations. There is little formal mathematics in the contributions here and what there is does not relate specifically to combinatorial optimisation – the main branch of mathematics employed in scheduling theory (French, 1982; Morton and Pentico, 1993; Pinedo, 1994) and the cornerstone of an extensive research literature on scheduling since the classic book by Baker (1974). The reason for the lack of prominence of mathematics is not to make the subject more accessible - many of the concepts across the range of contributions here are equally, if not more, challenging. However, this issue gets to the heart of what makes this book different.

Even a cursory study of the realities of planning and scheduling in any non-trivial industrial situation is likely to reveal the gulf between mathematical theory and industrial practice. This phenomenon has been noted by many and has been much discussed over the last two decades (e.g. Buxey, 1989; White, 1990; MacCarthy and Liu, 1993). It is also alluded to by many of the authors in this book. The extensive literature covering the many, sometimes elegant algorithms developed to generate optimal or near-optimal sequences for the flow shops and job shops of the classical theory, help to provide some insights into the nature of real problems. Unfortunately they do not in any sense 'solve' real problems. If they did then the management of operations across business and industry would be quite different. To many it seems as if the classical theory and real industrial problems reside in quite different worlds. Indeed a number of contributors to this book challenge what planning and scheduling really are in practice. The studies reported here show that in practice planning and scheduling do not come packaged as discrete problems that can be 'solved', optimally or otherwise. Rather they are dynamic processes that need to be managed over time. People, individuals or teams manage the processes. It is from this direction that most contributors in the book approach the subject.

The book offers a reassessment of a domain that is critical to industrial performance and has applications more widely in supply chain management and logistics. The book is motivated by three principal ideas – firstly, the importance of planning and scheduling processes to successful manufacturing enterprises, secondly, the importance of adopting a holistic view that incorporates technical, organisational and human dimensions and thirdly, the need to understand the human contribution.

The remainder of this chapter discusses the industrial and business context that drives the study of this area and highlights some of the major questions that are central to business improvement. An overview is then given of the structure and content of the book.

1.2 INDUSTRIAL AND BUSINESS CONTEXT

A number of trends have coalesced in recent years to create the pressures in which contemporary manufacturing businesses operate - global markets, global supply chains, international manufacturing operations, increased levels of product variety and customisation, continual emphasis on quality, and of course the ever-present cost pressures. Against this background is the requirement in all sectors to be responsive to market demands.

Effective planning, scheduling and control processes are essential for success in the competitive marketplaces create by these pressures. However, although the criticality of these business processes is generally accepted, we need to know far more about how to design, improve and manage them more effectively. The contributions in this book address these issues head on. Here we identify some of the strong underlying themes arising from current industrial and business realities that are pertinent to planning and scheduling processes

1.2.1 The human contribution

Across industrial sectors large numbers of people are involved in managing and facilitating planning, scheduling and control processes. Be they high-level planners allocating capacity across international sites, customer champions with sales and marketing roles, production managers with a range of responsibilities that include aspects of planning and scheduling, or old-fashioned progress chasers, businesses continue to rely on key human decision makers in planning, scheduling and control roles. They are always with us. Indeed the number of people involved in these roles, particularly in large organisations, can come as a surprise to some, even to the organisation themselves.

1.2.2 Conflict and blame

These processes (and the roles held by people within them) are often subject to conflict and contention. The visibility of problems at the lowest level is often a source of conflict but may not be the source of the problems. Problems emanating from poor capacity planning decisions, unrealistic aggregate or higher-level plans, quality problems, material supply problems, labour absenteeism or unreliable plant tend to manifest themselves in poor conformance to prescribed schedules. A 'blame' culture can easily develop when people are asked to achieve the impossible. Avoidance of problems and ensuring some degree of production continuity may require practices that are hidden from higher levels. In a large complex business the real systems may deviate significantly from acknowledged procedures.

1.2.3 Replacing humans with computer-based systems

The 1980s dream of Computer-Integrated Manufacturing (CIM) and full automation also reached the planning and scheduling functions. The hope, if not the expectation, was that computer-driven solutions would 'solve' industrial planning and scheduling problems with minimal requirements for human intervention. This has proved, at best, a limited perspective. Re-engineering of information systems in businesses generally has proved difficult, frequently daunting, with the pain often felt most acutely at the sharp end, in managing operations (Davenport, 1998). The failure of IT and computer-based systems to deliver effective 'solutions' in many manufacturing environments has occurred, seemingly irrespective of expenditure or the latest incarnation of software. Without doubt the latest breeds of enterprise resource planning (ERP) systems do facilitate enhanced information provision across organisations but there is little evidence that they have changed the nature of planning and scheduling functions significantly. The Manufacturing Resources Planning II (MRPII) paradigm, with all its attendant problems noted over many years (e.g., Anderson *et. al.,* 1982; Melnyck and Gonzalez, 1985; Scott, 1995) is often still the basis for the manufacturing planning and control modules within ERP. Planning, scheduling and control remain human-intensive processes in most environments.

1.2.4 Supply chains and the extended enterprise

Planning, scheduling and control are not limited to the four walls of the factory. These issues now permeate across the supply chain. Few companies exist in isolation. Long-term success depends on managing the extended enterprise, taking into consideration your suppliers' suppliers and your customers' customers. It is argued that in some sectors supply chains rather than individual organisations compete (Christopher, 1994). Hence many of the contributions in this book reflect dynamic management and decision-making in materials procurement, supplier and supply chain co-ordination and logistics. Planning, scheduling and control involves the whole enterprise and may include multi-site planning, international supply chains, business-to-business and business-to-customer linkages, strategic outsourcing and subcontracting, and lean manufacturing flows. Thus the issues addressed in the book relate to a wide range of human-intensive processes, managing and co-ordinating across the extended enterprise.

1.2.5 Work organisation

One of the most important developments with respect to planning and scheduling has been the changing nature of work organisation in many businesses. At the enterprise planning level, centralised control and decision-making is advocated. However, lower down in organisations, decision-making is being devolved to empowered teams and autonomous workgroups, often in cellular manufacturing environments. These contrasting developments raise many issues on roles and responsibilities, authority and support. In parallel there is a gradual realisation that planning, scheduling and control are much more than technical functions. They are multi-faceted processes that necessarily involve people, organisational structures and computer and information systems. The latter are necessary conditions for effective management of operations but can never be sufficient. Only people can make decisions in ever-changing environments using information and knowledge about the current situation, perceptions about the future and balancing a range of pressures and demands. Social and organisational processes that enable individuals and teams to dynamically manage and co-ordinate are essential prerequisites for successful operations.

1.2.6 Continuous improvement

The philosophy of continuous improvement has had a big impact on manufacturing industry world-wide. The emphasis has moved from specifically product quality-related issues to operational performance including customer service levels, efficiency and productivity, all closely entwined with planning and scheduling functions. Performance measurement and performance and practice improvements are cases in point. It is a well-known phenomenon that the metrics selected to assess performance may influence both practice and performance, and not always positively when viewed from an organisational perspective. There is little existing theory or knowledge on how to define and measure performance in planning, scheduling and control.

1.3. KEY QUESTIONS

There is a strong desire in many sectors to improve planning, scheduling and control processes to meet competitive challenges in tough markets. However, in the absence of proven theories or a body of knowledge to help in the design of these key business processes, there is limited understanding on how improvements might be achieved and sustained. Many of the contributions in this book address these from the perspective of contemporary practice and require us to re-evaluate our assumptions, conceptions or understanding of the relevance and interaction of the issues. Here we note some of the key questions that are of general relevance across sectors and organisational types that are recurrent throughout the book.

1.3.1 What is the domain?

From a purely technical point of view, manufacturing planning and control can appear as being well defined with precise definitions for functions, activities and relationships. Textbook theory in planning, scheduling and control is notable for its 'neatness' (e.g. Browne *et al.*, 1996). One of the first things evident from a study of people in organisations is that they rarely fit neatly into pigeonholes. When a number of planners and schedulers are studied across businesses (and sometimes within businesses) the diverse roles and responsibilities that participants may hold become apparent. When viewed holistically within an organisation, mapping of the planning and scheduling domain, charting its boundaries and understanding its interfaces becomes problematic. Obtaining a general definition of scheduling as practised in industry is a particularly difficult issue.

The development of performance improvement strategies requires identification of types or classes of manufacturing environment that show similarities and differences. Manufacturing enterprises may be classified from many perspectives (MacCarthy and Fernandes, 2000) with different emphases and different levels of approximation. What makes manufacturing environments different with respect to planning, scheduling and control characteristics? What kinds of taxonomies can be developed to facilitate the holistic view of planning and scheduling incorporating organisational, technical and human aspects of these processes?

How can such taxonomies help in systems design, redesign, implementation and management? Complexity affecting the nature of planning, scheduling and control may emanate from many attributes of an extended manufacturing enterprise. The aerospace sector, for instance, has very long lead times but great product complexity. Much of the food sector, on the other hand, tends to have simple products and relatively simple, if imprecise, manufacturing processes, but very demanding time pressures. In the fashion clothing sector, although a relatively low-tech industry, pressures result from the number of permutations of styles, sizes, colours, garment components and materials that have to be accounted for, the labour-intensive operations and the vagaries of fickle customers.

1.3.2 How is it done?

The issues that arise with respect to how planning and scheduling activities are carried out by practitioners is truly vast and we note just a few related to the cognitive and job design dimensions.

Is scheduling a cognitively taxing task or is it largely routine with repetitive behaviour? It may seem reasonable to argue that schedulers use some form of internal representation or mental model of the system they are working with, and 'run' or manipulate that model to help make scheduling decisions. However, there is no great clarity about the notion of mental models, how they might be identified and represented, and subsequently applied in systems analysis and design. Wilson and Rutherford (1989) and Rutherford and Wilson (1992) classify different types of mental models. There is general agreement that mental models of any work task or process will be multiple, incomplete and unstable. This raises questions on how to exemplify and differentiate models of the manufacturing facility, the scheduling information system or display and the schedule itself.

What is the true nature of the production scheduler's job in terms of design, cognitive effort and load, learning strategies, scheduler stress and burn out? More fundamentally, what do people bring to scheduling - do we overestimate or underestimate their capabilities and performance?

1.3.3 How should we organise and support planning and scheduling functions?

We have already noted the fuzziness and fluidity of real manufacturing businesses and the difficulty of precisely delineating processes, functions and roles. The most profitable approach is to view planning and scheduling as a continuum of activities and roles where, at one end, high-level planning sets the constraints within which the business operates and at the other, low-level scheduling loads the manufacturing system by dictating what is actually produced in the immediate and short-term. In different environments the number of levels, the roles and responsibilities vary significantly. However, some hierarchy from high-level planning to shop loading is usually evident. Viewed from the perspective of roles, then planners plan at some aggregate level and schedulers attempt to realise the plan in practice. Therein lies the source of many problems and conflicts that are evident in practice. There are many difficult questions on allocation of functions, authority and responsibilities across the continuum.

Are there effective generic strategies for co-ordinating planning with scheduling functions to achieve long-term overall expected benefits? What is the most appropriate balance between central prescription and local autonomy? What is the appropriate balance between rigidity and flexibility in plans? In the extreme of distributed autonomy, decisions are made close to task execution on the shop floor. Should this be the machine operator, the shift supervisor or the department manager? How many levels away from execution can decision-making occur and still rank as devolved? Does overly rigid control and highly prescriptive planning and scheduling result in shop floor personnel failing to take responsibility and failing to make necessary decisions?

More generally, do planning, scheduling and control processes evolve into systems that ultimately work in the reality or evolve by default into systems that make it easy for people to operate? Should we attempt to design such systems or accept organic development that may occur from high levels of local autonomy?

What do we mean by support for planners and schedulers – is it just restricted to computer-based systems or should we interpret it more widely? When and how much should we try to automate planning and scheduling functions? Are some manufacturing environments suited only for human scheduling whilst some lend themselves to wholly automated computer-based scheduling? Or do co-operative human-computer decision support systems always provide the best approach to scheduling? How many planners and schedulers does a manufacturing business or facility need? How lean can it get? Is there an irreducible level of human involvement needed for almost any manufacturing planning and control systems? Given that manufacturing businesses are selling their abilities to handle complexity, how simple can we afford to make manufacturing systems, without compromising their ability to operate in competitive markets?

1.3.4 How do we measure performance?

Many leading companies see performance measurement as crucial to the maintenance of organisational control and competitiveness. However performance measurement generally in business raises many issues – metrics are difficult to specify and/or to implement successfully (Neely, 1999). Ensuring that performance metrics are aligned with business strategy raises very difficult issues in this domain. Further difficulties arise in the context of planning and scheduling. Indeed, the very basis of metrics can be debated. Problems arise with time lags between action and results, and the complicating issue of empowered action and control that may be needed in a responsive, customer-focused environment.

What should the metrics be and how can they be captured accurately? How should the 'quality' of the planning and scheduling process be measured? Is it by schedule adherence, schedule stability, penalties, late jobs, excess inventory, etc.? Schedule adherence may be easy to achieve in a facility that is under-utilised with few resource conflicts but may not be a true reflector of schedule quality. Schedule stability may be slavishly adhered to, regardless of other issues but may be inappropriate in a highly responsive situation. Scheduling decisions may be overridden for political or organisational reasons - how does this affect performance measurement? When is it a good plan? When is it a poor schedule? What are the attributes of a good plan or schedule? Under what conditions do these attributes change? Objective measurement in planning, scheduling and control must account for the process by which plans are generated and executed, the people who are instrumental in generating them as well as the actual realisation of plans and schedules over time.

More generally how can all of these factors be accounted for in performance measurement systems that add value to planning, scheduling and control processes and that have a beneficial effect on organisational performance over time? A wider question is the performance appraisal of individual planners and schedulers, or teams involved in planning and scheduling.

1.4 OVERVIEW

The origins of the book go back to a conference stream organised by the editors at a major international conference in April 1998[1], to which many of the authors here contributed. The motivation for the conference stream was to bring together researchers that were tackling some of the issues highlighted in this chapter. The stream has been followed by two international workshops, one in June 1999[2] and one in May 2000[3]. Our goal in selecting contributions for the book was to present as many major studies of industry practice as possible and to address as many as possible of the key questions highlighted in Section 1.3 of this chapter. We have also aimed for wide coverage of industrial sectors and to get perspectives on the issues with respect to computer and information systems and with respect to supply chains and the extended enterprise.

The contributors to the book have a diverse range of backgrounds including operations management and operational researchers, industrial engineering and manufacturing systems, human factors and work psychology, computer science and information systems. This provides a breadth of experience and viewpoints. The book contains reflective chapters, field studies, and theory and model development. The contributors address many of the questions highlighted in Section 1.3 above, explicitly in some cases, implicitly in others. Not surprisingly different perspectives and contrasting views on the nature of the domain are evident but there is also much agreement on some of the core issues.

This book is divided into a five parts. *Part I, The Re-emergence of the Domain*, contains two further chapters in addition to this one: a literature review and a retrospective chapter by one of the pioneers of the field. The early literature in this area is patchy and certainly scattered across a number of disciplines. It stretches back to the early 1960s and the seminal work of Dutton (Dutton, 1962, 1964; Dutton and Starbuck, 1971). Up until the review by Sanderson (1989), the first to focus on planning and scheduling from an explicitly human factors perspective, there was a surprisingly limited amount of published output. In Chapter 2, *Crawford and Wiers,* researchers who themselves conducted extensive field work in the area – critically review the literature. They analyse the major

---

[1] INFORMS/CORS Conference, *Human Performance in Planning, Scheduling and Supply Chain Management,* Montreal, Canada, 28th-30th April 1998.
[2] 1st International Workshop on *Planning, Scheduling and Control in Manufacturing: New Perspectives on People, Information and Systems,* University of Nottingham, UK, 7th-9th June 1999.
[3] 2nd International Workshop on *Planning, Scheduling and Control in Manufacturing: Putting the Human Back in Control*, Swiss Federal Institute of Technology, ETH, Zurich, Switzerland, 3rd-5th May 2000.

studies from a number of perspectives. They identify the need to establish a firmer basis for research in the area and develop more formal understanding of the human contribution to planning and scheduling. Kenneth McKay did a lot to re-establish the domain by conducting longitudinal field studies from the late 1980s onwards (McKay *et al.*, 1988, 1995a, 1995b). In chapter 3, *McKay* presents a retrospective view on his work, the nature of the domain and how his thinking has developed in relation to the domain.

Part II, Field Studies of Planners, Schedulers and Industrial Practice contains six studies. In chapter 4, *Webster* describes a field study of an engineering company and how an individual scheduler manages to cope with the complexity presented. The importance of bottlenecks resources, information management issues and anticipation are highlighted. In chapter 5, *Crawford* presents a detailed case study, again in an engineering environment and focusing on an individual scheduler. The impact of performance measurement on behaviour is highlighted. In analysing the nature of the schedulers job a distinction is made between the different types of tasks carried out and different roles held or assumed by the scheduler. The need for sound methodology and effective and efficient methods in conducting field studies are emphasised. In chapter 6, *Harvey* develops the theme of responsibility and in particular looks at the supervisor's role in relation to planning and scheduling. Based on an empirical study of twelve small and medium-sized enterprises from a wide variety of sectors with various characteristics the evidence from the field shows very variable practice and characteristics with respect to where boundaries lie, even in ostensibly similar environments. The importance of informal systems in making formal systems work is clear. In chapter 7, *Vernon* describes a field study in a textiles and clothing firm. Some of the conclusions are based on a larger study, of which this study forms part. The characteristics and performance of planning and scheduling are examined along with the people who carry out the roles, both before and after the re-organisation and the introduction of a new system. The issues of how far scheduling decisions can be devolved downwards and the consequences of flexibility and freedom for autonomous workgroups are discussed. In chapter 8, *McKay and Wiers* also address autonomy and organisational issues and the environments that place the highest demands on people in terms of scheduling. They present a typology of four types of scheduling situation and discuss the kinds of support necessary. A case study is presented in an environment that places a high demand on human schedulers and the development of a scheduling support system described. The allocation of tasks to computer and human is discussed.

Flexible manufacturing systems (FMS) have been advocated as automation systems that potentially combine flexibility, productivity and autonomous working (so-called lights out production). The reality has been somewhat different with many FMS requiring intensive human support. Much of the literature considers the problems associated with FMS as purely technical, particularly their planning and scheduling. In chapter 9, *Slomp* presents a detailed longitudinal case study of the adoption and use and development of FMS from a human factors perspective. The high level of human involvement is noteworthy.

Part III, Plans, Schedules and Computer Systems, contains five studies. In chapter 10, *Wiers* addresses the development of a scheduling system for a firm that produces sheet material. It operates in an environment characterised by a large amount of internal and external uncertainty. The sheer difficulty of capturing the knowledge is clear as well as the additional factors that have to be taken into account for successful implementation. In chapter 11, *Cowling* discusses the development of a computer-based scheduling system for hot rolling in the steel industry, from the viewpoint of a commercial systems developer. Issues discussed include how much functionality should be allowed for different classes of user and the way in which different users use the system. In chapter 12, *Webster* presents another system development – a system based around relatively simple rules but that gives appropriate functionality for the type of environment. The issue of developing customised scheduling tools using co-operative development methods is noted. The study underlines the importance of the human element in effective implementation. In chapter 13, *Higgins* addresses generic issues of architecture and interfaces for computer-based support for scheduling. He notes that perplexity characterises the working environments of many schedulers. He advocates a number of analysis tools to deconstruct the type of cognitive support needed. Giving the decision maker freedom of action and goal direction are important. The study is illustrated with detailed analysis of scheduling in a printing firm. In chapter 14, *Nakamura* addresses the problem of matching capacity with schedules generated from MRP. This is key to effective scheduling in MRP-based environments and it is implicit in most descriptions of MRP. The fact that it is necessarily human-intensive is highlighted and how it might be supported is described.

Part IV, Context and Environment for Planning, Scheduling and Control, contains five studies. In chapter 15, *Efstathiou et al.* look at measuring manufacturing complexity and its impact on scheduling in particular. A measure of dynamic information-theoretic complexity is presented. Two field studies are discussed – one in a press shop and one in PCB assembly - where attempts have been made to capture and measure complexity. In chapter 16, *Fransoo and Raaymakers* look at the batch chemical sector, one of the most difficult areas within the process sector with respect to planning and scheduling. It is characterised by planners making plans without perfect information and schedulers having to live with the plans and having to attempt to put them into action. A regression approach is advocated to capture the complexity of a scheduling task and identify the most significant influencing factors. A case study in the pharmaceutical sector shows the key ideas. In chapter 17, *Batchelor* addresses the changing nature of the automotive sector with continual increases in variety, requirements for higher levels of customisation and shorter delivery times. A long term study with a major automotive supplier illustrates the pressures on managing these conflicting demands across the enterprise. This is examined from the perspective of logistics and the design of effective logistics systems with the collocation of personnel considered.

The next two chapters examine aspects of socio-technical systems theory in design and re-engineering of planning, scheduling and control systems. Many of the issues noted throughout the book can be viewed from a socio-technical perspective. In chapter 18, *Slomp* interprets guidelines for socio-technial design in the context of production planning and how they can be adapted. This is illustrated with a case study for a firm that produces perforated sheet metal. Accurate estimation of lead times is identified as key to the approach. In chapter 19, *Wäfler* gives another perspective on the socio-technical contribution to planning and scheduling systems analysis and design and identifies some of the inherent contradictions. The concept of planning and scheduling as secondary work systems is introduced, where task execution must be defined carefully. A detailed case study in a company that produces high-end plumbing items shows how much an existing system can be at variance with socio-technical principles and the insights provided by a socio-technial perspective.

The editors conclude the book with a brief chapter. It highlights some of the generic issues of most significance for industry and the research community. The importance of practice is emphasised. The opportunities for interdisciplinary research are highlighted, as is the need for research geared to improving the design and organisation of planning, scheduling and control processes to sustain competitiveness in demanding markets.

1.5 REFERENCES

Anderson, J. C., Schroeder, S. E., Tupy, S. E. and White, E. M. (1982). 'Material requirements planning: The state–of-the-art', *Production and Inventory Management,* 23, 4, 51-67.

Baker, K. R. (1974). *An Introduction to Sequencing and Scheduling*, New York, Wiley.

Browne, J., Harhen, J. and Shivnan, J. (1996). *Production Management Systems: An Integrated Perspective*, Addison-Wesley, Harlow, UK.

Buxey, G. (1989). 'Production scheduling: practice and theory', *European Journal of Operational Research,* 39, 17-31.

Christopher, M. (1994). 'Logistics and supply chain management', *Financial Times,* Irwin Professional Publishing.

Davenport, T. H. (1998). 'Putting the enterprise into the enterprise system', *Harvard Business Review,* July – August, 121-131.

Dutton, J. M. (1962). 'Simulation of an actual production scheduling and workflow control system', *International Journal of Production Research,* 4, 421-441.

Dutton, J. M. (1964). 'Production scheduling: a behaviour model', *International Journal of Production Research,* 3, 3-27.

Dutton, J. M. and Starbuck, W. (1971). 'Finding Charlie's run-time estimator', In J.M. Dutton and W. Starbuck (Eds), *Computer Simulation of Human Behaviour*, New York, Wiley, pp. 218-242.

French, S. (1982). *Sequencing and Scheduling : An Introduction to the Mathematics of the Job Shop*, Chichester, Ellis Horwood.

MacCarthy, B. L. and Fernandes, F. C. (2000). 'A multidimensional classification of production systems for the design and selection of production planning and control systems', *International Journal of Production Planning and Control*, 11, 5, 481-496.

MacCarthy, B. L. and Liu, J. (1993). 'Addressing the gap in scheduling research: a review of optimization and heuristic methods in production scheduling', *International Journal of Production Research,* 31, 1, 59-79.

McKay, K. N., Safayeni, F. R., and Buzacott, J. A. (1988). 'Job shop scheduling theory - What is relevant?' *Interfaces*, 18, 4, 84-90.

McKay, K. N., Safayeni, F. R. and Buzacott, J. A. (1995a). ' "Common sense" realities of planning and scheduling in printed circuit board production', *International Journal of Production Research*, 33, 6, 1587-1603.

McKay, K. N., Safayeni, F. R. and Buzacott, J. A. (1995b). 'Schedulers and planners: what and how can we learn from them?' In D.E. Brown and W.T. Scherer (Eds), *Intelligent Scheduling Systems*, Boston, Kluwer, pp. 41-62.

Melnyck, S. A. and Gonzalez, R. F. (1985). 'MRPII: The early returns are in', *Production and Inventory Management,* 26, 1, 124-137.

Morton, T. E. and Pentico, D. W. (1993). *Heuristic Scheduling Systems: with Applications to Production Systems and Project Management,* New York, Wiley.

Neely, A. (1999). 'The performance measurement revolution: why now and what next?' *International Journal of Operations and Production Management,* 19, 2, 205-228.

Pinedo, M. (1994). *'Scheduling: Theory, Algorithms and Systems,'* Englewood Cliffs, N.J, Prentice Hall.

Rutherford, A. and Wilson, J. R. (1992). 'Searching for the mental model in human-machine systems. In Y. Rogers, A. Rutherford and P. Bibby (Eds), *Models in the Mind: Perspectives, Theory and Application*. London, Academic Press, pp. 195-223.

Sanderson, P. M. (1989). 'The human planning and scheduling role in advanced manufacturing systems: an emerging human factors domain', *Human Factors*, 31, 635-666.

Scott, B. (1994). *Manufacturing Planning Systems*, London, McGraw-Hill.

White, K. P. (1990). 'Advances in the theory and practice of production scheduling', *Control and Dynamic Systems*, 37, 115-157.

Wilson, J. R. and Rutherford, A. (1989). 'Mental models: theory and application in human factors'. *Human Factors*, 31, 617-634.

CHAPTER TWO

From Anecdotes to Theory: A Review of Existing Knowledge on Human Factors of Planning and Scheduling

Sarah Crawford and Vincent C.S. Wiers

2.1 INTRODUCTION

Production scheduling has been studied intensively since the 1950s, mainly within the realms of operational research, management science and operations management. From the 1960s onwards, the disciplines within which production scheduling has been studied have expanded, with research being carried out in artificial intelligence, business and management, ergonomics/human factors and the social sciences. Concurrently within industry, the focus also broadened. The development of information technology in the 1980s saw manufacturing industry attempting to reduce the human input into the planning and scheduling process by automating the decision making functions. This task proved more onerous than expected and industry has now largely accepted that removing the human element from the process was not an appropriate course of action. It is now widely accepted within academia and industry that human planners and schedulers are essential elements for an effective planning and scheduling process.

The change in attitude and understanding as to the nature of the planning and scheduling process has been instrumental in the development of the human factors of planning and scheduling domain. However, the findings made by researchers in academia, and the practices carried out in industry often develop separately. Academia and industry often work without input from one another; this 'gap' between theory and practice has been widely noted (Buxey, 1989; MacCarthy and Liu, 1993; MacCarthy *et al,* 2001; Wiers, 1997b).

With this gap between the information held by researchers in various disciplines in industry and practitioners as a backdrop, the main feature of this chapter is its focus on collating the existing knowledge on human factors of planning and scheduling. Without a bounded research domain the information available will remain fragmented, duplications of findings and practices will remain the norm, and a lack of dissemination will limit structured developments. To begin to address these problems the authors present a re-assessment of the available information in the form of a number of research issues to underpin future studies and developments within the domain of human factors of planning and scheduling.

The chapter starts with an overview of how both field research and industrial practices have developed within the domain, and follows with a general problem statement to bring together the knowledge available within the academic and industrial domains. Subsequently, the authors present a meta-analysis of previous theoretical and applied studies. This meta-analysis provides a basis for a research issues framework, based on the information already available in the data collated from previous research and industrial practices.

Some of the discussion in this chapter may appear critical of previous studies. This is not our intention. The authors feel that this critique of previous research identifies many positive points. It demonstrates that researchers have been very successful in generating findings and solutions to questions about both scheduling theory and practice. It has also shown that there are many areas within the domain of human factors in planning and scheduling that justify focused research. Eight topic areas are evident by the meta-analysis. These are discussed below.

2.2 SCHEDULING STUDIES CONDUCTED BY ACADEMIA

There has been a wide range of research carried out under the heading of the human factors of planning and scheduling. In order to analyse the types of research undertaken, information on previous studies has been collated into a summary table (see Table 2.1). This allows the relevant data to be presented in a concise format. The table comprises of all the findings from the individual studies reviewed. It identifies the domain within which a study was carried out (or the particular problem that the researchers wanted to address), and the 'explicit' study findings that were the focus of the research. Table 2.1 also presents other issues raised by the study that can be directly inferred from the information provided in the papers. This summary therefore goes one step further than Sanderson's review (1989) by including results and findings that were not the initial focus of the studies.

The table presents the range of information available in the theoretical and applied studies on the human factors of planning and scheduling. A meta-analysis of these data has revealed three generations of research development in the planning and scheduling studies conducted by academic researchers. These generations will be discussed in the next section. A simple classification of the data included in Table 2.1 also generates eight topic areas within which many of the findings fall and factors that influence planning and scheduling in businesses. These topics will be further elaborated upon in Section 2.4.

Table 2.1 Summary of implicit and explicit data from previous studies

Study type	Domain/ Problem	Goal of study	Theoretical points	Other issues raised
Sanderson & Moray (1990) THEORY	Discuss human factors of scheduling behaviour	**Use scheduling models to understand how time pressure affects tasks. Use human performance models to understand how humans schedule**	Humans often perform better than dispatching rules because they are more flexible. Humans exploit opportunities for action. Humans look no more than 3 steps into future. Human decision making is simpler under time pressure; more use of 'intuition'. Difference between expirable & non-expirable tasks i.e. some tasks have fixed time limit. Macro-ergonomic factors influence scheduling. Decision Support Systems (DSS) must support appropriate recognition, decision & action behaviour, & interventions.	Need to understand subjective parameters that affect human scheduling performance. Building blocks of a Model Human Scheduler (MHS) are recognition, decisions & actions. Parameters influencing human scheduling behaviour are not plant specific. Need to understand error-prone behaviour & interruptions in relation to human scheduling. Need to understand when & why humans intervene if scheduling under computer control.
Sanderson (1991) THEORY	Theoretical perspective on modelling scheduling behaviour	**Develop framework for a Model Human Scheduler (MHS)**	Planning & scheduling span cognitive field. Need models of human planning & scheduling to support design decisions for Advanced Manufacturing Technology (AMT) systems. Must answer questions about optimal information display & how to aid decision making. MHS composed of sub-models of recognition, decision & action behaviour.	Need to understand nature of possible tasks when scheduling & human strengths & limitations i.e. need a global model. Attempt at model for discrete process control. Macro-ergonomic factors influence departure from 'best' scheduling procedure. MHS is theoretical basis for allocation of function decisions.
McKay *et al.* (1992) THEORY	Theoretical perspective on interdisciplinary nature of scheduling practice	**Discuss 'what is scheduling?' in terms of AI and OR. Introduce modelling paradigm to study scheduling**	Scheduling follows cognitive skill paradigm. Scheduling is making predictive decisions. Expert schedulers use 'broken-leg' cues i.e. trigger event that alters certainty of standard determinant event. Scheduling is feedback & control function.	Scheduling involves social interaction & role aspects e.g. trust, respect. Scheduler reduces problem space quickly & uses abstract models to problem solve 4 types of schedule: political, optimistic, idealistic & inevitable. Scheduler must totally understand domain.

Table 2.1 contd. Summary of implicit and explicit data from previous studies

Study type	Domain/ Problem	Goal of study	Theoretical points	Other issues raised
Nakamura & Salvendy (1994) THEORY	Discussing roles of human planners & schedulers in Advanced Manufacturing Systems (AMS)	**Develop mechanism for design of human roles in planning & scheduling functions of (AMS)**	If disruption to schedule, it is human scheduler who takes control of DSS because DSS do not offer 'good' decisions. Key factors of planning & scheduling system are dynamically changeable.	Sometimes scheduler has many attempts before finds satisfactory decision. Automated decisions for scheduling not possible as they need judgement & discernment. Being responsible for schedule is an issue.
Dessouky *et al.* (1995) THEORY	Using scheduling theory as basis for study of strategic behaviour	**Develop taxonomy of scheduling systems**	Cannot evaluate scheduler behaviour without definition of optimal or acceptable strategy. Schedulers try to reduce complexity of problem by simplifying it; oversimplification of system creates infeasible/suboptimal schedules. Must research way scheduling problem is presented/ represented to scheduler.	Several important reasons why there are problems finding optimum schedules. Must understand scheduling problem within its environment. Need to specify measures of schedule within a conceptual, underlying framework.
Stoop & Wiers (1996) THEORY	Review factors creating complexity of production scheduling in practice	**Summarise factors that make scheduling theory techniques infeasible to scheduling practice**	Scheduling techniques based on simplified representations of scheduling process. Strengths & weaknesses of human scheduler's cognition not considered in theoretical 'solutions'.	In practice, disturbances to schedule create deviations from planned performance. Human factors affecting scheduling success include scheduler's cognitive abilities & how they use theoretical scheduling techniques.
McKay & Wiers (1999) THEORY	Discussion about unifying Production scheduling: theory & practice	**Develop framework that aids understanding of requirements for extended definition of production scheduling**	Scheduling theory not successful in job shop environments or in industries with inherent uncertainty. Scheduler's problem involves information collection & validation, & determining environment. Scheduling is problem solving rather than computational task.	Scheduling is often manual task involving human judgement. Content of scheduling tasks varies across organisations. Human recovery in organisations is often lowest control level i.e. scheduling.

Study type	Domain/ Problem	Goal of study	Theoretical points	Other issues raised
Thurley & Hamblin (1962) FIELD STUDY	5 mini case studies in variety of manufacturing environments	**Importance of technological, human & organisational variables in behaviour of production supervisors**	Greater variety of operations means more complex planning behaviour needed by experienced personnel. Major function of production supervision is dealing & preparing for 'contingencies' i.e. breakdown of production plans.	Importance of supervisor's shopfloor knowledge & experience. Supervision must be understood in context. Performance measurement of supervision function. Supervisor selection & training criteria.
Dutton (1962) FIELD STUDY	Organisational decision making *Box-making plant*	**Production of corrugator scheduling program. Computer simulation of program**	Corrugator program is highly routinised Is an 'informal' program is stable; buffered by work flow control centre. Scheduler's goals measured by output as scheduling process not specified.	Implicit importance of concepts of devolved & team decision making. Different functions of scheduling & work flow control centres within plant. Decision making system adapts to situations; it is not static.
Dutton (1964) FIELD STUDY	Study of human behaviour in complex manufacturing situations *Box-making plant*	**Behavioural model & computer simulation of forming machine program**	Forming scheduling is goal directed, rule ordered behaviour. Main cognitive problem is combinatorial; 8 types of order combinations used. Model was early version of decision aid.	Scheduling & schedule revisions 'solved' by successive approximations. Access to & co-operation of plant personnel. Social system influences decisions. Schedule to meet production constraints.
Fox & Kriebel (1967) FIELD STUDY	Empirical study of decision making *Shoe-box manufacturing*	**Programme decision making of production scheduler**	Scheduler's decision can be represented by linear equation. Scheduling goals: timely delivery & low set-up costs on machines. Program replaces FCFS rule with SOR.	Advantages of combining descriptive & normative analysis of decision making. Consider empirical studies as guideline in analysis of mathematical simulations of scheduling rules.
Hurst & McNamara (1967) FIELD STUDY	Study of production scheduling decision behaviour *Woollen mill*	**Study planner's scheduling decision behaviour. Develop model of scheduler's average behaviour**	Transposed planner's priority decision variables into linear equation. Planner must implement schedule on non-scheduled operations after carding. Max-max heuristic used to decide job/machine combination for next order. Planner created weekly formal schedule. Model suggested as type of decision aid.	Assumption: scheduling is programmed decision behaviour. Decision criteria are embedded in decision situation so not able to model them. Certain decision variables not included in model because not considered objective. Planner aware of aspects not included in information sources used by researchers.

Table 2.1 contd. Summary of implicit and explicit data from previous studies

Study type	Domain/ Problem	Goal of study	Theoretical points	Other issues raised
Dutton & Starbuck (1971) FIELD STUDY	Computer simulation of factory scheduler *Textile plant*	**How scheduler specifies production sequences & predicts machine runtimes**	Objectives: maximise profit & utilisation of fabric width; minimise waste. Main performance measure is keep low average backlog of orders. Scheduler uses 2 non-linear relations to generate 1 simple linear relation.	Scheduler uses mental model of plant to assess plant state of affairs. Main scheduling task is prediction of order production time. Scheduler uses 'look-up' table of data.
Bainbridge (1974) FIELD STUDY	Study of human process control *Electricity distribution for steel-melting furnaces*	**Understand how operator organises attention on several continuous variables to make control decisions**	Optimum control is to balance current power usage vs. possible future events. Operator's behaviour described by approx. 12 'routines' or sequences of activity. Large part of behaviour is making predictions. Operator holds 'mental picture' of process.	Operator makes intermittent rather than continuous control actions. External interruptions not disruptive for operator. Operator's activities described using information processing approach. Inexperienced operators behave differently.
Beishon (1974) FIELD STUDY	Study of human process control *Continuous baking ovens*	**Understand cognitive aspects of operator's skill. Develop model & simulation of operator's control behaviour**	Operator uses 'routines' to deal with different phases of baking process. Operator uses number of 'look-up' tables. There is executive routine (ER) that directs operator's main procedure(s).	Operator plans 1/2 or 1 hour ahead. Operator uses perceptual processes for cues; difficult to simulate perception. Data represented as information-processing model of operator's behaviour. Attempt to test model using 'control game'.
Grant (1986) FIELD STUDY	Study application of AI to scheduling of repair jobs *RAF aircraft operations*	**Build prototype scheduling aid for scheduling aircraft repair jobs using AI techniques**	AI approach models human expertise with emphasis on knowledge engineering. OR systems approach 'downgrades' subjective element of human activities. RAF schedulers spend 80-90% of time determining constraints of environment. Scheduler uses many untaught heuristics.	Scheduling aid should help determine most effective relaxation of scheduling constraints. Scheduler works with large problem space. Scheduler uses schedule improvement rules & rules for applying dispatch & improvement rules; OR not studied these rules enough.

Study type	Domain/ Problem	Goal of study	Theoretical points	Other issues raised
Duchessi & O'Keefe (1990) FIELD STUDY	Knowledge elicitation with experienced production planner *Garden products factory*	**Develop a prototype KBS to support production planning activities**	Produced abstract model of human's knowledge to incorporate into KBS. KBS focuses on a credible production planning approach rather than optimality. KBS faster and more effective than planner at overall plan production.	State-space representation is natural framework for defining planning problem. Effective Knowledge-based Systems (KBS) must use heuristics demonstrated by human. Importance of working directly with planner to develop prototype.
Norman & Naveed (1990) FIELD STUDY	Development of 5 Expert Models (EM) *Rotary cement kiln*	**Compare performance of all EM with human kiln operators**	Actual kiln operation different from operators' descriptions & operating rules. Actual operation not the most effective method of kiln control.	Problem of human & EM comparison: human performance not optimal & decisions vary by operator. No evidence that EM would supervise kiln better or worse than operators.
Hardstone (1991) FIELD STUDY	Case study of Computer Aided Production Management (CAPM) System implementation *Jobbing print production unit*	**Understand CAPM scheduling module in context of firm's technology & organisation**	Must consider cognitive, technical & social aspects of module implementation. Module must integrate with features of production system e.g. quality etc. Visibility & accessibility of information are salient features of scheduling. Temporal aspects of scheduling noted.	Need to consider differing priorities of system's clients e.g. cost, ease of use, control. Management & organisational structure. Affect how new module perceived & used. Need to cater for various types of knowledge. Address issue of whether technological solutions are transferable.
Vera, Lewis & Lerch (1993) FIELD STUDY	Study of decision making in complex environment *Mail sorting facility*	**Design decision support tools to improve allocation of resources & deliveries**	Supervisors rely on environmental cues to make resource allocation decisions. Behaviour is independent of supervisor's level of expertise. Differences between expert & novice performance & information search behaviour.	No computer-based information source. Information gained from visual scan of floor & verbal interactions. Experts more efficient at information searching & use upstream information. Experts use processed data as information.

Table 2.1 contd. Summary of implicit and explicit data from previous studies

Study type	Domain/ Problem	Goal of study	Theoretical points	Other issues raised
McKay *et al.* (1995a) FIELD STUDY	Study of schedulers' decision making in rapidly changing industries *Electronics firm*	**Develop methodology to study scheduling practice & lack of technology transfer in rapid change industries**	Need to use hybrid behavioural & cognitive science approach to study scheduling. Difference between study of integrated & distributed decision making. Framework for studying scheduling task. Description of scheduling objectives, constraints & heuristics. Description of how scheduling task adapts to perceived risks.	Type of information needed for design of computer support systems & making recommendations for change in factory Possible techniques for scheduler performance & quality measurements Need researcher with suitable knowledge. Importance of methodology & methods for studying 'what is scheduling in real world?' Problems associated with real world research.
McKay *et al.* (1995b) FIELD STUDY	Study of instability & its relation to the planning & scheduling process *Printed Circuit Board (PCB) assembly plant*	**Applied goal: improve scheduling process, & discover if some decisions can be automated Theoretical goal: develop framework to capture nature of scheduling, & how to deal with potential instability**	Difficult to automate PCB assembly because manufacturing situation unstable Major differences between formal & actual scheduling procedures Scheduler used multiple schedules Used non-routine heuristics for exceptional conditions i.e. when has to use judgement Scheduling is a cognitive skill. Field data analysed using methods from skill & expertise area of cognitive science Scheduler uses own 'information network' Scheduler's task & role is as problem anticipator & solver, not a sequencer.	Scheduler ignores normal problems & concentrates on changes in the environment. Tasks included more than task assignments e.g. changes in capacity, build options etc. Some heuristics & decision processes direct link to company's performance metrics. Performance criteria affect scheduler behaviour & creates alternative schedules. Enriched data essential to scheduler. Issue of measuring scheduler performance. Issue of selecting & training schedulers.
Wiers (1996) FIELD STUDY	Quantitative field study of scheduler decision behaviour *Truck manufacturing plant*	**Investigate all aspects of human scheduling decision behaviour using quantitative model**	Performance criteria, actions & disturbances are 3 elements of quantitative model Performance of units not directly controlled by quality of the schedule. Performance not important factor in decision behaviour; cannot relate feedback to behaviour Schedulers show nervous decision behaviour.	Having good schedulers does not guarantee that order requirements will be fulfilled. Decision behaviour is different between schedulers because of the importance of individual differences. Quantitative method has disadvantages of oversimplification & non-causality. Scheduler decision process is 'hidden'.

Study type	Domain/ Problem	Goal of study	Theoretical points	Other issues raised
Webster (Chapter 4 of this book) FIELD STUDY	Case study of scheduling practice *Machine tool manufacturer*	**Reduce gap in scheduling research by describing actual scheduling practices in comparison to theory**	Schedulers emphasise early identification of problems as part of scheduling task. Typical on-going scheduling activities involve identifying & correcting errors in system data.	People responsible for scheduling carry out more than just scheduling task Schedulers must work round machine operators 'nervousness' of empty work stations
Halsall *et al.* (1994) SURVEY	Survey of planning & scheduling in UK SMEs	**Investigate companies decision support needs in planning & control**	Very low amount of work on application of scheduling theory in industrial practice, 64% of companies had staff member with main responsibility for production scheduling. Many companies had stability problems from internal & external sources.	Many companies frequently had to adjust or override published schedules Any new system developed must support - not replace - human scheduler's decision making.
Kenworthy *et al.* (1994) SURVEY	Case study survey of short-term scheduling	**Investigate influence of short-term scheduling in production control systems**	Short-term scheduling either not done (expediting), done manually with limited aids or on finite scheduling systems (MRP modules). Multi-functional software package not prerequisite to best scheduling practice.	Scheduling needs high calibre production control personnel, appropriately supported Must reduce material & capacity shortages at high level planning to reduce scheduling complexity so can be done 'in the head'.

2.2.1 Generations of research

The first set of studies were motivated more by researchers' general scientific interest in the area of human factors in scheduling rather than being inspired by the 'gap' between scheduling theory and practice. The second generation, which coincides approximately with Sanderson's (1989) informative review of the human role in planning and scheduling, contains studies that were motivated more by actual industrial planning and scheduling issues, and this is also applicable to many of the studies being carried out at present. However, the authors propose that a third generation of research is emerging - planning and scheduling research as a structured discipline that provides answers for both academic researchers and industrial practitioners (for example, Crawford *et al.*, 1999; McKay and Wiers, 1999; MacCarthy *et al.*, 2001). It is this generation of research that has made it apparent that the domain needs to be based upon a structured research framework. The three generations of research are discussed in detail below.

2.2.2 Generation 1: Motivated by scientific interest

Early studies in operations and manufacturing management acknowledge the human input to the scheduling function (for example, Emerson, 1913; Coburn, 1918; Gantt, 1919). The first contemporary studies on human factors in scheduling were carried out in the 1960s and continued until the late 1980s. Sanderson (1989) reviewed much of the work that we refer to as Generation 1. We will not repeat her discussion. Rather, this review will focus on the aspects of the theoretical and field studies that have received relatively little attention since her review.

The common theme of the early studies carried out lies in the motivation of the research. In the period of the 1960s to 1980s, the human factors aspect of scheduling was largely ignored by manufacturing industry. If researchers did study schedulers, it was simply to gain a deeper insight into the scheduling task, rather than being motivated by the explicit goal of improving scheduling performance. An important contributory factor was that the information technology needed to substantially aid human scheduling performance was not readily available.

Thurley and Hamblin's study (1962) was of five mini case studies conducted in a variety of manufacturing situations. The research focused on the fact that technological, human and organisational factors all influenced the behaviour of production supervisors who also carried out planning and scheduling tasks as part of their roles. The paper emphasised the importance of the supervisor's knowledge and experience of the shopfloor, and noted that the supervision task could only be understood in context. Furthermore, other related factors such as performance measurement, supervisor selection and training criteria were also mentioned. On re-examination it is interesting to note that this early work already pinpointed areas of interest that have been included in many other papers since. (See also Harvey, Chapter 6).

Following on from Thurley and Hamblin's work, comparisons can be drawn between the studies of Dutton (1962, 1964), Fox and Kriebel (1967), Hurst and

McNamara (1967), and Dutton and Starbuck (1971) regarding the goal of the studies: to describe the scheduler's decision behaviour using mathematical models. However, the scheduling tasks studied were relatively narrow compared to other Generation 1 studies; in two cases, the scheduler's task was to minimise material waste in a two-dimensional cutting problem (Dutton, 1962; Dutton and Starbuck, 1971). A further observation about the aim of these papers is that the studies generated similar theoretical statements. The papers do not appear to build on each other or on the results of previous research. They did include statements about dynamic decision making, scheduler responsibilities, team decision making, the social system, and other similar issues. However, these aspects fell outside the scope of the intended research and were reported on in an anecdotal manner and not elaborated upon.

Process control studies of this era did not deal with discrete production scheduling, but there are similarities between the behaviour exhibited by a process control operator and by a scheduler, so these studies should be used as an early frame of reference for subsequent scheduling field research. Beishon (1974), Bainbridge *et al.* (1968), and Bainbridge (1974) used verbal protocol methodology and observation of the participant operator in the field to study the process control task. Such studies reported new findings about the process control task; however, they focused on the cognitive processes of the operator and so little attention was given to the actual context within which the tasks were studied. This descriptive research approach has remained popular with both production and process control researchers, although many of these knowledge elicitation studies are now carried out under simulated or controlled conditions (Tabe *et al.*, 1988; Lopez *et al.*, 1998) or as limited field studies (Duchessi and O' Keefe, 1990; Norman and Naveed, 1990). The recent studies do report many interesting results; however, the findings tend to be specific to particular research conditions and therefore cannot be applied to aid scheduling practice generically.

As noted by Sanderson (1989) the research from the 1960s to the 1980s 'generally looked at highly experienced, relatively unaided individuals as they perform their tasks...who developed a high level of expertise based on heuristics built from experience and observation'. The other focus of these studies was an attempt to explain scheduler performance by understanding general problem-solving strategies rather than the scheduler's specialised domain knowledge. All of these earlier studies on the human factors in planning and scheduling appeared to be carried out in a 'random' manner and therefore could not provide a firm foundation for a definitive theory of human factors in planning and scheduling. This Generation 1 type of study practically ceased to exist as researchers moved towards Generation 2.

2.2.3 Generation 2: Motivated by industrial and business issues

At the end of the 1980s it became evident that businesses were considering production control as a competitive tool. Companies extended their focus beyond minimising costs and maximising efficiency. Customer service received more attention and delivery performance became part of the product itself. During the

1970s, many companies started to implement Materials Requirements Planning (MRP) systems, but by the end of the 1980s there was a growing realisation that the promises of MRP for production planning and control were not being fulfilled. MRP systems were useful at making data available to the organisation, but the tools failed to transfer this functionality into a competitive advantage. Moreover, because scientific and academic research on scheduling focused on greatly simplified situations, no relevant techniques were being made available to industrial practitioners.

In this period, information technology became more suitable for building simple, graphical scheduling tools. The combination of the lack of applicable scheduling techniques from academia with the demand for scheduling tools in practice, led to the development of commercial scheduling systems. These industrial systems focused on the presentation and manipulation of information using electronic Gantt charts (e.g., Kanet and Adelsberger, 1987; Kanet and Sridharan, 1990; Köhler, 1993). Some scheduling systems also incorporated scheduling techniques to automatically generate schedules. However, successful implementations of these scheduling systems in practice remained scarce (King, 1976; Graves, 1981; McKay *et al.,* 1988; 1989; Rodammer and White, 1988; Buxey, 1989; MacCarthy and Liu, 1993; Halsall *et al.*, 1994; Pinedo, 1995; Wiers, 1997b).

The problems faced by both industry and academic researchers in developing effective and competitive scheduling solutions in practice, motivated a new wave of research in human factors of scheduling. The authors refer to this period as Generation 2. It represented a shift by researchers towards the investigation of specific scheduling problems highlighted by industrial practice and was initiated by the work of Kenneth McKay. McKay's PhD thesis (1992) initially set out to evaluate the hierarchical production planning (HPP) paradigm in dynamic and uncertain production situations. It attempted to explain how the human scheduler was crucial to the decision making function and proposed a framework that captured the type of control necessary. In subsequent papers many issues were raised about the human factors of scheduling, all of which deserve further research (McKay *et al.*, 1995a; 1995b). However, in retrospect, it may be that this research was also attempting to cover too broad an area, the same problem faced by earlier scheduling researchers.

Other researchers had also been motivated by industrial and business issues. In 1996, Wiers used a quantitative descriptive approach to study scheduling behaviour in practice. The study served as a methodological experiment in Wiers' PhD thesis (1997a). The complexity of the scheduling situation resulted in many findings that were difficult to generalise. Subsequently, Wiers' work (1997b) focused on the question of why many scheduling systems are not used by human schedulers in practice. Again, a broad range of influences became evident including organisational factors, cognitive aspects, and information technology. The study generated a number of useful criteria for the design of scheduling systems but many questions about the human factors of scheduling remained open for future research.

Webster (Chapter 4 in this book) used a qualitative and observational approach to study scheduling practice in a machine tool manufacturer. His case study shows that scheduler experience, job knowledge and cognitive factors are important aspects that have been frequently overlooked by both industry and academia, and must be researched further in order to understand the gap that has opened up between 'classical' scheduling theory and complex real world scheduling practice.

A research programme at the University of Nottingham is working across a range of manufacturing sectors in the UK to study such questions about planning and scheduling practice. The research has taken a multidisciplinary and structured approach to planning and scheduling field studies (Crawford *et al.*, 1999; MacCarthy and Wilson, 2000a; MacCarthy *et al.*, 2000b). The overall research goal is to understand planning and scheduling from an organisational and a human factors perspective in order to underpin the development of more effective planning and scheduling processes. The project's findings have been enlightening in relation to the assumptions held by researchers and by industry about the actualities of scheduling practice and the scope of the planning and scheduling processes within manufacturing industry. The main problem with investigating planning and scheduling is in the field, the general lack of research integration within the domain. Some of the issues are explored in detail in other chapters contained in this volume (See Crawford, Chapter 5; Vernon, Chapter 7).

This second wave of studies shows a clearer motivation and justification to carry out such research, with researchers such as McKay and Wiers pushing at the limits of human factors knowledge in this domain. This type of study added vital information about the 'soft' system elements and provided evidence that humans were still a necessity in planning and scheduling systems. Although these studies often did not have defined methodological structures (Crawford *et al.*, 1999) and were carried out independently of each other, they provide useful insights into issues of relevance to contemporary manufacturing industry. It is from this level of research that researchers will be able to progress to Generation 3: building on previous research results.

2.2.4 Generation 3: Building on research results

Sanderson's review (1989) concluded with the observation that more and better co-ordinated research was required. The research reported in her review was widely dispersed over a variety of research journals and such research was often carried out in isolation. She also noted that a common research question addressed in much of the reviewed literature - which is better, humans or algorithms? - was no longer relevant. Humans and algorithms appeared to have complementary strengths that could be effectively combined. To be able to achieve this, it was clear that a sound understanding of the human scheduler was needed.

This lack of understanding of the human scheduler is the motivation behind studies in Generation 3. It is worth asking questions about whether current research is producing new and innovative findings about the performance of human schedulers, or whether researchers are simply repeating previous research because previous findings have been overlooked. Earlier human factors studies in this domain have always included data on the planning and scheduling task and the human scheduler carrying out these tasks, so by undertaking new research what knowledge will be generated that was not available to the scheduling researchers 35 years ago? Current researchers have realised the need to use and build upon previous results in order to generate a research framework for this domain.

As well as utilising findings available in previous research, researchers have also become acutely aware of the lack of integration between the findings of scheduling researchers in academia and the manufacturing planning, scheduling and control practitioners in industry. So, what body of knowledge that is utilised by industrial practitioners may have been overlooked by academic researchers? This question must be addressed to ensure that existing work in both arenas can be built upon effectively by the new emerging studies. An overview of emerging trends in business and industry that are having an impact on current industrial practices is presented next.

2.2.5 Industrial scheduling practice: from MRP to Supply Chain Management

During the 1990s, the performance of supply chains has received much attention from companies. In many businesses, Supply Chain Management (SCM) projects, consultancy firms and software vendors are promising 'solutions' to advance companies to new heights of logistical performance. A number of prevalent conditions have set the stage for the adoption of such SCM strategies and for industrial research into scheduling practices:

1. *Advanced Planning and Scheduling.* A new wave of information technology has emerged which is referred to as Advanced Planning and Scheduling (APS). In short, the strength of APS systems originates from sophisticated algorithms, powerful modelling abilities, the integrative nature of the solutions offered, and the effective 'hard sell' marketing of APS products!
2. *Intense and global competition.* Competition in industry has become fiercer than ever, which increases the need for cost reduction and the use of the supply chain as a competitive weapon. There are many success stories of SCM projects that have indeed returned on their investment in as little as a few weeks. However, the lack of documented failures is probably more related to the highly competitive nature of the APS systems market, rather than to the absence of such failures.
3. *Maturity of IT and availability of data.* The implementation of an Enterprise Resource Planning (ERP) system ensures the availability of reliable and complete data. At the same time, ERP implementations have failed to improve business performance in many cases. Many industrialists believe that APS technology could finally bring the investments made in ERP systems to fruition.

It is expected that in the coming years many companies will implement APS systems. Moreover, companies are considering or have already adopted ideas such as e-business, supply chain integration and network resource planning. Therefore, from an academic perspective, the following question has to be asked: how can academic researchers integrate their knowledge about scheduling techniques, algorithms and processes with the practitioners' knowledge that is being utilised in practice?

From a scientific point of view, the Supply Chain Management approach could also be criticised. However, it might be as appropriate to direct some of the criticism at academia as it has had limited success in influencing scheduling practices. If scheduling researchers want to produce results that can be applied to practical problems, a paradigm shift has to be made. Two main factors can help researchers regain practical relevance.

1. Researchers must catch up with practitioners by participating in supply chain management case studies.
2. Researchers must reorganise their research approaches, structure their domain, and use and build on each other's results.

2.3 THE GENERALISATION PROBLEM

Two sources of information for the human factors of planning and scheduling researchers have been presented: previous research and current industrial practices. However, even if current researchers utilise this information as a basis for their planning and scheduling studies, they are then faced with the major problem of a lack of focus for the application of the results of these studies.

In order to develop such a focus for the application of research results, a new perspective on the human factors of planning and scheduling is proposed. The authors believe that it will enable researchers to concentrate on sections of a research framework and produce 'bounded' findings which are more specific, studied more intensely, replicable, generalisable and therefore likely to be more applicable to practice.

So why is a new perspective of human factors of scheduling necessary? The amount of work available on planning and scheduling is vast and this fact *per se* has been widely documented (Sanderson, 1989; MacCarthy and Liu, 1993; Dessouky *et al.,* 1995). Although there are substantially less papers that focus specifically on the human factors of scheduling, there have still been numerous papers on this subject over the past 30 years (see Sanderson, 1989). It appears that there are three main problems with this body of work:

1. *Lack of co-ordination.* The research on the human performance in planning and scheduling is widely spread across a variety of research communities, disciplines and journals, including psychology, management science, OR, production management, HCI and human factors. Sanderson (1989) reviewed many of these dispersed papers, but even her eloquent summary failed to jump-start the planning and scheduling research community into formulating a common research framework.
2. *Lack of focus.* Many of the previous studies have been too broad and unstructured and have attempted to cover the large human factors domain from studies of the organisational aspects of planning and scheduling to studies of the cognitive performance of the human planner and scheduler. In many instances, studies discuss the same range of factors, which provides a summary of the issues but lack definitive results on a specific element.

3. *Lack of building on results*. A major problem with the available body of knowledge is the lack of generalisations made from previous work. Once the studies have been collated, and the findings extrapolated, there is no framework upon which to shape any generalisations. Without generalisable findings, theory and knowledge cannot be generated. How can academic planning and scheduling knowledge aid and influence planning and scheduling practice when as a research community we cannot offer generalised findings about human performance in this area?

There is an irony in this 'generalisation problem'; the more that researchers are aware of the gap, the more work is carried out to build the bridge between theory and practice. This is good for the domain as additional knowledge and information can only improve the situation. However, without an underlying framework each researcher needs to be an expert across all the areas covered by the 'human factors of planning and scheduling' in order to understand the results. These areas include HCI, organisational behaviour, mental workload and information processing skills. If a limited number of factors were investigated within each study, they would be understood in more depth so that the results could be more easily applied by scheduling researchers and practitioners.

The authors believe that the concepts and information needed to address this generalisation problem are already available in previous work but have often been overlooked in the overall study findings. Or they remain as implicit rather than explicit results, left loose and not tied up as definitive conclusions. It is these implicit findings, presented in Table 2.1, which can be re-assessed to form a useful research issues framework for future studies.

Here we present a meta-analysis of the previous work summarised in Table 2.1, in which the anecdotal, non-formalised, implicit findings are brought together. This review does not make assumptions about any of the studies. It presents a wider range of knowledge and information about the scenario investigated. That is, the wider data were included in the paper but were not presented as explicit findings because the data were not perceived as the main motivation for the study. This collection of implicit findings demonstrates how varied the scheduling function can be.

2.4 REFLECTING ON FINDINGS FROM PREVIOUS STUDIES

Eight topic areas are evident from the meta-analysis. These are discussed below.

2.4.1 Cognitive issues

Cognitive psychology and cognitive science cover a very large area. It is necessary to look at what cognitive factors appear to be important within both field and theoretical studies of planning and scheduling. In order to define our field of reference, relevant cognitive issues are considered as relating to how schedulers represent information internally. (This section does not cover experimentally-based work; a number of reviews are available that offer such insights (Sanderson,

1989; Sanderson and Moray, 1990; Moray *et al.*, 1991; Stoop and Wiers, 1996; Wiers, 1997b)).

Some researchers have documented that human schedulers look no more than three planning or scheduling steps into the future, or that they plan their scheduling behaviour only ½ to 1 hour ahead (Beishon, 1974; Sanderson and Moray, 1990). In psychological terms, a plan is a mental event that appears to specify a sequence of actions. An immediate question is whether the three-step limit is a human strategy generally, or whether schedulers are better or worse at planning because of the environment they operate within. The issues of the influence of external factors, and of aiding plan generation in production environments, will be discussed.

It has been noted that schedulers often use a 'mental model' or a 'mental picture' of the plant and of the production process, to assess the state of affairs or the problem that needs to be solved (Dutton and Starbuck, 1971; Bainbridge, 1974). This raises the issue of how a scheduler assesses the current situation but situation assessment in scheduling has not been widely studied. The topic of situation awareness, which was developed within defence research, could possibly aid researchers studying the scheduling function.

The scheduling problem is sometimes considered primarily as a combinatorial cognitive problem i.e. the scheduler must provide a satisfactory scheduling solution by dealing with a combination of real-time production parameters (Dutton, 1964). It has also been reported that the information needed to solve the problem is held in a mental 'look-up' table (Dutton and Starbuck, 1971; Beishon, 1974). Again, this raises questions as to the best type of problem representation and decision aid for production schedulers. If a cognitive psychology framework is used, the scheduler's activities can be described using an information-processing view but as pointed out by Beishon (1974), schedulers also make use of their perceptual processes to provide cues for scheduling activities, making simulation of such activity very difficult for researchers.

2.4.2 Decision making

A particular set of cognitive issues that is central to understanding scheduling behaviour relate to decision making. A range of factors can influence the decision-making process. There appears to be a general consensus that a main scheduling task is making predictive decisions of some sort (Dutton and Starbuck, 1971; Bainbridge, 1974; McKay *et al.*, 1992), although Wiers (1996) makes the point that the actual decision behaviour is different between schedulers and hence the researcher must not overlook the issue of individual differences. Dutton states that researchers must also realise that the social system within which scheduling takes place affects scheduling decision making (Dutton, 1964). The actual act of making the decision has been studied in previous research. Many researchers have stated that the cues that trigger decisions are embedded in the surrounding environment (Hurst and McNamara, 1967; Vera *et al.*, 1993), mainly in the form of 'enriched' data (McKay *et al*, 1995b). Human decision makers also exploit opportunities for action (Sanderson and Moray, 1990) and expert schedulers use 'broken-leg cues' for their decision making trigger events or information that alters the certainty of a standard event (McKay *et al.,* 1992). It becomes quite evident that the decision making system adapts to situations and is not static (Dutton, 1962) contrary to the implicit assumptions made by many IT designers, which then transfer into scheduling software.

Some research has proposed that there is an 'executive routine' or overall driver that directs the general decision making process (Beishon, 1974). This fits in with findings that schedulers use untaught and non-routine heuristics in many circumstances, but especially in conditions they consider exceptional (Grant, 1986; McKay *et al.,* 1995b). This leads onto the area of how best to aid scheduling decision making. Duchessi and O'Keefe (1990) reported that an effective knowledge-based system (KBS) must use human heuristics to support the operator. This sentiment was backed up by Halsall *et al.* (1994) who found that any new decision support system (DSS) must truly support and not replace human decision making. It must support appropriate human recognition, decision and action behaviour and scheduler interventions. System developers need also to understand how interruptions affect scheduler decision making (Sanderson and Moray, 1990) although other researchers have reported that external interruptions may not be disruptive (Bainbridge, 1974).

The different styles of decision making in scheduling have also been studied. There is an implicit understanding of the importance of devolved and team decision making in planning and scheduling in Dutton's work (1962; 1964), whilst McKay *et al.* (1995a) warn that different approaches are needed to study integrated and distributed decision making in companies. A final issue is that of performance measures. Wiers (1996) found that performance is not an important factor to the scheduler in terms of human decision making, because the scheduler is not often able to match the feedback to the appropriate decision making behaviour. However, studying scheduler decision making will always be a difficult task because, in scheduling practice, the actual decision process is hidden.

2.4.3 Environmental factors

From internal cognitive factors it is a natural progression to consider environmental factors that affect human scheduling. A number of researchers have found that macroergonomic factors affect scheduling - the social system that scheduling is a part of influences scheduling decisions (Sanderson and Moray, 1990; Sanderson, 1991; Dessouky *et al.*, 1995; McKay *et al.*, 1995b). There are also findings stating that these higher level factors influence the departure of the scheduler from 'best' scheduling practice. The issues of scheduling performance and performance measures are topics that cannot be done justice here. Many previous papers have reported findings on many aspects of scheduling performance measures (see Stoop, 1993; Dessouky *et al.*, 1995; Gary *et al.*, 1995; McKay *et al.*, 1995a /b; Stoop and Wiers, 1996; Wiers, 1996).

Apart from the influence of organisational issues, scheduling is also effected by such aspects as roles, trust, respect and interpersonal interaction (McKay *et al.,* 1992). It is not enough to consider scheduling as merely an organisational or management function; scheduling is also defined by how the scheduler interacts with other functions e.g. planning, purchasing, sales etc. This then brings in to play the role of the scheduler and the question of how the scheduler is perceived by peers, supervisors and other functions. Is the scheduler considered a dynamic problem-solver, a high level decision-maker or a shopfloor operative with decision-making responsibilities? The perceptions and representations of the scheduler within the organisation affect how the schedule is used and perceived by members of the company. The role of the scheduler also raises questions of how much production authority and responsibility the scheduler is given or is allowed (see Crawford, Chapter 5), which in turn affects how the schedule is generated and implemented.

Environmental factors must also be considered in terms of the implementation of planning and scheduling systems. The centrality of the scheduling function means that organisations must consider not only the technical aspects of a scheduling module implementation but also the cognitive and social aspects of such an introduction. For instance, the company must understand the differing priorities of a production system's clients, such as the cost of the implementation, the ease of use of the modules etc. Most importantly, the scheduling module must integrate with current features of the production system such as quality control and delivery routines. Finally, it must be remembered that the management and organisational structure of the company itself will also affect how a new scheduling system is perceived and used (Hardstone, 1991).

No matter how aware a researcher is of the influence of environmental factors, taking a holistic view of the scheduling function means that certain points will be overlooked because of the sheer size of the domain. The researcher often realises too late that the scheduler is aware of aspects that were not explicit research topics within the study (Hurst and McNamara, 1967). This in itself is not a failure; in the field of scheduling research the researcher must be aware that there are certain problems associated with real world research and should only attempt to cover as many factors as is feasible.

2.4.4 Domain-free and context-based factors

Closely related to environmental factors is the issue of whether scheduling is a domain-free or context-based task. Some researchers have found that the parameters influencing human scheduling behaviour are not plant specific. There are two separate strands here.

1. Where researchers attempt to define the factors that affect scheduling performance as *objective domain-free parameters,* e.g. the number of jobs, the number of machines that will have to be scheduled by the scheduler. However, this approach is considered ineffective by the majority of researchers because such parameters only indirectly influence the scheduler's performance.
2. Factors beyond the level of individual scheduling situations that influence scheduler recognition, decision-making and action are *subjective domain-free parameters* e.g. time available to examine the system, familiarity of system

state, and the number of plausible action alternatives. Certain researchers suggest that there is a need to understand these types of parameters to develop a generic understanding of human scheduling performance (Sanderson and Moray, 1990).

The antithesis of the previous approach is to view the scheduling task in context. Some researchers state that we must study the scheduling problem within its environment i.e. scheduling can only be fully understood in context. This argument is particularly relevant to findings about schedulers themselves. Previous studies have found that schedulers must understand their domain intimately, that is they can only work effectively when they have a thorough understanding of the process and product that they are being asked to schedule. This is to allow the scheduler to make decisions based on an ability to recognise hidden relationships and identify possible alternative resource assignments (McKay *et al,* 1992). The need for scheduling to be carried out by personnel with first-hand and intimate knowledge of shopfloor work has come about because of the complexity in scheduling created by demands from customers for customisation of products and improved performance overall (Thurley and Hamblin, 1962; Hurst and McNamara, 1967; Hardstone, 1991; Webster, Chapter 4 of this book).

2.4.5 Instability and complexity of the environment

So what type of scheduling environments do schedulers really have to deal with? Mostly schedulers have been documented as dealing with dynamic, complex and relatively unstable production situations. Instability exerts a major influence on scheduling and schedulers. A manufacturing environment that is unstable is prone to many changes, in production processes, in the materials used, in personnel etc. Human performance under these changeable conditions needs to be understood to aid the design of scheduling systems.

Many manufacturing companies often have to adjust or override their published or agreed schedules because of unstable conditions, which obviously adds to both the instability and complexity of a scheduler's working environment. Halsall *et al.* (1994) surveyed a large number of Small to Medium Enterprises (SMEs) and found that instability was mainly due to:

- *Internal uncertainty*: companies appeared to have reasonable certainty about the stages in their production process, but actual set-up and process times of operations were less well known. Companies also suffered disruption to the schedule because of breakdowns in the system.
- *External uncertainty*: many companies began orders before the customer requirements were fully known. Another problem was that customer delivery dates would change once the order had been started. Other factors include late or incorrect supplier deliveries to the company.

One solution to the problem of instability may be to try to automate the more certain types of manufacturing environment. However, some production environments are difficult to automate because of their inherent instability. In a longitudinal field study McKay *et al.* (1995b) attempted to encode the decision processes of the scheduler to try to automate the scheduling process in an assembly environment. The conclusion was that a human solution was reasonable until the manufacturing situation exceeded the decision capabilities of a human scheduler. One way in which human schedulers try to reduce the complexity of the scheduling problem is by simplification, for example by aggregating or eliminating variables or constraints. However, a simplified scheduling model leads to the oversimplification of the real system to be scheduled, and this in turn creates unfeasible or sub-optimal schedules (Dessouky *et al.*, 1995).

In order to minimise the level of risk and uncertainty that a human scheduler has to deal with, it has been proposed that a higher level solution needs to be formulated. The material and capacity shortages that afflict the scheduler at the production level must be reduced at the higher planning level. This in turn reduces the scheduling complexity so that the human scheduler is able to manage the scheduling situation 'in the head' (Kenworthy *et al.,* 1994).

2.4.6 Temporal and production constraints

We have looked at the scheduling function and the scheduling environment but are there specific constraints that affect how humans perform their scheduling tasks? One type of constraint that is implicitly included in the majority of previous research is time. Obviously, the concept of time is inherent in scheduling; if there were no time constraints, there would be no scheduling problem! However, the temporal influence is only made explicit in a limited number of papers. Hardstone (1991) makes the important point that the success of computer technology for scheduling can only be accurately predicted when it is considered in the context of real-time, i.e. how a system will actually be used by an organisation in practice. ' 'The worst problems were encountered when the [ideal] future became the present, and the system had to cope with real production' (Hardstone, 1991, p.18).

Another important temporal issue is raised by Sanderson and Moray (1990) who distinguish between expirable and non-expirable scheduling tasks (tasks that cannot be late or they become worthless). Although the examples of tasks used are related to other skilful, human behaviour such as driving and flying, it is interesting to consider that the definition used by the authors can be applied to manufacturing scheduling tasks.

Temporal boundaries are not the only limitations that human schedulers have to deal with. There are other production constraints to be considered (Dutton, 1962; 1964). The process by which schedulers actually scheduled work orders was not understood; the schedule was generated to meet specific goals, rather than generating the schedule by a specified means. For example, the scheduler had to meet the objectives of maximising material usage and minimising wastage. These constraints bounded the task of schedule generation but the constraints themselves did not provide pointers as to the actual scheduling process.

One would assume that human performance in scheduling was now more established, but recent findings suggest that schedulers spend 80 to 90% of their time determining the constraints of the environment that will affect their scheduling decisions (Fox and Smith, 1984; Grant, 1986). This means that only 10 to 20% of schedulers' time is spent on the generation and modification of schedules. In a wider context, these findings mean that when the complexity of manufacturing environments increases, production constraints may no longer be obvious to the scheduler. In turn, schedulers are not able to remain fully aware of all the constraints that they are working under. Therefore, an apparent solution to this problem of complex and conflicting production constraints is the development of effective scheduling aids that determine the most appropriate relaxation of constraints for schedulers.

2.4.7 Information issues

The issue of information in scheduling is often simply related to the design and implementation of information displays and raises questions about optimal information display (Sanderson, 1991). However, other considerations are the necessity and importance of information, which may be only alluded to implicitly in research. Increasing product variety and the need for greater levels of tracking and control also highlight these issues.

One area of discussion is the accessibility of information for scheduling. As Hardstone (1991) noted, the visibility and accessibility of information are salient features of scheduling. Schedulers need to be able to see and access relevant and critical information in order to be able to carry out the scheduling task. The issue of access was central to the work of Vera *et al.* (1993) when they studied the decision-making behaviour of supervisors who allocated resources and deliveries in a mail sorting facility. They found that the supervisors rely on environmental cues to make resource allocation decisions, and that relevant information was gathered by visual scans of the production area and from verbal interactions. This ease of access to necessary information was very important in this scenario because there was no computer-based information source underlying the scheduling decisions. The findings also found differences between expert and novice information search behaviour, and found that experts are more efficient at searching for relevant information. Experts also tend to include information from upstream functions within the planning system in their search behaviour.

From a slightly different perspective, McKay *et al.* (1995b) found that schedulers establish and use their own 'information networks' from which they glean necessary information, '...each contact point had the potential for providing some piece of key information needed to solve the planning and scheduling puzzle'. The research also discovered that 'enriched data' is essential to the scheduler for them to carry out their scheduling tasks. The authors define enriched information as 'that data not normally found in analytical models, MIS decision support tools, or the formal paperwork that is associated with work that is to be produced...[enriched data] includes aspects such as historical, organisational, cultural, personal and environmental'.

It can be seen that the understanding of the role of information in scheduling is not as straightforward as it is perceived to be in much of the work on human performance in planning and scheduling.

2.4.8 The scheduling function

Now that we have discussed the influence of cognitive and external factors on scheduling and the human scheduler, we can finally tackle the scheduling function itself: what is scheduling in practice?

There is a very limited amount of work on the application of scheduling theory in practice. In one particular survey, 64% of small to medium enterprises (SMEs) had a member of staff specifically responsible for production scheduling (Halsall *et al.,* 1994). But what is it that they should be doing? It appears to be well documented that scheduling tasks include more than simply assigning jobs to machines. Depending on what industry is studied, scheduling itself can be a variety of things. Primarily, scheduling can be considered a feedback and control function within the overall planning and scheduling system; however, there are major differences between formal (planned) and actual scheduling procedures. For instance, McKay *et al.* (1995b) found that formal scheduling procedures in a specific plant assumed that the capacity available was always 80% of the theoretical optimum. The scheduler working within this system, however, did not assume a static, set capacity, but worked on the assumption that the available capacity would fluctuate depending on the dynamics of the production situation. Similar findings had previously been documented in the early scheduling fieldwork of Dutton (1962). He stated that scheduling is primarily an informal function as the scheduling procedures have often simply evolved over time or been passed on by example or word of mouth. Therefore, there are significant differences between the formal procedures as perceived by the management and the informal procedures actually followed by the schedulers themselves.

There does not appear to be a consensus on the main function of the scheduler working within the scheduling environment. Some research states that the major task is to prepare for and deal with the technical contingencies that happen in production scheduling i.e. the breakdown of plans (Thurley and Hamblin, 1962). Other work proposes that the main task is the prediction of order run time, so that the scheduling of machines, especially continuously running machines, is an efficient and effective use of resources (Dutton and Starbuck, 1971).

There does not appear to be such a disparity in the way that previous research has characterised the scheduling task itself. Although it has been described in a variety of contexts, the findings are very similar. Scheduling is considered to be a highly routinised task. It is understood to be programmed and goal-directed, rule-ordered decision behaviour. In this context then, the scheduler has been found to make intermittent rather than continuous control actions, and appears to ignore 'normal' problems and concentrate mainly on changes in the environment. This can be explained as the scheduler only making interventions or changes to the current situation when there is a problem e.g. machine breakdown, late delivery of materials, scrapped items etc. If scheduling is a routine, goal-oriented task then the scheduler will keep a current mental picture of the state of affairs that is updated sporadically not continuously; the scheduler will review the production situation and act only when there is a situation that is outside of normal tolerance.

2.5 DISCUSSION AND NEXT STEPS

Our review has demonstrated that a structured approach to the domain of the human factors of planning and scheduling does not mean dismissing past work and replacing it with new research. On the contrary, the authors acknowledge that past studies have provided many findings on many aspects of a very large area; the only problem is that these findings have not been synthesised into a definitive research framework that can incorporate previous findings and act as a guide to the next generation of research and industrial practice. Our goal is to provide a conceptual, structured framework of research issues that provides a boundary to the domain of the human factors of planning and scheduling. It is within this context that the authors propose that the issues identified previously should be used by researchers as a guide in prioritising what needs to be investigated in future human factors of planning and scheduling research.

The results of the meta-analysis have shown that there are many areas within the domain that justify focused research. Below is a summary of a number of implications raised by the review.

Firstly, *cognitive factors* are prevalent in previous scheduling work. The issues raised from an analysis of the type of cognitive behaviour displayed by schedulers will be instrumental in laying the groundwork for better, and more appropriate, planning and scheduling aids. Researchers need to study the cognitive processes of industrial schedulers to discover how schedulers represent certain information internally. For example, how do schedulers 'plan'? Is there a generic strategy used by schedulers, or do individual schedulers plan differently because of the specific manufacturing environments they work within? What are the best types of information representation and decision aids for production schedulers? Although experimental studies of schedulers' cognitive processes will be of great value in generating a theory of cognition in relation to scheduling, research must also be undertaken in companies to validate experimental findings in practice. However, researchers must be aware that studying cognitive behaviour in this field as opposed to control room studies is made much more difficult by the variety of spatial and temporal dimensions that schedulers have to deal with.

Secondly, the review has shown that scheduling cannot be considered as a 'stand alone' function within today's dynamic manufacturing enterprises. *Environmental factors* such as management structure, organisational culture and performance metrics all influence the scheduling task, and the scheduler, within a business. Future research needs to acknowledge this fact and develop valid, reliable and practical methods to study, analyse and implement the findings from detailed, ethnographic research.

Thirdly, within the review papers there appears to be a disagreement as to whether scheduling is a *domain-free or context-based* task. A fuller understanding of this situation is necessary because of the implications for system design, job design, decision support systems and performance metrics if planning and scheduling are perceived incorrectly within organisations. This issue also raises questions about whether findings as to the structure of production control systems are as generalisable as many of the software houses believe they are. If planning and scheduling tasks are context-based, does this mean that the underlying production control system is also context dependent? If so, can effective ERP and APS systems be applied 'off the shelf', or will production control modules have to be customised to the planning and scheduling environments of individual businesses?

Fourthly, the majority of previous studies have documented the *instability and complexity of the scheduling domain*. There have been various suggestions as to how these factors can be minimised but it appears that at present, an answer is more likely to lie with the human scheduler rather than a technological, computer-based solution. It appears that the human is the only element that is equipped to cope with the differing demands of an unstable manufacturing system and this may be more so in the dynamic, responsive and competitive environments within which schedulers operate across manufacturing sectors. This reliance on the human aspect of the planning and scheduling task raises questions as to how to train an effective scheduler, and how the performance of planners and schedulers can be measured appropriately against the unstable and complex backdrops of manufacturing businesses. Other research issues also need to be addressed, such as whether instability and complexity are inherent attributes of the planning and scheduling functions? Would a reduction in the instability and/or complexity add to or detract from the overall performance of a business? Assumptions as to the 'optimal' manufacturing environment must be addressed before more effective planning and scheduling systems can be built and personnel trained.

Fifthly, a research issue linked to the previous discussion is that scheduling will always be subject to certain constraints, most of which are inherent to the scheduling problem. However, like instability and complexity, *temporal and production constraints* have not been researched enough and researchers, as well as practitioners, appear to have overlooked the distinction between theoretical and real-time issues. With increasing complexity and instability occurring within manufacturing environments, there is a need for the issue of production and temporal constraints and the assumptions made about them to be researched in more depth.

Sixthly, the review of previous work has shown that the *issue of information* in human planning and scheduling is much wider than the often researched information presentation and display topics. Information to schedulers is not just received or distributed through formal or computer-based systems; the information needed to generate and implement a schedule is also embedded in the environment. All of this needs to be considered in the design and implementation of information and production control systems, as well as information representation formats and displays. The knowledge that information is central to the planning and scheduling function is not the issue; the issue that must be addressed is how can this knowledge then be best applied in practice?

Finally, the fact that there is no definitive statement about what the *scheduling function* actually entails is not in itself a negative conclusion. Indeed, this signifies that there is no one correct way to study scheduling. Researchers and practitioners must ensure that they do not hold too narrow a view of the scheduling function and its position within an organisation. Assumptions as to what is meant by the terms 'planning' and 'scheduling' can only constrain the domain of the human factors of planning and scheduling. The goal for researchers must be to challenge these previous assumptions and produce valid and practical definitions and applications that can be utilised by researchers and practitioners alike.

The following chapters in this book will include reference to many of the research issues mentioned previously. The task for researchers now is to develop the findings, links and interactions of the issues and themes made explicit in this book into a sound and structured research framework for the study and understanding of human performance and practice in planning and scheduling. Only when the planning and scheduling processes are understood from a deep rather than broad perspective will research be able to make significant contributions to practice.

2. 6 REFERENCES

Bainbridge, L. (1974). Analysis of Verbal Protocols from a Process Control Task. In E. Edwards and F. P. Lees (eds), *The Human Operator in Process Control*. Taylor & Francis, London, pp. 146-158.

Bainbridge, L., Beishon, J., Hemming, J. H. and Splaine, M. (1968). A Study of Real-time Human Decision-making using a Plant Simulator. *Operational Research Quarterly*, 19: 91-106.

Beishon, R. J. (1974). An Analysis and Simulation of an Operator's Behaviour in Controlling Continuous Baking Ovens. In E. Edwards and F.P. Lees (eds), *The Human Operator in Process Control*. Taylor & Francis, London, pp. 79-90.

Buxey, G. (1989). Production Scheduling: Practice and Theory. *European Journal of Operational Research*, 39: 17-31.

Coburn, F. G. (circa 1918). Scheduling: the Coordination of Effort. In I. Mayer (ed.), *Organizing for Production and Other Papers on Management 1912-1924,* Hive Publishing, Easton, 1981, pp.149-172.

Crawford, S., MacCarthy, B. L., Wilson, J. R. and Vernon, C. (1999). Investigating the Work of Industrial Schedulers through Field Study. *Cognition, Technology & Work,* 1: 63-77.

Dessouky, M. I, Moray, N. and Kijowski, B (1995). Taxonomy of Scheduling Systems as a Basis for the Study of Strategic Behaviour. *Human Factors,* 37(3): 443-472.

Duchessi, P. and O'Keefe, R.M. (1990). A Knowledge-based Approach to Production Planning. *Journal of the Operational Research Society*, 41(5): 377-390.

Dutton, J. M. (1962). Simulation of an Actual Production Scheduling and Workflow Control System. *International Journal of Production Research,* 4: 421-441.

Dutton, J. M. (1964). Production Scheduling: a Behaviour Model. *International Journal of Production Research,* 3: 3-27.

Dutton, J. M. and Starbuck, W. (1971). Finding Charlie's Run-time Estimator. In J.M. Dutton and W. Starbuck (eds), *Computer Simulation of Human Behaviour*. John Wiley & Sons, New York, pp. 218-242.

Emerson, H. (1913). *Twelve Principles of Efficiency*. The Engineering Magazine, New York.

Fox, P. D. and Kriebel, C. H. (1967). An Empirical Study of Scheduling Decision Behavior. *The Journal of Industrial Engineering,* 18 (6): 354-360.

Fox, M. S. and Smith, S. F. (1984). ISIS - A Knowledge-based System for Factory Scheduling. *Expert Systems*, 1(1): 25-49.

Gantt, H. L. (1919). *Organizing for Work.* Allen and Unwin, London.

Gary, K., Uszoy, R., Smith, S. P and Kempf K. (1995) Measuring the Quality of Manufacturing Schedules. In D.E. Brown and W.T. Scherer (eds), *Intelligent Scheduling Systems.* Kluwer, Boston, pp. 129-154.

Grant, T. J. (1986). Lessons for OR from AI: A Scheduling Case Study. *Journal of the Operational Research Society*, 37(1): 41-57.

Graves, S. C. (1981). A Review of Production Scheduling. *Operations Research,* 29: 646-675.

Halsall, D. N, Muhlemann, A. P. and Price, D. H. (1994). A Review of Production Planning and Scheduling in Smaller Manufacturing Companies in the UK. *Production Planning and Control,* 5(5): 485-493.

Hardstone, G. (1991). A Simple Planning Board and a Jolly Good Memory: Time, Work Organisation and Production Management Technology. *Edinburgh Programme on Information and Communication Technologies (PICT), Student Paper No. 4.*

Hurst, E. G. and McNamara, A. B. (1967). Heuristic Scheduling in a Woollen Mill. *Management Science*, 14(2): 182-203.

Kanet, J. J. and Adelsberger, H. H. (1987). Expert Systems in Production Scheduling. *European Journal of Operational Research*, 29: 51-59.

Kanet, J. J. and Sridharan, V. (1990). The Electronic Leitstand: A New Tool for Shop Scheduling. *Manufacturing Review*, 3(3): 161-169.

Kenworthy, J. G., Little, B., Jarvis, P. and Porter, J. K. (1994). Short Term Scheduling and Its Influence on Production Control in the 90s. In K. Case and S. Newman (eds), *Advances in Manufacturing Technology VIII.* Taylor & Francis, London, pp. 436-440.

King, J. R. (1976). The Theory-Practice Gap in Job-Shop Scheduling. *The Production Engineer*, 55(March): 137-143.

Kohler, C. (1993). New Technological and Organisational Solutions to Shop Floor Control. *Computer Integrated Manufacturing Systems*, 6: 44-52.

Lopez, P., Esquirol, P., Haudot, L. and Sicard, M. (1998). Co-operative System Design in Scheduling. *International Journal of Production Research*, 36(1): 211-230.

MacCarthy, B. L. and Liu, J. (1993). Addressing the Gap in Scheduling Research: A Review of Optimization and Heuristic Methods in Production Scheduling. *International Journal of Production Research,* 31(1): 59-79.

MacCarthy, B. L. and Wilson. J. R. (2000a). *Effective Decision Support in Scheduling: A New Approach Combining Scheduling Theory and Human Factors*. Final Report to the Engineering and Physical Sciences Research Council, May 2000.

MacCarthy, B. L., Wilson, J. R. and Crawford, S (2000b). *Toward World Class Performance in Planning, Scheduling and Control, From Progress Chasers to Responsive Teams.* EPSRC, Innovative Manufacturing Initiative Project (IMI).

MacCarthy, B. L., Wilson, J. R. and Crawford, S. (2001). Human Performance in

Industrial Scheduling: A Framework for Understanding. *International Journal of Human Factors & Ergonomics in Manufacturing*. In press.

McKay, K.N. (1992). *Production Planning and Scheduling: A Model for Manufacturing Decisions Requiring Judgement*. PhD Thesis, University of Waterloo, Ontario, Canada.

McKay, K. N. and Wiers, V. C. S. (1999). Unifying the Theory and Practice of Production Scheduling. *Journal Of Manufacturing Systems,* 18(4): 241-255.

McKay, K. N., Safayeni, F. R. and Buzacott, J. A. (1988). Job-Shop Scheduling Theory: What Is Relevant? *Interfaces*, 18(4): 84-90.

McKay, K. N., Buzacott J. A., and Safayeni, F. R. (1989). The Scheduler's Knowledge of Uncertainty: The Missing Link. In J. Browne (ed.), *Knowledge-based Production Management Systems*. North-Holland, Amsterdam, pp. 171-189.

McKay, K. N., Safayeni, F. R. and Buzacott, J. A. (1995a). Schedulers and Planners: What and How Can We Learn from Them? In D. E. Brown and W. T. Scherer (eds), *Intelligent Scheduling Systems*. Kluwer, Boston, pp. 41-62.

McKay, K. N., Safayeni, F. R. and Buzacott, J. A. (1995b). 'Common Sense' Realities of Planning and Scheduling in Printed Circuit Board Manufacture. *International Journal of Production Research,* 33(6): 1587-1603.

McKay, K. N., Buzacott, J. A., Charness, N. and Safayeni, F. R. (1992). The Scheduler's Predictive Expertise: An Interdiciplinary Perspective. In G. I. Doukidis and R. J. Paul (eds), *Artificial Intelligence in Operational Research*. Macmillan, Basingstoke, pp. 139-150.

Moray, N, Dessouky, M. I. and Kijowski, B. A. (1991). Strategic Behaviour, Workload and Performance in Task Scheduling. *Human Factors,* 33(6): 607-629.

Nakamura, N. and Salvendy, G. (1994). Human Planner and Scheduler. In G. Salvendy and W. Karwowski (eds), *Design of Work and Development of Personnel in Advanced Manufacturing*. Wiley, New York, pp. 331-354.

Norman, P. and Naveed, S. (1990). A Comparison of Expert System and Human Operator Performance for Cement Kiln Operation. *Journal of the Operational Research Society*, 41(11): 1007-1019.

Pinedo, M. (1995). *Scheduling: Theory, Algorithms and Systems*. Prentice Hall, Englewood Cliffs: NJ.

Rodammer, F. A. and White, K. P. (1988). A Recent Survey of Production Scheduling. *IEEE Transactions on Systems, Man and Cybernetics*, 18(5): 841-851.

Sanderson, P. M. (1989). The Human Planning and Scheduling Role in Advanced Manufacturing Systems: An Emerging Human Factors Domain. *Human Factors,* 31(6): 635-666.

Sanderson, P. M. (1991). Towards the Model Human Scheduler. *International Journal of Human Factors in Manufacturing,* 1(3): 195-219.

Sanderson, P. M. and Moray, N. (1990). The Human Factors of Scheduling Behaviour. In W. Karwowski and M. Rahimi (eds), *Ergonomics of Hybrid Automated Systems II*. Elsevier Science Publishers, New York, pp. 399-406.

Stoop, P. P. M. (1993). *A Model for the Short Term Logistic Shop Floor Performance Evaluation and Diagnosis*. BETA, Eindhoven, The Netherlands.

Stoop, P. P. M. (1993). *A Model for the Short Term Logistic Shop Floor Performance Evaluation and Diagnosis*. BETA, Eindhoven, The Netherlands.

Stoop, P. P. M. and Wiers, V. C. S. (1996). The Complexity of Scheduling in Practice. *International Journal of Operations & Production Management,* 16(10): 37-53.

Tabe, T., Yamamuro, S. and Salvendy, G. (1988). An Approach to Knowledge Elicitation in Scheduling FMS: Toward a Hybrid Intelligent System. In W. Karwowski, H.R. Parsaei and M.R. Wilhelm (eds), *Ergonomics of Hybrid Automated Systems I.* Elsevier Science Publishers, New York, pp. 259-266.

Thurley, K. E. and Hamblin, A. C. (1962). The Supervisor's Role in Production Control. *International Journal of Production Research,* 1: 1-12.

Vera A. H., Lewis, R. J. and Lerch, F. J. (1993). Situated Decision-Making and Recognition Based Learning: Applying Symbolic Theories to Interactive Tasks. *Proceedings of the 15th Annual Conference of the Cognitive Science Society*, June 18-21, Institute of Cognitive Science, University of Colorado-Boulder, pp. 84-95.

Wiers, V. C. S. (1996). A Quantitative Field Study of the Decision Behaviour of Four Shop Floor Schedulers. *Production Planning and Control*, 7(4): 383-392.

Wiers, V.C.S. (1997a). *Human-Computer Interaction in Production Scheduling: Analysis and Design of Decision Support Systems in Production Scheduling Tasks*. PhD Thesis, BETA, Eindhoven, The Netherlands.

Wiers, V. C. S. (1997b). A Review of the Applicability of OR and AI Scheduling Techniques in Practice. *Omega*, 25(2): 145-153.

CHAPTER THREE

Lessons from the Factory Floor

Kenneth N. McKay

3.1 INTRODUCTION

This chapter is a retrospective attempt to summarize the insights and opinions the author has accumulated since being challenged by John Buzacott in 1985 with the question: What is Scheduling? In a sense this paper is an update to McKay *et al.* (1988). The author has spent fifteen years studying the production control problem, attempting to bridge the gap between theory and practice. Simply stated, the aim has been

- to develop an understanding of what is and what is not scheduling and what is needed to help scheduling and planning in the real world.

During the time span of this extended research agenda, a variety of activities have transpired: scheduling systems have been built and deployed; quantitative models have been built and studied; the scheduling task looked at upside down and back to front; many discussions held with scheduling vendors and research colleagues, and many days spent in manufacturing facilities, watching and working with schedulers. While the subject matter has always been production control, the research methodology has been multi- and inter-disciplinary. Organizational behavior, sociotechnology, cognitive science, operations management, and information systems have been the domains from which field methods, modelling concepts, and inspiration have been derived. The research agenda has been described as unique, exploratory, inspirational and revolutionary by some, foolhardy and a waste of time by others. The agenda is not complete yet and the understanding is far from in a desired state – the end is not even in sight. As schedulers say, time will reveal all.

When the 1988 paper was written, personal computer based tools were in their infancy, AI was a promising area, and much work was being done on scheduling heuristics and algorithms. Many researchers were optimistic that the new and improved scheduling heuristics and related search concepts would be successfully embedded in the computer-based tools, which would then in turn be widely used in job shops: computers would generate the *optimal* or *near-optimal* schedules which would then be executed. Such an outcome would have justified the thousands of research efforts and supported the thousands of claims of relevancy. While this has happened in certain sectors such as process industries, single large machine problems, and at different levels of planning such as supply chain analysis, this has not happened for job shops. In this chapter the job shop will be the focus, as will the scheduling task. Most job shop style factories are still scheduling and planning the old-fashioned way. With few exceptions, the majority of job shop scheduling researchers are also caught in the past, researching a potentially insignificant part of the larger problem.

During the past decade, the author has been repeatedly approached by colleagues and asked several questions. Firstly, what has changed since 1988? Secondly, what is the solution to the scheduling problem? Thirdly, what can we do with the research results of the last three decades? Before addressing these three questions, the next section will first review the author's research and what key lessons were learned from each major activity. The paper will conclude with a number of reflections about factory scheduling in general.

3.2 RESEARCH ACTIVITIES AND LESSONS

There have been three distinct periods of the author's own research and roughly twelve major activities. Between 1985 and 1991 there were five activities which centered on scheduling in the arena of real factories, real work, and real people:

- research on the information flows involving the scheduler;
- co-authorship and design of Watpass (Waterloo Planning and Scheduling System with N. Henriques and J.B. Moore);
- development of a decision model for rapidly changing situations requiring human judgement;
- development of Watpass-II prototype (a time-sensitive scheduling system);
- specification of a scheduling 'factory from hell' benchmark and scheduling survey.

The period 1992 to 1994 was one of:

- quantifying some of the qualitative heuristics observed in practice (O'Donovan *et al.,*1999; McKay *et al.,* 2000a,);
- researching historical perspectives of planning and control to better understand the implicit assumptions and how we got to where we are today (McKay and Buzacott, 1999; McKay and Buzacott, 2000);
- developing a cyclic or bio-rhythmic view of manufacturing patterns.

1995 to the present saw renewed field activity and sociotechnical analysis of the problem:

- working on a task view of scheduling (Wiers and McKay, 1996; McKay and Wiers, 1999; McKay and Wiers, 2000);
- development of an extended field study site where an anthropological approach could be taken to the study of the customs and habits of schedulers;
- understanding dynamics and interrelationships in the middle of a supply chain
- development, deployment, and evolution of a scheduler's information system based on a sociotechnical view of the problem (now entering its fourth year of use at the field site).

Each of the above will be briefly discussed in the following subsections.

3.2.1 Research on the information flows involving the scheduler

McKay (1987) was a study of six job shops and documented the information flows across the schedulers' desks using an Organizational Behavior approach. In the study, the interpretation, attenuation, and amplification of the data was documented and categorized. The study showed the degree of human involvement required for dynamic situations where the information was incomplete, ambiguous, or context sensitive. The schedulers viewed the factory as an organic whole and seemingly used holistic reasoning to partition the scheduling problem and apply appropriate solutions.

3.2.2 Co-authorship and design of Watpass (Waterloo Planning and Scheduling System with N. Henriques and J.B. Moore)

Watpass was a simple interactive job shop scheduling tool having a self contained database describing the work to be done and the resources which were available. It had several different Gantt chart views and attempted to provide a seamless interface between the data, the schedule, and schedule manipulation. Watpass was started in 1985 and was designed for relatively small situations which might not have other systems such as Materials Requirements Planning (MRP). It was later transferred to a third party who enhanced it and marketed it as Tactic. Watpass served as an excellent vehicle for learning about what was needed and why generic tools such as this are very hard to develop, deploy and see sustained use. We worked closely with a number of sites and were provided invaluable feedback from the schedulers about features, functions, and failings. We quickly learned that i) the computer had to support the human (and not the other way around), ii) the human had to have almost complete control and be able to specify scheduling assignments regardless of what the database said was possible, iii) the human had to intuitively understand what the scheduling system was doing and why a decision would be recommended, and iv) the scheduler was under extreme pressure and the system had to be faster and better than the old-fashioned way if we wanted it used.

3.2.3 Development of a decision model for rapidly changing situations requiring human judgement

McKay (1992) was a logical extension of McKay (1987) and the lessons being learned from Watpass about the dynamic nature of scheduling decisions. Using field methodology borrowed from Cognitive Psychology and Action Analysis, two scheduling situations were studied for six months using a longitudinal field study design. A two-stage decision model had been formulated that incorporated many aspects of the interpretation and control observed in practice and a number of propositions made about its existence and utility. The focus of the model was on rapidly changing situations and the decision adaptability required if production was to be 'smooth' and 'efficient'. Over two hundred special decisions were captured at one site and analyzed using episodic encoding. The scheduler's actions fit the two-stage control model and clearly illustrated the type of feed forward and proactive control that was possible when the decision mechanism was sensitive to the context. The other site did not have a two-stage decision model and had all of the maladies and activities predicted. In both cases approximately 10-15% of the decisions to be made required special logic or handling and were not *mechanical.*

When recognized as such and treated with adaptability and judgement, the dynamic decisions were documented as improving the situation in the majority of cases. When the special situations were treated mechanically and not reacted to, extreme expediting, fire fighting, and unnecessary costs resulted. The field studies illustrated the extent to which the future can be predicted and how proactive policies can be used to provide stability and avoid or minimize future problems (McKay *et al.*, 1995a; McKay *et al.,* 1995b).

3.2.4 Development of Watpass-II prototype (a time-sensitive scheduling system)

The requirements analysis and design of Watpass-II was started in 1988 and a prototype shown in 1990. This was an object-oriented scheduling tool where every data element and record and every algorithm was time-sensitive. Furthermore, each data element had 'close enough' specifications to allow for constraint relaxation. The time-sensitivity was two dimensional and the tool knew when it was operating in real time (e.g., knew it was making decisions in January) and knew about the relative future (e.g., knew that the decision was three months on the time horizon). An aggregate resource model was used and the prototype was able to dynamically configure itself. For example:

> schedule with all of the constraints at the finest detail level for the next few days and require tight thresholds on finite resources; for the following two weeks use the work cell for loading at a finite level and allow a 10% threshold for calculations; and then for decisions beyond that, use infinite loading at the cell level.

This type of time sensitivity was important because we had observed schedulers dealing with jobs differently in the winter versus summer and if the decision was in the next week versus the next month or six months out. At the time, it was generally thought that separate databases and scheduling engines would be required for hierarchical decision modelling and Watpass-II served to show that it was possible to make those types of decisions in a monolithic system. We believe that Watpass-II was the first scheduling system to include the meaning of time (absolute and relative) and to apply these principles pervasively throughout the tool. Other scheduling systems then and now view each day through the year equally and with no special context. For example, they do not incorporate knowledge about humidity affecting the casting process in the summer but not in the winter when determining estimates for processing time, and they do not include the impact on quality of switching from heating to cooling during May. Systems use calendars to control resource availability, not resource capability! Some of the issues driving the design of Watpass-II were summarized in McKay *et al.* (1989a), McKay *et al.* (1989b), McKay *et al.* (1990), and McKay and Murgiano (1991).

3.2.5 Specification of a scheduling 'factory from hell' benchmark and scheduling survey

Created for the Intelligent Manufacturing Management Program (IMMP) of CAM-I (McKay and Moore 1991), the scheduling benchmark was the reverse engineering of the Watpass-II requirements in the form of a fictitious factory description. The benchmark documented over three hundred modelling issues and several hundred other requirements gathered by the author. The benchmark and its concepts were approved by the IMMP members as being representative of their needs and it was then used as the basis of a scheduling survey for CAM-I. The survey included the sponsors and it was interesting to note the concepts considered important and mandatory by the sponsors and not considered in the same light by the vendors. There was a distinct mismatch between what the sponsors considered to be the problem and the solutions being offered by the vendors. One of the vendors called the case study *The Factory From Hell* and the name stuck (McKay, 1993). The survey did not say what should be implemented and how it should be done – it was a survey of capability and understanding. In fact, in the Watpass-II design which was driven by the same requirements, only about 20% of the concepts and features were considered to be hardwired while the rest would be supported by a scheduling language and grammar. Since its creation, *The Factory From Hell* continues to be regarded as a comprehensive requirements document for scheduling and has been used by several organizations as a concept benchmark and challenge document. The case study and survey clearly illustrated the magnitude of the problem faced by generic tool providers.

3.2.6 Quantifying some of the qualitative heuristics observed in practice

In researching what schedulers did, a number of qualitative heuristics and 'rules of thumb' were encountered. For example, they would choose a 'special' job to schedule next on a machine after the machine was repaired. Or, they would decide to run a small batch of a part prior to running the whole batch because something about the process or situation had changed since the last time. We often hear that the human schedulers do not make smart choices and are incapable of generating good schedules. Were the qualitative heuristics used by the observed schedulers an asset or liability? To probe this further several of the heuristics have been formulated and experiments performed. The model most refined is called Aversion Dynamics (O'Donovan *et al.*, 1999; McKay *et al.*, 2000a) and clearly shows the wisdom of the scheduler and that the momentary sub-optimization on a local level yields global optimization results when the bigger picture is accounted for. The key to many of the heuristics is the temporal nature of the problem - a trigger activates the special logic and the special logic is active for a limited time and its power to make special decisions decays over time. A major part of research has been in trying to get the quantitative heuristic to behave like the observed rule of thumb. The preliminary research results suggest that when the 'scheduler' is right about what might happen, the results are significantly better and when the 'scheduler' is wrong about the event, the sub-optimization for a limited time is not terrible and the overall schedule is not severely punished. Work continues on this class of dynamic and context-sensitive heuristics.

3.2.7 Researching historical perspectives of planning and control to better understand the implicit assumptions and how we got to where we are today

McKay and Buzacott (1999) present an analysis of the underlying assumptions used by the prevailing production control methods and models in use today and discuss the inherent weaknesses in the approaches for today's manufacturers. For 150 years, authors have been discussing factory design and production control methods (McKay *et al.*, 1995c). Many, if not all, of the *new* concepts advocated in the last twenty years can be found in the literature from the turn of the century. It is interesting to see how standardization, specialization and mass production paradigms came into being and what the original authors warned and debated about. There have always been the handful of very successful factories where everything was synchronizd and which all the other manufacturers tried to copy. The copiers would visit the star's factory and then try to achieve the same result with a single silver bullet or theme of the month. Needless to say, this technique was as successful then as it is now. But, what of the basic control paradigms? None of the paradigms implicitly imbedded in our MRP-II and Enterprise Resource Planning (ERP) systems or in our hierarchical production planning models are designed for rapidly changing situations where human judgement is required. The current techniques are designed for relatively mature and stable situations and not where change is sustained and pervasive. For example, centralized production control for all of the product line is unwieldy and is not conducive for learning about the results of decisions. Conversely, techniques such as focused factories, with the possibility of feedback and learning, are better suited for changing situations and there are a number of techniques that can be used by firms to manage during change.

3.2.8 Developing a cyclic or bio-rhythmic view of manufacturing patterns

The manufacturing world is not made up of independent random events. The past and future are not simple time series where day 10 is inherently the same as day 11. Almost everything in a manufacturing situation can be viewed as having a life cycle and knowing where any object is in its life cycle provides valuable insights into the scheduling problem and possible solutions. For example, knowing that a machine is close to its decommissioning can provide clues about its potential operation under stress conditions. Knowing that a part has had almost all of its operations already performed indicates how much value added has also been done and what the risk is if the part is now damaged in process. Knowing that it is the last week of the fiscal quarter will provide clues about what constraints will be relaxed in order for certain objectives to be met. In addition, weeks start off in a certain way, have a plateau mid week and wind down towards the end - another miniature life cycle pattern. There are many of these patterns and they interact in many interesting ways. Understanding these patterns, what triggers them, what they mean, and how they can be exploited is part of what makes a good scheduler great. It is how they avoid doing the 'dumb' things and minimize the risk to processes and material. The Aversion Dynamic model mentioned above is an example of a minor life cycle for a machine and its meaning - the cycle between repairs and what happens at a repair point.

3.2.9 Working on a task view of scheduling

A sociotechnical view looks at the degree of control the scheduler has in determining the final sequence and how much uncertainty there is in the situation (Wiers and McKay, 1996; McKay and Wiers, 2000). A topic specifically addressed is the gap between the theory and practice of scheduling and a number of principles and concepts for a unified theory of scheduling that attempts to bridge the gap have been formulated (McKay and Wiers, 1999).

3.2.10 Development of an extended field study site where an anthropological approach could be taken to the study of the customs and habits of schedulers

For the past four years, the author has been working closely with a firm to overhaul its production control processes (McKay, 2000). This has been done in such a way that the researcher has been accepted by the rank and file workers and is seen to be one of them. The ability to sit, observe, and work with the dispatchers and schedulers on a daily basis provides an environment wherein the results of prior decisions can be monitored, information flows observed, and an intimate knowledge of the plant, products, and processes obtained. It also provides a better appreciation of what it means to be at the factory at 5:45am and what work gets performed prior to a shift change. Over time, subtle relationships reveal themselves and these fine points would not be obvious to an occasional tourist or inexperienced researcher. The author remains amazed by the schedulers' ability to keep track of so much data in their heads and to bring it to bear as needed. Considering the amount of data and the complexity of the usual scheduling problem, it is amazing that they make so few mistakes. However, not all schedulers are good, and not all schedulers are good at all scheduling tasks. Determining the strengths and weaknesses of a scheduler is almost impossible to do via short visits and superficial analysis of schedules and results. It requires a longitudinal field project where events and conditions can be tracked and recorded - clearly showing what went wrong and if it was indeed the scheduler's fault or an artifact of the system itself. The information flows and decisions observed during the study were used in the design of the scheduler's information system (see below, McKay and Buzacott, 2000). It was important that the system matched the 'good' habits and built on the existing culture and processes. Working closely with the schedulers also allowed us to improve the situation manually first, prior to the design and implementation of a scheduling tool. The intimacy and awareness of the scheduler's task was important since the schedulers had to believe that the researcher understood their problem, their situation, and what their constraints were. They also had to believe that the system ultimately developed by the researcher was designed to help them and to reduce stress and work and not add work. In addition to the intimate knowledge about this specific factory, the author obtained confirmation about the dynamic nature of the problem and the existence of common themes across manufacturing sectors and industries.

3.2.11 Understanding dynamics and interrelationships in the middle of a supply chain

The extended field site is an automotive supplier and is situated between the steel mills and its customers' assembly plants. As such, it has been an excellent site from where to understand the demand end of the assembly plants, observe the first and second tier impacts within the plant and the third tier demand directed at the steel mills. Of particular interest has been the impact on scheduling at the second tier within the plant and how the first tier can operate to provide better stability upstream. The basic lessons in this case relate to i) exploiting what downstream operations can do to improve upstream scheduling and ii) what upstream operations do to create future problems for themselves. For example, the first tier operations can use their inventory banks intelligently and stabilize the demand on the second tier. In certain cases, they do not have to ripple back changes from the assembly plant immediately and can absorb the fluctuations. They can also use their inventory banks to take the pressure off the second tier in a crisis situation and plan for future recovery in an orderly fashion. Such co-operation between two different areas of a large plant is not to be taken for granted and the system view must be consciously applied. In a repetitive job shop using raw materials that come in bulk form (e.g., steel coils) the quantities produced rarely match the demand, and production sizes can be under or over. This uncertainty results in the next orders out on the horizon moving and giving any long lead time items a degree of nervousness. Similarly, if a job is pulled short or is run longer than it should, this will also create problems in the supply chain. Often schedulers and dispatchers are unaware of the impact they have on the whole system and this is an area for improvement and further research.

3.2.12 Development, deployment, and evolution of a scheduler's information system based on a sociotechnical view of the problem (now entering its fourth year of use at the field site)

Three years ago, the senior scheduler took the prototype version of PPS (Production Planning System) for his first test spin. He was basically computer illiterate and had just seen the system do several hours of his manual work in several minutes. He was able to understand it and able to run it. He was able to understand when things had to be done and he was able to 'see' his shop as he put it. He only required several hours of training. How had this been done? The author had built the system while sitting with the junior scheduler who had been re-assigned for the purpose. A rapid prototyping style of software development was used and the first version was ready in a month once we decided that a custom system was needed. The system was designed specifically to fit the task and to fit the schedulers' model of the problem (Wiers and McKay, 1996; McKay, 1997). It was designed to facilitate partial decision-making with partial information and to match the information flows crossing the scheduler's desk at various points of the day. It is extremely custom built and uses the plant terminology and meta-objects. For example, the schedulers and plant use shifts as a basic unit of measure, so many of the features and functions use shifts as the focus: slide work to the next shift, split at shift boundary, etc. The tool generates shop floor paperwork and the schedulers can intercept the reports before they are printed and edit any data as necessary - shop floor paperwork must reflect what will be done and not what the computer's database says will be done. The system is hooked to the well known ERP system SAP and obtains demand and inventory information in a batch fashion. Due to its secondary database and information nonexistent in SAP, the scheduling tool now generates almost all of the operational reports for half of the factory (i.e., the second tier) and its reports are used by approximately 75% of the planners, supervisors, managers, and so forth.

The approach has been to work with the schedulers, encouraging the better practices and where necessary eliminating the undesired practices through training or function removal. The schedulers perceive that they are in control of the system and that the system exists to help them. The development and research strategy had been laid out in advance and the author had *carte blanche* from the plant manager to do it the 'right way'. The general opinion at the plant is that the scheduling tool has been successful and that production control is now ready for the next stage in its evolution. The author does not claim that the results are transferable or general, or that this development approach is quick and inexpensive. However, the difference in use between a custom system like PPS designed with a task view and fitting the information flow and a generic tool like Watpass is astounding. It is difficult to obtain supporting evidence, but a common discussion among people involved with generic scheduling tools is how they seem to make more revenue from the follow-up customizing and consulting than they do from the primary tool itself!

3.2.13 Summary

To summarize, there have been twelve major research activities and each has contributed to a better understanding of what scheduling is, or is not. What is scheduling? Scheduling is not something you can define in a single sentence or even two. Scheduling is an activity that involves a wide assortment of inputs and a large number of constraints with an equally large set of rules for constraint relaxation and enforcement. There are many trade-offs involving local issues and global system issues. The author is still not sure what scheduling is. The following three sections address the questions noted in the introduction: what has changed since McKay *et al.* (1988) appeared? what is the solution? what can we do with research results of the last three decades?

3.3 WHAT HAS CHANGED?

Since 1988 some things have changed for the good, some for the worse, and others have remained the same.

Firstly, what has changed for the worse? The scheduler's job is more stressful, as inventory is removed from the system and the system is leaner and meaner. Personnel formerly used to help verify data and to verify options has been downsized. Personnel used to back-fill during training and vacation has also been reduced, which in turn adds to the scheduler's workload several times during the year. The schedulers are expected to be computer literate and be able to use spreadsheets, word processors, e-mail, ERP systems and do this with little or no training. There also appears to be even more pressure than before in hitting the fiscal month and quarter targets for the stock markets. The processes and products are continuously changing and the schedulers are finding it harder to keep up with what can be done, where, and to what level of performance. Because things are changing so rapidly, the Industrial Engineering estimates, ever suspect, are more suspect. There seem to be fewer researchers investigating the scheduling problem and assuming that scheduling research has some utility, this is not a positive change. It is the author's sense that many scheduling researchers are now looking at the supply chain and larger issues and no longer focusing on the mundane level of dispatching and sequencing. Detailed planning and dispatching may be considered mundane by some, but if you have ever had the luxury of sitting with real dispatchers in the heat of battle, mundane is not a word that comes to mind. The problem still exists and is as real and as challenging as ever before – possibly more so.

Secondly, what is largely the same? The contextual nature of the scheduling decision is largely intact and has not changed since 1988. Humidity is still with us, as are the weekend effects, vacation effects, first snowfall effects, the state-of-the-art risks of new technology, and so forth. The general state of academic scheduling results used in industry remains the same - rare events which border on statistical noise. The fact that scheduling algorithms exist in scheduling tools does not mean that they are used or that they generate schedules which are feasible and capable of being executed. Outside of the process situation or single-machine type of situation or general logic for controlling work release into a complex situation like wafer-fab, there are no well-documented and well-known examples of a true job shop being automatically scheduled with traditional scheduling heuristics and where the executed plan resembles the original computer-generated plan. Computer algorithms are used to sequence work in a due date order or basic priority order and then the human scheduler is usually expected to tune the schedule manually by moving the Gantt chart bars about. The basic problem definition used by the vast majority of academic researchers remains the same and does not embrace any real world phenomena.

Thirdly, what has changed for the better? There is an increasing awareness of the scheduler and the need for an enhanced and extended view of the scheduling problem. This is illustrated by this book which would have been far thinner in 1988. The inclusion of some real-world phenomena in mathematical modelling is slowly becoming accepted and, while it is still scarce, it is happening and there are signs that it is becoming easier to do (see McKay and Wiers, 1999). In addition, many more senior academics are publicly recognizing that there is a problem. Programs such as the Leaders For Manufacturing Program at MIT have exposed faculty to situations previously unexplored and the research community is stronger as a result. The exposure is too recent to have had a major impact yet, but it is expected that this awareness will start filtering down into graduate work, the reviewing of papers, and so forth. The computers are faster and graphics nicer, but are the generic scheduling solutions better today than yesterday for the job shop problem? Some will claim yes, but the author believes otherwise. Many of the finite scheduling tools introduced in the late 1980s and early 1990s are no longer on the market and others have taken their place. The fundamental assumptions still exist and the vendors find new and unique challenges at each factory when they try to install their generic system. Some will say that ERPs such as SAP or Baan have improved the scheduler's life. This is partially true. The scheduler using an ERP has a nice database system that supports queries, tracking, and integration of functions. However, the ERPs do not incorporate finite modelling and do not directly support scheduling and schedule manipulation of the kind needed. In addition, the MRP logic today is basically the same as yesterday and is still best suited for mature, stable, mass-production situations. A better situation exists for planners orchestrating the supply chain: excellent tools exist today for supply chain management and for aggregate modelling of the situation. Several of these tools started off their lives as detailed finite schedulers in the process industry or in highly regimented situations. While they were not widely successful at detailed job shop scheduling, they have provided excellent value at the higher levels of planning where many constraints can be ignored, minor instabilities assumed away, and the decisions are 'bigger'. Another positive change is associated with the personal computer. It is relatively easy today for factories to build small Excel tools to help track shop data and analyze what is happening.

To summarize, there have been some changes which have improved the scheduler's life but most of these do not relate to sequencing tools and automated scheduling systems. From what we hear about experiences in the field, the level of difficulty in installing scheduling tools has not decreased during the decade and we are far from a turn-key application which is truly generic. There has also not been any significant and widespread improvement in the transfer of scheduling technology or knowledge from the world of academia to industry. As an optimist, the author would point to this book and the efforts to build context-sensitive heuristics as signs that we have turned the corner and there is light at the end of the tunnel. As a realist, the author points out that we are not at the corner yet and cannot see the end of the tunnel but that there is a reasonable chance that there is indeed a light there. We will eventually see the light if we work with a realistic problem definition and are rooted in real factory experiences.

3.4 WHAT IS THE SOLUTION?

The answer depends on what the problem is. The question has been posed in several contexts. Firstly, how do we make scheduling research relevant? Secondly, what are models that will work? Thirdly, how do we get industrial practitioners to use the scheduling research?

We can make scheduling research relevant in several ways. The first challenge is to develop a better understanding of what the problem is and how the problem changes within a factory and how the problem changes between factories. We have only begun in this area and there is much work to be done. Once we have a better understanding of what scheduling is, we can then start to build frameworks and taxonomies that support research results. While we are not sure what scheduling is, we are beginning to understand what scheduling is not. Scheduling is not a single heuristic applied day in and day out for the complete time horizon. Scheduling is not a single decision using only primary data (machine, job data) but it also involves knowledge about secondary support functions and impacts. Scheduling is not a static control model that does not learn or adapt to new processes, materials, or events. Scheduling is not a task absent of human judgement. These observations suggest that to make scheduling research truly relevant, an integrated and interdisciplinary framework is needed that supports a variety of algorithms and concepts at various levels. Without a framework, it is difficult to know where any little piece fits or what its potential relevance is. The scheduling research community must also invite and work with researchers from different disciplines to cure the whole patient and not just a finger nail.

It is not clear what models will work. McKay *et al.* (2000b) present an overall analysis of the current state of scheduling research and how this relates to the practice of scheduling. The author has also proposed a two-stage control model that adapts and learns in its environment and can handle perhaps 60-70% of the special cases we have seen humans address (McKay, 1992; McKay *et al.,* 1992; McKay *et al.*, 1995d). The remaining percentage will still require human skills - identifying the situation and creating solutions to match the context. This type of sophisticated control is probably inappropriate for almost all manufacturing situations today and would fit only the best shops. Even in the best shops, it might be better and more cost effective to hire university students to function as the intelligent control agent using previously generated rules of thumb to identify situations requiring non-mechanical decision-making. In many shops, until the bigger problems are solved, fine tuning the scheduling via computer software is meaningless. It is more effective to train the schedulers and production control supervisors to think about proactive control and to be aware of their environment. It has been suggested that the two-stage control model and dynamic heuristics such as Aversion Dynamics be imbedded in a large scale computer network for scheduling distributed work. The computer environment is a clean and pristine situation compared to most real factories and we hope that this will serve as a vehicle to demonstrate the benefits of context-sensitive heuristics. Once the first implementation in a clean situation has been achieved and dynamics analyzed, it might then be appropriate to create a factory scheduling tool using the same principles and determine its utility in a less clean environment.

Why don't factories use our results? The answer is fairly obvious and does not start with the factories. Industrial practitioners are constantly looking for tools and ideas which will help them manage their chaotic environment and make them more competitive. The author has yet to meet a scheduler who loves chaos, stress, and people yelling at them. They will rapidly accept and use a tool that helps them - they will be grateful and be your friend. However, the tools must match the problem at hand, be understandable, and be up to the task of being used daily in a hostile environment. Furthermore, the scheduling tool cannot constantly generate *silly* schedules and force the scheduler to make (what are to them) obvious corrections. There is another challenge to getting research results used in practice. A scheduling algorithm, regardless of its optimality, cannot be used in a real factory without all of the other bits and pieces that go into a fully fledged scheduling system. A researcher may believe that their special algorithm will be the best thing to ever happen to the plant and to industry at large, but how is someone to use it? Even without real time inventory and demand interchange, sophisticated modelling of the bills of material, electronic data interchange, etc., scheduling systems start out at about 50,000 lines of code and go up from there. Building such a tool takes significant time and to make it robust and of high quality requires a substantial quality assurance effort. Unless the scheduling tool is interfaced with the plant's existing system and provides the necessary representation and manipulation and provides the necessary outputs, the tool will NOT be used in an operational setting in a sustained fashion. There are not too many researchers who can or will build an industrial strength scheduling tool to house their favorite research results and then invest the time and energy to deploy the tool in a real setting. If the researcher does not build the scheduling tool, can they work with a scheduling vendor? This is difficult for a number of reasons: i) there are many more researchers than vendors, ii) vendors have little interest in academic studies and experiments which are time and resource consuming, and iii) most vendors understand the limitations of the myopic models typically worked on by researchers. Scheduling tool vendors have discovered that at the detailed job shop level, almost any simple algorithm is good enough since the problem will be changing in the next half hour if not sooner. They have invested their research time and energies into the supply chain level, process industries, and software structures which lend themselves more easily to customization. Unless someone is funded to produce a public domain, industrial strength scheduling tool for researchers to work with, it is doubtful if the majority of researchers will ever see their results used by a real factory. Some researchers will be lucky enough to work with a scheduling vendor or find a plant willing to fund a speculative project, but these will be rare.

3.5 USING THREE DECADES OF PRIOR RESEARCH

There are three ways in which we can find value in the vast body of research results which are not directly relevant for job shop use.

Firstly, an abstract analysis of the problem is useful for establishing theoretical bounds for performance. The bounds will never be reached, but progress or digression can be measured against the ideal. In this case, it is more important to have a rich representation than to have a quick algorithm since setting bounds is not an operational activity and would not be continuously done in a factory.

Secondly, if we can identify periods of stability and simplicity, the models can be used for those periods. There are many analogies for this approach. For example, a captain of a sailing vessel can use an auto-pilot quite safely when on the open sea but should use caution and turn off the auto-pilot when close to shore in dense fog. This approach is a basic swap - one set of logic for another depending on the state. Conceptually, the human would make the sequencing decisions when the system was not stable. The computer system could do a rough plan and have the human tune it.

Thirdly, there are the types of heuristics similar to the Aversion Dynamics noted above. These heuristics are more of a control theoretic approach to the problem. They are heuristics which are built into the main scheduling heuristic but which normally have no effect. When triggered, they modify problem parameters to reflect the meaning of the event and this in turn alters priorities and the ultimate decision choice. The heuristics use exponential decays to reduce their effect over time. The decay matches the event and can be easily tailored. For example, it is easy to determine the exponential decay to restrict the power of the heuristic to within one or two jobs, or one or two shifts. This approach is very different from the one outlined in the paragraph above, where patterns of stability have to be recognized, logic swapped, and human intervention invoked. In this third approach, the assumption is that the system is relatively stable most of the time and needs special logic for the unstable periods. The traditional heuristic is used as the base and is used in its pure form when the system is stable and in a modified form during instability. This does not dramatically alter the structure of the schedule since the same underlying rationale is used for both periods.

3.6 THE IMPORTANCE OF TIME – ABSOLUTE AND RELATIVE

It is natural and necessary in problem solving to abstract, simplify, reduce, and make assumptions about the problem. This is how general models and theories are derived and evolved. The power of general models lies in their ability to be applied to different contexts and explain basic real-world phenomena in hindsight (descriptive) or in foresight (prescriptive). The trick of course is to pick the right elements to abstract, simplify, and make assumptions about. It is difficult to pick the elements that make up the salient nature of a phenomenon without studying the phenomenon first. This is not a new observation. In the early days of Operations Research, it was recommended that the problem be understood prior to mathematical modelling. The author believes that this step was not done properly for the scheduling problem. As a field, we have been solving part of the scheduling problem, but not the scheduling problem itself.

Throughout this retrospective, the reader will notice that time and context are repeated themes. It keeps coming up and it does not go away. It is not possible for a foundry manager to wish away summer. It is not possible for a manager to ignore that production on the third shift is not and never will be the same as production on the day shift. It is not possible for the manager to ignore that repairs or upgrades are not always 100% and there might be a period of instability following such an event. However, in *all* of the traditional research on scheduling in job shops, these types of factors are ignored and assumed away. Time is always modelled in an abstract way with each time unit independent of preceding and succeeding units.

Time is used to measure duration but it is not used to guide sequencing, batch sizing, or special actions. Time has no meaning to scheduling researchers. Time has been assumed away!

If you assume that time has no meaning, you will not include it in models. It is also likely that you will look at the problem or situations with this set of glasses and not see the importance of time even if it is staring you in the face. When time and context is assumed away, the problem looks simple and without any further probing of the problem, a math model can be created and experimented with. This is a nice, safe and secure world to perform research in and the math is challenging enough without added complications. To include the meaning of time would require the researcher first to take the time to discover the meaning of time and that would take away from the modelling activity and require different training and skill sets. The academic model of research and advancement is not friendly to this scenario.

The fact remains though that time and the meaning of time is the essence of scheduling in the real world. It defines the problem and it defines the set of possible options that the scheduler can choose from. Without including the essence of the problem, it is not unexpected that the resulting solutions will not be relevant.

Time has been constantly at the center of the author's research. In the beginning, it was 'what do schedulers do with their time?' as a way to understand what is scheduling. Then, it was 'why is the scheduler making that decision now, when in a seemingly similar situation before, the scheduler made a different decision?' and 'how does the scheduling decision-making adapt to the situation.?' Recently, it was 'how can a scheduling tool be designed to fit what a scheduler needs to do and when they need to do it?' More than ever before, the author believes that the understanding of time is a key component missing from the research activities. The challenge is to define the meaning and to develop appropriate techniques to include the meaning in the modelling.

3.7 REFLECTIONS

It is difficult to perform research on a single topic for over a decade and not develop opinions. The opinions may not be supported by quantitative results, but they are important nonetheless. The following three subsections expand upon themes and issues about which the author has developed opinions.

3.7.1 Importance of scheduling

Is scheduling an issue worthy of research? The answer depends on what you define scheduling to be. If scheduling is simply sequencing work at a machine independent of all context, the answer is probably no in all cases. It is more important to manage the tactical decisions, to orchestrate the supply chain, to ensure products are designed for manufacturability and so forth. If you do not have good overall management and a system designed for success, better scheduling of any form will not help you a great deal. Poor scheduling may be the straw that breaks the camel's back, but the camel is doomed in any event. If the factory and products are well-designed, then the quality of scheduling becomes a key factor in the process. Good scheduling will not occur where events are not reasonably predictable and consistent.

Quality scheduling is very important for lean manufacturing, nimble processes, and where a few 'profit points' separate success from failure; for good scheduling can reduce flow times, inventory costs, overtime, subcontracting, and many other indirect costs. But, what constitutes quality scheduling? It is the author's opinion that quality scheduling must address the whole system and relies upon the context in which the decisions are being made - when, how, what, why, and by who. Quality 'scheduling' must encompass the decisions a scheduler makes and we need to understand what these decisions are and then perform research on them. Sequencing is just one such decision and is a minor one.

3.7.2 Value of lateral thinking

Assuming that scheduling is important, is scheduling research in the classical view of sequencing at the machine level important anymore? If it is, it is not as widespread as it once was. Consider the lean and mean manufacturing process. There are small inventory buffers and the work is pulled or flowed through the system. It has been our experience that many of these situations are so lean that due dates are the single driving force and there are few times that any amount of tardiness is acceptable. In these cases, the scheduler's first priority is how to avoid being late at all costs. The scheduling task becomes one of seeking and making alternatives when there are conflicts for resources. The second priority is to avoid doing things to make the situation worse; things that would damage machines or material. In fact, in early descriptions of the scheduler's task, a main part of the scheduler's job was to anticipate the future and discount problems prior to their happening. The third priority is to reduce costs associated with unnecessary setups and inventories. These priorities and actions suggest that for these situations, the most important entity to manage is the inventory bank and keeping it in harmony with the resources. That is, instead of taking the quantity and due dates as given, these can be variables (to some degree) and be used to alter the job load to avoid conflicts. For example, if a resource is used to make two parts, when it starts one part there should be enough inventory of the other to cover the first's build cycle. If there is not, there will be a conflict when the second part's quantity hits the reorder point. The problem is compounded when there might be dozens or hundreds of parts crossing the same resource. Perhaps we should be looking at how to change the problem dynamics to create smooth loads instead of accepting the problem parameters and attempting to minimize total tardiness?

3.7.3 Value of job shop scheduling tools

It is the opinion of the author that many generic scheduling tools are purchased by desperate people attempting to stop the bleeding associated with fatal wounds unrelated to the scheduling task itself. The systems sold look pretty and have excellent demonstrations that illustrate to higher management how easy it will be for the schedulers to use. Upper management, often not understanding what schedulers do, naively believe that the scheduling tools can be used to create plans which can be executed with little or no human intervention. If they are lucky, production control personnel can be reduced, or at least all of those 'scheduling mistakes' will be eliminated. The author has encountered many senior managers who fit this description. Instead of fixing other management, process, and system problems and managing the whole plant in an effective and efficient manner, they turn to a scheduling tool and abdicate their responsibility. It is a rare plant manager who has a true appreciation of production control and the associated challenges. In too many cases, the scheduler has little input and is told that a solution to all of their ills will be delivered shortly.

The author has had numerous discussions with scheduling vendors over the years and a common topic centered around the benefits associated with scheduling tools. In some cases, the tool never leaves Industrial Engineering or the manager's desk and gathers dust. When put into practice, the first and in some cases the only benefit comes from trying to install such a system. The installation of an *expensive* system frees up the necessary resources to clean up the databases, update the process descriptions, and provide some order in the midst of chaos. After that, it is suspected that most of the systems are used for reports and for tracking, but not for scheduling *per se*. Each day, the schedulers either adjust many of the future jobs to fit a view of reality or do not bother because it will be different tomorrow – they really only set the true sequence or schedule for the next 24-48 hours.

In one case the author knows of, the major benefit came from having all of the supervisors and shifts use the same rules and policies - this dramatically reduced the down-time associated with shift changeovers and yielded a 10-15% production gain. It was not the generated schedules. On a technical level, a $5,000 solution is probably good enough for most factories to get the first benefit but an expense of that magnitude would be insufficient to get management's attention. Many scheduling tools cost in excess of $100,000 when you add in all of the direct and indirect costs and this level of expense will get management's attention. A gain of 10-20% will likely arise from the initial clean up and sorting out of production – regardless of tool choice. The difference between solutions will be at levels beyond this and the author is not aware of any data-supported results in job shops for generic tools where higher gains have been made.

3.8 CONCLUSION

The author is a self-acknowledged plant rat. Each and every day in a plant has provided new insights or reinforcement to maturing ideas. Talking to people involved in production control is almost always interesting and engaging. Hearing about it second-hand from a graduate student is not the same as being there and living it. Working alongside the schedulers for months on end, from 5:45am till the end of their shift has been an education unattainable elsewhere. The time and energy invested has provided deep knowledge about subtle interrelationships that exist in constraint relaxation and production dynamics. Spending time in the factory takes away time from building models or writing up results. There are few outlets for publishing such research once it is written. The response of the research community once it is printed is also interesting. Papers such as McKay *et al.* (1988) created as much negative reaction as positive encouragement.

The original question that started this research agenda remains largely unanswered: What is scheduling? This question is too large for any single researcher to answer or to understand (yet). The invitation to write a retrospective has forced the author to stand back and take stock. We may indeed be close to a turning point and hopefully we are building momentum. Research collaborations and interdisciplinary teams seem to be part of the answer. Looking for patterns in the workplace, understanding what a scheduler does, and understanding why special decisions are made are good starting points for researchers interested in this question.

What next? The author hopes to continue working with the field site and participate in the evolution of production control in a real way. The work to quantify the qualitative heuristics will continue as well. The impact of the types of manufacturing found in a supply chain on production control is a new topic that looks very interesting. The timing of this volume marks a turning point for the author and it is suspected that the field work will decrease slightly and new collaborative research partnerships will be formed to look deeper into the scheduler task and the scheduler-computer interface. Planning versus scheduling is also a new theme that bears analysis and research.

It is time for the research community interested in the human scheduler to compare notes and to work together. It is a relatively small group, but it is full of researchers with a passion for the topic and there are exciting times ahead as we begin to understand the problem better. The opportunity exists for us to formulate descriptive and prescriptive theories which will help schedulers do their task and that is our challenge. As individuals, it is unlikely progress will be made, but together there is a high probability we can make an impact.

3.9 ACKNOWLEDGEMENTS

The author would like to acknowledge all past and present co-authors for insightful discussions and research collaborations. Professors Buzacott and Safayeni provided the opportunity to do non-traditional research and the intellectual stimulation that continues to provide motivation. Of equal value have been all of the schedulers and planners who generously shared confidences and explained what they really did and not what their managers thought they did. The research activities have been graciously supported by Canada's NSERC agency.

3.10 REFERENCES

McKay, K. N. (1987). *Conceptual Framework For Job Shop Scheduling*, MASc Dissertation, University of Waterloo, Department of Management Sciences, Ontario, Canada.

McKay, K. N. (1992). *Production Planning and Scheduling: A Model for Manufacturing Decisions Requiring Judgement*, PhD Dissertation, University of Waterloo, Department of Management Sciences, Ontario, Canada.

McKay, K. N. (1993). The Factory from Hell - A Modelling Benchmark, *Proceedings of the NSF Workshop on Intelligent, Dynamic Scheduling*, January 1993, Cocoa Beach, Florida, USA. pp. 97-113.

McKay, K. N. (1997). Scheduler Adaptation in Reactive Settings - Design Issues for Context-Sensitive Scheduling Tools, *Proceedings of IEE Conference Practice and Applications*, November 1997, San Diego, USA, pp. 1283-1288.

McKay, K. N. (2000). *Improving Production Control via Action Science: A Study in Patience*, Working Paper.

McKay, K. N. and Buzacott, J. A. (1999). Adaptive Production Control in Modern Industries, in *Manufacturing Systems Modelling: From Strategic Planning to Real-time Control*, A. Villa and P. Brandimarte (eds), Springer-Verlag, pp. 193-215.

McKay, K. N. and Buzacott, J. A. (2000). The Application of Computerized Production Control Systems in Job Shop Environments, *Computers in Industry*, **42**, pp. 79-97.

McKay, K. N. and Moore, J. B. (1991). Planning & Control Reference Case Study - Version 1.3, *CAM-I IMMP Technical Report R-91-IMMP-01*, Arlington, Texas, USA.

McKay, K. N. and Murgiano, C. J. (1991). Decision Support Requirements for Planning and Scheduling Within The Automated Factory, *Proceedings of Autofact '91*, November 1991, Chicago, **10**, 1, pp. 10-19.

McKay, K. N. and Wiers, V. C. S. (1999). Unifying the Theory and Practice of Production Scheduling, *Journal of Manufacturing Systems*, **18**, 4, pp. 241-255.

McKay K. N. and Wiers, V. C. S. (2001). Decision Support for Production Scheduling Tasks in Shops with Much Uncertainty and Little Autonomous Flexibility, in *Human Performance in Planning and Scheduling,* B. MacCarthy and J. Wilson (eds*)*, Taylor &Francis, London, pp. 165-177.

McKay, K. N., Safayeni F. R. and Buzacott, J. A. (1988). Job Shop Scheduling - What is Relevant?, *Interfaces*, **18**, 4, pp. 84-90.

McKay, K. N., Buzacott, J. A. and Safayeni, F. R. (1989a). The Scheduler's Information System - What is Going On? Insights for Automated Environments, *Proceedings of INCOM'89*, September 1989, Madrid, Spain, pp. 405-409.

McKay, K. N., Buzacott J. A. and Safayeni, F. R. (1989b). The Scheduler's Knowledge of Uncertainty: The Missing Link, in *Knowledge-Based Production Management Systems*, J. Browne (ed.), North-Holland, Amsterdam, pp. 171-189.

McKay, K. N., Buzacott J. A. and. Safayeni, F. R. (1990). The Scheduler's Desk - Can it be automated?, in *Decisional Structures in Automated Manufacturing*, A. Villa and G. Murari (eds), Pergamon Press, Oxford, pp. 57-61.

McKay, K. N., Buzacott J. A., Charness N. and Safayeni, F. R. (1992). The Scheduler's Predictive Expertise: An Interdisciplinary Perspective, in *Artificial Intelligence in Operational Research*, G. I. Doukidis and R. J. Paul (eds), Macmillan Press, London, pp. 139-150.

McKay, K. N., Safayeni, F. R. and Buzacott, J. A. (1995a). Schedulers & Planners: What and How Can We Learn From Them, in *Intelligent Scheduling Systems*, D. E. Brown and W. T. Scherer (eds), Kluwer Publishers, Boston, pp. 41-62.

McKay, K. N., Safayeni, F. R. and Buzacott, J. A. (1995b). 'Common Sense' Realities in Planning and Scheduling Printed Circuit Board Production, *International Journal of Production Research*, **33**, 6, pp. 1587-1603.

McKay, K. N., Safayeni, F. and Buzacott, J. A. (1995c). A Review of Hierarchical Production Planning and its Applicability for Modern Manufacturing, *Production Planning & Control*, **6**, 5, pp. 384-394.

McKay, K. N., Safayeni, F. R. and Buzacott, J. A. (1995d). An Information Systems Based Paradigm For Decision Making In Rapidly Changing Industries, *Control Engineering Practice*, **3**, 1, pp. 77-88.

McKay, K. N., Morton, T. E., Ramnath, P. and Wang, J. (2000a). Aversion Dynamics – Scheduling when the System Changes, *Journal of Scheduling*. **3**, 2, pp. 71-88.

McKay, K. N., Pinedo, M. and Webster, S. (2000b). A Practice-Focused Agenda For Production Scheduling Research, *Journal of the Production and Operations Management Society*. In press.

O'Donovan, R., Uzsoy, R. and McKay, K. (1999). Predictable Scheduling and Rescheduling on a Single Machine in the Presence of Machine Breakdowns and Sensitive Jobs, *International Journal of Production Research*, **37**, 18, pp. 4217-4233.

Wiers, V. C. S. and McKay, K. N. (1996). Task Allocation: Human Computer Interaction in Intelligent Scheduling, *Proceedings of the 15th Workshop of the UK Planning & Scheduling Special Interest Group*, November 1996, Liverpool, UK, pp. 333-344.

PART II

Field Studies of Planners, Schedulers and Industrial Practice

CHAPTER FOUR

A Case Study of Scheduling Practice at a Machine Tool Manufacturer

Scott Webster

4.1 INTRODUCTION

As a research area, the topic of scheduling is something of an enigma. It has attracted researchers from a range of disciplines (e.g., see Coffmann, 1976) and has been the subject of over 20,000 journal articles (Dessouky *et al.,* 1995) and more than 20 books. Studies have found that different scheduling methods can result in large differences in system performance, yet the results of scheduling research have had relatively little impact on scheduling practice (Graves, 1981; Buxey, 1989; White, 1990; MacCarthy and Liu, 1993). One means towards reducing this gap is through descriptive studies of scheduling practice. Such work is especially relevant today. With advances in information technology and increasing industry interest in computer-based scheduling tools, the opportunity to reduce the gap between theory and practice has perhaps never been greater. Descriptive studies play a useful role in laying the groundwork for taking advantage of this opportunity.

Early descriptive scheduling research can be traced back to the 1960s with the work of Dutton (1962, 1964), Dutton and Starbuck (1971), Fox and Kriebel (1967), and Hurst and McNamara (1967). Few studies of scheduling practice appeared in the literature from the early 1970s to the early 1990s. Recent years have witnessed increased activity in this area (e.g., Halsall *et al.*, 1994, McKay *et al.*, 1995; Wiers, 1997; see MacCarthy *et al.*, 1997 or Wiers, 1997 for a review). This paper builds on this activity by describing the scheduling function at a machine tool manufacturer and assessing implications for research.

In summary, scheduling orders for cutting tools at the plant is largely a manual process with most of the decision-making performed by two individuals. The logic is complex, systematic, and being continually refined. Some data-gathering and decision-making activities are conducted on a cyclic basis while other activities are ongoing. Over 30 different statistics regarding system status are monitored, and depending on the values of these statistics, over 100 separate rules for guiding scheduling decisions are employed. A major emphasis is on early identification of problems. The application of this logic has been successful, with fill rates and lead times improving over recent years.

The next section contains essential background information on plant operations. Section 4.3 presents a description of the scheduling function. Section 4.4 discusses the main strategies used by the schedulers and considers implications for scheduling research. Section 4.5 concludes the paper.

4.2 BACKGROUND

The plant produces cutting tools from bar stock (see Figure 4.1). About 60% of the orders are for custom product, which range from slight variations in existing product to cutting tools designed from scratch, and the remaining orders are for standard product. Approximately one-half of the standard product is made-to-stock, or in other words, about 80% of the orders are either engineered-to-order or made-to-order. Work-in-process averages 1,500 orders and flow times from order release to final inspection average four weeks.

Figure 4.1 Examples of finished product.

The plant, which is illustrated in Figure 4.2, has 37,500 square feet of manufacturing space and employs about 115 people working three shifts and five days per week. There are 29 workcenters that perform various machining functions (i.e., turning, grinding, boring, milling, and drilling). Twenty of these workcenters are unique. A cutting tool requires an average of ten operations but there are two operations that are especially relevant for understanding the scheduling function. These are *turning* and *pocketing*. Turning is the gateway operation; all products are processed on a lathe immediately after order release. Pocketing, which is typically the fifth operation, is where the cutting tool is shaped to hold the carbide inserts that act as the blade of the cutting tool. (See Webster, Chapter 12 in this book for a case study of the plant that produces the inserts). Relative to other machining operations, pocketing is generally much more intricate and time-consuming. It is also the bottleneck operation and a focal point for scheduling activities.

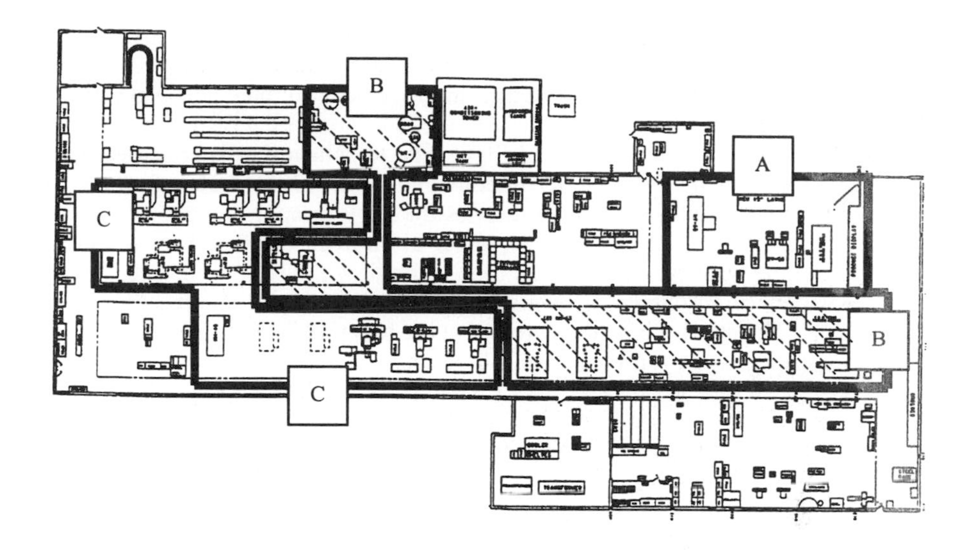

Figure 4.2 The plant layout. The lathes (gateway operation) are located in the work area labeled A and the pocket machines (bottleneck resource) are located in the work area labeled C. Operations between turning and pocketing take place in the shaded work area labeled B. Product flow up to and including the bottleneck resource is generally from right to left, moving through work areas A, B, and C.

4.3 THE SCHEDULING FUNCTION

Scheduling is primarily the responsibility of two individuals who we will refer to as Pete and Steve. Pete's title is production manager. He has over 20 years of experience. In general, he is responsible for developing and co-ordinating production schedules, and supervising and allocating manufacturing personnel, material, and equipment. Among other specific activities, he is the one who schedules the pocketing machines. Steve is an assistant to Pete, who has the title of master scheduler. He is responsible for order release and scheduling the lathes.

An understanding of the scheduling function was gained through interviews and task observation. Pete, and to a lesser extent Steve, regularly reviewed and edited the evolving document. Once the initial document was complete, three orders were tracked from start-to-finish to validate the logic. In total, 90 hours were spent on site over a period of three months ending in May 1997.

The main source for day-to-day scheduling logic was the schedulers themselves. Given the pressing and important concerns of effectively getting product through the plant, it can be a challenge to find schedulers with the patience to carefully help document the way they go about scheduling. We were fortunate that Pete was very willing to commit time and resources to the project. Over the years, he had refined a comprehensive and systematic approach to scheduling that seemed to be working well. There is perhaps some pride in seeing one's creation on paper, but more significantly, Pete viewed the project as a means towards clarifying the scheduling logic in his own mind and opening the logic to others who might see opportunities for further improvement or take on some of the responsibilities in the future.

The basic approach that Pete and Steve take to scheduling is consistent with the Drum-Buffer-Rope concept of Goldratt and Cox (1992). The drum is the bottleneck resource and its schedule dictates the flow through the plant. Careful attention is paid to the buffer, or queue, of the drum to ensure that there is enough work to keep the bottleneck busy. The rope is the flow time between order release and the bottleneck resource. Orders are released to keep the length of the rope fairly stable. The approach is implemented at the plant through two main activities that are described in the following two subsections.

4.3.1 Order release and first-stage scheduling

A key information source for Steve is the *unreleased orders file*. (Prior to 1996, Steve manually reviewed the order entry database to identify candidate orders to release.) Recommended orders for stocked items are generated whenever finished inventory drops below a reorder point. For each recommended order, the inventory control group checks if the routing needs to be modified, gets the routing updated if necessary and, once the order is ready to be released to the shop floor, transfers the order to the unreleased orders file. Several steps take place before a *non-stock order* (i.e., product that is not made-to-stock) is entered into the system. These steps include product design (if the product has not been produced before), cost and lead-time quotation, and customer confirmation. The engineering department reviews non-stock orders in the order entry database. After verifying the route and writing NC programs (if appropriate), the order is ready for release and is transferred from the order entry database to the unreleased orders file.

Another key source of information for Steve is the *open order file*. This file lists all orders awaiting processing at a pocketing machine and the associated pocket machine processing time requirements.

Each morning, Steve estimates pocket machine daily capacity (capacity varies due to machine breakdowns) and selects orders to release from the unreleased orders file. He releases about one day's worth of pocket machine work, but this can vary with the priority of unreleased orders and the workload of the pocket machine buffers. For example, if there is between five and ten days of work in the buffers, Steve will set a release target of one day's worth of pocket machine work. If the buffer length is outside these limits, he will increase or decrease the target by about 10%. Once the workload target is established, he considers job priority, which hinges in part on the order due date. Normally Steve uses the quoted promise date as the order due date. However, sometimes customers request earlier ship dates, and this may be used as the order due date when shop workload is low. An order is classified as priority *A* if it is behind schedule. More specifically, an order is priority *A* if it is a special order (e.g., rework, warranty item, or coded as expedite), is due within four weeks, is a stock item with less than four weeks of supply, or requires more than 12 operations. Priority *B* is assigned if an order is on time (i.e., due within four to six weeks, or a stock item with four to eight weeks of supply). Priority *C* is used for all other orders. All priority *A* orders are released. Priority *B* orders, and perhaps some priority *C* orders, are released until the workload target is met. This logic is adjusted if the first-stage buffers are especially high or low.

When an order is released, an automatic email message is generated and sent to the purchasing department. With few exceptions, no raw material is maintained. Purchasing alerts the appropriate suppliers and raw material arrives within two to four days (raw material may be ordered prior to order release if the design department anticipates long raw material lead times). In the meantime, Steve prepares a routing packet for each released order; the packet contains detailed routing information and the priority code. Once the raw material arrives, material handlers consolidate the material with the routing packet and move to the lathe. Lathe operators process orders according to priority code (i.e., all priority *A* before priority *B*, and all priority *B* before priority *C*) but attempt to save set-ups when sequencing orders with the same priority.

4.3.2 Pocket machine scheduling

The logic for pocket machine scheduling is more involved. Compared to other operations, pocketing requires longer set-up and processing time, involves a more complicated machining function, and requires a higher level of operator skill. As noted earlier, pocket machines are the bottleneck resource in the plant and their schedule significantly impacts plant performance. Pete's main objective is to ship every order on time. However, he is also interested in achieving short flow times, low machine idle time, high stability in flow times, and flexibility to respond to change (e.g., machine breakdowns, absenteeism, rush orders). Naturally, conflicts arise between objectives; the following logic builds on over 20 years of experience and attempts to strike a reasonable balance.

One way to outline Pete's basic strategy for pocket machine scheduling is through an analogy from warehouse operations. Suppose, for example, that someone named Pat is the manager of a warehouse with a dozen loading docks for outbound shipments. Loading is very time-consuming, and as a consequence, the loading docks are the bottleneck resource. Trucks are loaded 24 hours a day and whenever a full truck leaves, an empty truck is available to take its place. Pat makes use of three levels for staging product to be loaded. When a product is picked, it is moved to the level 1 staging area. A packing slip must be prepared and attached to product before it can be moved to the level 2 staging area. The third level of staging corresponds to areas in front of each loading dock. About three days of work is lined up in front of each loading dock and product is sequenced in the order it is to be loaded on a truck.

With the exception of truck loading, all material handling takes place from 4 a.m. to 8 a.m. Product is picked and moved to the level 1 staging area, and product in level 1 with a packing slip is moved to the level 2 staging area. Some product is also moved from level 2 to staging areas in front of the loading docks in accordance with the staging plan developed by Pat the previous day. Pat arrives at

8 a.m. and uses the morning to get an accurate read on status (e.g., product in the staging area) and identify potential future problems (e.g., product that is behind on picking or packing slip preparation). Lunchtime provides an important break from the usual rush of activities that, if necessary, she can use to absorb the ramifications of new information and begin to generate ideas on a plan. Among other managerial activities, the afternoon is spent developing a staging plan that identifies product to be moved from level 2 staging to the level 3 staging areas in front of the loading docks. A number of factors are considered during this decision-making process including the amount of work in front of each loading dock, urgency of a shipment, specialized trucking requirements, and customer regions.

4.3.2.1 Routine scheduling activities – the Tuesday-Thursday cycle

Loading docks in the preceding example correspond to pocket machines. Pete's routine scheduling activities are composed of an assessment stage, which takes place every Tuesday and Thursday, and a decision-making stage that takes place every Monday, Wednesday, and Friday. The main purpose of the Tuesday-Thursday cycle of activities is to lay the groundwork for updating the pocket machine schedules by identifying orders that should be scheduled in the near future. The process is facilitated through staging areas, which in this case, are data files. Every Tuesday and Thursday morning, Pete executes a program that searches company databases and generates a file called *PeteStuf.* (Pete used to identify these orders manually; the Management Information Systems (MIS) group has since written computer programs that automate his logic.) *PeteStuf* contains a list of orders that satisfy the following conditions:

1. Order has been released, and
2. a non-stock order with less than 12 operations and due in less than three weeks, or
3. a non-stock order with 12 or more operations and due in less than six weeks, or
4. a stock order and finished goods inventory is less than four weeks of supply, or
5. order has not been released and is due in less than six weeks.

Each record in *PeteStuf* contains data stored in the system that are relevant for pocket machine scheduling. This information includes order code, customer ID, order promise date, customer request date, pocket machine assignment according to the routing, estimated pocket machine set-up and processing time, indicator for whether the pocket machine Numerical Control (NC) program has been written, indicator for whether the order has been released, and codes that identify which operations have been completed. The first step after generating *PeteStuf* is to find and remove problem orders. For example, if an order has not been released or raw material has not arrived, then this order is removed from *PeteStuf* after follow-up (e.g., determine if system data are accurate, and if so, track down reasons for delay and take appropriate corrective actions). Once *PeteStuf* has been reviewed and cleaned up, it contains candidate orders for pocket scheduling in the near future. The next step is to partition the orders in *PeteStuf* into those that are ready to pocket and those that cannot be pocketed until other operations are complete. Pete manually identifies orders to transfer to a file called *NotThere* (orders that are not ready to pocket) and to a file called *Next* (orders that are ready to pocket). In point of fact, there are actually two data files that contain orders that are ready to pocket and two files that contain orders that are not ready to pocket. This additional level of refinement is used to segregate orders according to special machining requirements. The distinction is not essential for gaining a basic understanding of the scheduling logic and is therefore omitted from the description.

The status of each order is updated in the system as it moves through the plant so, at least in theory, it is a simple matter to check whether all operations prior to pocketing are complete. Pete takes the time to manually partition the orders for two reasons. First, there are occasional errors in the system data (e.g., operator neglected to sign-off on an order), and second, the process reinforces his sense of where orders are in the plant.

NotThere corresponds to the level 1 staging area in the warehouse analogy. All the orders in this file have been released but require additional operations before being ready to pocket. Pete reviews the orders in *NotThere*, identifies those that are not progressing at a suitable pace, and follows up accordingly (e.g., check status of previously planned corrective actions, gather information on causes of delay, formulate plan to resolve a problem, take corrective action). *Next* corresponds to the level 2 staging area in the warehouse analogy. All orders in this file are available to be scheduled on a pocket machine. During the Tuesday-Thursday cycle of activities, Pete has identified current or potential future problems (and perhaps taken corrective actions) and he has gathered information on orders that can be scheduled. He has the opportunity to reflect on this information before making scheduling decisions the next day.

4.3.2.2 *Routine scheduling activities – the Monday-Wednesday-Friday cycle*

Orders in *Next* are scheduled every Monday, Wednesday, and Friday. There is a file containing the sequence of orders to be processed at each pocket machine. This set of files corresponds to the level 3 staging area in the warehouse analogy. Pete's task is to add new orders to these files. He begins by looking for system errors that affect the calculation of scheduled workload at each pocket machine. For example, a pocket machine operator may have neglected to sign-off on the completion of an order, or a 100-hour pocket machine operation may be 90% complete while the system is unable to store anything other than 100 hours of work remaining. Both of these errors overstate scheduled workload.

Once Pete arrives at a reasonably accurate estimate of scheduled workload, he selects pocket machines that need work. Selection criteria are based on machine workload and the maximum due date among orders scheduled on a machine. For example, a machine is always selected when there is less than three days of scheduled work. When there is more than three days of work, a machine with a higher maximum due date is favored for selection over a machine with a lower maximum due date.

After machine selection, Pete selects orders in *Next* to be added to the machine schedules. Rather than taking the time to examine every order, he usually sets a due date window (e.g., three weeks into the future) and only considers scheduling orders due within this window. The machine assignment decision depends on many factors including urgency of an order, difficulty of the operation, machine capability, skill level of the machine operators, and opportunities to save set-up time. Some machines can process nearly all orders while the range of other machines is more limited. To increase scheduling flexibility, Pete focuses on scheduling the least flexible machines first. Once orders are assigned to a machine, they need to be sequenced. The order due date, the operation processing time, and opportunity for set-up savings are the main factors that influence sequencing decisions. In some cases, Pete will contact the sales department for input on the relative importance of different orders. In general, tardiness is not much of a problem as long as it is small (e.g., within a couple of days of the target); the impact on customer relations can increase dramatically once outside of this band.

Several steps take place before an order in a machine schedule is processed. Each day, the NC programming group examines the pocket machine schedules to make sure that the machine assignments are feasible (orders are reassigned if necessary) and that NC programs are ready. The pocket machine assignment in the routing developed prior to order release can be wrong as much as 80% of the time, and as a result, the NC program - if already written - may require modification. The error rate is high because Pete's decision on machine assignment is significantly influenced by relative machine workload - information that is not available many weeks earlier when the route is selected. The NC programming group indicates that the NC program for an order is ready by changing the value of a flag variable in the order record of the pocket machine schedule file. Once the NC program is ready, Pete sends an email message to Steve who moves the order from a storage shelf to the specific pocket machine queue.

At the beginning of each shift, the employees responsible for tool setting at the pocket machines will print out the schedule for each machine. The tool setters, who understand the changeover time implications of a sequence better than anyone, use their judgement to modify the sequence according to the following guidelines. The order at the top of the list is always first. Orders from elsewhere in the list can be sequenced ahead of the second order if it helps to reduce changeover time. The same process is repeated for inserting orders between the second and third position in the original list, and so on. Tooling is prepared and is set up on the pocket machines according to the specified sequence.

From the tool setters' perspective, an important element of the scheduling approach is the forward visibility of several days worth of work scheduled on each pocketing machine. Forward visibility is important because of the time it takes to prepare tools and, more significantly, the time required to identify and resolve tooling conflicts (e.g., make sure that a set of tools is not needed in two places at once). The importance of forward visibility is one reason why Pete emphasizes schedule stability (i.e., the relative priority of orders do not change much as they flow through the shop). He relates a story of when, as a scheduler at a different company, he witnessed a bottleneck machine remain idle for four days; the operator was told to continually change the set-up because some new job was especially hot.

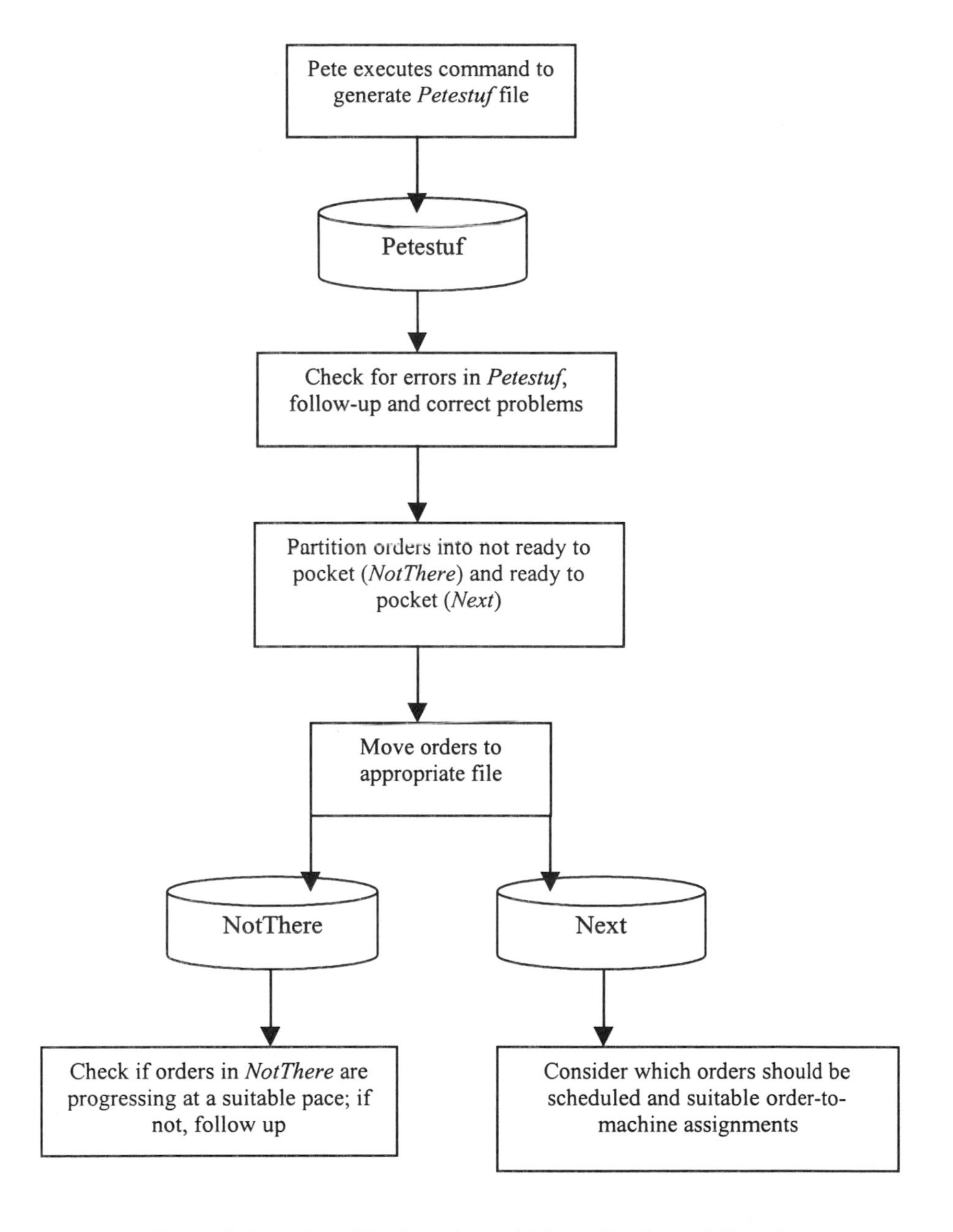

Figure 4.3 Overview of Pete's routine activities on Tuesday and Thursday.

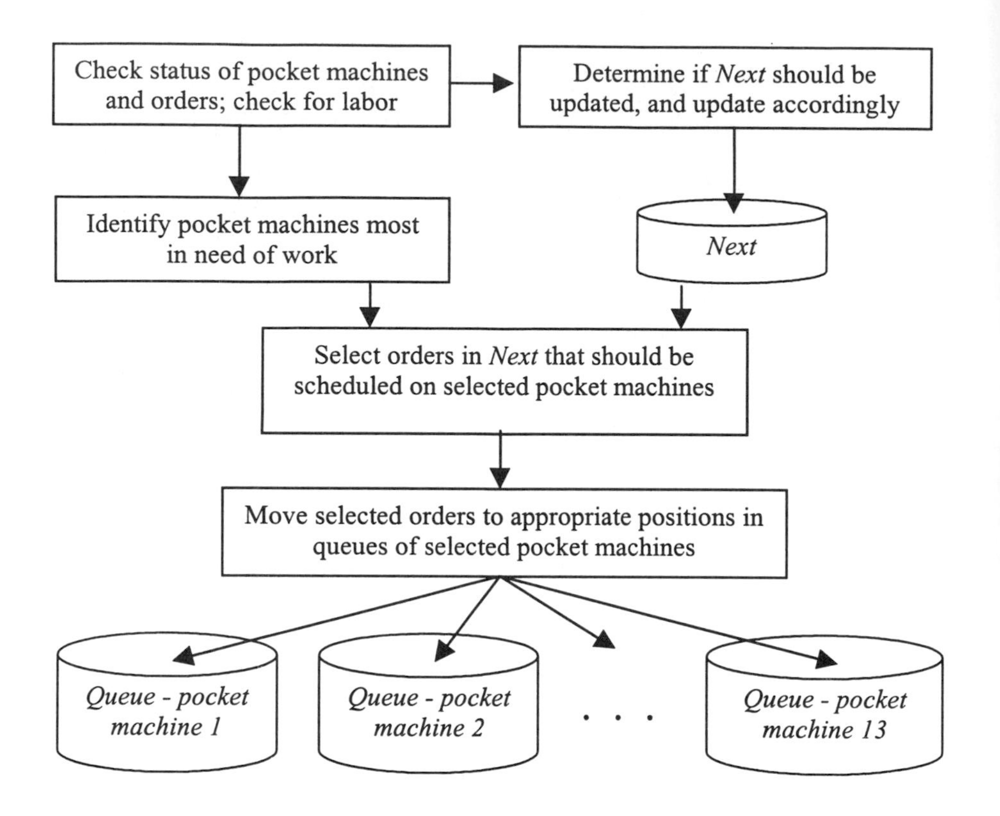

Figure 4.4 Overview of Pete's routine activities on Monday, Wednesday, and Friday.

4.3.2.3 Non-routine scheduling activities

We have described the routine scheduling activities that take place on a Tuesday-Thursday cycle and a Monday-Wednesday-Friday cycle (See Figures 4.3 and 4.4 for flowcharts). There are many scheduling activities that take place as the need arises or on an ongoing basis as time allows. These activities typically involve identifying and correcting errors in system data, tracking changes in the environment, and if appropriate, responding to these changes. Examples of error checking include reviewing pocket machine set-up and processing time estimates (average error is 30%), verifying the order code (the code contains information that can affect pocket machine assignment), and comparing orders in *Next* with the orders in the shelf storage area in front of the pocket machines. Pete also periodically reviews the progress of orders in *NotThere*. He believes that one key to effective scheduling is early identification of orders that are not properly moving through the system. To paraphrase, 'orders in motion will tend to stay in motion, but a slow-moving order will tend to be a problem throughout the system unless identified and corrected'.

As orders become ready to pocket, Pete will either move the order to *Next* or directly to a machine schedule file. There is not enough pocket machine capacity at the plant. Consequently, Pete is also regularly looking for orders in *Next* that are good candidates to be outsourced (e.g., make-to-stock orders with low urgency). As these orders are identified, Pete transfers the orders to a file called *Contract*. Orders in *Contract* are sequenced in priority order. Every Thursday a subcontractor picks up and drops off about 225 hours worth of these orders. There is also the occasional need to devise and implement policies that impact on the scheduling function. For example, in May 1997 Pete dedicated a particular pocketing machine to a new product line.

Pete plays close attention to queues as he is evaluating how work is progressing out on the shop floor. Operators tend to feel very good and are more apt to relax when their out-station is full; it gives a signal that they have been working hard and have accomplished a lot. On the other hand, an empty in-station can be very threatening as a lack of work may raise questions on the importance of their job. Consequently, Pete manages the material handling function to keep in-stations nearly full and out-stations nearly empty.

4.4 ANALYSIS AND DISCUSSION

We have described how scheduling is done at one machine tool manufacturing plant. The approach, which builds on 20-plus years of experience, is systematic, complex, and appears to work well. Steve, Pete's assistant, is primarily responsible for order release and first-stage scheduling. He refers to 12 statistics and employs 33 separate rules for governing order release and priority decisions. He estimates that these activities consume about 30% of his time. Pete takes responsibility for scheduling the bottleneck resource. He refers to 20 statistics, uses 69 separate rules, and spends about 60% of his time on this activity. (We did not specify all the rules and statistics as this level of detail is unnecessary for a basic understanding of Pete and Steve's methodology.) Both Pete and Steve regularly interact with a wide variety of personnel (e.g., foreman, operators, stock room personnel, inside sales, stock production manager, purchasing manager, design engineer, manufacturing engineer, part programmer, proposal engineer, quality engineer, insert plant manager, receiving clerk, product manager, computer programmer, R&D personnel, and upper management). Relatively little management attention is paid to scheduling beyond the bottleneck resource. After an order leaves the bottleneck, it is generally processed on a first-come-first-served (FCFS) basis as the remaining operations are completed. Operators have flexibility to deviate from FCFS in the interest of efficiency as long as orders are completed within about five days (actual work time after the bottleneck is about two days). Over the last few years, order flow times have been decreasing and the percent of orders shipped on time has been increasing.

Pete and Steve's approach to scheduling has six salient characteristics that are identified and discussed below:

1. *Pareto scheduling* - scheduling activities are concentrated on a few key operations.

2. *Schedule guideposts* - the schedule leaves some flexibility for operator discretion.
3. *Nip in the bud* - a focus on early identification and resolution of problems.
4. *Holding tanks* - use of virtual staging areas to help structure the chaos of 1,500 orders on the shop floor.
5. *Quiet time* - time separation between assessment and schedule generation.
6. *Queueing psychology* - recognize human nature in the management of in-stations and out-stations.

Pareto scheduling refers to the use of the Drum-Buffer-Rope concept; attention is focused on effectively scheduling the bottleneck and using this schedule to dictate the flow elsewhere in the plant. Management scheduling concerns are primarily focused on the two areas that have the biggest impact on shop performance. These two areas are the lathe machines (i.e., gateway operation) and the pocket machines (i.e., bottleneck resource). Pete and Steve's routine scheduling decisions center on i) determining which orders to release and their associated priority and ii) assigning and prioritizing orders at the pocket machines.

Scheduling guideposts refer to Steve's use of three priority classes at the gateway operation and the tool setters' authority to rearrange pocket machine queues. Lathe operators at the gateway operation can select any order in their queue that doesn't violate the class priority scheme. Pocket machine tool setters, while generally following the sequence identified by Pete, are allowed to move a lower priority job ahead in the sequence if it will save set-up time. The main reason for management's use of scheduling guideposts is their interest in low set-up time. Pete and Steve establish a scheduling framework consistent with plant-wide concerns and delegate decision-making authority within this framework to those with the best understanding of set-up requirements. In so doing, management is explicitly recognizing special expertise of shop laborers and, as a consequence, the use of scheduling guideposts may also have a beneficial impact on worker motivation.

Nip in the bud refers to the fact that very little of the time that Pete and Steve spend on scheduling is dedicated to creating a schedule. The majority of their 'scheduling' time is spent trying to anticipate and avoid future problems. Being able to do this well depends to a large degree on having an accurate pulse of the plant. Consequently, much time is spent checking and correcting inconsistencies between different information sources, and gathering new information from a variety of sources.

One interesting element of Pete's approach is his preference for manually moving orders from the file *PeteStuf* to two different files depending on order status. This process could easily be performed automatically by a computer program, but Pete finds the exercise worth the time because it enriches his sense of shop status. The approach highlights the idea that while automating aspects of scheduling logic can offer benefits it may also make shop status less transparent to management, thus inhibiting a potentially key human component of shop scheduling - problem identification. In Pete's case, many scheduling problems are not evident from information in the computer system (due to a combination of inaccurate and incomplete information), and thus are not amenable to automatic detection.

There are two main reasons why *nip in the bud* is used at the plant. First, the forward visibility provided by a schedule is essential for effectively managing set-ups at the pocket machines and, as noted above, set-up time is a significant concern. Tool setters rely on the schedule to plan and complete the offline tool preparation work and resolve tooling conflicts so that there are no (or minimal) delays when tooling is needed at a pocket machine. Naturally, there is little forward visibility if a schedule is regularly changing due to unanticipated problems. Second, Pete's 20 years of experience have led to a belief that orders in motion tend to stay in motion and orders at rest tend to get delayed throughout the production system unless the source of delay is eliminated. A natural management consequence is a focus on early identification and resolution of problem orders.

Holding tanks refer to the partitioning of shop orders according to status. Pete periodically identifies orders that should be scheduled on a pocket machine in the near future and loads each of these orders into one of two computer files -*NotThere* or *Next*. These files are essentially virtual staging areas that are used to monitor order progress and isolate orders that can be scheduled. *NotThere* lists orders that require additional work before being assigned to a pocket machine. Pete monitors these orders and takes corrective action if an order is not moving at a suitable pace. *Next* lists orders that are ready to be scheduled at a pocket machine. Pete is able to focus attention on a relatively small set of orders when making routine scheduling decisions or in response to unanticipated events (e.g., tool breakage eliminates the possibility of processing orders currently scheduled on a machine so Pete must quickly identify other orders to process on the machine).

Quiet time refers to Pete's approach of collecting and organizing information relevant for pocket machine scheduling, then having the evening to mull it over before making schedule decisions. Pete's job environment is hectic. He accumulates and assesses large quantities of sometimes contradictory information from multiple sources, and is regularly faced with urgent matters requiring immediate attention. Pete finds that the opportunity for more relaxed consideration of shop status is especially useful for detecting important missing information, identifying key problems related to scheduling, and considering alternative solutions - activities that ultimately result in more effective plans.

Finally, *queueing psychology* refers to the motivational effects of in-station and out-station queue lengths, and the use of material handling strategies that take advantage of these effects. Pete believes people tend to work slower as their out-station queue becomes full and feel anxious if their in-station queue remains nearly empty. Pete uses this insight to his advantage by managing the material handling function to keep out-stations nearly empty and in-stations nearly full.

Pete and Steve face a dynamic and information-intense environment. For example, the high degree of change in the environment is the main reason why Pete will not schedule an order on a pocket machine until all preceding operations are complete. The first four characteristics - *Pareto scheduling*, *schedule guideposts*, *nip in the bud*, *holding tanks* - are tactics that help manage the uncertainty and complexity through decomposition. These tactics help sort out information most deserving of management attention as a means towards early recognition of problems, rapid formulation of schedule adjustments in response to change, and ultimately, efficient use of resources combined with short and predictable lead times.

Of the six characteristics, the third is perhaps the most interesting because it is the most at odds with the prevailing scheduling literature. In general, the research literature takes a *problem solving* orientation. Many properties and algorithms have been developed that lead to more effective solutions to a variety of different scheduling problems. The third characteristic is oriented towards *problem identification*, or in particular, early identification of situations that may negatively impact performance. This is consistent with an observation by Graves (1981) on future directions of scheduling research as well as a finding by Fox and Smith (1984) that 80% to 90% of a human scheduler's time is spent in problem identification. McKay *et al.* (1995) also found this characteristic in their study of scheduling practice at a printed circuit board plant.

Pete and Steve have continually refined their approach and, with lead times and on-time deliveries improving in recent years, it seems to be working well. The dominant theme is a set of tactics that appear to be a reasoned response to the desire for forward visibility when the environment is dynamic and information-intense and the formal system contains incomplete and sometimes inaccurate information. The case study highlights the importance of the human element in scheduling. Many of the data collection and scheduling activities were performed manually in order to compensate for weaknesses in the formal information system. It is this last point that is perhaps most meaningful for researchers when considered in light of a general trend towards increasing capabilities and performance-to-cost ratio of data collection and storage technology. Given that this trend continues to the point where formal information systems can more nearly match reality in the majority of organizations and assuming that the dynamic nature of a scheduling environment is typical and unlikely to change, the case study suggests that computer-supported scheduling diagnostics as well as methods for handling dynamics will become increasingly important areas for research. Example activities include research on effective user interfaces (e.g., to facilitate quick assessment of shop status), diagnostic heuristics that help highlight potential future problems, rescheduling methods that control change between old and new schedules, and methods for achieving stable schedules.

4.5 CONCLUSION

This paper is motivated by the gap between the research literature and scheduling practice. We have described the day-to-day scheduling activities at one plant, summarized the main tactics, and have found that one of the most pressing concerns of the schedulers is largely absent from the research literature.

As with any individual case study, the extent to which the main lessons extend to other settings is questionable. However, when taken as a whole, these studies play a useful role in directing research to areas of high impact. The thin case study literature combined with increasing industry interest in scheduling systems underscores the need for additional descriptive work on scheduling practice with an emphasis on assessing implications for research.

4.6 ACKNOWLEDGEMENTS

I would like to acknowledge Ta-Yao Wei, a graduate student at the University of Wisconsin-Madison, who spent many hours collecting data and documenting the scheduling function at the plant. I would also like to acknowledge the significant time and effort of the production manager and the master scheduler at the plant.

4.7 REFERENCES

Buxey, G. (1989). Production scheduling: practice and theory. *European Journal of Operational Research*, 39, pp. 17-31.

Coffman, E. G. (1976). Introduction to deterministic scheduling theory. In *Computer and Job Shop Scheduling Theory*, edited by Coffman, E.G., New York: John Wiley & Sons, pp. 1-50.

Dessouky, M. I., Moray, N. and Kijowski, B. (1995). Strategic behavior and scheduling theory. *Human Factors*, 37, pp. 443-472.

Dutton, J. M. (1962). Simulation of an actual production scheduling and workflow control system. *International Journal of Production Research*, 1, pp. 421-441.

Dutton, J. M. (1964). Production scheduling: a behaviour model. *International Journal of Production Research*, 3, pp. 3-27.

Dutton, J. M. and Starbuck, W. (1971). Finding Charlie's run-time estimator. In *Computer Simulation of Human Behaviour*, edited by Dutton, J.M. and Starbuck, W., New York: John Wiley & Sons, pp. 218-242.

Fox, P. D. and Kriebel, C. H. (1967). An empirical study of scheduling decision behaviour. *Journal of Industrial Engineering*, 18, pp. 354-360.

Fox, M. S. and Smith, S. F. (1984). ISIS: A knowledge-based system for factory scheduling. *Expert Systems*, 1, pp. 24-49.

Goldratt, E. M. and Cox, J. (1992). *The Goal*, 2nd revised edition. Great Barrington, MA: North River Press Inc.

Graves, S. C. (1981). A review of production scheduling. *Operations Research*, 29, pp. 646-676.

Halsall, D. N., Muhlemann, A. P. and Price, D. H. (1994). A review of production planning and scheduling in smaller manufacturing companies in the UK. *Production Planning and Control*, 5, pp. 485-493.

Hurst, E. G. and McNamara, A. B. (1967). Heuristic scheduling in a woollen mill. *Management Science*, 14, pp. 182-203.

MacCarthy, B. L., Crawford, S., Vernon, C. F. and Wilson, J. R. (1997). How do humans plan and schedule? *The Third International Workshop on Models and Algorithms for Planning and Scheduling Problems*, Queens College Cambridge.

MacCarthy, B. L. and Liu, J. (1993). A recent survey of production scheduling. *International Journal of Production Research*, 31, pp. 59-79.

McKay, K. N., Safayeni, F. R. and Buzacott, J. A. (1995). 'Common sense' realities of planning and scheduling in printed circuit board production. *International Journal of Production Research*, 33, pp. 1587-1603.

Wiers, V. C. S. (1997). Human-computer Interaction in Production Scheduling: Analysis and design of decision support systems in production scheduling. Ph.D. Thesis, BETA, Eindoven, The Netherlands.

White, Jr., K. P. (1990). Advances in the theory and practice of production scheduling. *Control and Dynamic Systems*, 37, pp. 115-157.

CHAPTER FIVE

Making Sense of Scheduling: The Realities of Scheduling Practice in an Engineering Firm

Sarah Crawford

5.1 INTRODUCTION

Scheduling occurs in many environments; the focus of this case study is scheduling in manufacturing industry. A general description of scheduling is the allocation of resources over time to perform a collection of tasks (Baker, 1974). This 'classical' approach to scheduling problems has been formalised into the underlying framework referred to as scheduling theory. Scheduling in practice, however, varies greatly between different companies with no definitive model of 'real world' scheduling available to aid manufacturing industry. Overlooking the realities of scheduling has resulted in many industries being far from successful in their development of planning and scheduling systems, or in their understanding of the role of the human scheduler within these systems. Instead, businesses have exploited the increasing trend for IT solutions to automate planning and scheduling decision- making processes. However, these systems approach production scheduling as a well defined mathematical problem in contrast to the reality of people carrying out scheduling as a dynamic process. The business' simple, implicit assumption is that the human scheduler will 'solve' most of the problems that arise.

The majority of previous scheduling research has ignored the presence of the large numbers of people in manufacturing organisations that are involved with, or can impact upon, scheduling decisions. Human decision-making is not usually considered in the development and implementation of planning and scheduling systems. Therefore, there is a need to understand the role of the human scheduler and the nature of scheduling practice as a basis for developing more effective systems. This case study was conducted as part of an EPSRC funded project[1] whose aim is to study and capture the reality of planning and scheduling across manufacturing sectors and to explain what actually happens in 'real world' practices. The study looks at the nature of scheduling in an engineering firm, specifically investigating how a scheduler works and how the scheduler's performance is measured.

The case study covers three main areas. Firstly, there is an overview of the firm where the fieldwork took place and an introduction to 'John', the participant

---

[1] Effective decision support for production planning and scheduling: a new approach combining scheduling theory and human factors (GR/L/31364).

scheduler in the study. The issues of selecting schedulers for study, gaining access to companies and carrying out fieldwork within companies are also discussed. The paper then deals with data analysis and presents the planning and scheduling information system in use at the firm and the different types of behaviour and performance that John displays within his role. Finally, we look at the topic of performance measures and the difference between formalised, institutional measures and informal, unrecognised measures. This section also discusses how performance measures influence the behaviour of the scheduler, and other personnel, within the firm. The paper concludes with a discussion section about the points raised in the case study and suggests ideas for future human factors research in production planning and scheduling.

5.2 BACKGROUND

The fieldwork was conducted in the machining department of a large engineering firm based in the UK. There are approximately 2,500 personnel employed by the firm, which is organised so that manufacturing, sub assembly and assembly take place within four 'departments'. These departments are considered autonomous, and each regards the other as their customers. The machining department studied produces made-to-order machined parts for other departments on site, and also for other sites within the engineering firm. The department also has to account for 'replacement' orders that arrive as six month batches from another firm. These have to be planned into the machining plan as smaller batches of orders. Within the machining department there are five Cells that each manufacture or finish different parts and items. John works within the Engineered Products Cell (See Figure 5.1). There are seven workcentres within this Cell and a number of 'facilities' or individual machines within the workcentres. The workcentres perform a variety of machining jobs i.e. boring, frazing, grinding, part marking, drilling, inspection, turning, vertical, horizontal and conventional milling. Two of the milling workcentres are considered prime plant and are afforded extra vigilance from the Cell management, the Planners and the Shop floor Supervisor.

There are three Planners within the Cell of whom John is one. He is in his mid thirties and has worked for the firm for 16 years. He has been in his current role of Cell Planner for four years and his previous job was as a maintenance engineer at the same firm. John started working in scheduling after it was suggested by a manager 'who knew what I could do' and 'who saw that I got bored easily and needed a more interesting job'. John holds definite views on his role, stating that schedulers have to be 'practical' especially when it comes to their initial training, which he describes as 'sink or swim...on-the-job training'.

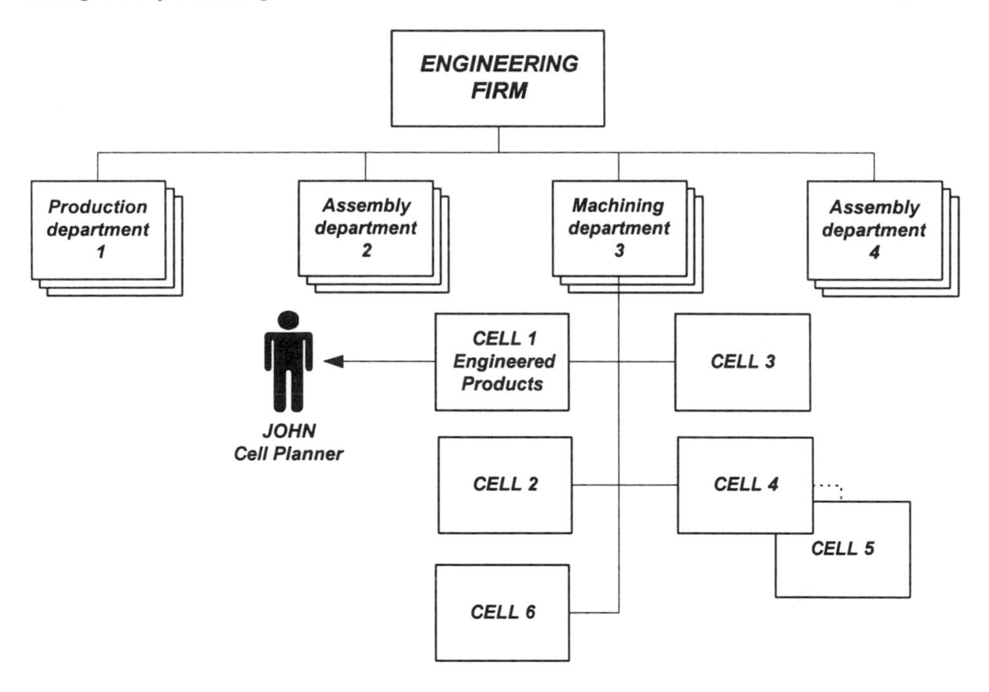

Figure 5.1 John's position in the engineering firm.

5.3 METHODOLOGY

But how was John selected as the scheduler to be studied? The engineering industry in Britain is vast and to gather relevant data on scheduling practice in the engineering industry an appropriate participant scheduler had to be found. The process of selecting participant schedulers is discussed later. Discussions had taken place with the industrial collaborators at the beginning of the project and the engineering firm had volunteered its time. A contact name of a Planning Manager at the firm was made available and, after preliminary discussions about the research, the manager arranged access to the firm in order to find appropriate participants. These preliminary visits are important; the researchers need to meet the managers and possible participant schedulers and vice versa, so that all those who might be involved know what the aims and time frame of the fieldwork are. The discussion and design of visits within companies must not be overlooked as fieldwork relies on the co-operation of both parties. (See Crawford *et al.*, 1999 for a full discussion of the methodological issues and methods involved in studying scheduling practice).

The researcher met with John at his Cell to briefly explain what the field visit would entail. It had become evident from pilot studies carried out before the main field visits that the selection of participant schedulers was very important. Previous participants had often performed very different roles to the 'classical' scheduling function. Therefore, a quick and accurate interview method of selecting participants had been devised, asking about the overall scheduling task, who does the scheduling, how it is done, what systems are used and in what format the schedule is released to the shop floor. Analysis of this interview shows if the participant is a suitable candidate for the study of scheduling in that department. John carried out the Scheduling Task Interview and subsequently a start date for the study was agreed upon. John was also happy to participate in the follow-up visits that would be needed in order to validate the information gathered during the main visit.

The preliminary visit had paved the way for the field visits, but what information would be needed in order to gather empirical data about John's role and how he executed scheduling-related tasks? And how would this information be recorded and validated? To objectively gather data on John's scheduling practice, documented field methods for gathering and recording information in dynamic industrial situations were studied. The data-gathering and recording methods chosen were adapted from applied social sciences research methods and formed a 'toolkit' of six main techniques to capture relevant data. The methods were:

1. Observation
2. Structured and semi-structured interviews
3. Verbal protocols
4. Task analysis
5. Retrospective decision probes
6. Attributes analysis.

The methods noted above had already been piloted so the next issue to consider was how to approach the field visits at the engineering firm. From the researcher's previous experience of the scheduling domain, and from the pilot study findings, it was assumed that John has certain days in the week when he undertakes large amounts of 'scheduling', for example when a new plan is produced. Certain tasks may follow cyclical patterns, updating the schedule at a shift change or regenerating the schedule after a Materials Requirements Planning (MRP) run, and certain episodes would be unplanned and random, machine breakdowns or stock shortages. Therefore, the visit was designed so that John was observed for five days to gather information on his scheduling behaviour across a whole working week.

At the end of the visit John agreed to two follow-up visits to gather information on any non-planned, random events that had not occurred during the visit. These visits were then finalised by a feedback visit where John and the researcher worked together off-line to validate the data collected. This fieldwork design was chosen to allow an intensive study of the nature of scheduling followed up by visits over a longitudinal time frame to gather information on episodes such as introduction of new products, processes or computer systems, within a context of studying other schedulers at several other companies as well. John validated the data from the information fed back to him and was able to make any changes or ask any questions that he felt were applicable. Once he was happy with the information, the data were considered valid and could be used for analysis.

So the participant scheduler had been selected, the field visits had been carried out and the data had been collected; the next stage was therefore to analyse the data.

5. 4 ANALYSIS OF JOHN'S JOB

The main objective of the analysis was to accurately describe John's scheduling practice from the large amounts of information gathered during the field visit. The author's basic research questions were:

1. How does a scheduler schedule, that is what is John's scheduling behaviour?
2. How well does a scheduler schedule, that is how is John's performance measured?

The answers to these research questions were not easy to extract from the fieldwork data. The amount of data itself was not a significant problem; it was the interrelationships between the data that made it difficult to describe specific motivators of John's behaviour and performance. The framework proposed by Crawford and Wiers in Chapter 2 of this volume, offers a structured approach to the analysis of such data. This framework proposes that John was working within the influence of five main factors: organisational; IS/IT system; people; manufacturing processes and performance measures. The case study analysis must account for the influence of these main factors, and also for the influences generated by the interactions between the factors.

If we consider John's behaviour within this wider context and not just a 'planning and scheduling' context we can conjecture that his job does not consist solely of scheduling and sequencing work orders. The main aim at this stage of analysis is to document John's main tasks and also approach the issue of performance measures. To fully appreciate the nature of John's job we need a brief description of the department's planning and scheduling information system as John relied heavily on the data and information from this system to carry out his main tasks.

5.4.1 The planning and scheduling information system

John tended to use four modules of the overall information system.

5.4.1.1 Module A - Manufacturing control module

This MRP-based module acts as an interface between the engineering data, materials and purchasing data, capacity planning data, shop floor control data and data from other relevant manufacturing modules. It is the engine of the planning system with the overall aim of balancing supply and demand within the department. Module A receives the bill of materials (BOM) information and forms the basis of the department's manufacturing plan.

5.4.1.2 Module B - Short term capacity planning module

This capacity planning module provides a detailed picture of workload and capacity and highlights any incompatibilities. This allows John to see possible capacity problems and gives him a limited modelling function to try out alternative scenarios. Module B is also the basis for the work-to-lists used on the shop floor.

5.4.1.3 Module C - Shop floor management and control module

This takes the production plan from Module B and inputs from scheduling, and provides each workcentre with a prioritised work-to-list. It also tracks the location of machined items through the Cell and the department. John can print off Cell documentation and inspection lists using this module, and create reporting groups (categorised work-to-lists) so that product types and certain operations can be reorganised into more suitable groups to fit the functions of the Cell.

5.4.1.4 Module D - Material requirements system

This is the main materials module used by the machining department and it registers, collates and downloads the requirements for individual orders to the suppliers so that material deliveries can be made on time. It is also linked to the Cell's main supplier so that John can access their stock levels and vice versa.

5.5 STRUCTURE OF JOHN'S JOB – PRELIMINARY ANALYSIS

A preliminary analysis of the fieldwork data indicated an underlying structure to John's job. It appeared to be comprised of two main areas: tasks and roles. John did not carry out all tasks in the same way all of the time and his roles also changed over time. This was due to the factors mentioned previously and also to the inherent instability and uncertainty of the scheduling environment. John carried out his scheduling responsibilities within the boundaries defined by the tasks and the roles. It is only when John's job is considered using this type of holistic viewpoint, that his scheduling behaviour and performance can be fully understood. Figure 5.2 presents the interactions between types of tasks and roles that John carried out as his job as a scheduler within the Cell.

The task and role areas should not be thought of as equal or balanced; each area will change and adapt as the environment demands different responses of John as a scheduler. The model should therefore be considered simply as an illustrative, schematic device to demonstrate that John's job comprises interactions between different tasks and roles.

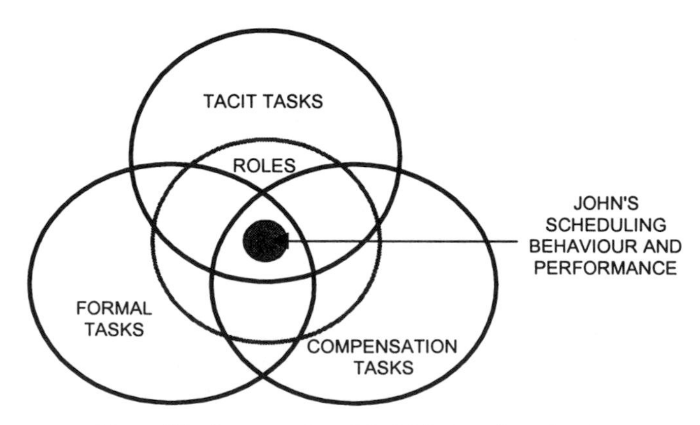

Figure 5.2 Conceptual model of John's tasks and roles.

The information system supported the majority of John's main tasks. However, he is not only subject to the demands of the computer system. He must also aggregate the influences of all the other factors in order to fulfil his job. We begin by looking at John's formal tasks, but it is only when all of a scheduler's tasks and roles are considered as interdependent that we begin to fully understand the richness and depth of a scheduler's behaviour and performance. It then becomes evident that a scheduler does not solely schedule and sequence orders in the 'classical' sense.

5.5.1 John's formal tasks - the building blocks of John's job

Tasks are categorised as formal because the firm has specified that these types of tasks are expected to be carried out by employees in John's position; these tasks have been made explicit and are formalised within the firm's job specifications.

5.5.1.1 Firm and cut planned status orders

Every day John checks the electronic mailbox messages sent to him from Module A that tell him the status of the production orders. Within the department an order must pass through seven sequential stages from its start to its completion, with its status changing through:

1. Planned
2. Firm
3. Cut
4. Released/ Picked
5. Start
6. Complete
7. Closed.

Planned orders are downloaded onto Module A at a level previous to John's involvement. Once the orders have been allocated as planned, Module A sends the status messages to the mailboxes when orders are due to be firmed and cut, based on final due date requirements. The module recognises which mailbox to send these firm and cut messages to, because each Cell Planner has their own planner code. There is an automatic firm and cut facility within this module that allows the system to release planned orders without manual intervention; in this set-up, only auto-flagged orders will be brought to the attention of the Cell Planner. However, within John's Cell the Planners have taken the decision to override this function and release all planned orders manually. John feels that manual release allows more control of orders and provides him with more up-to-date knowledge about what orders are being released to the shop floor at what time.

5.5.1.2 Order materials using the material requirements module

Once the orders have been firmed and cut, this triggers John to call in the materials for the production orders using Module D. He checks the status of the orders on the system many times each day and the module gives him one of three material status messages:

1. WA (Waiting Allocation) - the stores have not had time to check the storage areas for the appropriate stock. This means that the order is delayed until the stock is checked and allocated.
2. WM (Waiting Material) - there is no stock. John then checks the bill of materials for the ruling dimensions to see if there are any alternatives available. If no alternatives are available or if allocating alternative stock affects other orders, John has to decide if he should make any changes to the order or just wait for new stock.
3. WR (Waiting Request) - there is material available and the system is waiting for it to be allocated to an order, i.e. this order has the correct materials available.

If Module D states WR, John books the orders onto Module C which then generates the workcentre reporting groups that are used by the Shop-floor Supervisor and machine operators. At this stage the order packs or dockets stating the operations needed on that order, and its production route, are also run off.

5.5.1.3 Negotiate with customers

John checks the status of the order book (the overall MRP - generated plan) constantly during the day. If he finds that the order book is not balanced he will contact the person responsible to find out why. He can trace the progress of the orders by using the planner codes as a reference of who 'owns' the order within the department from its inception to delivery due date.

John visualises how the orders have been affected using a 'mental' Gantt chart:

1. If he 'sees' that the orders have moved to the right, to John this means that the lead-time has increased and so he leaves the orders until they become current on his order book. He therefore deletes any messages from the mailbox that apply to changes affecting that order.
2. If he 'sees' that the orders have moved to the left by approximately four or five days, John will overlook the query generated by Module A, and firm and cut the planned orders anyway because he feels that in most circumstances the machining department can normally cope with a slightly reduced lead-time. He then deletes any messages from the mailbox that apply to the queried orders.
3. If he 'sees' that the orders have moved to the left by more than four or five days then John will phone the customer and try to negotiate an agreement as to the changed lead-time. He will leave the mailbox message until the problem is sorted out, which is normally on the same day. Once the problem has been solved, John firms and cuts the order and deletes the mailbox messages. However, if John feels that he will not be able to negotiate an agreement about the order with the customer, he will put the queried order back to its full normal lead-time and show the customer how changing the due date without giving the machining department the proper notification affects the overall machining department plan. Often this triggers the customer to be more flexible with their demands so that John is able to re-negotiate the lead-time with the customer. John knows that it is important that any rogue changes in due date are dealt with because if they are left they will affect the machining department's schedule adherence and performance measures.

5.5.1.4 Adjust scrap and 'lost on shop' figures

Every day John collects the scrap and 'lost on shop' figures from the inspection personnel on the shop floor. He then adjusts these figures on Module A to keep the system up-to-date in 'real time'. If the scrap or lost item is a production item, then he contacts the customer to inform them. If the scrapping of an item has adjusted the plan, then he must renegotiate the lead-time with the customer. If it is a replacement order, he contacts George as the machining department's contact for replacements. So John has three options when he adjusts the scrap and 'lost on shop' figures:

1. He can reduce the customer order to match the amount left on the shop floor. The customer often accepts such a reschedule especially if the parts are needed immediately.
2. He can ask the customer to redate the order, if the order is a replacement order. Normal made-to-order work cannot be redated or the Cell's schedule adherence and performance measures will be negatively affected.
3. He can bring an order of a similar item in line behind the affected order to ensure that there are replacement parts available in the near future. However, such an action loses some or all of the adjusted order's lead-time. In this case, John has to ensure that he is scheduling effectively and not affecting the overall planning and performance of the shop floor.

John communicates any changes to the customer so that the customer knows what is happening with the order. This allows John to build up a co-operative relationship with the customer who in turn often allows John to produce more flexible schedules to try to give the shop floor an 'achievable' schedule.

5.5.2 John's tacit tasks - the influence of environmental factors

John's tacit tasks are performed because a combination of certain factors creates a need for the tasks to be undertaken. John assumes a tacit responsibility for the tasks, as there is no formalised justification that they should be carried out by Cell Planners.

5.5.2.1 Monitor shop floor production and problem solve

John monitors the shop floor throughout the day. For example, he is expected to follow up concessions[2], to deal with any alternative materials substituted by the supplier or to ensure that the preferred method of production is being followed. Any situations that mean that the materials he would normally order would not be acceptable or would need to be changed 'triggers' John to solve the problem. As John himself states, he is responsible for all orders in the Cell 'from birth to death'. This feeling of responsibility motivates him to provide solutions to a myriad of problems. If there is not enough time to satisfy the complete customer order because of certain manufacturing process variables, John tries to offer the customer

---

[2] A concession is a check needed to ensure that an item can be put into production because it is different from the original specifications, for example, if the item has an extra hole, or will be produced from an alternative material etc.

a compromise, such as delivery of at least half of the order on time by splitting batches. Or John will change an EBQ (Economic Batch Quantity) in order to solve a production problem because of a change of raw material delivery type i.e. sometimes reducing the EBQ will allow material billets to fit the machines better.

As part of monitoring shop floor production, he has to request material for the orders ten days before the material needs to be delivered. Even though the delivery agreement between the department and the suppliers states a three day lead-time, it has become easier for the production unit to accept a ten day lead-time as this ensures a more accurate delivery service from the suppliers. In these types of situation it has become John's responsibility to plan ahead and ensure that such manufacturing process factors do not affect the schedule.

Another way to avert shop floor problems is to 'clean' Module A data. He removes the orders that have already been started on the shop-floor from the reporting groups so that the machine operators can see what work is still available to be started. This is to keep the shop-floor picture up-to-date and to keep the operators happy, as they become 'nervous' if they cannot see a supply of future orders. Certain technical factors mean that the information system will eventually catch up with the shop-floor picture, but the work-to-list generated by Module C will be longer than John's manually adjusted list and the dynamic nature of the shop-floor means that some data will always be slightly behind the actual production situation.

A more formal way that he monitors production is to print off reports from Module A that tell him what is happening on the shop-floor. These reports take the form of:

- Due starts
- Failed starts
- Due completes
- Failed completes
- Due closed
- Failed closed
- Critical backlog report.

These performance measures factors, and their influence on John's scheduling behaviour and performance, are discussed in the next section.

5.5.2.2 Monitor and achieve Cell performance measures

The Cell has a number of key performance indicators (KPIs) that are used by all the personnel to measure the department's schedule adherence and delivery performance. However, a full understanding of the KPIs' influence on John is important because their impact is twofold. Firstly, the Cell KPIs are an example of performance measures as a factor that influence and motivate John's behaviour. However, secondly, because of their interrelationship with organisational, people and manufacturing process factors they are also used by the Cell as an informal measure of John's performance.

Due starts. Each order is allocated a start date on Module A to allow the monitoring of an order's progression against the original plan. Therefore, John knows when his orders are due to start on the shop-floor and, working back from this date, when he should be firming and cutting planned orders. If John does not ensure that the order starts on time, then the order shows up on the *failed starts* report.

Due completes. Each order is allocated a completion date on Module A so that they will be delivered to the customer on time. Therefore, John knows when his orders are due to be completed. This due date is downloaded onto Module C to generate the work-to-list so that John knows at what stage in its operations an order should be at any point. If John does not ensure that the order is completed on time, then the order shows up on the *failed completes* report.

Due closed. Each order is allocated a closed date on Module A. The agreement set by the department for the time allowed between the complete and closed dates is three days, i.e. an order has three days to be booked out of the Cell and booked into stores. Therefore, John knows that if he does not ensure that the order is booked into stores within three days of its completion, the order will show up on the *failed closed* report.

Critical backlog report. This report is made up of *actual* and *potential* backlog data. It shows John the amount of backlog that there is within the department and therefore what needs to be 'pushed' through the shop as priority orders. Although John does refer to this report to give him an overall picture of the loading situation within the Cell, this report is used more by the Shop floor Supervisor who uses it as his plan of priorities to sequence late jobs and orders through the shop floor.

By combining all of this information, John can generate *predicted* and *actual* reports for the week. These show John's predictions of what orders should be started and completed for the following week and his justifications for these predictions. These reports, and all of the others, are used by the Cell Manager who, as an example of a people factor, expects John to prepare this information to enable the manager to justify the Cell's performance at the department's review meeting, which is attended by all the departmental Cell Managers and the department's Manufacturing Manager. Further examples of the influence of performance measures on John's behaviour and performance will be discussed below.

5.6 IN-DEPTH DATA ANALYSIS

5.6.1 Analysis of data

The previous section presented two types of tasks within a conceptual model of John's scheduling behaviour and performance. We can use this model as a framework to provide further insight and interpretation. Generally, detailed findings are not evident purely from a preliminary analysis or from studying John's observable behaviour. The methods used to study John and his work provide data which must be analysed appropriately to allow insight into those factors that influence a scheduler's behaviour and performance. The methodology of analysing qualitative data is outlined below.

The majority of the data collected were qualitative, that is, the data were usually in the form of words rather than numbers. As described by Miles and Huberman (1994):

> 'Qualitative data...are a source of well-grounded, rich descriptions and explanations of processes in identifiable local contexts. With qualitative data

one can preserve chronological flow, see precisely which events led to which consequences, and derive fruitful explanations'.

However, the use of qualitative data is not without its problems. It is easy to gather reams of data that are difficult to compile into a user-friendly format, and therefore are difficult to analyse. The task of analysing qualitative data is iterative. One must analyse a piece of data and record preliminary findings which can then be validated by referring to the findings of data collected by a different method. Once all data have been analysed and findings cross-referenced, final results can be presented. It is imperative therefore that analysts understand the need to triangulate their qualitative data analysis.

There is no one definitive method of analysing qualitative data; the volume, richness and quality of such data mean that there can be no formalised analysis rules. However, several analysts have presented analytic guidelines (Strauss, 1987; Miles and Huberman, 1994) which were used during the analysis of the data from the field visit.

All the field data were written up as soon as possible after the field visit. This ensures that notes made are understood and that data are not forgotten. From there the first main task is to *code* the data. This, in essence, means applying a single summarising notation to a 'chunk' of data. For example, this is the coding process on a section of the observation data:

10.52 a.m. *Checking overall finished reporting groups (work-to-lists)*

This appears to be an example of John checking the status of orders within the Cell.

11.40.a.m. *No time to check each order in detail-just check order book will be balanced*

This appears to be an example of John checking the status of the overall plan for the Cell.

This process continues as the observation data are read. Preliminary notes are made about what is happening (*descriptive* coding) or about what John's behaviour means (*interpretative* coding). From there, a third class of more explanatory codes becomes discernible (*pattern* coding). After further scanning of the data, a preliminary master list of codes is drawn up. This is revised accordingly during further analysis. The above example was added to by further analysis so that an *assessment of the situation* code became evident, as did various sub codes:

AS (assessment of the situation) : *ST* (status of) / *or* (orders)
pl (plan)
mats (materials) etc.

Once all of the data had been reviewed and coded they can be collated into 'four...summarisers: themes, explanations, relationships among people and...theoretical constructs' (Miles and Huberman, 1994). The following section presents the findings extrapolated from the coded data summaries using this method.

5.7 JOHN'S TACIT TASKS - THE NATURE OF SCHEDULING PRACTICE

The detailed analysis showed the extent to which factors are interrelated and demonstrated how this interdependency generated many, sometimes conflicting, motivators for John as a scheduler within a dynamic machining department. The following sections should be considered as further examples of John's tacit tasks; the important point to note is that these tasks only became evident when more detailed analysis was applied to the fieldwork data.

5.7.1 Routine and clerical work

The majority of John's job as a Cell Planner appears to be mainly routine and clerical work. 'Routine' is defined as something that John does every day, that is not done by the planning and scheduling information system, and something that must be carried out on a regular basis. However, this does not imply that routine or clerical work is an easy or negative aspect of the scheduling task. Often the routine work is important to keep John aware of what is going on.

Another important point is that routine work needs a full understanding of manufacturing process factors. John needs to know about the Cell products, processes, constraints and short cuts to efficiently complete his tasks. For example, John faxes material requirements to the suppliers at least once a day. Although, theoretically this could be done by computer via a modem, the information system input is too limited. John needs to know the availability of stock, the accuracy of stock figures, the lead-time of special materials, whether there is transport available and a number of other small, but important factors to ensure that materials of an acceptable quality arrive in the right place at the right time. All this commences with the routine task of collating and faxing material orders to the supplier. So the job of Cell Planner relies on system and job knowledge, therefore someone who is not knowledgeable about the systems, products, processes, machining techniques and personnel could not perform this job. This has important implications for the selection, training and job design of schedulers. The idea held by many in the firm that Cell Planners are 'swappable' between different departments, without much training, is incorrect.

5.7.2 Adjust scrap and 'lost on shop' figures

John must work through the amendment sheets every day. If it is not done, Module A becomes out-of-date as order quantities are not adjusted. However, this task also represents a chance for John to communicate with the shop floor. As he goes out to the inspection department he has a chat with various people and a chance to catch up on any news. It also represents an opportunity for John to update his mental model of the shop floor as he is made aware where there may be problems as certain items are scrapped or lost. He will then know of potential problems before they become actual shortage and customer delivery problems. This tacit task therefore allows John the opportunity to act on these potential problems and reduce the possibility of scrapped or lost items affecting customer delivery.

5.7.3 Check and re-check

The job of a Cell Planner relies heavily on checking. Organisational and manufacturing process factors impose the need to constantly check materials, suppliers' delivery promises, orders, dockets, the overall plan, the status of the work-to-list, the status of the shop floor and stock figures. John must be proactive; he needs to be aware of his situation. He cannot simply react to production problems and disturbances because of the tight lead-times demanded by customers. John needs a good memory, attention to detail and the ability to disaggregate large amounts of data from reports, performance and capacity graphs, work-to-lists, routings and engineering drawings. Again, such issues are important for the selection and training of schedulers.

5.8 JOHN'S COMPENSATION TASKS – 'FILLING IN THE GAPS'

We have already analysed a number of aspects of John's behaviour as a Cell Planner in this engineering firm's machining department. The analysis has demonstrated that planning and scheduling takes place within a framework of interrelated factors, and that John's scheduling behaviour is best explained using the concepts of tasks and roles. In this case study, compensation tasks and John's compensatory behaviour are considered as different from the behaviour demonstrated in formal and tacit tasks. The compensation tasks described in this section encompass all the functions of John's job. We define compensatory behaviour as the behaviour John demonstrates while attempting to make up for some level of failure, for example, in the information system, in the organisational structure or the production unit layout etc. In other companies compensation tasks could be linked to a number of other factors. Here they are driven by the performance measures factor because this aspect is so pervasive in this firm. This type of task is a good example of Bainbridge's (1982) ironies of automation; when attempting to automate a system, the most difficult and complex tasks are inevitably assigned to the human operator because they cannot be performed as effectively by a computer.

5.8.1 Formal performance measures

There are very definite performance measures in place in the form of KPIs, as described in Section 5.5.6.2. It is by these measures that the Cell is measured, both internally and externally. It is John's responsibility, and that of the Cell as a whole, to minimise the number of failed orders and the amount of backlog against the plan. In practice, these performance indicators appear appropriate to the Cell; they are translated into weekly reports and graphs that are used at various meetings, and Cell personnel feel that the measures are visible and defined. However, the severity with which the measures are pursued means that the measures also generate compensation tasks for John.

5.8.1.1 'Beating the system' tasks

As well as trying to achieve the system-defined targets by carrying out formal and tacit tasks, John's behaviour indicates that he often tries to 'beat the system'. Three examples of 'beating the system' tasks are presented here in order to explain this concept further:

1. The firm's service agreement allows three days between an order being completed and booked out of the Cell, and booked into stores. John knows from experience that in most situations it only takes a day to complete and close an order. Therefore, when needed he knows he can 'borrow' two days of the service agreement to use as extra production days to increase his chances of achieving the scheduled due dates.
2. Orders completed by the Cell on a Thursday or Friday may not be booked into the treatment plant by treatment personnel until the following Monday because of the lag of Module A over the weekend. However, John can book an order into the treatment plant himself on a Friday; this will 'beat' the Module A weekend update so that it appears as though the orders are already in treatments. In reality, the orders are still in the Cell's despatch area but John will have achieved the performance target for closing the order on time.
3. The Cell Manager needs the performance measures reports by 8a.m. on a Thursday morning to assess them before his meeting. Instead of just running the reports directly off Module B, John goes through and 'checks' the status of the orders. Where possible he updates them, that is, he books them further through the operation statuses instead of allowing the system to update the statuses. John's adjustments allow the Manager to present better statistics at the Cell review meeting. This is a non-value-adding exercise for John because Module A updates system information on a Friday. Therefore, John wastes valuable planning time 'beating the system' to present more positive KPIs.

This type of behaviour raises a number of issues. For example, it could be that within this firm, the KPIs are not realistically achievable and so to achieve them in practice employees have to use 'short cuts', or maybe John does not have the appropriate knowledge, training or skills to achieve the targets set by the Cell.

So it can be seen that as well as generating compensation tasks that John is expected to undertake as part of his job, the KPIs can be used as measures of John's behaviour. There are other measures or means by which John is judged, which for the purposes of this paper have been defined as 'informal' performance measures. These measures are implicit at firm and machining department levels and there is little awareness of their influence. These measures are discussed next.

5.8.2 Informal performance measures

5.8.2.1 Visibility of tasks

John's tasks are very visible. 'Visibility' is defined as when it is apparent to many people, such as management, operators and supervisors and suppliers, how John is performing in relation to departmental expectations. Visibility means that it can be seen when an order is late against its planned due date or when materials have not arrived from the supplier; it can be seen when a machine breaks down and production stops. John is not the only person who is subjected to this type of scrutiny; the Cell Supervisor and machine operators are also judged against the same measures. However, John is the only indirect or support staff subject to these measures. If the Cell Engineer sets up a cell that is not efficient, this can only be inferred over time from the performance indicators used to measure that Cell. John's role is such that physical objects and problems are seen directly, not inferred. If this measure is applied to the jobs preceding John's role, an unfair comparison is drawn. For example, the Cell cannot see if suppliers have no materials available; if the suppliers state they have, the Cell cannot see this until the materials fail to arrive on time, at which point the non-appearance of materials will then be John's responsibility. If the plan generated by Module A is not feasible, it is only when visible factors present themselves, for instance, overloaded machines or backlog problems, that 'John's failures' are made visible.

5.8.2.2 Suggest and solve tasks

Cell personnel expect John to make suggestions or give solutions to production or logistics problems that go beyond his job description. John is expected to give appropriate, effective and rapid solutions to problems that may be someone else's responsibility, for instance about machine capacity, cell layout, the effectiveness of Cell performance measures, reduction of backlog. This is not necessarily an unfair performance measure, but it raises the important question of where boundaries should be drawn around John's role. This has implications for job grading and pay, and what kind of training and experience John needs before Cell personnel consider him competent and reliable enough to 'solve' such varied production and scheduling problems. Such performance measures have important implications for the selection and training of schedulers within the engineering industry.

5.8.2.3 Equilibrium tasks

John is expected to successfully balance the flexibility demanded by the customer, the Cell environment, and the uncertainty of the machining department as a whole. For example, John has to order material from the supplier ten days before its delivery date, yet shop floor personnel want the appropriate material in place for each order, and management want John to minimise inventory levels. The department's Manufacturing Manager asks John to offload approximately five days worth of backlog orders but still guarantee that his scheduling will achieve customer due dates. The shop floor prefer that John issues them with a work-to-list that gives them up to ten days visibility yet Module C only produces a five day work-to-list. The question is why does John have to carry out these tasks? How and why are uncertain and unstable elements arriving at this planning and scheduling level?

5.9 JOHN'S ROLES

We have already discussed the type of tasks that John undertakes, but there are also certain 'roles' that John holds within the Cell. A role is best defined in a sociological sense, that is:

> 'Each status in society is accompanied by a number of norms which define how an individual occupying a particular status is expected to act. This group of norms is known as a 'role'...Social roles regulate and organise behaviour. In particular, they provide means for accomplishing certain tasks.' (Haralambos and Holborn, 1991)

If we use this perspective to analyse John's behaviour within the machining department then it becomes evident that the interplay of factors impose many different 'roles' on John's job within the Cell. If we consider these roles in relation to the tasks already extrapolated from the fieldwork data it can be seen that we are moving closer to answering the questions of how this particular scheduler schedules and how well he performs.

5.9.1 Dynamic problem predictor and problem solver

John holds the role of a dynamic problem predictor and solver within the Cell, which means that John's planning and scheduling tasks are in the form of problems that he has to solve before he can carry on with other related tasks, and that John must try to anticipate or pre-empt possible problems. He is dynamic in the sense that he is proactive rather than reactive. Evidence of such behaviour can be seen in the fieldwork data. For example, the Cell Engineer asks John to check the status of an order going through the Cell. John decides that the order might cause a problem because of its wrongly allocated Economic Batch Quantity. It was not John's responsibility to find out about changing the EBQ because the order was on schedule at that time but he altered the EBQ so that when the order arrived at the Cell, it would cause less problems for the shop floor and production rate overall.

5.9.2 Information hub

John acts as an information hub within the Cell. The concept can be described using an analogy with the American airports' major and minor hub system, where certain airlines use certain airports as a major hub for their business, and in turn the major hubs are linked to a number of smaller minor hubs across the country. This infrastructure provides a co-ordinated service for each airline's customers. In order for a customer to reach the destination they require, they must travel via a series of hub transfers.

5.9.2.1 Dissemination of information

If John is considered analogous to a major hub within the Cell, that is, as the point of contact for internal and external enquiries, then other Cell personnel are analogous to minor hubs. John has information for these minor hubs from hubs outside of the Cell. Information is therefore disseminated across John's Cell and other Cells via this network. However, it is not a one-way service; the minor hubs that John speaks to also have relevant and current information for him and John constantly collates this information to ensure updated information is being disseminated.

5.9.2.2 Information use

As well as updating and passing on information, John must also use the information given to him for his own job needs. Or John can do both: use the information, and then update it and pass it on. In this role John can be considered as an 'information provider', an 'information investigator' and an 'information user'. When these behaviours are combined, John's role as an 'information hub' becomes evident.

This information role assumes that John cannot successfully carry out his job without a transfer of information, and that the people connected with him and his job cannot successfully perform their tasks without the information. This has many implications for job design, work design, information quality, teamworking initiatives, system design, and interface design. The information role imposed on John suggests that he cannot be considered purely as someone who plans and schedules work within the Cell; the various factors' influence on information and therefore on John's behaviour must be fully understood, as must the necessity, timing, quality, amount and adjustment to information passing through the 'hub' systems within departments. Without such an in-depth analysis of the areas of John's behaviour, major factors influencing scheduling practice in this machining department will be overlooked.

5.9.3 Decision filter

John also fulfils the role of a decision filter. To explain this role fully, the interactions between John, the Senior Cell Planner, George, and the Shop floor Supervisor, Dave, are included in the description.

5.9.3.1 Senior Cell Planner filter

George looks at orders planned by Module A using Module B. Module A generates a plan using its programmed algorithms but as the information system cannot incorporate all the current constraints, the plan often translates into an overload or underload situation in relation to capacity. This 'approximate' plan is visually presented to George as either a graph or a table via Module B. It is then George who makes the adjustments to the graph or figures to 'make them fit' the system-generated capacity line. That is, the information system aggregates the data within the department, but it is the human who produces the final 'best fit' of the data to the real-world constraints and dynamics of the production situation. George knows which machines can cope with an overload for a certain amount of time, he knows who to ask if he needs to include extra overtime. There are too many nuances and informal pieces of information that cannot be added to the system's data to enable the system to be able to generate a feasible, 'do-able' plan; that is the role of the human filter, George. He sifts through his constantly updated mental model to apply the best information to the generation of an appropriate production plan.

5.9.3.2 Cell Planner filter

So George 'filters' unfeasible plans and workloads from progressing to the next level by making the final capacity-loading decision. If George then assesses the planned workload as being feasible, these orders are fed back into Module A, as planned status orders to be worked on by John.

John assesses the overall picture of planned orders shown to him by Module A and acts as another filter. In the past, Module A had been programmed to firm and cut orders automatically, i.e. the planning and order release process was totally system generated. However, now all orders are firmed and cut manually to allow John more control over the orders being released to the shop floor, and to allow him to develop a more accurate mental model of what the shop floor 'looks like' at a particular moment in time.

Sometimes there is a problem with a particular planned order that is not evident from Module A, but John knows about it from his different sources of information. He can act on the problem before allowing the order to proceed to the next stage. If the order were allowed to progress automatically, the problem would have moved one step nearer to creating uncertainty at the shop floor level. If John knows that there is no material available at the suppliers for a certain order because he spoke to them earlier in the day, then he can start chasing the material while the order is still at planned status and buy himself some time to troubleshoot before the order becomes 'live'. This minimises the chances of a shortage stopping the progress of an order. Even when the orders are firmed and cut, they do not all smoothly progress through the next stages of material allocation and kitting. If there is a problem, such as late or incorrect deliveries, it is John's responsibility to filter out the problems before the order progresses to the next stage, the shop floor.

5.9.3.3 Shop floor Supervisor filter

On the shop floor, the order meets another human filter, the Shop floor Supervisor, Dave. Once John has placed orders in reporting groups using Module C and kitted them up with their appropriate parts, the orders do not just get placed at appropriate machines as is assumed by the information system. Dave first assesses the current shop floor situation, and evaluates the information on the backlog report, and from the work-to-lists before allowing the orders to be physically placed at machines. He must also consider the implications of the non-system-generated information such as changes in customer demands, and the management and planning of jobs that have to go to and come back from the external treatment plant. Dave integrates all this extra information and data into an assessment of the overall shop floor situation before he finally places orders at machines to be put into production.

Even at this final stage the underlying assumption made by the information system is that the production situation is amenable to the released orders, that is, that all materials are available, that there are operators capable of carrying out the jobs, that there have been or are no machine breakdowns. If any of these situations, or countless others, occur it is people like George, John and Dave who filter the problem through their current situation mental databases, and implement a solution. The value of the information systems within the scheduling process is often purely to aggregate the data input by humans and update any data linked to it in other integrated systems. That is, there are two methods of data flow in this manufacturing scenario: the information system data update, and the transfer and 'filtering' of knowledge and information by the human element to the next level of the planning and production process.

5.9.4 Production responsibility and authority

A final role to explain is that of production responsibility and authority. Most of the time, it is relatively easy for people within the Cell to know where to go for certain information and help; this comes with experience and knowledge of the Cell's working structure. For example, when the Cell Manager wants an order pushing through the Cell because it is late, he will go to Dave if the order is on the shop floor, or to John if the order is due to go into production because it is John's 'production responsibility' role to get the order ready for production. The Manager tells John to get the order ready on time; he can do this because he has the authority imparted to him via the firm's hierarchical organisational structure. John must then 'tell' people to get the order ready. For example, he 'tells' stores to prepare the materials needed or he 'tells' the suppliers to deliver the materials early. However, John has no 'production authority' to tell people what to do; the department's organisational structure means that he is not 'above' these people. In some cases, they may be equal to or higher up than him. So how does John gain a production authority role? There are a number of factors that account for other people allowing him to hold this role:

- *Respect*: for John and his ability to do his job well.
- *Trust*: that John is asking them to do it out of real need.
- *Friendship*: if the person asked knows John and likes him.
- *Favours*: as a bargaining tool.

The *favour* concept appears to be important to a dynamic role such as John's. Favours are used as bargaining tools. If John asks someone who does not have to obey to do something, then it is often done as a favour with the, sometimes implicit, understanding that this favour will be reciprocated when they ask John to do something for them. For example, sometimes when John checks Module A's mailbox messages in the morning, the other Cell Planners Robert and Michael have not had time to go through their messages. John will go through all of the messages and clear them on the understanding that he "knows" that Robert or Michael would do it for him if he was short of time. Therefore, people and organisational factors have interacted to produce informal communication and bargaining networks within John's domain.

5.10 DISCUSSION

The generation of a valid conceptual model of a scheduler's behaviour and performance within a business was not the only aim of this intensive, observational case study. A variety of other points were noted from the in-depth, ethnographic-based data collection techniques. Some of the other issues raised during the case study are discussed below.

Much of the data presented above raise an interesting question of how the department within this study measures performance within a Cell. If John and other members of the Cell can 'work around' the measures that have been set, then perhaps the measures are themselves a problem; that is, maybe the department should be measuring *how* something gets done to reach a 'good' standard rather than just measuring *what* is done against a performance measure. The alternative way to look at this point is that it does not matter *how* the Cell Planner achieves the performance criteria, only that the criteria specified by the department are achieved.

Another related issue is that it is important that the department ensures John is measured against targets that are within his domain. For example, failure of suppliers to deliver materials on time is partly John's responsibility because he should be monitoring ongoing delivery. However, the converse of such a 'failure' is that the department is allowing the supplier's failure to be translated onto John; the supplier should also be measured as a separate factor. The most common way to address this problem is to introduce vendor measures. In this case, this study has demonstrated that when informal measures are made explicit, they can be used by a business to assess if their formal performance methods of rating and measuring personnel are valid.

The major finding from using this type of data analysis was that John does not schedule in the 'classical' sense of 'the allocation of resources over time to perform tasks' (Baker, 1974) or at the level of sequencing and routing jobs to specific machines. The generation or updating of the plan or schedule by John is done at the order level, not at an operations level. However, even changes at the order level are minimal because it is the MRP system that generates the plan from the data input at a higher operations planning level. By the time the plan reaches John it has been transferred to Module C, which compiles work-to-lists using an earliest due date to customer algorithm. John is only authorised to make changes to the work-to-list as a last resort, so that 'queue jumping' does not worsen the Cell's customer order backlog problem. Therefore, it is not John's job to schedule, according to the department it is not anyone's job to schedule as the system is relied upon to generate the plan and schedules! What John is charged with, *is an overall responsibility to realise the plan and to deal with any problems that affect the plan and schedules as efficiently as possible to minimise disturbances to production.*

All of these findings have also been apparent in the other case studies conducted as part of the overall research project. Six other schedulers, across a range of UK manufacturing sectors, have been observed using the same research methodology and data collection methods. The most interesting comparisons are that all of the other case studies demonstrated schedulers carrying out a variety of tasks and roles as an inherent part of their job. In particular, the information hub and decision filter behaviour have become evident as important roles in all of the other case studies. The transfer of information and the filtering of information in order to allow up-to-date and accurate decisions to be made at the operational level appear to be generic scheduler roles.

It is also interesting to note that other schedulers also do not schedule in the 'classical' sense. In these cases, it is evident that the participant schedulers undertake broader tasks and roles and act as shop floor control 'facilitators' or general 'manufacturing controllers'. This evidence, that the responsibilities of certain schedulers lie more with ensuring that the planned schedule is physically implemented and 'pushed' through shop floor production rather than building the schedule itself, has important implications for the study of scheduling within industry, and also for industrial practitioners.

This case study has raised many previously overlooked issues about the human performance of planning and scheduling. However, it appears that there are still two major questions that human factors practitioners within this domain must address: what is production scheduling and what is the role of the human scheduler in planning and scheduling environments? It is apparent that valid answers to these questions can only come from reliable in-depth, applied studies of planning and scheduling, where the reality of the role of the human scheduler can be recorded, and the totality of human planning and scheduling performance can be placed into context.

5.11 ACKNOWLEDGEMENT

The author would like to thank all the participating schedulers and the companies that have enabled this research programme to take place.

5.12 REFERENCES

Bainbridge, L. (1982). Ironies of Automation. *Proceedings of Conference on Analysis, Design and Evaluation of Man-Machine Systems*, Baden-Baden, W. Germany. pp 129-135.

Baker, K.R. (1974). *An Introduction to Sequencing and Scheduling*. Wiley, New York.

Crawford, S., MacCarthy, B.L., Wilson, J.R. and Vernon, C. (1999). Investigating the Work of Industrial Schedulers through Field Study. *Cognition, Technology and Work,* 1: 63-77.

Haralambos, M. and Holborn, M. (1991). *Sociology: Themes and Perspectives.* Collins Educational, Harper Collins, London, UK.

Miles, M.B. and Huberman, A.M. (1994). *Qualitative Data Analysis: An Expanded Sourcebook.* Sage Publications Inc, CA, USA.

Strauss, A.L. (1987). *Qualitative Analysis for Social Scientists.* Cambridge University Press, Cambridge, UK.

CHAPTER SIX

Boundaries of the Supervisory Role and their Impact on Planning and Control

Howarth Harvey

6.1 INTRODUCTION

Many activities carried out by staff in production planning, scheduling and control contribute to the formulation and outcome of the shop floor schedule. Each of these contributions may be divided into activities which are formally prescribed by the system of planning and control and those activities which are not prescribed or are informal. Appreciation of the position of these boundaries and the degree to which the formal system is unable to cope with reality is critical to improving the co-ordination of these activities. In the short-term, it may be possible to realise greater improvements in operational performance through co-ordination that is specific to a manufacturing environment than through development of commercial planning software. Commercial planning software is generally considered as part of the formal system and is often seen as inadequate or inadequately implemented.

This contribution pursues this idea through analysis of an empirical investigation into planning and control in small to medium, Make-To-Order (MTO) companies. In this analysis of the empirical data, which was part of a wider study, it was decided to focus on the role of the supervisor, although choice of other roles would have been equally valid. Deepening the empirical knowledge of activities which impact on planning and control will facilitate improved co-ordination. It will also help define where the boundary should be between formal and informal activity and thus also assist the formal modelling process and its implementation.

This introduction is completed with a brief overview of the motivation and methodology of the source investigation for this contribution and a review of other empirical studies into the supervisory role. Section 6.2 describes the diversity of line management structure and functional links with other indirect staff found in the source investigation case studies. Presentation and analysis of the empirical data for planning and control activities of the supervisor follow in Sections 6.3 and 6.4 respectively. Finally, there are conclusions and suggestions for future research.

6.1.1 Motivation and methodology of the source investigation

This contribution has been drawn from an investigation which was initially motivated by the difficulty in identifying and selecting context specific ways of effecting production planning and control functions. Appreciation of this difficulty was gained through the FORCAST project (Harvey, 1993).

The above difficulty was combined with a realisation that the descriptions of industrially practised systems were inadequate. Nominally, the same system can be implemented in a variety of ways (Giesberts, 1991; Burcher, 1992). Very different scheduling systems have been selected for the same environment (Harvey, 1993). Nominally the same system has been found equally successful and unsuccessful in the same and different environments (Sridharan, 1992). There have been well-argued cases for the unsuitability of some scheduling tools as currently structured for use in particular environments (Hendry, 1991).

There was a need to acquire improved descriptions of current practice, not to enable mimicry, but to improve the understanding of how incomplete planners and planning and control models (Rodammer, 1988) were completed in practice. It was considered that this could only be done through investigating how such systems were combined with the organisation, the manufacturing processes and the external environment to the company. A case study methodology was adopted.

An experimental structure was devised to determine relationships between parameter choices in the manufacturing system under a particular set of constraints. The environment chosen was Make-To-Order (MTO), small to medium enterprises manufacturing discrete end items. These end items were not customised on the receipt of each order. The customer would either be making repeat orders of their own particular variant of a product type or would be choosing a variant within the company's portfolio of standard products.

A set of 12 case studies were selected from an original set of 56 companies where face to face interviews were carried out for preliminary data to ensure that they would conform to the case study constraints. Each case study company was subject to four or more further interviews, spread over a few weeks. Much emphasis was placed on collecting quantitative data which was amenable to comparison and which, in some cases, would serve as partial proxies for qualitative attributes, e.g. the number of part types put together in final assembly was a partial indicator of the complexity of the product.

The set of 12 case studies was divided into four groups, with between two and five companies in each. Each company was investigated through the impact of its 'choices' on a single product. Each group consisted of members who put the same number of part types together in Final Assembly (FA), for a range of demand in annualised averages of units of the particular end item. Figure 6.1 shows the distribution of the case studies; the product was not selected on any criteria of optimisation in respect to its fit with the company's systems. It did have to be in continuous production. A brief outline of each of the companies is given in Table 6.1.

The objective of the research was to observe how, within a single group of companies, a product 'performed' under what was expected to be a changing set of choices in system parameters as the demand increased, e.g., layout, product decomposition, resource sharing. Further detail of the methodology can be found in Harvey (1998a).

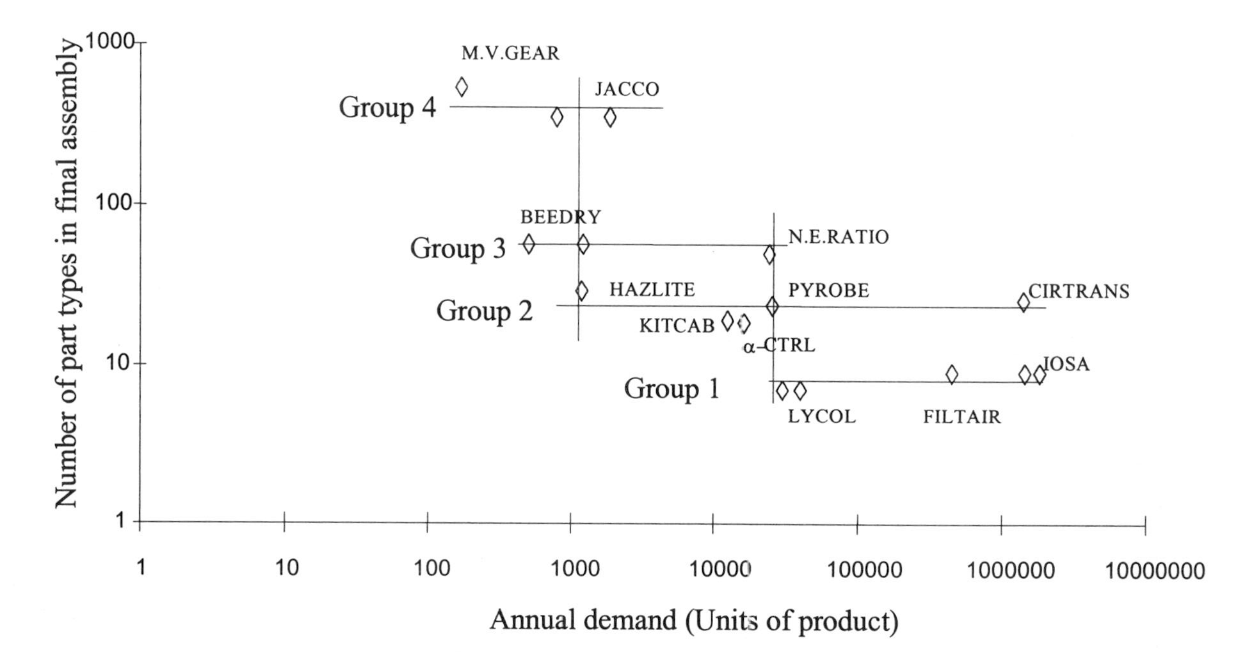

Figure 6.1 Distribution of case study groups in the domain.

Table 6.1 Selection of the case study products and their grouping

Group	Average number of part types added in Final Assembly	Company[1]	Product	Number of employees	Customers for product
1	8	LYCOL	Lighting column	123	Local authorities, infrastructure contractors
		FILTAIR	Air filter	450	Original Equipment Manufacturers (OEMs) and industry
		IOSA	Bimetallic switch	240	OEMs
2	23	HAZLITE	Lamp unit	160	Independent electrical distributors, process sites
		KITCAB	Kitchen unit carcass	74	Building contractors (70%) and the public
		'α' CTRL	Accelerator control	135	OEMs and own distributors
		PYROBE	Wardrobe	630	Public and retail
		CIRTRANS	Central heating pump	200	OEMs
3	57	BEEDRY	Building dryer	300	Plant hire companies
		N.E.RATIO	Truck gearbox	130	OEMs
4	325	M.V.GEAR	Medium voltage switchgear	600	Utility companies and large industrial electricity users
		JACCO	Loom equipment	375	Cloth manufacturers

[1]The names given to the companies are fictitious in compliance with the agreement on confidentiality

6.1.2 Motivation for selecting the issue of the supervisor's input to planning and control

There were a number of reasons for selecting the role of the supervisor in planning and control as an issue for this contribution. Firstly, the supervisor role contains a number of activities that impact on the quality and outcome of the schedule, 'external' to the scheduler or planning department. The supervisor role appears to be under-reported in relation to the emphasis placed on differences in outcome between types of commercial software for planning and control (e.g. Sridharan, 1992; Stoop and Wiers, 1998). It is also distinct from empirical work on the scheduler (McKay, 1995). In nearly all the source investigation case studies (Harvey, 1998a), the supervisor was found to be intimately involved with planning as well as control/plan-executing functions at shop floor level, irrespective of the formal system employed.

The second reason for selecting the supervisor role is that the allocation of staff appears to be becoming an increasingly important element in planning and control. Traditionally, this has been a function of the supervisor but greater interchangeability of staff through multi-skilling and the changing ratio of direct staff to work stations in the implementation of lean practice (Riat and Prickett, 1995), has added to the impact of this function on planning.

The third reason for selecting this role is that the activity of the supervisor is often incorporated into the measure of success of the schedule. Unless there is some improved understanding of the supervisor's contribution, it will not be possible to improve appropriate measures of performance.

6.1.3 Other empirical investigations into the supervisory role

If the continuing number of 'manuals' on the role of the supervisor is any indication (e.g. Bittel, 1990; Savedra, 1990; Chapman, 1992; Cusins, 1994), it is clear that the role of the supervisor is perceived to remain a key one. In a more investigative approach, Senker (1994) makes a comparison between the different emphasis on recent training practice for supervision between Britain, Japan and Germany. However, none of these recent contributions describe supervisory practice in specific manufacturing contexts or planning and control systems.

In an empirical study of cultural differences in industrial relations, Dore (1973) makes an interesting comparison in supervising practice between Japan and Britain, i.e. before the Japanese were seen as the source for improvement in manufacturing operations. However, the focus was not on the supervisor's role in planning or the technicalities of production methods employed. It was focused more on the relationship between direct staff and supervisor. It contrasted the difference in the relationship between the 'market orientation' of British industry and the 'organisation orientation' of the Japanese. This study was drawing on evidence from much larger companies than the current source investigation.

Other studies of supervisor practice, as well as being from larger organisations, have largely focused on the automotive sector (Melman, 1958; Grimm and Dunn, 1986; Lowe, 1993).

In very general terms, it can be said that the role of the supervisor in the early stages of industrialisation at the end of the 19th century was one of being completely responsible for planning, i.e., on occasions almost acting as sub-contractor to the owner. In such cases (Stone, 1973; Melling, 1980), the supervisor would be an expert in their craft, might select the number and type of staff to work in the team, decide the ratio of wages between the staff, as well as decide how the job should be done, how long it should take etc. With changing skill levels and larger organisations, the role concentrated much more on the control of people rather than planning; already in the 1920s, Ford in the USA had introduced the $5 day. Taylor (1947) at the same time was laying down the 'scientific' basis for work study and emphasising the need to separate planning from execution.

Melman (1958) describes an automotive plant in the early 1950s, where fully developed planning, work study and personnel departments had removed several functions from the supervisor. However, he describes the role of the supervisor moving away from monitoring individual workers to ensuring the flow of materials and in particular the meeting of product output targets. On pages 99 to 102 (ibid.), a supervisor gives a description of a daily routine which in planning and control terms would be hard to distinguish from the current practice exhibited at the case study companies of the current source investigation. Even later than this, a move away from control of the rate of work by piece rates remained a concern (CDEEA, 1968); this report suggested that the move to Measured Day Work would require the levels of supervision to double from 50:1 to between 25 and 30:1.

It may be that things have moved on in the larger organisations of today (Grimm and Dunn, 1986; Alasoini, 1993), where some of the supervisor's responsibilities have been returned within a cellular framework, e.g. maintenance, continuous improvement of techniques and team building. However, although these authors also suggest that some of the technical content of the role is increasing in respect of the equipment that has to be operated, this is not equivalent to planning the skilled work of others. Lowe (1993), who admits the limitations of employing idealised models in studying the changes from mass to lean production, produces anecdotal empirical counter-evidence of any real change in the role of the supervisor, even in large automotive plants.

It is clear from even this limited amount of empirical data that a number of nominally different supervisory roles can be operating in different companies at the same period of time. It is not possible to say if any evolution has taken place. The evidence reported from this current investigation indicates there is currently much diversity in the role of the supervisor for the size and category of company investigated. Some of this practice seems little different from that of 50 years ago and this is despite the introduction of a wide range of software for planning and control in the latter part of that period.

6.2 DIVERSITY IN SUPERVISORY STRUCTURE AND FUNCTION DISTRIBUTION FOUND IN MTO SMES

Before focusing on the specific impact of the supervisor's role in planning and control, it was considered useful to situate this role with respect to some other features of the case study companies that might influence that role.

6.2.1 Supervisory level observed

There is inevitably some problem with the nomenclature of the supervisor's position. Therefore, a choice was made to investigate the lowest level of supervision that took on some planning activity (formal or informal), irrespective of the name given to that role (for a definition of planning activity, see Section 6.3).

6.2.2 Hierarchical level of the selected supervisory position

The first issue investigated was the position of the selected level of supervision in the overall hierarchy of line management (see Table 6.2) and how this might be related to the number of employees at the company. It is not uncommon that indirect staff task division increases with the number of employees. However, no particular pattern was obvious when the number of levels above and below the selected position were plotted in the domain of number of employees versus demand. Neither was there a pattern with respect to the total number of levels. Demand was chosen as the other axis of the domain because formality in control was known to increase with demand (see Section 6.4 on control activities of supervisors).

PYROBE and M.V.GEAR, companies with the most employees in the selected case studies, had more sub-divided hierarchy in line management than most of the others. However, LYCOL, one of the smaller companies, had even more levels than these and, even ignoring this as an exception, there seemed to be no obvious gradation top to bottom or diagonally across the domain (see Figure 6.2).

A second related issue is the density of supervision, the number of direct staff (those carrying out manufacturing operations) for each supervisor (the numbers of staff per supervisor at the selected level are shown in Table 6.2). This also proved to be a very variable factor within the chosen domain. JACCO, the case with probably the most normative version of cellular layout, had the densest supervision with a range of four to eleven staff per team leader. There was a wide range of case study companies where the ratio is between 1:20 and 1:30, from the lowest demand at M.V.GEAR to almost the highest at CIRTRANS. Density of supervision does not seem to be related to the number of employees.

It might be argued that the density of supervision could be dependent on the skill level of the direct staff. In general, the level of skill was higher in component manufacture than in assembly. The figures for component manufacture commonly show a lower density of supervision where the capacity requirement is for more staff. At M.V.GEAR, component supervision was divided on skill type, between

Table 6.2 Position and number of lowest level of supervisory staff (italicised) making scheduling decisions

Group	Company	Component supervision	Sub-assembly supervision	Assembly[1] supervision
		Level above *Subject supervisor* Level below	Level above *Subject supervisor* Level below	Level above *Subject supervisor* Level below
1	LYCOL	←	Operations manager	→
		Shop superintendent (19) *Section foreman* (?) Chargehand	No sub-assembly but component goes for outside processing	Shop superintendent (16) *Section foreman* (?) Chargehand
	FILTAIR	←	Production director	→
		Separate site- same structure as assembly ? (?)		Production manager *Team leader* (22) None
	IOSA	←	Production manager	→
		Team manager (shared) (?) Team leader None	Team leader(coding) None	*Team manager (shared)* (?) Team leader (8) None
2	HAZLITE	← ←	Operations manager Production manager	→ →
		Shop manager (15) None		*Shop manager* (17) None
	KITCAB	←	Works manager	→
		←	*Shop manager* (28)	→
		Senior operator None	? None	Senior operator none
	'α' CTRL	←	Production Manager	→
		Cell leader (5) None	*Leaders for 2 cells* (21 & 2) None	*Cell leader* (9) None

2	PYROBE	*Machine shop manager* (?) Section manager 1/shift Team leaders: 2/day-shift Chargehands: 2/shift		*Ass. man.*(6 lines: ≈11/line) Section man: 6 lines Team leaders: 3/ 6 lines Chargehands: 3
	CIRTRANS	←	Production manager	→
		←	*Shop supervisor* (36)	→
		Team leader None	Team leader None	Team leader None
3	BEEDRY	←	Production manager	→
		Fabrication supervisor (30) None		*Line supervisor* (5 - 9) None
	N.E.RATIO	←	Manufacturing manager	→
		Shop supervisor (35) [2] None		*Shop Supervisor* (25) None
4	M.V. GEAR	←	Manufacturing manager	→
		←	Module leader	→
		Machine Cell leader (33) [3] None		*Assembly Cell leader* (35) Section leaders
	JACCO	←	Managing Director	→
		Zone manager *8 Cell Leaders* (78) [4] None	Zone manager *7 Cell Leaders* (78) [4] None	Zone manager *2 Cell Leaders* (23 - 28) None

Notes:

1. If there is not a separate supervisor for sub-assembly, the single supervisor for sub- and final assembly is included in this column. If there is no sub-assembly, this is indicated in the previous column. Figures in brackets indicate number of direct staff supervised.
2. There is a single supervisor for each of the 3 shifts
3. There are two other cells in component manufacture: fabrication and process. Machining cell also has 3 technical support staff.
4. These cells are not divided between component manufacture and sub-assembly. One zone deals primarily with mechanical components and their sub-assembly. The other zone deals with electrical parts and their sub-assembly.

machine shop, fabrication and process cells. JACCO, on the other hand, with its normative cellular structure, had lower ratios of staff to supervisor in the feeder cells than in final assembly. The cell leaders would also be part of the operating team which was not the case elsewhere with the exception of α-CTRL and LYCOL in component manufacture and, in exceptional cases, at FILTAIR in assembly.

6.2.3 Links between supervisors at the same hierarchical level

Another issue that impacts on the role of the supervisor is the nature of the links that the supervisor has with other staff/functions within the planning and control systems. Although the methodology adopted for this investigation precluded coming to an understanding of the 'negotiation' that might occur through these links, it was able to establish in most cases what links existed on a formal or informal basis.

The links were looked at from two perspectives. The first was the links between functions *within* line management, in particular links that arose through material routings crossing boundaries of responsibility from raw material to completed end product. The case study companies fell into the following groups based on the number of boundaries:

Number of boundaries	Case study company
0	KITCAB, CIRTRANS
1	FILTAIR, IOSA, PYROBE, BEEDRY, JACCO and HAZLITE
1/2	α-CTRL (Component: 1; sub-assembly: 2)
2	N.E.RATIO, M.V.GEAR
5	LYCOL

These groupings do not follow those of the case study structure or the groupings based on the scheduling technique (MRPx - a type of MRP or MRPII - and non-MRPx).

The two examples where there were no boundaries had a single supervisor for component manufacture and assembly (see Table 6.2). For the six cases in which there is one boundary, that boundary is between component manufacture and assembly. This can be significant, particularly where there are no links and where the planning decisions can be quite different, e.g. between job and flow shop scheduling. The links between supervision for component manufacture and assembly are shown in Table 6.3, last column. The exceptionally large number of boundaries at LYCOL mainly derive from a grading/payment scheme for direct staff rather than supervisory functions. N.E.RATIO and M.V. GEAR were considered to have two boundaries because they sent material outside the company for processing. This also accounted for one of LYCOL's boundaries. The random distribution of the differences did not appear responsible for disrupting any observable trends.

Table 6.3 Links between supervisor and other indirect staff

Group	Company	Complimentary functions of scheduler for lowest level of schedule	Links between Selected[1] scheduler and supervisor	Other operational functions carried out in supervisor's location	Material: responsibility boundaries	Links between component and assembly supervision
1	LYCOL	Also operations manager to whom supervision and dispatch answer. Higher level schedule/forecasting.	INFORM(2)	None in component manufacture or assembly	5	No links (?)
	FILTAIR	Smooths order entry by volume and to financial targets	NO LINK(?)	No supplementary information acquired	1	No contact
	IOSA	No other functions volunteered in replies	NO LINK	Cell team leaders check quality, train 'solve problems'	1	No links
2	HAZLITE	Scheduling and purchasing share same office; release of works orders; check kitting	FORM(Twice weekly work-to-list reviewed)	None in component manufacture or assembly	1	Daily meetings
	KITCAB	Sales order processors build up the single level schedule	INFORM(2) (Frequent late Inserts from sales order processing; lack of specs on schedule issue)	Shares tasks with factory manager who visits main site 2/3x per week	0	Single manager for components and assembly
	'α' CTRL	Customer order processing, schedule completion, resource variation	INFORM(2)	None	1 for component and 2 for sub-assembly	Informal contact
	PYROBE	Sales order processors build up single level schedule modified by assembly. manager & dispatch	FORM: daily meeting of senior departmental managers	No supplementary information acquired Daily departmental meetings	1 + (?)	Daily meetings
	CIRTRANS	Processes customer orders; purchases materials; issue dispatch paperwork	FORM (monthly meeting) & INFORM (2)	Yes: ?	0	Single supervisor for components and assembly

3	BEEDRY	Material planner + 4 schedulers who deal with routine purchasing and material shortages	Component: FORM (weekly) & INFORM (2) Assembly: Only FORM with case study assembly	No supplementary information acquired	1	Weekly meetings with planner
	N.E.RATIO	Order processing, assists & signs off higher level schedule	Component: NO LINK but see next column. Assembly progress chaser and scheduler do Build Plan	**Daily** meeting of production manager, shop supervisors and assembly progress chaser on Build Plan	2	Informal contact
4	M.V.GEAR	Module planner + senior module planner: ordering materials, machine loading; performance measures	FORM (weekly meeting) Assembly: INFORM (2)	Cell controller, planner, 1/2 engineers, section leaders (assembly only). Weekly meeting with other cell leaders and production manager	2	Daily meetings
	JACCO	Weekly meeting with sales, monthly with dispatch and higher level schedule review which includes MD & zone managers	NO LINK	Daily meetings with zone manager and material controller.	1 (Final assembly is split in 2)	No contact

NOTES:

1. The selected supervisor is the italicised level in the supervisory structure in Table 6.2. The scheduler is the one producing the lowest level of schedule and whose other functions are described in the third column.

FORM: Formal links through regular meetings or reports; INFORM(1): Regular informal contact through sharing office; INFORM(2): Regular informal contact by individual initiative; NO LINK: No direct communication exists on a regular basis.

6.2.4 Links of the selected level of supervisor with indirect production staff

The second perspective on links were those between the selected level of supervisor and indirect staff not in line management but having an impact on the schedule or its execution. This mainly focused on the functions carried out by the scheduler producing the lowest level of formal schedule but it also looked at others who the selected supervisor had contact with either formally or informally (see Table 6.3).

It was fairly straightforward to pick out extremes of 'connectedness' between functions. For example, at CIRTRANS, the selected level of scheduler had the functions of material ordering, customer order processing and making out the dispatch paperwork for completed items. The scheduler had formal and informal links with the supervisor who in turn was responsible for both assembly and component manufacture (see Section 6.1.3, for the supervisor's role in planning). This was considered to demonstrate a high degree of 'connectedness'. However, analysis of the allocation of functions and the hierarchy of responsibility was not always sufficient to judge the value of 'connectedness'. At LYCOL, the very large number of functions invested in the role of the Operations Manager appeared excessive. This company was also troubled by having a particularly anarchic set of parallel routes of authority to the shop floor.

In a qualitative sense, it was possible to point to examples of a close connectivity between functions or a loose connectivity. It was not possible to find examples where there was close connectivity and poor performance or loose connectivity and good performance. Isolation of the contribution of connectivity needs to be the subject of further work.

6.2.5 Supervisory role and the level of technology

Finally, the technology of the processes involved in manufacture is considered for its impact on the role of the supervisor. This remains an issue that is hard to index in a quantitative way, particularly if it is to be viewed as more than the obverse of the skill content in the job. In a qualitative sense, it was again possible to pick out two extremes. At one extreme lay CIRTRANS, where a special machining carousel was employed to carry out the complete set of machining operations on the pump body. This company also had conveyor-fed assembly operations, many of which were completely automated. The opposite of this was the assembly at FILTAIR, also at the high volume end of the case study set, where all operations were completely manual. In fact, FILTAIR claimed to have dispensed with any production engineering in this area and arranged assemblers on the same operations around circular tables. It had the appearance of home-working away from home. The supervisory roles were in many ways quite similar with respect to planning. On the control side, the production manager at FILTAIR claimed that his main pre-occupation was that assembly staff would produce consistent quantities of completed product at a particular operation. At CIRTRANS, the shop manager expressed this as getting the 'standard minute time' down (in this case, actual quantity of product produced divided by actual hours worked on a sectional basis). This does not seem very different.

For the remainder of the group, component manufacture showed some variation in the number and complexity of Computer Numerical Control (CNC) machines employed, otherwise there was not a great difference between case studies.

There was a sub-set of companies where the number and form of raw materials was limited, i.e. those companies dealing with a select number of sheet materials. These materials often went through simple machines dedicated to single operation types (e.g. cutting, bending, drilling and finishing). This sub-group consisted of LYCOL, FILTAIR, IOSA, KITCAB, PYROBE and BEEDRY. However, it was not possible to attribute this group with specific supervisory characteristics.

6.2.6 Concluding remarks on diversity of supervisory structure and function

The above observations are made outside the case study structure, and the group of 12 companies is too small for any statistical generalisation to be made. However, the hierarchical position, functions and linkages of the selected level of supervision appears diverse within the chosen domain as does the density of supervision. These results might be considered to be confirmatory of the lack of theoretical underpinning to current industrial practice. However, it might also indicate that the boundaries of the domain of investigation have been wrongly chosen or are insufficiently constrained. Given the key role assigned to supervision it would seem worthwhile resolving this dilemma.

It was recognised that supervision could be involved in additional functions to those impacting on planning and control (e.g. health and safety, maintenance and tooling) but these roles were not pursued in the original investigation.

6.3 PLANNING ACTIVITIES OF SUPERVISORS

As stated in the introduction, the main subject of the original investigation was the industrial practice of planning and control for Make-To-Order companies. In order to situate that investigation, it was necessary to state the selected meanings of the terms planning and control since there are no commonly accepted definitions.

Planning (and scheduling) was defined as the pre-operational choice of what and how much to make, with which resources and when. Control was seen as the means of ensuring that execution of the schedule conformed with the schedule 'instructions', e.g. in terms of times for starts, finishes and durations (acting or not acting as a feedback loop to planning). In some cases, there is also a more static element to control involving decisions on inter-station buffer sizes to take account of process variability and line imbalance.

One of the potential sources of confusion that arises with definitions of planning and control in practice relates to where the boundary is set between formal and informal planning. Informal planning itself can be sub-divided into that which is accepted as being not amenable to formal planning or not required to be formally planned and that planning which is not officially recognised by the organisation. The reasons for the latter include the failure to appreciate the limitations of the current model of planning and control or ignorance of the required operational knowledge.

However, provided the function of planning is not seen to be vested in a particular person, section or department but that it is commonly going on in several places and needs to be seen as a whole, then the above distinction between planning and control would appear to suffice.

Some practical examples of a supervisor's planning and scheduling activities are:

- allocation of staff to jobs or work stations;
- rebatching or resequencing of material;
- local replanning to deal with production disturbances.

As will be discussed later, the area of responsibility and the planning system itself can have an impact on the planning decisions made by the supervisor.

6.3.1 Allocation of staff to jobs and work stations

This is commonly seen as an activity of a supervisor but is not commonly regarded as a planning or scheduling activity by the producers of commercial scheduling software.

Choices of how to allocate staff when there are more machines than staff can have a big impact on the outcome of the schedule. In the case where staff changes are a consequence of change in the product, then decisions on when to change the sequence or when/how to change the sequence in the event of a disturbance can also be critical. Component manufacture and assembly do not often have the same degree of flexibility to change product sequence. It is therefore interesting to view this with respect to the links between supervision in these two areas when they are separately supervised (see Table 6.3, last column). Examples were not encountered of any support given to the decision processes on staff allocation, although it was quite common for output targets to be set.

An attempt was made to categorise the difficulty of managing different staffing changes with change in product model that might be encountered in assembly. The different staffing changes are shown in Table 6.4, and ordered with respect to difficulty from 1 (no changes and easiest) to 6 where the total number of staff in the area change; the number of staffed stations change; and there are changes to the number of staff on one or more stations. The case study companies were assigned the permutation number that most closely fitted their particular organisation for final assembly. Although there was simplification in the allocation requirement with increasing demand for the first of the four case study groups and for the first three cases of the second group, this trend was not followed for the remainder of this group and the remaining two groups. When all the companies were viewed as a single group, no pattern was discernible either (see Figure 6.2).

The sampling questionnaire (56 companies) for the original source investigation showed that in three quarters of companies, staff were shared across products in both assembly and component manufacture. The case study companies showed that most of the allocation issues occur within assembly where the number of defined skills are fewer and the time to acquire those skills is less. The trend to greater interchangeability between direct staff is likely to be ongoing and with the objective of increasing productivity. Decision support to supervision in this area and its integration with the formal planning and scheduling would therefore appear to be a critical area for development.

Table 6.4 Categorisation of direct staff allocation requirements.

(a) Categorisation of staff change requirements on model changes

Permutation number	Total staff in area	Number of staffed stations	Number of people per station	Practicality of permutation
1	No change	No change	No change	Possible
2	No change	No change	Change	Possible
3	No change	Change	Change	Possible
4	Change	No change	Change	Possible
5	Change	Change	No change	Possible
6	Change	Change	Change	Possible
7	Change	No change	No change	Impossible
8	No change	Change	No change	Impossible

(b) Results of the categorisations for final assembly

Group	Company	Changes with model Permutation no.	Interchangeability of skills	Number of grades
1	LYCOL	5[1]	No	1
	FILTAIR	3	Partial	1
	IOSA	1	'Theoretically'	1
2	HAZLITE	3/6	Semi-skilled: Yes	2
	KITCAB	5	Yes	1
	α-CTRL	1	Yes	1
	PYROBE	3[2]	Partial	2
	CIRTRANS	2/3[3]	Grade 2:Yes Grade 3: No	2
3	BEEDRY	2/3[4]	Partial	1
	N.E.RATIO	3/6[5]	Partial	2
4	M.V.GEAR	Constant staff	Limited within skill	1
	JACCO	6	Partial	1

Table 6.4 NOTES:

[1] In this case work stations are not fixed but assemblers go to the job, however, they were said to stay with the same set of tasks not matter what the product variant.

[2]It is being assumed that the sub-assembly area in the assembly shop which feeds directly into the final assembly is a single unit.

[3]A large percentage of the final assembly is automated. There are several sets of stations looked after by a single person. There are two specifications of pump that require to have additional work done to them on separate lines of equipment but these specifications represent around 3% of the volume only. One sub-assembly workstation is operated with between 2 and 4 staff. It was not determined how this change was accommodated.

[4]The categorisation will depend on the volume being produced. When the cell is staffed at the minimum, final assembly stations will be left vacant while sub-assemblies are made up by members of the same team.

[5]It was not established how the significant increase in staff was achieved when changing product family.

6.3.2 Rebatching and resequencing of material

Decisions on batch sizing and sequencing of batches by supervision can be divided into three categories.

1. Batch sizing and sequencing is formally the responsibility of the supervisor.
2. Batch sizing and sequencing is informally modified by the supervisor.
3. Supervisor adheres to formal planning output on batch size and sequence.

These three categories are distinct from the activity required in the event of a schedule disturbance, e.g., a machine breakdown or the non-availability of staff.

6.3.2.1 Category 1

The first category of company is where the decisions on batch sizes and sequences were formally left with supervision. In these cases, it was common for the supervisor or cell leader to work directly from a final assembly end item requirement. Choices of batch size and sequence for sub-assemblies or components were made by the supervisor to comply with that final assembly list for a given period. In the case of sub-assemblies or components using shared equipment, some of these scheduling problems were complex and there was no obvious decision support. Examples of this were observed at CIRTRANS and the sub-assembly of PCBs at α-CTRL.

In the case of CIRTRANS, the shop supervisor receives a weekly programme of end-item requirements on Thursday afternoon or Friday for the following week's

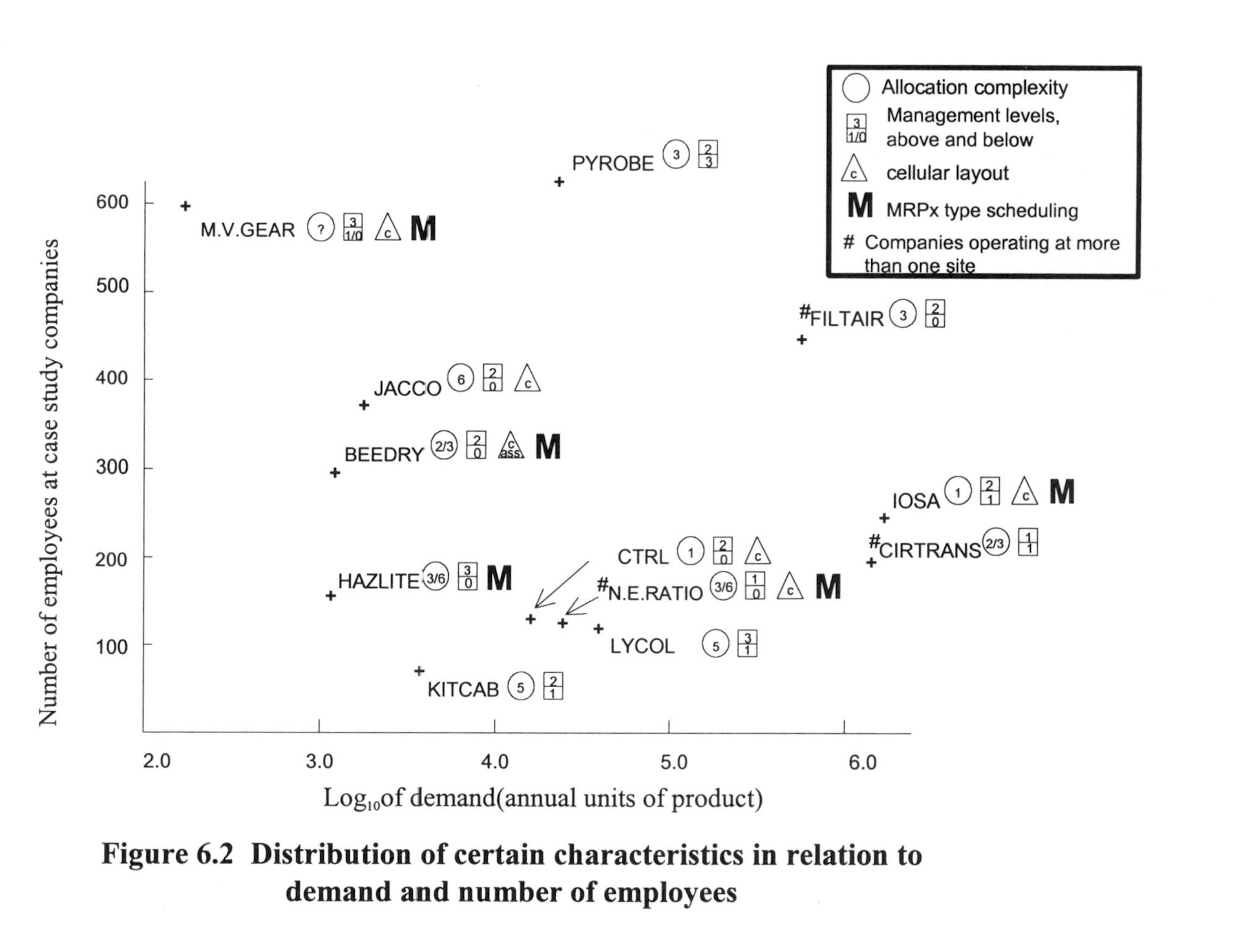

Figure 6.2 Distribution of certain characteristics in relation to demand and number of employees

production. Production days are normally allocated to particular customers with their own sub-set of variants.

The shop supervisor will then determine the quantities that will be manufactured for each component type and sub-assembly required to meet the respective set of orders so that they arrive in sequence at final assembly. This is normally carried out in a particular order related to bottlenecks and minimising changeovers.

There are approximately 200 variants of this product but the number of commonly requested variants is much smaller. There are only four types of pump housing and a single type of stator housing. Four of five changes of variant per day were not uncommon.

The schedule that the supervisor produces also has to take into account certain stock level targets for each of the principal items for all variant types. However, the main target that the supervisor is required to meet is the 'standard minute time' for the completed pump operations. At this company this means the actual hours worked divided by the quantity of pumps produced.

This planning process was done without any form of decision support apart from the supervisor's own experience. The main difficulties were said to be keeping track of changing customer requirements and unpicking previous instructions.

At α-CTRL there is a sub-assembly operation which consists of populating Printed Circuit Boards (PCB). The team leader of this 21 strong section receives a weekly schedule of end-item requirements split into individual days. This schedule indicates the board specification number for each variant of those end items for six separate final assembly cells. The investigation being reported was considering the products of only one of those final assembly cells. The boards for that product family shared the automatic insert machines with other product final assembly cells. Manual insertion of components was dedicated for this particular product family. The product also shared the cropping, flow soldering and the final section where components that could not be flow-soldered were added and testing took place. Daily quantities for variants of this product family could vary from 1 to 100.

The batching of product through the shared facilities was quantified and sequenced by the team leader of the sub-assembly cell. Set-up time on the automatic insertion machines was a significant issue in how to batch the work at this stage. It was common for weekly amounts to be made at these stations whereas the other operations in this cell might be divided up into daily quantities.

However, this will depend on a number of factors and clearly could become complex to resolve, and be planned and executed with varying effectiveness. As in the case of CIRTRANS, there were no planning tools to assist the supervisor of this section.

Responsibility for the replenishment of stock for sub-assemblies and components could be considered a sub-category of the formal attribution of batching and sequencing to the supervisor. Instances of this activity were found at α-CTRL and KITCAB component manufacture. In certain respects, the team leader at FILTAIR had this role in final assembly through having to regulate the long pull system to the Kanban buffer for final identity marking of standard stock. This requirement had to be balanced with the need to insert orders (Kanbans) to meet the orders of particular OEMs that required additional variants.

6.3.2.2 Category 2

The second category of decision by supervisors was where batch sizes were defined by the scheduling system. In some cases, these systems would also employ a rule to determine the sequence of priority at given work centres or individual machines. These cases were generally running MRPx systems (i.e. any type of MRP or MRP-II). The form of decision in these cases was to deliberately override the choice of the system because the supervisor did not perceive the planning output conformed with actual constraints or performance targets. This is being regarded as informal scheduling by the supervisor. In these cases, the batches may be aggregated over periods, split or changes in priority taken.

The methodology employed in this investigation was not able to distinguish how much of this was being done because of material shortages or machine breakdown and how much was attributable to the lack of confidence in the system output.

One example of this kind of practice was found at N.E.RATIO. In this case study a layout of equipment dedicated to particular components, each of which would have a number of variants, was compromised by the combination of formal and informal systems of planning. It also has to be said that the rapid growth in demand and the apparent difficulty of their suppliers to accompany this growth was also an important contributory factor to the operational difficulties.

The role of the shop supervisor for component manufacture included the manual filling-in of Work Priority Sheets at the beginning of each shift. This sheet showed the current and following job for each work centre for which that shift supervisor was responsible and these areas of responsibility cut across the dedicated component flow lines. The input to this Work Priority Sheet was the output from the MRPx system which came in two types of report. The first was the Operation Machining Programme which showed weekly aggregates by component variant together with their arrears. In other words this indicated what needed to be launched that week and had a horizon of 6 months. The second report, the Shop Schedule, was a daily update of where material was currently in the shop across all work centres.

The supervisor overrides this formal output with the Work Priority Sheets. An example was given of one particular component line where there was currently an arrears of one of the component variants due to a raw material shortage. This shortage had been rectified that morning; following an enquiry by an operator about what to do next, the supervisor decided to launch the complete quantity of incoming raw material which covered the arrears, the current week and part of the following week's requirements. A material requisition would be issued and the information downdated on the system by the supervisor.

In this case, it was said to be an easy decision because there were no other outstanding requirements for other variants of this component. It was said that decisions were not always this obvious and would on occasions require physical inspection on the shop floor if the calculation was 'tight'.

Launched batches were said to become progressively smaller as the particular material approached its final operation before assembly. Typically, a batch of 1500 would be divided into five or more intermediate batches in order to meet immediate assembly requirements. In other words, any initial launch sequence was lost and timings on make-span could be quite variable despite a relatively simple layout.

Additionally, and in effect a permanent requirement for supervisory planning amongst all MRPx companies (IOSA, BEEDRY, HAZLITE, M.V.GEAR and N.E.RATIO), was how to reincorporate backlog. All these companies ran with considerably more backlog than the non-MRPx companies. It is possible that if backlog is not automatically reported, there is more incentive to operate without it as it cannot be tracked and used as a source of 'control'.

The issue of backlog begs the question of what policy there is with respect to the size of disturbance which will trigger rescheduling. Rescheduling is being defined here as the decision to withdraw a current formally issued schedule and replace it with another which includes the material in progress from the previous schedule. It was found that such a policy was only initiated in exceptional circumstances and only in two out of the 12 case studies. In non-MRPx companies, the commonest policy was to attempt to return to the original schedule as soon as possible with a backstop of modifying the up-coming schedule to accommodate overrun between periods. In MRPx companies, there appeared to be a permanent rolling backlog. In both these situations, it is the supervisory function that would appear to be largely responsible for planning these scenarios. Even if it is considered to be a separate issue from the execution of the formal schedule, it is hardly likely to be without impact on that schedule.

6.3.2.3 Category 3

The third category of company is where the batching and sequencing was not changed from the formal decision of the system except in the case of a disturbance. The company that came closest to this was BEEDRY. This company had a system that could be said to conform most closely with a Period Batch Control system. In this company a single component manufacturing area is feeding 14 assembly lines. Once the order load is reviewed by the assembly manager for a given set of end items for a 2-day period, this list is exploded into its component parts. This list of components is taken over by a production control clerk who optimises use of panel materials and issues a daily machine shop programme. This cycle repeats itself and is considered uninterruptable. In other words, it can take up to 4 days to re-introduce a product. Although there is not much resequencing or rebatching *per se*, the system would appear to require a lot of staff to monitor progress. Even in such formally adhered to system outputs, there can still be undefined elements like staff allocation described in Section 6.3.1.

The above examples give some idea of how common it was in this group of companies for supervisors to make decisions on batch size and batch sequence. Given the way that schedule performance is currently measured (Harvey, 1998b; Gary *et al*., 1995), determining the source and magnitude of the effects on planning remains difficult to assess. However, it would be surprising if choice on batch size and sequence were insignificant.

6.3.3 Local replanning to deal with production disturbances

Local planning in the event of machine breakdowns, absenteeism, etc., was not dealt with by the original source investigation; the variable nature of these events would be hard to investigate with the methodology employed. This is not intended to de-emphasise the importance of this supervisory role. It is mainly the consequence of it not being part of the main focus of the original investigation.

6.3.4 Impact of the scheduling system on the supervision decision framework

The final planning issue to be dealt with concerns the impact of the scheduling system itself on the choices made by the supervisor. For the sake of simplification, the scheduling systems at the case study companies were divided into two basic types: MRPx and non-MRPx.

The decision to assign a case study to one type or other was based on the following understanding of the systems:

Companies with MRPx systems would basically back schedule any dependent demand from the due date. This generally requires the definition of operation and set-up times at work centres that may consist of one or more machines or stations. Depending on the system, it may also require the estimate of queue and move times in between operations. This is much harder to estimate and becomes more so the greater the interdependency of the routings through shared facilities. Essentially, the same issue arises in bucketed systems, although the information is not as explicit.

The significant issue here is that it is very hard for the supervisor, or anybody else for that matter, to know what the consequences are in the rest of the network when making changes at individual workcentres. This limited visibility can make it harder to insist on complying with the issued schedule particularly if the model does not represent reality that closely (see Scott, 1994, pp 209-211 for non-universal assumptions of MRP systems). The work centres can often only be assessed as individual islands. Although it is possible to peg orders, this was not found to be practised in this group and, as a consequence, the destination of components to end items was normally obscure.

This situation contrasts with the non-MRPx systems where, with simple layouts, it was commonly possible to issue sub-assembly or component manufacturing supervisors with the sequence of final assembly end items. From this sequence, the supervisor would determine the requirements of dependent demand, i.e. there was a fairly direct association with particular end items. This does not mean that there are not other complexities as has already been alluded to in the section on rebatching and resequencing.

The distribution between MRPx and non-MRPx amongst the case study companies can be seen in Figure 6.2.

In order to penetrate the differences dictated by the systems a bit further, it was decided to break down the planning and control activities of the chosen level of supervisor into a number of categories. For example, getting materials to operations, allocating staff or jobs to machines and maintaining progress. The answers from the protocol were interpreted and generally it was possible to determine whether or not the supervisors carried out these activities. These data are presented in Table 6.5. The incidence of 'YES' was then calculated for the two types of system.

The largest contrast occurred for material call up. In non-MRPx companies, in both component manufacture and assembly, supervision is much more likely to be responsible for this task than in MRPx companies. There was also a distinct difference in the requirements to allocate staff in component production. There was no requirement for this in a sample of five non-MRPx companies whereas in the MRPx companies, the incidence of this activity was 60%.

Although the above gives an indication of the impact of the system on the decision framework of the supervisor, an investigation with a more specific focus in this area would be useful in order to improve the understanding of the role of the supervisor in scheduling. There now follows a description of the supervisor's role in control.

6.4 CONTROL ACTIVITIES OF SUPERVISORS

Although control as defined at the beginning of the previous section is a more commonly accepted role of supervisors, there is still a need to differentiate the forms control takes, the dependence of these forms on other choices and the interaction these activities have with the form of planning adopted. Without such an analysis, it will be problematic to identify the contributions of different parts of the system to the overall outcome in a particular manufacturing context. Graves (1995), has also noted the limited amount of investigation that has taken place on the interaction between scheduling and material flow control.

Some preliminary steps were made in this direction within the framework of the case study structure being described. Two aspects were considered: the 'formality' of control with change in demand and the subject of control, i.e. whether the system of control was based on conformance with individual durations or was based on achieving target rates of production.

The scale of 'formality' was defined between two poles:

- decisions on any element are being reviewed on each instantiation and on a localised basis;
- complete sets of decisions are effected through a centrally, predetermined set of rules or when decision-making is eliminated by 'hard wiring' out any alternative in the combination of system and layout.

For example, the movement of material between work stations could develop from requiring an individual instruction from the supervisor; to the incorporation of this task with other tasks of the operator; to dedicating material movement to a particular person not directly involved in manufacturing operations; to mechanising the movement of material between stations.

In general, within the groups of case studies the tendency was for formalisation of material movement to increase with increasing demand. However, there were some exceptions to this which were not obviously attributable to other factors on which evidence was collected. Moreover, when comparison was made between groups, between companies of similar demand but different product complexity, important exceptions were found to what should have been similar levels of formality. This result was confirmatory of others that showed that, despite operating under similar constraints, industrial practice appears to have arrived at many choices through default rather than design.

With respect to the subject of control, a distinction was made between those companies that tried to control manufacture through conformance with the sum of individual durations of activities and those which looked at target rates of production for complete sets of operations. This latter subject of control is commonly expressed in units of product per unit time.

Table 6.5. Schedule execution: task division at supervisor level

Company	Area	Transform	Material call up	Allocating direct staff	Allocating jobs to machines	Monitoring progress
LYCOL	**Component**	Yes	?	No	By exception	Yes, "continuously"
	Assembly	No	Yes, one material only	Yes	Yes	Yes
FILTAIR	**Component**	?	?	?	Yes	?
	Assembly	Yes, Manager & Team Leader	Yes, manager by exception	Exceptionally	Not applicable	Yes
IOSA	**Component**	Yes	No	Yes	No	No
	Assembly	No	No	No	No	Yes
HAZLITE	**Component**	Yes	No	Yes	Yes	Yes
	Assembly	Yes	No	Yes	No	Yes
KITCAB	**Component**	Yes,not formally scheduled	Exceptionally	Exceptionally	No	Yes
	Assembly Same supervisor	No	Exceptionally	Yes, with change in demand	No	Yes
α-CTRL	**Component**	No	Yes	?	?	?
	Sub-assembly	Yes, 2H/wk	Yes, mins to hrs	Exceptionally	Yes	Yes
	Final assembly	No	Yes	No	No	?
PYROBE	**Component**	No	Yes	No	Exceptionally	Yes
	Assembly	No	Yes	Yes	Exceptionally	Yes
CIRTRANS	**Component**	Yes	Yes	No	Yes	No
	Assembly Same supervisor	Yes	No	No, only seasonal change	No	No
BEEDRY	**Component**	Yes	No	No	No	Yes
	Assembly	No	No	Exceptionally	No	No
N.E.RATIO	**Component**	Yes (4 to 6 H/ week)	Yes	No	No	Yes (12H/ week)
	Assembly	Yes	No	Yes	No	No

M.V.GEAR	Component	No	No	Yes (26H/week load smoothing)	Exceptionally	No
	Assembly: cell & section leaders	Yes	No	Yes	Yes	(Much firefighting)
JACCO	Component	Yes	Yes	Exceptionally	No	Yes
	Assembly	No	Yes	Yes (mainly in Final Assembly (FA))	No	No Paced FA, sub-ass. synchronised

"Yes" replies (%)							
	MRPx	Component	80	20	60	20	60
		Assembly	60	0	60	20	60
	Non-MRPx	Component	57	60(out of 5 cases)	0(out of 5 cases)	33	80
		Assembly	28	57	57	16	67

NOTES:
The shaded rows indicate companies employing MRPx type scheduling.
'Exceptionally ' is counted as a 'No'
Column headings: Each column heading represents an activity of supervision with respect to the execution of the schedule:
'Transform' indicates that supervision will be spending some time on producing an informal schedule and/or modifiying the sequence or the priority of the lowest formalised level of schedule.
'Material call up' indicates the supervisor initiates the move of the material to one or more of the operations.
'Allocating direct staff' covers the requirement of supervision to make decisions on the distribution of labour across the various tasks e.g. because of change of requirements with product change or the sharing of labour across a number of tasks.
'Allocating jobs to machines': This supervisory task will occur in those cases where there is a possibility of adapting item routings to suit machine loadings.
'Monitoring progress of work' will register an entry when it is the supervisor's job function to ensure the movement through the shop or register completions in some way.

In general, it was MRPx companies that adopted durations as their subject of control although there were two companies out of this group, NE.RATIO and IOSA, who used a hybrid system: sums of durations for component manufacture and rates of production for assembly. In the MRPx group of companies, although durations of operations were recorded on occasions as starts or finishes with respect to real time, the reports were invariably in fairly aggregated forms and on a weekly or monthly frequency. This made analysis for control purposes retrospective and often hard to disaggregate to useful operational units. In some cases it would be possible to tell whether individual operators were consistently not achieving durations of operations but these times can often represent quite a small proportion of the time an item remains within the manufacturing unit and may well only represent disconnected islands in terms of material flow.

For those companies that used target rates of production, it was almost invariable that no account would be taken of the range of rates for the different variants of the product mix. In fact, a variation of 25% or more was typical of this group in relation to the product variant taking least staff hours. Commonly, it was anticipated by the companies that variations in the product mix would balance themselves out within the accounting period. In practice, this may not always work out and can be a major influence on the supervisor in their scheduling decisions.

The one exception in this second group was JACCO. The final assembly operation at this company could mean a variation of between 22 and 28 members of staff on the line in order to maintain the same cycle time for the different variants. The exceptional nature of this company was that this change in requirement was taken into account in the scheduling rules, e.g. there could not be more than one variant change per day and there were also a number of what were described as softer constraints on sequence changes between particular variants. It was suspected that this understanding was a contributory factor to the company's superior performance relative to a majority of the other case studies.

In both groups, MRPx and non-MRPx, there are potential areas for improvement in the mechanisms and measurement for control that could assist the operational practice of the supervisor.

6.5 CONCLUSIONS AND FUTURE WORK

The results of the source investigation have deepened the reported understanding of the current involvement of the shop supervisor in shop scheduling. The prosecution of that role and the structure within which it takes place, exhibits a diversity which in part may be a reflection of individual company development and in part the lack of underpinning theory in which that development takes place.

There is still a need to improve the understanding of the relation between the formal and informal and between the contributing functions of schedule planning and execution. For comparability, it will be important that such work is situated within an identifiable context in order to develop manufacturing theory. This will also require improvements in performance measurement which should also benefit operational practice.

With the intention of focusing future empirical investigation the following observations are highlighted:

Although in some cases there appears to have been a simplification of the scheduling task, particularly in the non-MRPx companies, this seems to have been accompanied by a more formal assignment of the task of shop floor scheduling to the supervisor. In most cases, there would appear to be little decision support in this planning work. It might be that advances that have taken place in scheduling theory could assist in some of these simplified sub-sets of scheduling problems.

However, some elements of planning are making a greater impact. Greater interchangeability of staff - multi-tasking and minding of more machines per member of direct staff, means that the staff allocation issue has become more critical to the outcome, particularly if the multi-skilling is introduced for the purpose of increasing productivity.

In other cases, mainly MRPx companies, the shop floor scheduling remains as complex as previously. It doesn't appear that this role has changed a great deal for the supervisor with the exception of having ordered task lists for every workstation. The degree to which this helps will depend on how close this model is to reality.

Where the complexity of task of shop floor scheduling is constrained to reviewing individual operational islands within a path to completion, the value of collecting additional empirical evidence on this activity would appear limited. Such evidence might be valuable in an investigation of individual coping strategies in a behavioural framework of organisation. It would seem unlikely to contribute to the development of scheduling theory except in the sense of understanding the boundaries between theory and practice.

In some instances where simplification has taken place, the gap between scheduling theory and practice is either non-existent or small. In such cases, the potential for localised decision support for the supervisor or team leader would seem to be positive. However, great care will need to be taken that any simplification has not been made at the expense of displacing the complexity of the problem elsewhere. There were examples from the source investigation where the chosen boundary of the individual company or manufacturing unit excluded a more balanced perspective of the scheduling of material transformation as a whole.

6.6 REFERENCES

Alasoini, T. (1993) 'Transformation of work organisation in time based production management', *The International Journal of Human Factors in Manufacturing*, Vol. 3, No. 4, pp 319-333.

Bittel, L.R. (1990) *What Every Supervisor Should Know*, McGraw Hill.

Burcher, P., (1992) 'Master production planning and capacity planning: the link?' *Integrated Manufacturing Systems*, Vol. 3, No. 4, pp 16-22.

CDEEA: Coventry and District Engineering Employers' Association (1968) *Working Party Report on Wage Drift, Work Measurement and Systems of Payment,* Coventry, UK.

Chapman, E.N. (1992). *First Time Supervisor: A guide for the newly promoted*, Kogan Page.

Cusins, P. (1994). *Be a Successful Supervisor*, Kogan Page.

Dore, R. (1973) *British Factory, Japanese Factory: The origins of national diversity in industrial relations* with author's 1990 afterword, University of California Press, Berkeley, USA.

Dunkerley, D. (1975) '*The Foreman, Aspects of Task and Structure*', Routledge & Kegan Paul, London.

Gary, K., Uzsoy, R., Smith, S.P. and Kempf, K. (1995) 'Measuring the quality of manufacturing schedules' in *Intelligent Scheduling Systems,* editors: D.E. Brown and W.T. Scherer, Kluwer Academic Publishers, Boston. pp. 129-154.

Giesberts, P.M.J. (1991) 'Master production scheduling: a function based approach', *International Journal of Production Economics*, Vol. 24, pp 65-76.

Graves, R.J. (1995) 'Literature review of material flow control mechanisms', *Production Planning and Control*, Vol. 6, No. 5, pp 395-403.

Grimm, J.W. and Dunn, T.P. (1986) 'The contemporary foreman status', *Work and Occupations,* Vol.13, No. 3, pp 359-376.

Harvey, H. (1993) 'Report on the impact of Lucas methodology for factory re-design on manufacturing, planning and control systems', *Sub-project 1, FORCAST/DTI Project*, DTI, London.

Harvey, H. (1995) 'Selection of scheduling, control and manufacturing systems: can there be all these recipes without any relationships?' *Proceedings of the International Conference on Advanced Manufacturing,* University of Sunderland, UK, September, 1995.

Harvey, H. (1998a) '*Scheduling and Control Activities in Industrial Manufacturing System Frameworks*', Unpublished PhD Thesis, University of Sunderland, U.K.

Harvey, H. (1998b) 'A case for improving the industrial measurement of schedule adherence', *Proceedings of the 14th National Conference on Manufacturing Research*, University of Derby, UK, pp 243-250.

Hendry, L. (1991) 'A decision support system for job release in Make-To-Order companies', *International Journal of Operations and Production Management*, Vol. 11, No.6, pp 6 -16.

Lowe, J. (1993) 'Manufacturing reform and the changing role of the production supervisor', *Journal of Management Studies,* Vol. 30, No. 5, pp 754 -758.

McKay, K.N. (1995) 'Common sense' realities of planning and scheduling in printed circuit board production', *International Journal of Production Research,* Vol. 33, No. 6, pp 1587 -1603.

Melling, J. (1980) 'Non-commissioned officers: British employers and their supervisory workers, 1890 – 1920', *Social History,* Vol. 5, No. 2, pp 183-221.

Melman, S. (1958) '*Decision-making and Productivity*', Blackwell, Oxford, UK.

Riat, A. and Prickett, P.J. (1995) 'The under-utilisation of the production employee', *Proceedings of the International Conference on Advanced Manufacturing*, University of Sunderland, Sunderland, Section G1.

Rodammer, F.A. (1988) 'A recent survey of production scheduling', *IEEE Transactions on Systems, Man and Cybernetics,* Vol. 18, No. 6, pp 841-851.

Savedra, M. (1990) *Supervision,* Macmillan, London.

Senker, P. (1994) 'Supervision in manufacturing organisations', *Journal of General Management*, Vol. 20, No. 1, pp 44-60.

Scott, W. (1994) *Manufacturing and Planning Systems*, McGraw-Hill Book Company, London, UK.

Sridharan, V. (1992) 'Manufacturing planning and control: is there one definitive answer?', *Production and Inventory Management Journal*, First Quarter, pp 50-54.

Stone, K. (1973) 'The origin of job structures in the steel industry', *Proceedings of the Conference on Labor Market Segmentation,* Harvard University, March 1973.

Stoop, P.P.M. and Wiers, V.C.S. (1996). 'The complexity of scheduling in practice', *International Journal of Operations and Production Management*, Vol. 16, No. 10, pp 37-54.

Taylor, F.W. (1947). *Scientific Management*, Harper and Bothers, New York and London. This volume included: 'Shop Management', 1903; 'Principles of Scientific Management', 1911 and 'Hearings before the Special Committee of the House of Representatives to Investigate the Taylor and other Systems of Shop Management', 1912.

CHAPTER SEVEN

Lingering Amongst the Lingerie: An Observation-based Study into Support for Scheduling at a Garment Manufacturer

Caroline Vernon

7.1 INTRODUCTION

There are many planning and scheduling papers whose opening lines, quite rightly, extol the importance of effective production planning and shop floor control to the success of a manufacturing company. A vast body of the literature on planning and scheduling focuses on scheduling theory (a mathematical approach seeking optimal solutions to the scheduling problem) or on artificial intelligence (e.g. modelling human expertise and integrating that expertise into a computer system) with the ultimate objective of incorporating these algorithms or elicited knowledge into computer-based scheduling systems. Other literature focuses on the organisation of planning and scheduling, looking at how layers of control within planning and scheduling should be organised (e.g. Bauer *et al.*, 1994; Doumeingts *et al.*, 1995; Williams *et al.*, 1994). A small but significant number of papers focus on the role of the human in planning and scheduling and take an interdisciplinary approach to research, borrowing concepts from the social sciences, cognitive psychology and organisational behaviour (see Crawford and Wiers, Chapter 2). Many of the studies that focus on the human in planning and scheduling have had the objective of developing some form of decision support system and only a few authors have attempted to examine other issues such as re-evaluating the hierarchical paradigm for planning and scheduling (see McKay *et al.*, 1995a) or examining CAPM systems introduction from a social and organisational perspective (see Hardstone, 1991; Mahenthiran *et al.*, 1999).

Unlike many other studies focusing on the human in scheduling, the research project on which this paper is based set out to examine six companies from different industrial sectors (Engineering and Physical Sciences Research Council Grant GR/L/31364)[1]. The overall objectives of the study were: i) to propose guidelines for supporting scheduling given the existence of different manufacturing environments and different social contexts, and ii) to propose guidelines for the design of human- centred planning, scheduling and control structures.

---

[1] Engineering and Physical Sciences Research Council (EPSRC Grant GR/L/31364): Effective Decision Support for Scheduling: Combining Scheduling Theory and Human Factors.

This chapter examines these issues in the context of an industrial case study, although discussion of some of the themes is in context of the wider project. The emphasis is on how scheduling in particular is a human process. A holistic approach is taken for the examination of scheduler support. The scheduling process is viewed as a social as well as a cognitive process. The support for schedulers therefore requires social and organisational considerations in addition to its cognitive and technological dimensions. The environmental and organisational issues impacting on the scheduling function and influencing the scheduling role are examined, as is the impact of the control approach on the scheduling role and interrelationships with other functions. The influence of technological support (or lack of it) on the scheduling role is discussed. Considerations for appropriate support are discussed including computer systems, task redesign, improvements in information quality, revision of performance measures, changes in communication practices and changes in the design of the control hierarchy.

Section 7.2 provides a background to the company and a description of the business and scheduling environment. Section 7.3 looks at how scheduling was undertaken from both a macro and micro perspective and Section 7.4 looks at the changes that have occurred to the process from the past to the present. Section 7.5 discusses the findings from the case study and the issue of appropriate support.

7.2 BACKGROUND

7.2.1 The methodology

The research approach undertaken for this study can be described as 'naturalistic enquiry' (Lincoln and Guba, 1985). With naturalistic enquiry there is a preference for the analysis to be inductive, for the theory to emerge from (to be grounded in) the data and for the research design to emerge from (known as 'Emergent Design') the interaction with the study.

The methods used within this approach included observation, retrospective decision probes, task probes and semi-structured interviews. The methods were piloted on three occasions, fine tuning them to suit our needs. Prior to conducting each field study we identified appropriate participants by asking managers to suggest the most appropriate people, who we then interviewed briefly to assess suitability.

The toolkit was designed to capture the tasks the participants were performing, the information used, decisions made, knowledge used, information system data used, and the different methods allowed for triangulation of the data. For a more detailed description of method development and use, see Crawford *et al.* (1999).

7.2.2 The company, the factory and the people

Alpha Garments Plc is a large garment manufacturer with a number of divisions. This case study focuses on Geoff, the Production Manager who schedules one of the underwear/lingerie factories. The factory is geographically distant from the head office where the planner for the factory resides. The head office is where planning, selling, buying and design occurs centrally for all the factories in Alpha Garments. An organisational chart for Alpha Garments in shown in Figure 7.2.

The shop floor of this factory is divided into 28 lines of machinists and the cutting department, which cuts styles for all of the divisions' factories. Each line or team consists of four or five women seated at sewing machines. The machines in the lines will be of different types to cope with all the operations on the garment that line has been given to stitch. There are three Line Managers and Supervisors each responsible for a number of lines of machinists (see Figure 7.1).

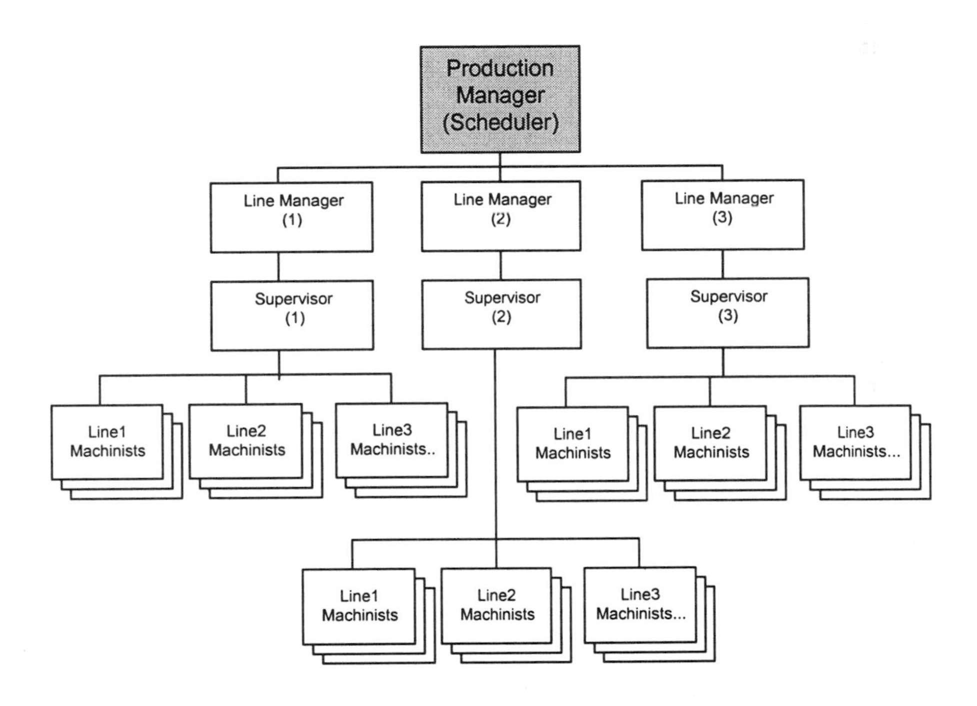

Figure 7.1 The factory organisational chart.

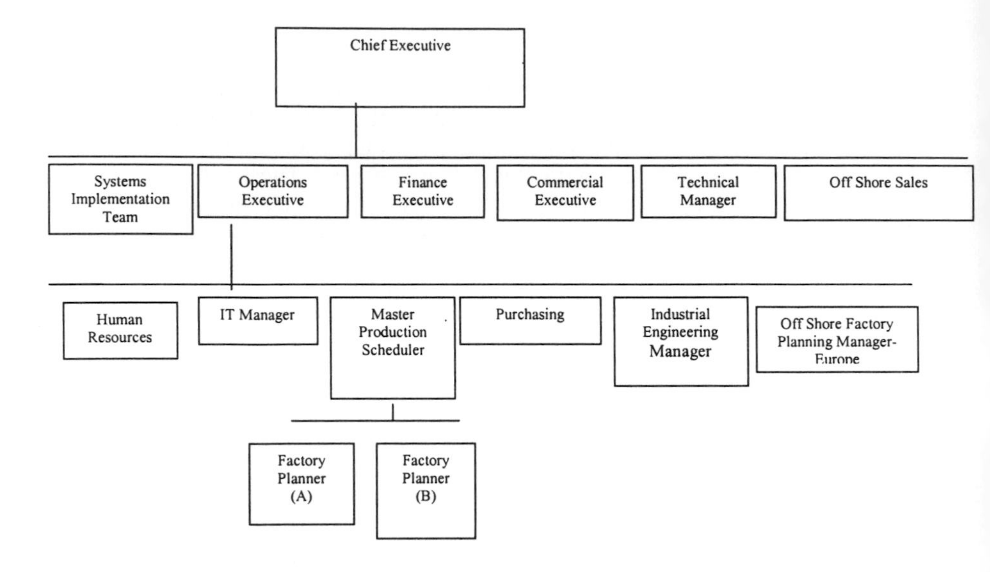

Figure 7.2 The Head Office organisational chart.

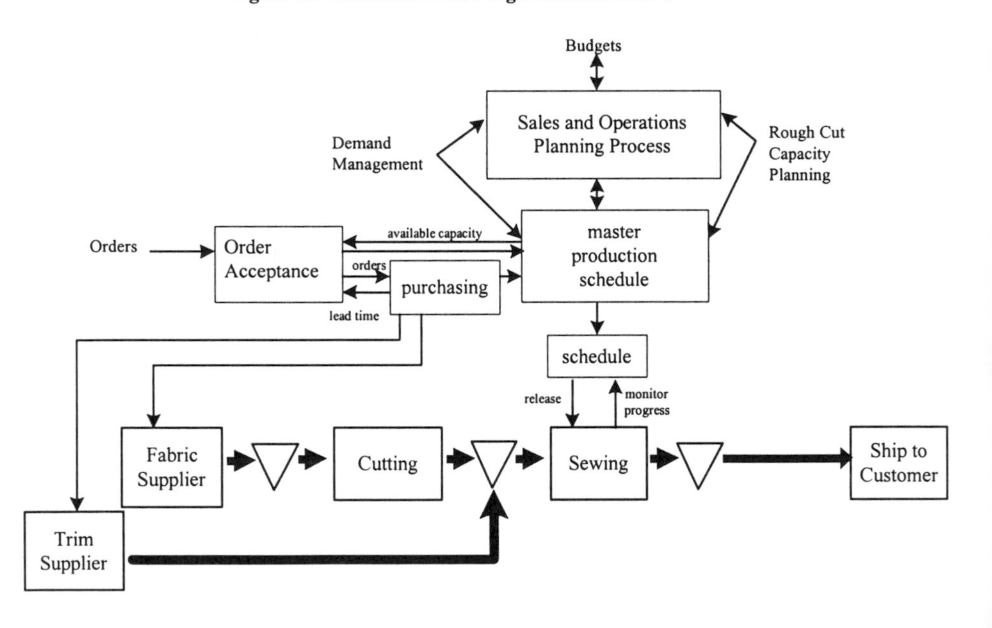

Figure 7.3 Production control, order management and material flows.

7.2.3 Order management

The company operates on a make-to-order basis with orders being placed at the style/colour level at the start of each season[2]. The garments are designed in-house under negotiation with the customer. The range that a customer buys each season is made up from newly designed garments and repeat business. The customer will have previously reserved capacity for the whole season's orders with the Master Production Scheduler and the orders that come in will be subtracted from the reserved capacity. The customer will specify a delivery date the order is required and the planner and the purchasing department will review the date based on available capacity and fabric availability. When the order is accepted it is added to the Master Production Schedule and confirmed to the Purchasing Department (see Figure 7.3).

If the order specifies the style and colour of the garment, then fabric can be ordered from the supplier. Size specific items cannot be ordered until the customer specifies the exact quantity required per size, referred to as 'sizing'. The initial style/colour order may specify quantities with staggered delivery dates which will require 'sizing up' at different times. 'Size ups' are required seven weeks after initial enquiry but may not be for the full contract quantity. If the product sells well in the stores then the customer may size up the balance of the contract and request for production to be brought forward. If sales are slow then the customer will not want to size up all the contract at one time and may wish to push the additional unsized contract quantities back in the plan or may even on occasions cancel the contract balances.

7.2.4 Complexity and uncertainty

Complexity within the business lies in the variety of product styles demanded by the customer and the responsiveness and flexibility that the customer demands. It does not lie within the manufacturing process. Although the basic product may appear simple, in order to compete in the marketplace and to excite retail buyers, products are increasingly made from new fabrics. The trend for garment manufacturers is to send basic volume garments overseas and use their UK manufacturing bases to produce the more complex and detailed, smaller volume garments. This has ramifications for the UK manufacturing bases, who are required to be increasingly responsive and flexible whilst being faced with a complex product mix. On the factory floor this has implications for the number of changeovers[3] required and associated set-up time. This has implications for throughput and it requires the workforce to be multi-skilled.

A critical path process describes the detail involved in the process of taking each style from conception to manufacture. The critical path document includes major milestones, which must be completed by the date set in order for production

---

[2] Typically there are two main seasons per year in terms of volume of product order, Summer and Winter . There are also two smaller seasons Spring and Autumn.

[3] A changeover may involve cleaning down the machine when going from a dark colour of thread to a lighter colour of thread or it may involve resetting the machine to work with a different fabric type.

to be able to start on time. If week one is the initial enquiry point then the critical path would include milestones such as:

Week 2: Approval from the customer as to the shade of fabric required
Week 7: Confirmation of the ratio of sizes of the garments required
Week 9: Customer approval of a sample garment - correct for fabric and make-up although shade and trims can vary. Issue of patterns to production.
Week 10: Issue of garment technical specifications to the factory. Bulk fabric and trims in house.
Week 12: Cut and kit
Week 13: Load and make
Week 14: Declare available to the customer.

If detail is not attended to, say labels are ordered with an incorrect fibre composition printed on them or a technical specification for a style does not contain correct packing information, these incidents could at worst cause garments to be rejected by the customer or at the very least cause a delay in production whilst the problem is rectified. So complexity inherent in the business processes prior to production, multiplied by the variety of product styles can, if not managed well, pose a problem.

Complexity is also inherent in the product itself. Producing fabric is not an exact science. Fabric properties can vary from one batch to the next which, whilst being in tolerance, can affect how well the fabric sews. Shading between batches can vary which again, whilst being in tolerance, can be very obvious if a cut panel from one batch of fabric is sewn together with a cut panel from another batch of fabric. Different fabric can require the sewing machines to be set very differently and so changeovers may involve intense effort on the part of the mechanics.

Uncertainty can come from many sources. Customers compacting the lead time to manufacture by placing orders late affects the ability of functions within the critical path such as the garment technology to produce patterns and samples on time. Customers, by sizing orders late, can affect the acquisition of size specific labels/packaging by the required date. Fabric rejections may be a result of the inability of the supplier to perform or can be an outcome of the quantity of new developments being put through. Disturbances can also be a result of human error in dealing with the detail at the front end.

Disturbances cause the plan to be changed. Moving styles forward and back in the plan has ramifications. In some industries or businesses uncertainty, in the form of rush orders, may be countered by the planner reserving some capacity in order to allow the business to be responsive and to allow orders to be moved around. Capacity for a garment manufacturer is made up of machinists all keen to be kept fed with work in order to earn their bonuses. Even though piece rate has been replaced with a flat rate per team, a bonus system still exists to encourage productivity. So the notion of saving capacity and keeping some machinists idle in order to cope with uncertainty is not a strategy that can be employed. The shop floor must therefore always be loaded to full capacity, less contingencies for sickness and absence. If a style needs to be pushed back, then another style needs to be pulled forward to fill the gap. Some styles may not qualify for pulling

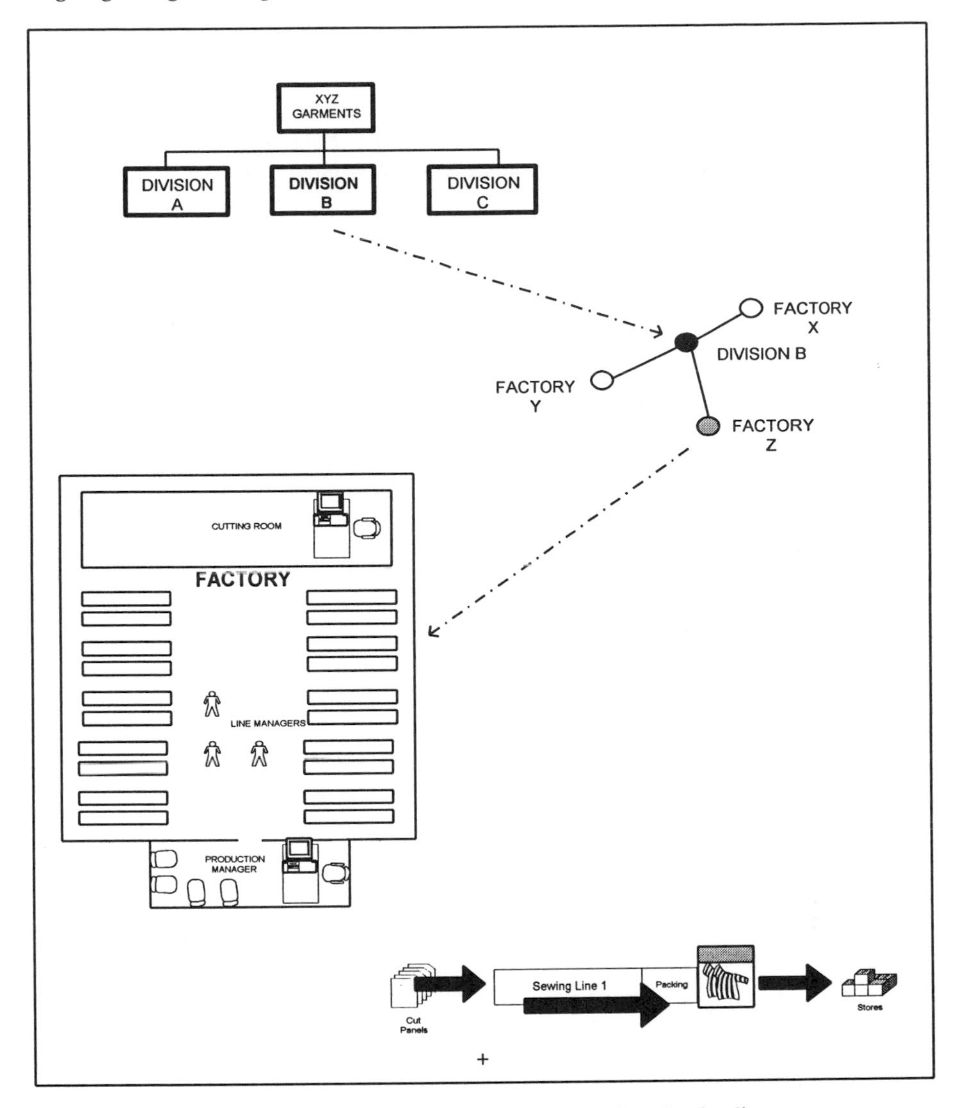

Figure 7.4 The company, division, factory and production line.

forward due to materials not being available, but not filling the gap will mean machinists running out of work. A customer who finds one of their styles is selling exceptionally well may wish for some production to be pulled forward. This can only happen if materials are available and if they can agree that other styles can be pushed back or if over-time is worked.

7.2.5 The scheduling process

Being a flow process with no transfer of work between lines means that the scheduling process is relatively simple. There are no bottleneck operations to single out for scheduling. The line managers plan balancing of the lines before a style of garment is due to be produced (Figure 7.4). This is based on the time for each operation and the machinists' abilities.

Sequencing is the main issue that needs to be tackled by the production manager and the term 'maintaining continuity' is commonly used. In order to maintain continuity the scheduler will try, for example, to follow a run of garments of one fabric type with a run of garments of a similar fabric type on a particular line to avoid the mechanics having to fully reset the machines. If a cotton garment is being followed by a cotton lycra garment then the handling properties of these two fabrics are very different so the technicians would have to do a full reset of the machines. Also colours are sequenced from light to dark and machines must be cleaned down when going from a dark to a light colour. Compiling a schedule involves deciding on the allocation of styles per team based on sequence constraints, changeover issues and team performance. The decision is bounded by the prescriptive factory plan in terms of quantities and delivery dates and by the grouping of the teams on the shop floor; i.e. a *brief* is allocated to one of the *brief* teams. In situations where the scheduler is attempting to schedule an entire factory in daily buckets and attempting to remember all the issues surrounding certain styles as well as the performances of the teams and their skills, then compiling the schedule can soon appear to be a daunting and time-consuming task. This case study illustrates how alternatives to automating the scheduling process can be simpler and more cost effective.

7.2.6 Summary

The scheduling function faces challenges as a result of many factors within the environment that impact upon it. Some of these factors are manageable, others are inherent to the type of business the company is in. Challenges for management in supporting the scheduling function lie in successfully tackling the manageable factors and devising procedures to best cope with the inherent factors. Examples of manageable factors that impact on the scheduling function here are:

- the product volume and mix versus the skill base of the machinists to maximise scheduling choices;
- customer pressure on lead times versus the capacity of front-end departments and the need for pilots and raw materials to be checked in sufficient time prior to production in order to minimise problems in production;
- the level of responsiveness demanded by the customer and the delivery reliability of suppliers versus the stability required by the shop floor to best utilise the resources available to set up machines and pilot styles in order to minimise last minute rescheduling and disturbances.

Factors that are considered as inherent to the business here but will also impact on the scheduling function include, for example, the responsiveness required versus the nature of capacity.

7.3 SCHEDULING THE FACTORY- THE PAST

Two extended observational visits were undertaken, three months apart. During that time much had changed with regards to the scheduling process and the roles and responsibilities of the people involved. This section concentrates on the past and is dominated by the following points:

- the existence of a traditional hierarchical production planning and control approach with prescriptive planning;
- the scheduling process being facilitated by 'in the head' expertise and knowledge of a single scheduler;
- the non-existence of computerised planning and scheduling;
- scheduling behaviour influenced by throughput based performance measures.

7.3.1 The macro view – manufacturing planning and control

7.3.1.1 Type of control approach

A traditional hierarchical approach to control was very much in evidence where:

- higher levels believe they know what is best for lower levels but in fact do not know the detailed workings of lower levels; and
- higher levels constrain lower levels and use aggregated constructs or models of lower levels (McKay *et al.*, 1995a, 1995c).

There is a strong management belief that the scheduling function should be driven by the planning function so the plan that the scheduling function receives is prescriptive i.e. the shop floor is required to follow the plan as closely as possible. The reason the shop floor is dictated to, rather than given a window in which to re-schedule to better suit the shop floor and to improve efficiency, lies in the delivery buckets quoted (in weeks) and the rates scheduled and resulting production run length (often more than a week, seldom one or two days).

In some companies the shop floor can be given a quantity of orders and be told to schedule them as they see fit providing the due dates are hit. The concept of the window only works when individual orders have short processing times e.g. a couple of hours or a couple of days, which span a duration shorter than that of the delivery quote bucket e.g. a week.

The benefits of rescheduling within a window are:

- the plan is more robust because it is less prescriptive and because disturbances can be dealt with at a low level without affecting the customer due date;
- less planning effort is required for the reason given above;
- this means that less feedback to the planner and therefore to the customer is required for a given level of disturbances.

A business that has a relatively long production run length (often more than a week, seldom one or two days) but a delivery bucket of say a week has more of a problem to manage as:

- the plan is required to be more prescriptive and therefore will be less robust and has more chance of going wrong;
- the planning effort will be increased as a result;

- the shop floor will be required to follow the plan strictly, which implies less opportunity to react to disturbances and less opportunity for empowerment at the lower levels with less opportunity to make scheduling decisions;
- more feedback will be required from the scheduler to the planner for a given level of disturbances.

The factory featured here is in the latter category (long production runs) and has the additional problem of a relatively high level of disturbances. Although it would be beneficial for this factory to adopt the scheduling window concept it would face a number of issues in attempting to do this. If the delivery time frames were increased, this would militate against the customer requirements for quick replenishment that dictate that deliveries are made at least once a week. Conversely, if the run lengths were shortened then this would imply that machinists would have less time to get used to a style and to build up their efficiencies. This would have a negative effect on throughput volumes which production is currently measured by. It would also mean that there would be more changeovers, which at present levels still need to be made more efficient, and would imply more load on the technicians. A step change in production policy would be required before this would be a feasible option.

7.3.1.2 Planning and scheduling interaction

There are several issues with respect to planning and scheduling where the planner and Production Manager have different views. One relates to the limited level of feedback received by the planner from the Production Manager, another is the lack of involvement of the scheduling function in generating the plan and a third relates to the lack of changes made to the plan even when a critical path milestone has not been achieved on time.

When talking about feedback it is important to distinguish two types:

- feedback to aid plan generation, i.e. on whether the plan is feasible;
- feedback on the progress of actual production versus the plan.

The planner complains of getting little feedback from the production manager as to whether the plan is feasible. Where new styles of garments are being introduced the planner does not include a learning curve for that style. He feels that in some cases this might be preferable, but does not receive enough feedback from production to allow him to do this. He also feels that he does not receive enough feedback on production progress, i.e. whether they are in front or whether they are behind.

The Production Manager feels that the plan is not generated with enough input from the scheduling function. He feels that the planner does not appreciate all the constraints he faces at shop floor level. A line supervisor corroborates this view, stating that there are situations where the plan could be better organised to take account of constraints on the shop floor without jeopardising customer due dates. Input from the scheduling function into the plan is perceived as undesirable by planning management who desire the plan to be prescriptive and in their view this excludes involvement of the scheduling function in generating a feasible plan.

Given that both the scheduling function and the planner want better communication and more collaboration, it is hard to comprehend why the scheduling function is not guiding and constraining the planner by providing more detailed and specific information about how the plan should be altered to make it more feasible. Basic collaboration, introducing learning curves to new styles could be achieved easily but is not currently being done.

It is clear that the loose coupling is not the result of a poor working relationship or lack of the opportunity to communicate despite the geographical distance. This suggests therefore that the issue may stem from organisational or cultural issues. We know from the work of McKay *et al.*, (1995b) that a scheduler will often have a number of different schedules, the political schedule, the schedule that he believes in and the schedule that he pushes the line to achieve. We also have evidence from decisions we captured that the scheduler is motivated by the need to meet performance measures and to keep the machinists in work (see Section 7.3.2.2). This can result in the scheduler bringing forward work contrary to that which has been planned in order to keep the line in work as a result of a disturbance affecting the loading of the planned style.

The scheduler's actions, driven by the need to meet his objectives derived from the organisational and cultural environment, conflict with the hierarchical production control paradigm demanding that he sticks to the plan which is also in place. It is not hard to imagine how a scheduler might limit certain feedback to the planner regarding actions he is taking because of this conflict of objectives. For example, the scheduler might not tell the planner about a contingency plan involving pulling forward a style not required for two weeks because another style has failed. Given the amount of disturbances that are evident it is hard to imagine the scheduler doing anything less than attempting to juggle work in order to keep lines going and throughput measures met. It is not our intention to point the finger of blame at the scheduler by highlighting the gap between *actual* and *expected* behaviour. We note only how the scheduler is often found to be in a position somewhere between a rock and a hard place.

The third issue of contention relates to the critical path process. The Production Manager attends both a production meeting and a critical path meeting on a weekly basis and it is in these meetings that he gets information regarding possible issues with loading. It is in these meetings that Geoff says he gets more information by 'reading between the lines'. When people say that fabric or trims will arrive on time, but show hesitation, he interprets this as a need to develop a contingency plan for that style i.e. to figure out what else could be loaded if say trims or components failed to arrive at the last minute. The reticence of the planner and others to move styles back even if there is uncertainty over delivery relates to the fact that capacity is based on people's throughput. As such if a style is moved back another style must be pulled forward, otherwise that capacity will be lost. This translates into disgruntled machinists sitting idle without work not earning their throughput bonuses. To pull forward a style can also involve a lot of work checking whether materials are available, and may involve negotiating with suppliers to bring them in earlier. This issue highlights how problems can come to light very close to the point of production commencement because of fears that moving a style which is failing critical path milestones may cause further problems.

The issues highlighted above are not the result of poor working relationships or a lack of opportunities to communicate. They appear to be a result of conflicting objectives. The planner is focused on planning to customer due dates and the scheduler is focused on meeting utilisation measures, keeping the teams in work and retaining some stability for the shop floor. Neither are formally encouraged to collaborate and negotiate their positions through the design of the planning process so it is not surprising that they both feel there is a gap in communication. A plan can still be followed to the letter, still satisfy the customer and yet have input from the shop floor. There is an argument to say that this is more likely to happen naturally and informally when the planner and scheduler are situated in the same building especially if they have a good and co-operative working relationship. In cases where the planner and scheduler are geographically distant it will be important to 'design in' lines for communication and negotiation over the plan.

7.3.2 The micro view – scheduling and facilitating the schedule

Very rarely can a person be found who holds the job title of ‘scheduler’. People who schedule have a variety of different job titles, have varying degrees of authority and power and also a variety of different primary or secondary tasks for which they are responsible other than scheduling. So people who schedule do not always spend the majority of their time generating a schedule. They may have other quality, people management or materials management responsibilities for example. We have come across people whose primary task is people management with scheduling only taking up a small percentage of their time. Others have people management responsibilities but find that generating and facilitating the schedule takes up more of their time than they would wish it to. Others again recognise that they could improve upon the work-to-list or plan that they have been given but prefer to concentrate their energies on people management and to deal with scheduling issues as they occur.

On first inspection it can appear that there are few similarities between the tasks and responsibilities of one scheduler compared to another. Even the jobs of people who schedule in different areas of the same business can appear quite different, with different task sequences, different priorities of task, different levels of authority and responsibility and different levels of computer interaction. But like process control, where operator behaviours can be categorised as monitoring, situation assessment, planning and trouble-shooting (Mitchell and Sundstrom 1997), ‘scheduling’ and ‘facilitating the schedule’ can also be categorised. The behaviours have been separated here into information collection (involving information sampling, receipt and monitoring) and trouble-shooting (involving situation assessment and decision making). For Geoff, who is a Production Manager who also schedules, the actual task of compiling a schedule is just part of his role that takes up a relatively small proportion of his time, with the majority of his time being spent information-gathering and trouble-shooting.

7.3.2.1 Information collection

Information collection in scheduling can be broken down into various components: monitoring, information sampling and information receipt. The strategy an individual employs for information collection will depend on the individual, the responsibilities of the individual (McKay *et al.*, 1995b) and the situation. Where or whom the information is collected from or accepted from can be a function of the perceived reliability and quality of the source, experience (Vera *et al.,* 1993; Randel and Pugh, 1996), the personality of the individual (i.e. is the scheduler a 'social animal' or does he prefer to extract information from a computer screen) and of the potential sources (i.e. if the usual source is a person, is that person approachable or is he or she known to be difficult?). The information collected will depend on the task, the responsibilities of the individual, the availability and format of the information, and the time constraints imposed.

- Monitoring is the formal process of routinely collecting information rather than the activity being triggered in some way. The most common example of monitoring that is undertaken by schedulers is that of checking actual production against planned production. Depending on the business, this could be done once a week or several times a day.
- Information sampling is an informal process that involves the collection of information by exception and focuses on the resources that need to come together at the right time to ensure that production will occur on time. Information sampling occurs as a result of a trigger or cue. It may be in response to a disturbance that has just occurred or it may be in response to a disturbance discovered that does not yet warrant action but requires monitoring. If the information sampled does not provide all the pieces to the puzzle required for the scheduler to assess the situation to his or her satisfaction then this will trigger additional sampling of information.
- Information receipt can be formal in the form of paper-based reports, information gleaned at meetings and meeting notes. It can also be informal, such as verbal notification of disturbances[4] or exceptions and information gleaned verbally that is of use to the scheduler's assessment of the situation.

Informal routes may be very important in dynamic shop floor environments. These information collection behaviours are one component brought to the scheduling arena by 'humans'. It is part of the process by which schedulers oil the wheels to ensure production proceeds smoothly. In a dynamic information-rich environment the 'human' cannot be outside the control loop because much of the rich, dynamic information that exists cannot be captured into historical reports or on computer (see McKay *et al.*, 1995b). Evidence from our work so far indicates that information sampling can be triggered immediately by cues such as the existence of an actual or possible disturbance or may occur after a period of delay according to the scheduler's assessment of the situation and the priority attached to that particular trigger. If a scheduler mistrusts formal information received via a report or computer system, or even informal information received from some individuals, the scheduler may well sample additional sources of information for verification, leading to duplication of effort. The need to sample some information points can be eliminated by effective management of certain resources at higher levels in the planning hierarchy. Figure 7.5 maps out the production manager's information

[4] Disturbances can be delays in raw material delivery, machine breakdowns or quality issues, for example.

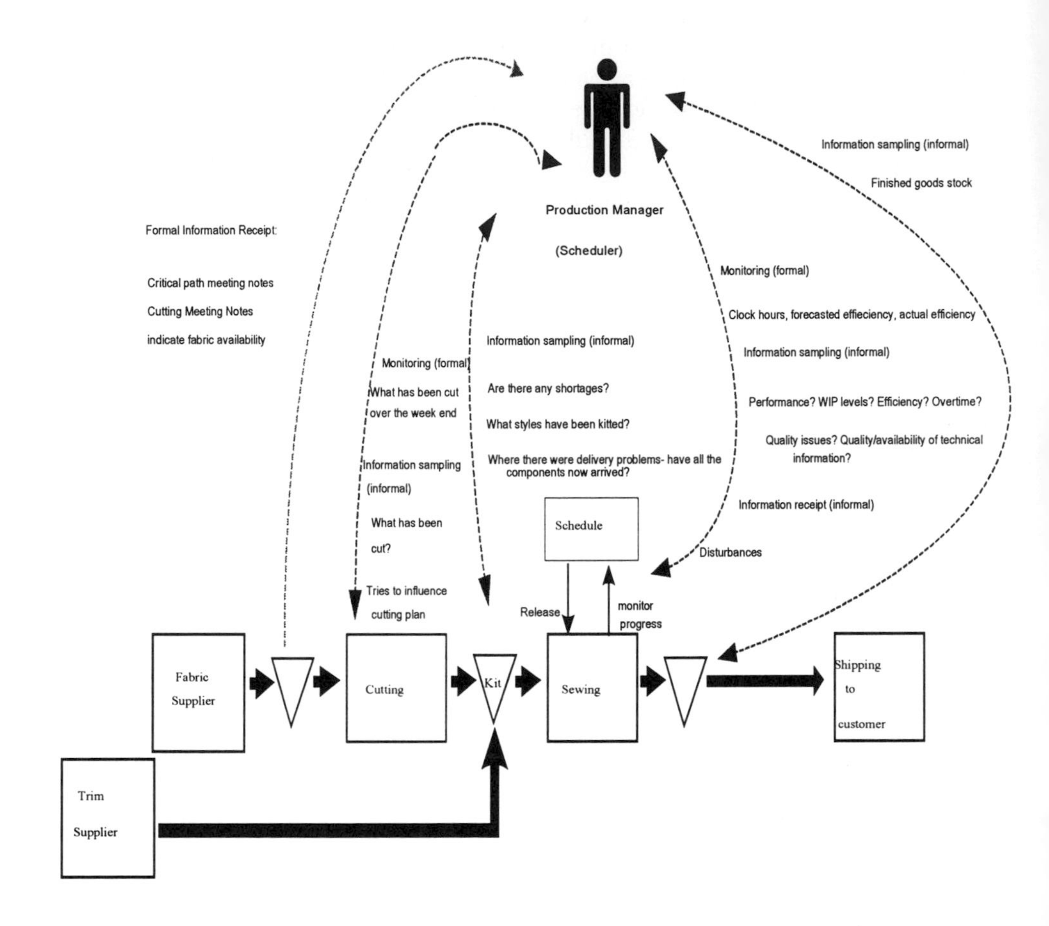

Figure 7. 5 Geoff's information collection map.

collection behaviour. There are certain resources that need to come together at the right time in order for production to start on time. These include:

- receipt of fabric
- receipt of trims
- cutting of fabric
- knitting of all components for a style.

Geoff may need to collect information about a style for example that has shortages which are threatening its production start date, or he may be collecting information about a style that he is hoping to use as a contingency if another style fails. He may therefore check on trim availability or whether a style has been kitted and also let the people in these areas know that the particular style he is asking about is a priority. He may also check whether a style has been cut and if not whether it can be brought forward in the cutting plan. Being a key department just before him in the supply chain, the Cutting Department can determine whether or not Geoff has work to feed the lines if the planned style has failed. There should be a cutting buffer of 5 to 6 days between the two departments most days but even so Geoff cannot always find a suitable alternative to load amongst the buffer styles if there is a disturbance. In this situation he may request the Cutting Department to bring other styles forward. Not being within his jurisdiction Geoff cannot order a particular style to be cut but can use his authority as a Production Manager to influence the situation.

Although Geoff samples a lot of information because of his position of authority he can also rely on a lot of information coming to him rather than having to seek it out. During the day Geoff receives constant interruptions. He encourages this as it allows him to receive information regarding the status of the shop floor and about disturbances affecting production. The downside of this policy is that not all information brought to him and taking up his time is relevant and many problems brought to him by the line supervisors are problems that they are perfectly capable of sorting themselves. Sampling information is an important part of facilitating the schedule for Geoff but he can also rely on a quantity of information coming to him due to his position of responsibility. Schedulers in other businesses without people working for them and with less authority often have to rely more heavily on sampling information as well as their bargaining and negotiation skills to get things done.

Information sampling and information receipt in terms of disturbances are key inputs to the scheduler and the scheduler's decision-making process, discussed further in the next section.

7.3.2.2 Trouble-shooting: situation assessment and decision making

Decision making in scheduling can relate to the thought process involved in the mechanical process of compiling the schedule or the decisions made in real time to facilitate the schedule in the event of disturbances, in other words problem-solving or trouble-shooting. Schedule compilation is an off-line task in that the scheduler is forced to take a snap shot of events and draw up a game plan for the next period of time. The process may or may not be time-constrained and it may or may not be combinatorially complex.

Facilitating the schedule in the event of disturbances involves real-time on-line decision making relating to dispatching, i.e. what needs to be loaded to the shop floor next or in the next time period. In Geoff's case this can include, for example, decisions on whether or not to accept borderline fabric to allow a style to be loaded in the sequence scheduled and Geoff's thought process may include taking into account who the customer is, whether there is a contingency plan in place if he suspects there might be a fabric issue or whether the line would be short of work, how much stock the customer has had delivered of that style recently, how urgent the style is and how *political* the style is (to mention just some of the factors).

The ability to analyse complex, ill-structured problems in dynamic environments and to search for and gather relevant information is part of the situation assessment process in naturalistic decision making (Salterio, 1996).

According to Smith and Sage (1991), situation assessment has several components: the inputs (described above as information gained by monitoring, sampling or information received), personal profiles, perspectives, stress levels (time constraints can affect this), decision strategies, learning and individual versus group decision-making considerations. We have observed a single scheduler in each case and so we are not examining the effect of individual differences in the same situation but instead focusing upon situation awareness and the decision-making process of the individual taking account of decision triggers, information used to make the decision, knowledge used, goals, options considered, time constraints and eventual outcome.

The following factors were key components of Geoff's situation assessment and decision-making process:

- the decision to respond now or to respond later;
- the process of contingency planning i.e. planning for suspected or predicted problems;
- Geoff's motivation behind his decision making;
- Geoff's influencing behaviour.

Geoff's decision-making process involved deciding whether to respond now to a cue or whether to respond later, deciding either that the situation was of low priority or that the situation required monitoring. Geoff knows that he operates in a dynamic environment and had learned from past mistakes that to respond too quickly can be detrimental. Sequentially occurring disturbances can cancel a previous disturbance or compound it. Geoff develops contingency plans but does not always enforce them straight away as he knows that the issues he faces can be so fluid that it can be politic to wait. He may set the wheels in motion to create the possibility of bringing the contingency plan into play but may wait to the last minute to make the final decision about what to do. One minute there may be the issue of a team appearing as if they will run out of work by a certain point in the afternoon; an hour later it may actually be the case that the machinists are not able to perform at the rate predicted on that particular style, or that a machinist falls sick and has to go home, or that the costed standard minutes are incorrect or an operation has been missed meaning that the work will last longer and there will no longer be a gap in production. Schaub (1997) describes this phenomenon as being inherent to complex dynamic systems. Unlike traditional decision-making tasks, a complex system has built-in dynamics which can result in it not being necessary to react at all in some cases as the system dynamics can result in the goal state arising independently as described above.

McKay *et al.* (1995b) describe their scheduler Ralph as exhibiting predictive expertise. He identifies triggers signalling potentially risky situations (e.g. length of time since last production; a machine upgrade; a procedure change; an engineering change on a product; a material change in a sub-assembly or component; or a change of supplier). He predicts the possible impact, whether direct, indirect, immediate or delayed (e.g. the process will take longer; the yield will be less; personnel will be taken off other tasks to help; or the next items will suffer variations in time or yield because of the current product). He decides what action to take to either minimise or avoid the impact (e.g. by choosing a different resource; by scheduling a test run prior to the main batch; or not scheduling a critical item close to a suspected problem). For example, if a product had not been made for a long time, say six months, then Ralph would organise for a special run to be made with a small volume several weeks ahead of when the product is required to be running at full rate. This allows for any anomalies in documentation, required equipment or manufacture to be ironed out and avoids unwanted losses in capacity.

Geoff exhibits similar behaviour. One of his main goals is to avoid capacity losses and to maximise throughput, as the main metric he is measured by is utilisation. Lost capacity equates to machinists being out of work. Both Geoff and Ralph make dispatching decisions based on their perceptions of not only the skill but the attitude and personality of the workers. Both Geoff and Ralph also condone pilot runs of potentially troublesome products. A significant aspect of Geoff's troubleshooting behaviour is his contingency planning. If he is told of a disturbance or predicts a disturbance will occur he forms a contingency plan where possible. As mentioned before, in some cases the prediction may be based on a suspicion from 'reading between the lines' rather than from what may be thought of as an established cue. It is critical that Geoff can keep the machinists in work and so contingency planning is an important part of his strategy. Geoff monitors what has not been cut on time via the computer system screens and what has not been delivered on time via the critical path report notes in order to get an advanced warning of potential problems. Some problems, such as material being found to be substandard only when it reaches the cutting table or an item being missed from the bill of material only being discovered as it is about to be stitched, also have to be dealt with by the scheduler and are hard to predict. These types of problems can be tackled through reorganisation, better procedures and attention to detail but in a pressurised business such as garment manufacture can be exacerbated under pressure from the customer to reduce lead times, to produce less volume of more styles and to be as flexible as possible. It is clear that the role of scheduler is to 'oil the wheels' and deal with disturbances but the questions that should also be asked are 'how much should the scheduler be responsible for?' and 'what realistically should they be expected to cope with?' and 'what problems could be avoided by other departments putting their own houses in order?'.

Targets and motivations that influence the scheduler affect the scheduler's behaviour. These influences can be structural, environmental and cultural. In the case of Geoff the main drivers are:

1. the principal performance measure, which is a utilisation metric (i.e. efficiency);
2. the motivation to keep machinists in work to achieve (1) and also to give the machinists a chance of earning their bonuses;
3. to retain credibility through shop floor stability, which is related to the fact that the scheduler needs to be respected in order to be able to ask for favours i.e. higher output to meet targets or overtime to catch back backlog, but also relates to (2) and (1).

All the motivations are interlinked and are related to organisational and/or production control goals and constraints. Keeping the teams in work ensures that throughput performance measures and targets are met, machinists have work to stitch and therefore have an opportunity to earn production related bonuses, and capacity is not lost because of machinists being idle. Retaining credibility through shop floor stability also relates to meeting throughput performance measures and targets. Instability implies changing styles more often, which also leads to increased set-up times. The more instability that exists the more disgruntled machinists and supervisors will be. Machinists have less of a chance to earn bonuses when the load is unstable and supervisors are faced with having to set up a team for one style having only just set it up for a different style. The happier machinists and supervisors are, the more favours Geoff is likely to be granted if overtime or increased production rates are required. An interesting point is that the motivations are strongly related to production performance and Geoff's domain, and indirectly rather than directly related to customer satisfaction. These motivations do not always induce behaviour congruent with management expectations. Below are three examples of behaviour which management discourage:

- *Resequencing sewing or cutting schedules in order to counter actual or predicted disturbances* e.g making changes to the schedule bringing it out of line with the plan in order to fill a gap in shop floor loading. Changes observed included: moving styles not required for up to four weeks. requesting changes to the cutting plan to provide a contingency plan against predicted disturbances.
- *Opting to maintain throughput momentum in the face of uncertainty* e.g machinists stitching in the face of 'borderline' quality, be it garment quality or fabric and trims quality, or instructing packing to continue when faced with uncertain packing instructions, rather than stopping whilst investigating what the correct method should be.
- *Opting to load styles without full component availability* e.g. Loading a style without the packaging being available.

The first of the three strategies is contrary to the expectations of planning management. The last is common to many of the manufacturing environments in the studied sample and a classic example of the 'trying to make it happen despite all odds' syndrome. The middle strategy is possibly the most contentious as it relies on subjective opinion of the effects on the aesthetics of the garment, of the strictness of the particular customer and of how discerning the consumer actually will be, weighed against the reality that lingerie/underwear manufacturing is a cost-conscious environment.

7.3.3 Summary

The ability to be flexible, to predict disturbances and to cope with uncertain, dynamic environments is what the human brings to the scheduling arena. Scheduling behaviour is complex and influenced by many factors and not readily automated. Schedulers are often described as being an important cog in the machinery to *make it happen*. But does the end always justify the means when the schedulers' efforts to *make it happen* are in conflict with organisational objectives and management expectations? Much has been made of the scheduler's predictive expertise but Geoff's actions as a result of predicting problems are not always popular. His contingency plans involve deviations from the plan by not just bringing forward the next style but bringing forward styles that will best meet the utilisation production performance targets. He is able to do this because of the authority afforded to him through his position as the Production Manager; not all schedulers have the same level of influence. Clearly the fault here does not lie with the scheduler but the design of the organisation and its objectives. There is a need to re-examine and realign business objectives and performance targets to ensure that the qualities the scheduler brings to the scheduling arena are used to best advantage.

Allowing the scheduler to have some flexibility to manage disturbances and best organise production within boundaries will also maximise benefits from the qualities brought to scheduling by the human scheduler. Constraining the scheduler and reducing his or her options in the face of a problem will limit the scheduler's ability to *make it happen*. Conversely, tackling the underlying issues that cause disturbances will reduce the need for the scheduler to get involved in trouble-shooting and making it happen.

7.4 SCHEDULING THE FACTORY- THE PRESENT

7.4.1 The macro view

A return visit to the factory was made three months after the initial visit. Although much had changed at the level of the individual it can be argued that little had changed at the macro level in the preceding three months with the prescriptive planning approach still being emphasised.

Changes were being driven by the need to change the business prior to introducing a new MRPII system. Under the banner of a 'business excellence' programme, team-working was being introduced which had changed the whole way the shop floor was scheduled and the role of Geoff as a scheduler. A transitional scheduling approach is currently in place to suit the devolution of scheduling decisions to the line managers. Once the new computer system is introduced the scheduling approach will be changed again.

7.4.2 The micro view – scheduling and facilitating the schedule

The line leaders are now more involved in the scheduling process. Product types are allocated to groups of lines and each line leader is responsible for a certain group of products. Long seam products which include slips and chemises are the responsibility of the first line leader, briefs are allocated to the second line leader and tops, vests and shorts are the responsibility of the third line leader. Some products sit in a grey area for allocation and could be given to the brief teams or the tops, vest and shorts teams. This decision rests with the Production Manager.

The task of *compiling a schedule* has now become a joint activity between the line leaders and the Production Manager. The schedule is no longer developed using an Excel spreadsheet but is drawn up on a white board. A style written in blue highlights that there is a complete style change between that style and the previous style. A style written in black denotes that the current style will change onto a different colour. The white board has the advantage of being very visual, allowing the Production Manager and the line leader to view all lines at once while sequencing styles. The only disadvantage is that it is fixed and not transportable. To counter this disadvantage the Production Manager produces an Excel document based on the schedule generated on the white board which lists the sequences of the styles allocated to each line per week rather than day and highlights whether each team will have enough work to fulfil their potential capacities. Another aspect that has changed is that the Production Manager deals mainly with top down disturbances and the line leaders deal with bottom up disturbances. As a result the decision load of the Production Manager is reduced. An interesting point to note is that the change in schedule generation is described as solely an interim measure prior to the new system implementation.

Devolving the responsibility for sequencing to line leaders has resulted in the reduction in the vertical division of labour thus giving them job enrichment (Spur *et al.,* 1994). Although this level of self-scheduling allows the line leaders to schedule their own lines and reduces the length of the scheduling hierarchy between scheduler and executor it does not allow the actual machinists executing the schedule to schedule for themselves. The concept of self-planning or to be more specific in this situation, self-scheduling, should include the executor in the scheduling loop with a view to driving proactive behaviour and improving the transparency of the schedule. So the measures that have been taken have enriched the jobs of the line leaders and by default the supervisors, but not the machinists.

Devolving the responsibility for sequencing has also meant that the Production Manager now has far less contact with his line leaders. Whereas before his office was seldom free of line leaders or supervisors reporting issues or asking how to solve a particular issue the office is now quiet with only the odd interruption. The Production Manager has openly admitted that this has left him with an 'uneasy feeling'. He no longer has an up-to-date mental model of the status of the shop floor and also is required to problem-solve and trouble-shoot less which has left him 'feeling empty'. Geoff believes that these feelings will pass and the changes that have been made are for the better and he is also thankful for the extra time available to actually manage the factory.

7.5 DISCUSSION

At the beginning of the study Geoff was scheduling the entire factory and he found the schedule compilation process taxing. However, the majority of his effort was spent facilitating the schedule as opposed to compiling it. He was heavily reliant on informal information sampling, his 'in the head knowledge' and the volume of information that came to him via his line leaders. Unlike schedulers with less authority, Geoff could use the power of his position to influence rather than negotiate and to ensure that a certain amount of information came to him without having to seek it.

This study also highlights Geoff's motivation behind his scheduling *behaviour* i.e. decision making and actions, and highlights how the structure of the organisation and environment and the design of the performance measurement systems result in behaviour that is not always congruent with management expectations.

By the end of the study Geoff's scheduling responsibilities have dramatically changed. Some of his scheduling responsibility has been devolved down to the line leaders and scheduling has been transformed from a process reliant on one person 'with in the head knowledge' to a more transparent process.

This case raises a number of questions with regard to supporting the scheduling function, for example, *what activities should be supported?* Scheduling software currently focuses on supporting the mechanical process of compiling the schedule. Many researchers advocate the 'interactive' or 'hybrid' approach to the issue of schedule generation proving that the former outperforms both fully automated and fully manual approaches (Haider *et al.,* 1981; Nakamura and Salvendy, 1988, 1994; Higgins, 1996, 1999; Baek *et al.,* 1999).

Clearly compiling the schedule is not always the part of the job that requires the most effort. So can the process of facilitating the schedule be supported? To facilitate the schedule, the scheduler needs to collect information (monitor, sample, receive), assess the situation and make decisions about any issues threatening production.

7.5.1 Information sampling for situation assessment

As previously mentioned, information sampling is part of the schedule facilitation process. It is often a time-consuming process with information typically being provided informally and verbally from multiple sources. Informal information may be sought if a formal source does not exist, is not available, or is not trusted. If the formal source is perceived as being 'sometimes' unreliable then both sources are often used with the informal information being used to double-check the formal information. Informal sources can pose problems for the scheduler as the people providing the information may not always be available and all the information received from one source may not be relevant and may require filtering. Conversely the information from one source may not be complete forcing the scheduler to seek another source.

Formal information from sources such as information systems, meetings and paper reports can also cause the scheduler a number of problems. The information may not be trusted by the scheduler and require double-checking against a different source. The data may be difficult to access, being spread across a number of different screens, systems or reports. The format may be unhelpful causing the scheduler to have to transpose the information into a more usable format. The information required may be buried in a mass of data in a report or on a screen. The information may be incomplete without reference to another source.

Schedulers have to try to integrate information from a wide variety of sources, both formal and informal. The difficulties associated with this, such as the time it takes to collect and integrate the information, leaves little time for analysis (Shobrys, 1995) and a lack of either timely or accurate data is one of the greatest obstacles to informed decision making (Bauer *et al.,* 1994).

Clearly if the scheduler is to facilitate the schedule effectively and avoid threats to production by effectively assessing situations and making appropriate decisions, then the information she or he requires needs to be portrayed in a manner that supports the scheduler's situation assessment and decision making process. Traditional systems analysis data-orientated techniques fail to achieve this (O'Hare *et al.,* 1994; Vicente *et al.,* 1995; Kaempf *et al.,*1996; Wong *et al.,* 1996; Klein *et al.,*1997).

How to present the information is not the only issue. There are other fundamental questions that should be raised:

- *Should we attempt to formalise all informal sources in order to support the scheduler?* What impact will it have on the social structure and the scheduler's level of job satisfaction if all the information the scheduler needs is on a screen in front of him? It will undoubtedly cut down on time spent seeking out information. But as she or he will no longer need to talk to people face-to-face, will the lack of personal contact affect the scheduler's ability to bargain and ask for favours? Will the scheduler miss other cues from the environment for example a wrong part being loaded onto a machine, if she or he is less mobile due to the effective information provision? These are not questions that researchers looking at decision making in ambulance dispatching or naval decision making have to face. However they are issues that need to be raised when examining the role of the scheduler. Common sense dictates that the final solution would be a compromise. The limitations of information presentable at the interface would ensure this. The situation is similar to that in McKay *et al.* (1995b), where the scheduler's heuristics could not all be included in a decision support system because some of the heuristics included data that are enriched. For example, qualitative and intuitive information, such as the workers' attitudes during a training session or the current level of politics between plants, could not - for obvious reasons - be included. This means that not all information used by the scheduler would be displayed, the scheduler would still be free to be mobile and justifiably seek out sources of information, however the increase in appropriate information via an information system would cut down on non-value-added activity.
- *Is all of the information sampling really necessary?* From the evidence we have seen, the answer is probably not. Obviously information sampling can be reduced through appropriate presentation at the interface. However there can be other underlying issues. Reviewing the information sampling behaviour of a scheduler can provide a means of performing a health check of the production planning and control system. Excessive sampling of information points and sampling of more information points than is necessary can indicate deficiencies in the formal information process or can indicate that there is a control issue in the organisation causing the scheduler additional work. For example, the scheduler may be being forced to sample information and make decisions that are not part of his role due to ineptitude of others. Alternatively, the scheduler might be forced to sample information and make decisions at a level that could be reduced through better management of disturbances of others. This also leads to the wider question of what to do if the scheduler's actual behaviour is in conflict with management expectations and higher organisational goals. In Section 7.3.2.2. the conflict between performance measures and goals at the scheduling level and the planning level is illustrated. If these conflicts remain unresolved it leaves the system designer with the dilemma of which behaviour to support- the 'actual' behaviour or the

'expected' behaviour. Supporting 'actual' behaviour would be unpopular with planning management and supporting 'expected' behaviour would result in the informal systems still remaining and the information system implementation would either be used in parallel with the informal system or not at all.

- *What if the geography of the organisation changes, what implications does this have for the scheduler's information sampling strategy and information requirements?* Proximity has a fundamental impact on the accessibility of information, especially where limited amounts of data are found on the computer system and the primary form of transmission is via verbal communication. In a lot of work roles, information acquisition can occur informally by the coffee machine, in the corridor or in the smoke room. It can even occur passively, overhearing conversations or observing others' actions in the periphery, such as in the cockpit of an aircraft or in a London Underground control room (Hutchins, 1995; Hutchins and Klausen 1996; Luff *et al.*, 1992). As Geoff is sited in a satellite factory and other functions such as buying, fabric and garment technology, sales and planning occur at the head office, he must rely on being informed of disturbances such as fabric and component availability, technical failures and changes to the plan in a timely fashion. This lack of proximity and the reduced possibility of informal face-to-face contact means that formal systems in these settings have to be efficient and effective and data flow has to be well designed. In cases such as these it becomes more important to improve the data flow to the scheduler and appropriate presentation of information at an interface has obvious advantages. However the dynamic nature of environments may mean this option is not feasible. This is where alternative technologies, for example the concept of computer-supported co-operative work, become more appropriate.
- *Is technology the only answer?* Technology is not the only answer. In the case of Geoff and the planner, communication is perceived to be lacking despite a good working relationship. Because the planner is focused on planning to customer due dates and the scheduler is focused on meeting utilisation measures, keeping the teams in work and retaining some stability for the shop floor, there is a conflict of interest. Neither are formally encouraged to collaborate and negotiate their positions through the design of the planning process so it is not surprising that they both feel there is a gap in communication, and geographical separation adds to this divide. Data flow can be inhibited as a result of the control structure and not just through a lack of formal data flow procedures or information systems. So the dilemma here is what information can reasonably be transmitted to and from the scheduler via an information system in real time in a dynamic discrete manufacturing environment, and what information will always have to be communicated informally and how can control systems be designed to encourage this communication.

7.5.2 Co-operation and communication

Relying on the 'in the head' expertise of a sole scheduler has its disadvantages. Scheduling can appear mysterious rather than transparent as the decision-making process is not made explicit to others. The majority of knowledge can be informal rather than formalised and scheduling strategies can be hard to make explicit and to pass onto others. Decision making load on just the one person can be high. Researchers to date have focused on supporting the sole scheduler, trying to eliminate problems by reducing cognitive load, providing cognitive aids and eliciting the scheduler's knowledge (rules and heuristics) to provide decision support or to make the scheduling process more transparent and to assist learning (McKay *et al.*, 1995b; Wiers, 1997; Higgins, 1999).

Another way of simplifying the scheduler's cognitive and decision-making load is to devolve part of the scheduling responsibility to the shop floor operators or supervisors. In this case in particular, the shop floor had already been divided into product areas. Devolving scheduling responsibility has meant that each line leader is now responsible for scheduling their own area. Decision making for sequencing of styles by teams within a product area transformed a centralised into a distributed system, although the decision making involved in managing the overall picture and reallocating resources remained with the scheduler. The decision-making process became hierarchical with line managers having local decision-making authority and with the scheduler retaining the power to override decisions if necessary and retaining the power to reallocate resources. This is a more cost-effective way of providing support for the scheduler than a conventional decision support system. In the case of Ralph (McKay *et al.*, 1995b), a single scheduler with 'in the head knowledge', the company decided to hire a graduate to assist Ralph rather than implement a decision support system. This 'human solution' was said to be a reasonable resolution as long as the number of product options Ralph was in charge of did not increase.

When removing all the decision making from one person's head and distributing the decision making amongst a team of people, the need for co-operation, communication and co-ordination of the decision-making process increases. According to Hoefer (1985) 'the creation of self-contained tasks aids in reducing complexity and uncertainty in task execution and it also eases the co-ordination of decision making' and in the same way decisions themselves need to be self-contained. If a person in charge of making a particular set of decisions has all the information required to make those decisions, is empowered to make that decision and does not have to refer to others before executing the decision then the co-ordination of decision making is reduced. Though reduced, decision making will still need to be co-ordinated at some level and information flows will need to be co-ordinated also.

When the emphasis moves from the single scheduler to a team of people this adds another dimension to the problem of supporting the scheduling function. It is necessary for the team to be well integrated to ensure scheduling performance does not suffer, information flows need to be well designed and facilitated and any physical separation of scheduling actors needs to taken into consideration. A well informed, integrated and co-ordinated scheduling team may be achievable through organisational design or it may require technological assistance depending on the situation. If we think of the scheduling function as a small team of people within a small factory then appropriate information systems support may well be beneficial, yet computer support for communication would seem unnecessary. If we think of the need to co-ordinate scheduling across numerous cells in, say, a large machine shop, then a system to co-ordinate the scheduling activity as well as provide appropriate information presentation would possibly be the most beneficial form of support. If we expand this scenario to include the need for co-operation between the scheduling function and the planning function, then an appropriate computer system would include the functionality to allow co-operative schedule/plan generation, provide appropriate, accurate well-presented information at the interface, and would facilitate communication.

Advances in display and multi-media technology and networking have resulted in the development of research into computer supported co-operative work (CSCW) and groupware. Scrivener (1994) describes CSCW as 'a generic term which combines the understanding of the way people work in groups with the enabling technologies of computer networking and associated hardware, software, services and techniques'. Groupware is technology designed to facilitate the work of groups. This technology may be used to communicate, co-operate, co-ordinate, solve problems, compete, or negotiate. The designer of a groupware system to support production control would not only be concerned with what information to present to the user and how to present it, how to support co-operative schedule generation but would also need to consider how to facilitate communications between users i.e. which methods of communication (audio, video, display) are best matched to different tasks (Harvey and Koubek, 1998; Klies *et al.*,1998). There is no doubt that introducing computer supported interaction changes the nature of social interaction by both constraining and facilitating that interaction. Computer supported interaction can be more inhibited, more blunt and less creative than face-to-face communication but where face-to-face contact is difficult, CSCW can facilitate communication and co-operative work (Wellman *et al.,* 1996).

Communication and information sharing is not guaranteed even with the existence of a communication system. Unless the organisational culture supports the information exchange and individual and organisational goals are aligned as described above, then the same problems can exist with a groupware system as between individuals without the technology (Barua *et al.,* 1997).

To summarise, any technological support for the scheduling function that proves to be appropriate and feasible will only be successful if due consideration is given to organisational design. The technology that is appropriate to support the scheduling function will depend on the situation and environment. Like other authors we have found that the production planning and control function will be heavily influenced in design and performance by uncertainty and complexity in the business and by the level of buffering against these that is possible (e.g. stocks, capacity etc.) (Little *et al.,*1994). By examining the organisational and environmental context of a particular company it is possible to identify inherent and manageable factors which constrain and influence how the production planning and control system does operate and how it is required to operate. By redesigning the organisation and the production planning and control system, simplifications can be made and factors that were influencing the complexity and uncertainty facing the scheduler can be reduced. Some uncertainty and complexity however will be inherent in the business, and this will influence the scheduler's activity and the proportion of schedule compilation time, information collection time and co-operation and communication that makes up the scheduler's role.

Such distribution of activities, as well as geographic issues, will define the type of support most appropriate.

The other major issue, of course, is commercial availability and supply of technology, which by definition will be customised rather than generic. What is being described above calls for more in-depth requirements analysis than would be undertaken for traditional IT solutions resulting in mixed media, interactive solutions with tailored interfaces and software customised for multiple rather than individual users. In the majority of company-wide planning system implementations companies buy generic systems; a large proportion of the cost is then in buying the software and the remaining cost is in customisation and implementation. Time for implementation and costs are the drawbacks of such systems, which demand that users adapt their ways of working to the system. Clearly a step change in technological solutions offering cheap and easy customisation would be necessary to be able to provide truly effective support for the scheduling function.

7.6 CONCLUSIONS

A key issue to resolve when reviewing support for schedulers is to establish whether support is required in the form of organisational remedies to reduce the complexity of the scheduling, improve communication and co-operation and reduce the effort involved in facilitating the schedule, or whether IT support is required to provide better integration of data, better task-technology fit and better scheduling functionality. The answer, in fact, is that the solution will be context specific and both types of support may be required. The solution will be dependent on the company's objectives, inherent structural features of the business, the complexity of the scheduling problem and the level of disturbances that the scheduler has to cope with in order for the business to be responsive. What the business needs to do is better understand the role of the human scheduler and to understand the interplay between factors within the environment that can impact on the scheduling role. In addition, the business needs to support value-added activity where required but also support the scheduler by taking measures to eliminate the non-value-added activity.

7.7 REFERENCES

Baek, D. H., Sang, Y. O and Wan C. Y. (1999). 'A visualized human-computer interactive approach to job-shop scheduling', *International Journal of Computer Integrated Manufacture*, 12, 1, pp 75-83.

Barua, A., Ravindra, S. and Whinston, A. B. (1997). 'Effective intra-organizational exchange', *Journal of Information Science*, 23, 3, pp 239-248.

Bauer, A., Bowden, R., Browne, J., Duggan, J. and Lyons, G. (1994). *'Shop Floor Control Systems: From Design to Implementation'*, Chapman & Hall, London.

Crawford, S., MacCarthy, B. L., Wilson, J. R and Vernon, C. (1999). 'Investigating the work of industrial schedulers through field study, cognition', *Technology & Work*, 1, pp 63-77.

Doumeingts, G., Vallespir, B. and Chen, D. (1995). 'Methodologies for designing CIM systems: a survey', *Computers in Industry,* **25**, pp 263-280.
Ferguson, R. L. and Jones, C. H. (1969). 'A computer-aided decision system' *Management Science*, **15**, pp 55-61.
Haider, S. W., Moodie, C. L. and Buck, J. R. (1981). 'An investigation of the advantages of using a man-machine interactive scheduling methodology for job shops', *International Journal of Production Research*, **19**, pp 381-392.
Hardstone, G. (1991). '*A Simple Planning Board and a Jolly Good Memory: Time, Work Organisation and Production Management Technology*', Edinburgh PICT Student Paper No 4.
Harvey, C. M and Koubek, R. J. (1998). 'Towards a model of distributed engineering collaboration', *Computers and Industrial Engineering*, **35**, 1-2, pp 173-176.
Higgins, P. (1996). 'Interaction in hybrid intelligent scheduling', *International Journal of Human Factors in Manufacuring,* **6**, 3, pp 185-203.
Higgins, P. (1999). *Job Shop Scheduling: Hybrid Intelligent Human Computer Paradigm*. PhD Thesis. Swinburn University of Technology, Australia.
Hoefer, H. (1985). 'Theory and practice of decentralised production management systems- A philosophy rather than methods', *Computers In Industry*, **6**, pp 515-527.
Hutchins, E. (1995). 'How a cockpit remembers its speeds', *Cognitive Science*, **19**, pp 265-288.
Hutchins, E. and Klausen, T. (1996). 'Distributed cognition in an airline cockpit' in Engestrom Y. and Middleton D. (eds), *Cognition at Work*, Cambridge University Press, pp 15-34.
Kaempf, G. L., Klein, G., Thordsen, M. and Wolf, S. (1996). 'Decision making in complex naval command and control environments', *Human Factors,* **38**, 2, pp 220-231.
Klein, G., Kaempf, G. L., Wolf, S., Thorsden, M. and Miller, T. (1997). 'Applying decision requirements to user-centred design', *International Journal of Human-Computer Studies,* **46**, pp 1-15.
Klies, K. J., Williges, R. C. and Rosson, M. B. (1998). 'Co-ordinating computer-supported co-operative work: A review of research issues and strategies', *Journal of the American Society for Information Science,* **49**, 9, pp 776-791.
Lincoln, Y. S. and Guba, E. (1985). *Naturalistic Inquiry*, Sage Publications, Newbury Park, California.
Little, D., Kenworthy, J. G. and Jarvis, P. C. (1994). 'Performance measurement of production control systems' in *Proceedings for the Euroma Conference, Operational Strategy and Performance*, June 1994, Cambridge University, UK, pp 93-99.
Luff, P., Jirotka, M., Heath, C. and Greatbach, D. (1992). 'Tasks and Social Interaction: the Relevance of Naturalistic Analyses of Conduct for Requirements Engineering', *Proceedings of the IEEE International Symposium on Requirements Engineering,* San Diego, California, USA, IEEE Computer Society Press, pp 187-190.
McKay, K. N., Safayeni, F. R. and Buzacott, J. A. (1995a). 'An information systems based paradigm for decisions in rapidly changing industries', *Control Engineering Practice,* **3**, 1 pp 77-88.

McKay, K. N., Safayeni, F. R. and Buzacott, J. A. (1995b). ' "Common sense" realities of planning and scheduling in printed circuit board production', *International Journal of Production Research*, **33**, 6, pp 1587-1603.

McKay, K. N., Safayeni, F. R. and Buzacott, J. A. (1995c). 'A review of hierarchical production planning and its applicability for modern manufacturing', *Production Planning and Control*, **6**, 5, pp 384-394.

Mahenthiran, S., D'Itri, M. P. and Donn, R. E. (1999). 'The role of organizational reality in implementing technology: A field study', *Information Systems Management,* **16**, 2, pp 46-57

Mitchell, and Sundstrom (1997). 'Human interaction with complex systems: design issues and research approaches', *IEEE Transactions on Systems, Man and Cybernetics- Part A Systems and Humans,* 27, 3, May.

Nakamura, N. and Salvendy, G. (1988). 'An experimental study of human decision-making in computer based scheduling of flexible manufacturing system', *International Journal of Production Research,* **26**, pp 567-583.

Nakamura, N. and Salvendy, G. (1994). 'Human planner and scheduler', in Salvendy, G. and Karwowski, W. (eds), *Design of Work and Development of Personnel in Advanced Manufacturing*, John Wiley & Sons, Inc., New York, pp 331-354.

O'Hare, D., Wiggins, M., Batt, R. and Morrison, D. (1994) 'Cognitive failure analysis for aircraft accident investigation', *Ergonomics,* **37**, pp 1855-1869.

Randel, J. M. and Pugh, L.(1996). 'Differences in expert and novice situation awareness in naturalistic decision making', *International Journal of Human-Computer Studies*, **45**, pp 579-597.

Salterio, S. (1996). 'Decision support and information search in complex environment: Evidence from archival data in auditing', *Human Factors*, **38**, 3, pp 495-505.

Schaub, H. (1997). 'Decision making in complex situations: Cognitive and motivational limitations', in Flin, R, Salas, E., Strub, M. and Martin, L. (eds), *Decision Making Under Stress: Emerging Themes and Applications*, Aldershot, Ashgate.

Shobrys, D. E. (1995). 'Scheduling', in Boddington C.E. (ed.), *Planning and Scheduling, and Control Integration in the Process Industries*, MacGraw Hill, New York, pp 123-156.

Smith, C.L. and Sage, A.P. (1991) 'A theory of situation assessment for decision support', *Information and Decision Technologies*, **17**, 2, pp 91-124.

Spur, G., Specht, D, and Herter, J. (1994). 'Job design' in Salvendy, G. and Karwowski, W. (eds), *Design of Work and Development of Personnel in Advanced Manufacturing,* John Wiley & Sons, New York.

Vera, A. H., Lewis, R. L. and Lerch, F. J. (1993). 'Situated decision-making and recognition-based learning: applying symbolic theories to interactive tasks', *Proceedings of the 15th Annual Conference of the Cognitive Science Society,* Institute of Cognitive Science, University of Colorado-Boulder, pp 84-95.

Vicente, K. J., Christoffersen, K. and Pereklita, A. (1995). 'Supporting operator problem solving through ecological interface design', *IEEE Transactions on Systems, Man and Cybernetics,* **25**, 4, pp 59-545.

Wellman, B., Slaff, B., Dimitrova, D., Garton, D. and Gulia, M. (1996). 'Computer networks as social networks: Collaborative work, telework and virtual community', *Annual Review of Sociology,* **22**, pp 213-238.

Wiers, V. C. S. and Van de Schaaf, T. (1997). 'A framework for decision support in production scheduling tasks', *Production Planning and Control,* **8**, 6, pp 533-544.

Williams, T. J., Bernus, P., Brosvic J., Chen, D., Doumeingts, G., Nemes, L., Nevins, J. L, Vallespir, B., Vlietstra, J. and Zoetekouw, D. (1994). 'Architectures for integrating manufacturing activities and enterprises', *Computers in Industry,* **24**, pp 111-139.

Wong, W. B. L., O'Hare, D. and Sallis, P. J. (1996). 'A goal oriented approach for designing decision support displays in dynamic environments', *Proceedings from OZCHI '96 Computer Human Interaction Conference*, Hamilton, New Zealand, pp 78-85.

CHAPTER EIGHT

Decision Support for Production Scheduling Tasks in Shops with Much Uncertainty and Little Autonomous Flexibility

Kenneth N. McKay and Vincent C.S. Wiers

8.1 INTRODUCTION

Computerized decision support systems have been advocated as useful tools for schedulers to use and there have been many systems developed (e.g., Hess, 1995; Melnyk, 1995; Friscia, 1997; Weil, 1997). Unfortunately, the number of shops using such tools is relatively meager and there has not been widespread implementation and sustained use (Wiers 1997a).

For a scheduling tool to be successfully used it should match the type of scheduling the scheduler does. If the tool does not support the scheduler's decision-making, the scheduler will be forced to revert to past behavior and thereby ignore any automatically-generated plans. In any scheduling situation, the decision support system must: i) support an appropriate model of the manufacturing system, ii) present the scheduler with an adequate representation of the supply and demand problem, iii) provide the automatic or manual functions with which to generate a schedule/plan, and iv) give the user tools to help evaluate the quality of the result. The specific nature of these requirements will depend on the scheduling task.

In this paper, we will firstly review types of scheduling situation: the smooth, social, sociotechnical and stress shops. Subsequently, we will discuss the requirements associated with what we consider to be the most demanding situation: reactive scheduling in a stress shop. A system designed for reactive scheduling will be used to illustrate the design issues inherent in such a problem. We will conclude with recommendations for further research into reactive scheduling support systems.

The paper is structured as follows: Section 8.2 explains a typology of scheduling situations based on characteristics of shops. Section 8.3 presents a cognitive model of information processing that will be used to categorize human decision behavior. In Section 8.4, a case study is presented where a decision support system is implemented for a specific type of shop. Section 8.5 gives the conclusions and outlines areas for future research.

8.2 HUMAN SCHEDULING AND DECISION SUPPORT

8.2.1 Scheduling situations

A scheduler or planner has two basic types of planning that he or she does: predictive and reactive. Predictive scheduling covers some kind of planning horizon and establishes the work release sequence and any key dispatching required by the shop for the horizon. The planning horizon a scheduler deals with is usually long enough for the plant to make any useful changes in manpower, accommodate necessary preventative maintenance, or deal with any other short-term capacity/supply issues. Reactive scheduling deals with the unexpected: a change in some status that affects the immediate - or near-term future. How much one must do of each - predictive and reactive scheduling - depends on the manufacturing situation.

In situations which are extremely stable or which have significant degrees of flexibility on the shop floor, it is possible to predict, and prescribe, and ultimately observe little or no reactive scheduling. That is, the shop has little or no uncertainty and everything is deterministic, or the workers on the floor can autonomously deal with the problem without significantly changing the schedule. Similarly, in some other ideal situations having pervasive redundancy, it is possible to eliminate predictive scheduling and solely react; there is sufficient spare material, resources, and capacity to deal with any request, any time. Consider a third situation between these two ideals: a plan is made and something happens and the scheduler is required to alter the short-term schedule: work releases, task assignments, or dispatch decisions. While we will concentrate on this third situation in this chapter, it is important to understand the range of situations implied by the interaction of autonomy and uncertainty.

Wiers (1997b) studied the use of scheduling techniques by human schedulers and described a number of concepts that should guide the design of scheduling decision support systems. One of these concepts, autonomy, describes the degrees of freedom at a certain level in an organization. Generally, in scheduling literature, all autonomy is assumed to be in the hands of the scheduler and the shop floor operators do not have any decision freedom regarding the schedule - the sequence and allocation of work. However, because of their close relation to the production process, shop floor operators can be faster and better able to react to disturbances than the scheduler. For example, knowledge regarding the determinants of flexibility and uncertainty within production units is often in the hands of operators and foremen.

In van der Schaaf (1995), the concept of human recovery is presented as the positive role that human operators can play in the prevention of system failures. Two conditions for human recovery mentioned by Rasmussen (1986) are: i) observability, i.e., the ability to detect possible system failures, and ii) correctability, i.e., the ability to correct a possible system failure. In production scheduling, the concept of human recovery is used to refer to the *ability* of the operators on the shop floor to use flexibility to compensate for uncertainty (Wiers, 1997b). Human recovery can be employed in an organization by allocating autonomy to the shop floor, i.e., the *authority* to act on disturbances. It is important to note here that autonomy is a different construct than human recovery: autonomy indicates that shop floor operators are allowed to perform certain corrective actions, and human recovery indicates that shop floor operators are able to perform certain corrective actions.

Certain disturbances within the production unit can and should be solved by the operators (Wiers and van der Schaaf, 1997). The allocation of scheduling autonomy depends on the following two characteristics of production units: i) uncertainty, and ii) human recovery. The concept of human recovery in a production scheduling setting can be described as an active form of flexibility: it refers to the ability of the operators on the shop floor to use flexibility to compensate for uncertainty. The two characteristics can be combined to create four scheduling situations. These are depicted in Table 8.1.

Table 8.1 Typology of production units

	No uncertainty	Uncertainty
No human recovery	**Smooth shop** Optimize	**Stress shop** Support reactive scheduling
Human recovery	**Social shop** Schedule as advice	**Sociotechnical shop** Schedule as framework

The names of the production unit types illustrate the nature of the scheduling task, resulting from a certain division of autonomy between the scheduler and the production unit. It should be noted that the above situations are stereotypical: in many companies, different parts of the supply chain have mixed characteristics.
In the

- *smooth shop*, there is neither internal nor external uncertainty and as a result there is little need for human intervention and problem-solving, or, different parts of human recovery. The shop is very stable and can be considered mature. In these shops, schedules can be carried out exactly, and in these cases it makes sense to generate schedules that are optimal or near optimal.
- In the *social shop*, the scheduler can lay out the basic schedule with sequences and timing, but allow for autonomy on the shop floor to tune the final work sequence at any resource. The scheduler may provide an optimized recommendation, but acknowledges that from a social and motivational point of view some human recovery might be preferred. Ideally, the schedule identifies the operation sequence, recommended timing, and possible bounds for advancing or delaying the work. In a social shop, the scheduler is tasked with giving a feasible amount of work and the shop has sufficient flexibility and capacity to achieve any due dates.
- In the *sociotechnical shop*, it is neither necessary nor possible to *a priori* imbed the necessary flexibility into the schedule to take into account uncertainty. In these production units, human recovery should be employed to compensate for disturbances in the production unit. This is achieved via excess capacity or capability utilized by the shop floor - either will absorb the uncertainty without significantly damaging the overall plans. Hence, schedules are only generated to provide a framework for production and are not

executed exactly; optimization therefore does not make sense. The goal of scheduling in a sociotechnical shop must be to create feasible loads and to insert slack intelligently in the plan so that the shop can react accordingly.

- In the *stress shop*, there is much uncertainty that cannot be compensated for by allocating autonomy to the shop floor because sufficient localized recovery is impossible. In these cases, the impact of the uncertainty affects multiple resources and tasks requiring the bigger picture to be taken into account. Therefore, disturbances have to be managed by the scheduler. To enable effective rescheduling actions, high demands are placed on the speed and accuracy of feedback from the shop floor.

This chapter will focus on the stress shop. Decision-making in stress shops is highly *context sensitive* in terms of what is known when certain solutions are feasible, and when the scheduler must make certain decisions. Because of this uncertainty and tight linkage with the shop floor, we propose that stress shops place the highest demands on the human-computer interaction (HCI) for the decision support system. The decision support system needs to match the types of decisions to be made by the scheduler in a quick, transparent and efficient manner. This paper will present a case study in a stress shop, where the scheduler is faced by pervasive uncertainty and hence the HCI issues focus on the if-then style of reasoning.

8.2.2 Decision support

The above typology of scheduling situations was initially aimed at providing guidelines for the division of scheduling autonomy. That is, using the framework and typology to predict and prescribe the type of suitable scheduling decisions, and guiding the types of decision-making to imbed in a decision support system. The stress shop is obviously the most demanding of the four scheduling situations for the human scheduler, because he or she is an intimate link in the shop floor execution and the demands are high for any decision support system. In a stress shop, reactive scheduling can account for 80-90% of the scheduler's daily work. This is in contrast to a smooth shop: a predictable and reliable situation where a schedule is generated in the morning or beginning of the week and followed with certainty.

The degree of reactive scheduling places a strain on the scheduler and the tools used. For example, the system must support the capability of having a timely and accurate picture of the manufacturing situation whenever decisions are made, and it must support what-if capabilities resembling those which can be ultimately executed. However, commercial scheduling tools have been forced to be generic and they attempt to satisfy as many different manufacturing situations as possible. This does not mean that there are no differences between scheduling systems; however, these mostly apply to modeling capabilities and not to functionality. This is illustrated by Figure 8.1.

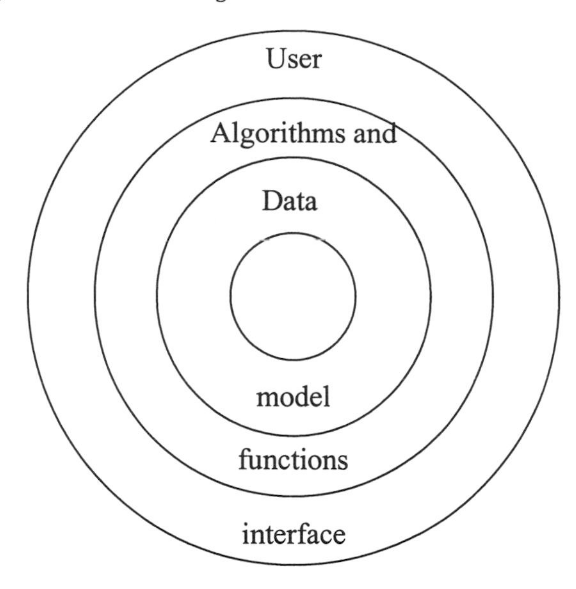

Figure 8.1 Structure of scheduling systems.

Most commercial scheduling systems differentiate their products by industry segment, such as oil and chemicals, customer packaged goods (CPG), high-tech, metals, food, semi-process, etc. When a supplier focuses on a segment, such as metals, this usually means that their product is able to model milling and casting operations, and use volumes in plans. However, the algorithms used in commercial packages are mostly similar across various suppliers from the viewpoint of the planner.

Hence, commercial systems functionality can be characterized as primarily context-free and makes them suitable for the smooth, social, and possibly sociotechnical shops. We propose that one of the reasons why scheduling systems have not been widely adopted and used in industry is that many situations are stress shops and context-free systems do not have adequate automatic, supporting, and enabling functionality. They are not easier to use than manual systems and they do not support the type of decision-making implied by the situation.

8.3 INFORMATION PROCESSING AND DECISION-MAKING

In this section, we introduce a cognitive model of information processing used to explain activities in the scheduling task and the focus on rule-based reasoning. The cognitive model used is the decision ladder of Rasmussen (1986).
Figure 8.2 depicts the GEMS model of Reason (1990), which is an adapted version of the decision ladder of Rasmussen.

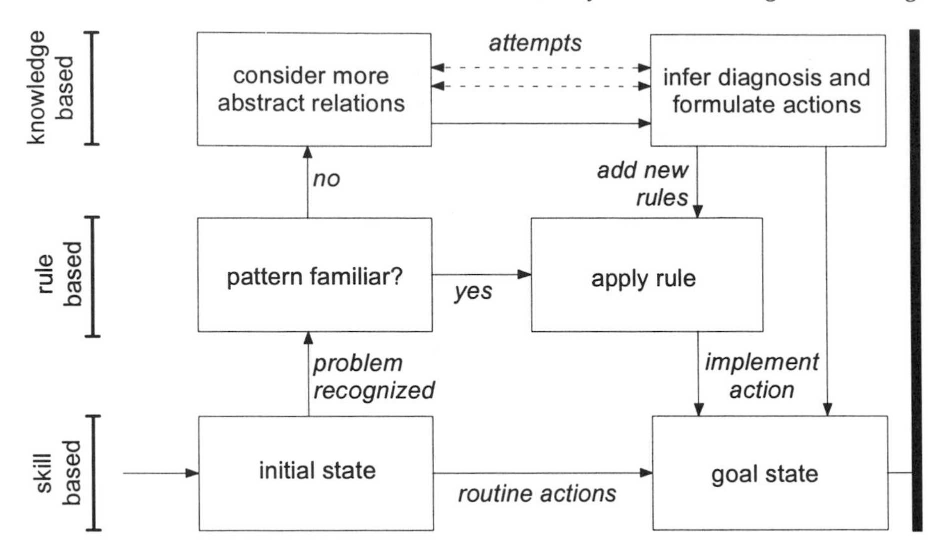

Figure 8.2 A model of human decision behavior (adapted from Reason, 1990).

The model distinguishes between three levels of reasoning: skill-based, rule-based and knowledge-based. At the *skill-based* level, actions are carried out almost automatically, i.e., without the need for conscious decision-making. It is a form of holistic reasoning where the solution is obvious to the scheduler and it 'just makes sense'. If a difference between the expected and real outcome is noted or a series of outcomes are consciously linked, control passes on to the *rule-based* level. At the *rule-based* level there are many *if-then* rules competing to become active. If there are no rules that match the environment, reasoning passes on to the *knowledge-based* level where arguments are made from first principles or detailed analyses (see also Higgins, chapter 13 in this volume).

The Rasmussen model has been used to explain human decision behavior in the production scheduling task (Sanderson, 1991; Wiers and McKay, 1996). Sanderson used the model to construct a model of human scheduling (MHS), which was never finished. A framework for the MHS was built that consists of twenty–seven production rules linking different types of scheduling activities. The model is also used by Wiers and van der Schaaf (1997) and Wiers (1997b) to explain human decision behavior in production scheduling tasks. However, in these studies, a distinction is made only between elements of the scheduling task that do not require attention of the human scheduler, and task elements that do require attention. Moreover, no attempt was made to link the production unit typology (i.e. smooth, social, sociotechnical, stress) to human decision behavior.

There are many cognitive models and one may claim that the choice for the GEMS model for the production scheduling task is more or less arbitrary. Indeed, much more detailed cognitive study is needed to validate this model in scheduling tasks. However, the use of the GEMS model in this domain is justified by the fact that the model captures a number of essential phenomena in the scheduling world. In particular, the notion of routine and exception in production scheduling, which was identified by McKay (1992) and Wiers (1997b), is well covered in the GEMS model. In previous research, the GEMS model is applied to the scheduling task by applying the notion of routine and exceptions; this chapter attempts to take the applicability of the model one step further by focusing on a specific reasoning level, i.e., the rule-based level and its requirements for decision support.

In Wiers and McKay (1996), the statement has been made that for timely, effective and efficient decision-making in a stress shop, the information processing supporting the rule-based reasoning level is critical. This includes how information is obtained, the structure of the information, the representation of the information, what can be done with the information, and the manipulation of the scheduling objects linked implicitly or explicitly to the information. However, no empirical evidence is given in Wiers and McKay (1996) to support the postulated relationship. To redress this, in the next section a case study in a stress shop is described, which shows that, in stress shops, the scheduling system should focus on rule-based reasoning processes.

8.4 A FIELD STUDY

8.4.1 The production unit

The field study was carried out in a production unit that can be characterized as a stress shop, with much uncertainty and little human recovery on the shop floor in terms of resequencing or alternative routings. The production unit is line oriented, in which each line represents one operation in the routing of products. In the production unit, there are more resources than people and the mobile workforce operates the lines in shifts. The operators can move from line to line; however, there is a reluctance to move people during the shift. A separate group of operators takes care of setting up the lines. The main control issues in this production unit are:

- allocation of parts to lines
- determining set-up changes
- determining run quantities
- start and end times of jobs
- allocating people to lines
- monitoring the number of people available per shift (labor usage).

The production unit contains over fifty primary resources (i.e., lines or combinations of lines), several hundred part numbers, and it has an average run duration of two to three days.

8.4.2 The scheduling system

A scheduling decision support system was developed in 1996 in close consultation with the human schedulers. The system is interfaced in a batch fashion with a manufacturing system for production requirements and information about the status of the shop floor. An interface has also been built to a fixture database for the purpose of obtaining timely data about fixture repairs and readiness.

In the following sections the scheduling system will be described by its functionality and user interface. The functionality of the system will be described according to its level of support for the user decision-making. The level of support lies in the possible variants of sharing responsibility between the human scheduler and the decision support system. Sheridan (1980) identifies ten possible levels in allocating functions to humans or to computers. At one extreme, the human acts as a principal controller, taking advice from the computer. The opposite extreme has the computer as the principal controller with the human performing corrections and adjustments. Interactive scheduling is located between these extremes (Sanderson, 1989; Higgins, 1996). Using the production typology we can be more precise and say that a smooth shop would be positioned close to the latter boundary with a stress shop towards the former. The social and sociotechnical shops would be between the two.

We will use three levels of support categories to describe the functionality of the system:

1. *automatic functionality*, that carries out actions without intervention of the human scheduler;
2. *supporting functionality*, that supports the scheduler in carrying out procedures that have to be carried out periodically; and
3. *enabling functionality*, that helps the scheduler in solving ill–defined problems.

8.4.2.1 Automatic functionality

The automatic scheduling functions of the system provide a starting point from which the scheduler starts altering the schedule. In other words, the automatic functions get the scheduler into the ballpark and then let the scheduler adjust accordingly. This approach also allows the scheduler to dynamically adjust any objectives without the system creating too tight a plan that would require too much alteration.

The general performance criterion of the scheduler is to produce orders on time. Hence, the due date of orders provides a guideline to construct an initial schedule. The task of sorting new orders in the schedule according to their due date is carried out by the system, and requires no intervention of the human scheduler. An important constraint in scheduling jobs is the available person capacity. The scheduling system guards the necessary capacity and provides automatic rescheduling functions to smooth capacity demand throughout the shifts.

While the smoothing algorithms might not be considered standard fare in scheduling systems, the basic scheduling capability is, and is usually more sophisticated than what was developed in this tool. This conscious decision to limit the mathematical algorithms was directly related to the stress shop categorization. The schedulers are initially driven by the ideal world in which every job meets its due date without conflict with other jobs, resource limitations, material issues etc. In the factory being studied, approximately 33% of the jobs are in conflict at any point in time. It should be noted that the causes of conflict were studied and, for the large part, they were outside of the schedulers' control and could not be addressed in the short-term. A great deal of effort could be put into scheduling algorithms and such, but with so many possible adjustments being necessary, it was considered paramount to quickly set the stage for the schedulers to view and interpret. In a sense, the automatic functionality establishes the baseline schedule for the if-then reasoning. The schedulers need to know immediately from the Gantt chart why jobs are where they are and previous experience with sophisticated algorithms indicated that human schedulers must be able to easily understand a generated schedule and it cannot be magic and mirrors.

The system contains a number of other automatic functions which are not directly associated with schedule generation: importing of data, comparing today's demand with yesterday's to pinpoint changes, updating a network version of the schedule, and so forth. As with the scheduling logic, these types of automated processing functions are not unusual in scheduling systems with the exception of the comparison and exception analyses. In a reactive setting, the schedulers do not have time to study today's requirements to see what is different in terms of dates or quantities and the automated analyses are a way to address this issue.

8.4.2.2. Supporting functionality

The supporting functionality of the scheduling system assists the scheduler in routine actions that have to be carried out recurrently. The system is structured so as to enable the scheduler to rapidly carry out certain actions and to eliminate wasted time and effort. The menu structure supports this work routine and has a start of day menu, morning menu (separated into 7.00a.m, 7:30a.m and 10.00a.m segments), afternoon menu (for 3.00p.m and 6.00p.m), and a weekly menu. In these menus, relevant functions and tasks are listed, and duplicated where necessary. For example, there are functions that progressively work through the lines where work is set and allows the scheduler to input what happened on the last shift. Another example is a function that progresses through the lines and asks if the line is manned, how large is the crew, and if not manned, how many shifts will the line be left idle.

These supporting functions are needed for the rapid establishment of the factory status and to support the if-then reasoning. For example, at 7.00a.m certain information is known and certain partial scheduling decisions are required to be made. The menus are supported by dialogs which are optimized based on the information flow that makes sense at the time. For example, in the 7.00a.m menu, only the information known at 7.00a.m is required. The schedulers' time is not wasted asking for irrelevant data or manipulating unnecessary levels of menus. Neither is the scheduler deluged with noise in the form of information, buttons, and the like, not relevant to the specific task. By having custom dialogs designed based on the information model, the scheduler can rapidly process information and start the if-then decision process.

The information flow analysis supporting this level of functionality was obtained through interviews, report analysis and a significant number of days spent sitting with the scheduler from 6.00a.m onwards. Living with the scheduler was invaluable to the design of the HCI: one knew the sources of information, the conditions under which the information was obtained, what was done manually with the information, what information was missing, and what time allowances existed for the scheduler to process the information.

The supporting and automatic functionality provide the pre-processing for the enabling functionality that is required by the scheduling task. This functionality, which is focused on reacting and adjusting, is described in the next subsection.

8.4.2.3 Enabling functionality

The system supports the human scheduler in many actions that directly affect the allocation of work to resources and the sequence of the work. This is the "if-then" part of the decision-making process. For example, if it is observed that too many jobs are being set up for a shift (and a few other conditions as well), jobs being set up close to the end of the shift will be delayed to the next shift. This example implies two levels of functionality: i) the ability to see the relevant parts of the 'if' (expanded upon in the next subsection), and ii) the ability to rapidly do something about it with enabling functionality.

As with many interactive scheduling systems, the schedulers can cut and paste jobs from line to line and can change the quantities, set-up requirements, processing times, and so forth. These are the context insensitive types of functionality and are quite useful in almost any shop type, from smooth to stress. However, in a stress shop, there are many context sensitive situations and the scheduling system needs to bring some of those to the forefront. Some of the functionality might relate to how capacity is viewed and what forms of capacity have the greatest elasticity. For example, while the lines and main running crews might be strictly guarded for finite capacity, it might be possible to overload (to a degree) the fixture crews. By allowing the schedulers to violate capacity constraints in much the same fashion as they can in the real world, the 'then' part of the decision-making has higher fidelity, which means a more truthful execution of the plan. It is typical of stress shops that they have a high degree of flexibility which is used to combat the uncertainty, but it is at the discretion of the schedulers and management when the degrees of freedom are used.

Another example of decision flexibility is the way set-ups can be manipulated independently of the running of the job. The fixture crews and running crews may be on different shifts and it is also possible for a line to be set one shift and then left idle for several shifts until a running crew is available. This pre-setting balances the fixture crews and ensures that the running crew has a line to work with. The scheduling tool enables this type of decision-making to be made quickly and actually supplies a function that splits a job between set-up and running and inserts a shift delay. After the job is split, the set-up and run segments can be manipulated independently if desired. As noted above, the shop works on a shift view and this was directly captured in the enabling functionality: functions to align work to shift boundaries, functions to enable/disable resources for shifts.

The enabling functionality was derived from the close inspection and analysis of the existing manual system. Not everything they did was supported in the system (e.g., bad habits) but the degrees of freedom were captured and replicated. The scheduling functions took on the noun and verb terminology of the schedulers and were mapped one-to-one whenever possible. It was observed that when several more generic functions were supplied that could be used in conjunction with each other to achieve one meta function, the schedulers inevitably voiced their concern. They would explain their real world function and expect that the scheduling system would have one function to match, thereby saving time and possibly avoiding errors.

8.4.2.4 Information presentation

The schedule is presented in a graphical way, i.e., with 'coloured magnets' on the screen. The normal generic user interface functions that would be expected of any modern interactive scheduling system are provided: click on the schedule, insert a job, inspect the detailed data that a magnet represents, find a job, and so forth. Like other systems, the scheduling tool uses color to represent late work or work that might be late if anything else happens to it (close to late). Beyond these typical features, the stress shop implies that the presentation of information is very important for the 'if' part of the decision process. *Context* is important in stress shops and the scheduling system must easily show the schedulers what they want to see. For example, it is important to know what jobs on the schedule have fixtures that might be in the tool room or leaving the tool room. It is also important to know what jobs are close to a shift boundary. This type of information is what the schedulers talk about between themselves and with others involved in the decision-making process. Therefore it is logical that the scheduling tool highlight or show the type of information they commonly ask about and search out. Similarly, the schedule objectives and constraints are shown at the top and bottom of the schedule for each shift, allowing the scheduler to see changes to the performance metrics and capacity as a change is made to the schedule.

8.4.3 Field feedback

The scheduling system has been used for close to a year by two schedulers working as a team across two shifts. The perception of the factory's management is that the scheduled area is running smoother and that there are less problems than before, and these perceptions are supported by the plant's internal performance tracking. The perception of the schedulers is that the tool is theirs and that it does things the way they think and the way they want it done. They like the way that the tool talks their language and that there are functions and features that match their daily job tasks. They like the way a number of manual tasks were automated with the chance for error eliminated. The schedulers are in a very stressful situation and the tool had to be faster and better than the old way: it had to provide added value to the decision task. They were not forced to use the tool, which means that if the tool was not giving them benefits, they would not use it.

8.5 CONCLUSION

A typology of scheduling situations has been employed to extend the use of the GEMS cognitive model to the scheduling task. Stress shops place high demands on the rule-based part of human information processing, which has consequences for decision support systems supporting a human in a stress shop situation. However, current standard software systems for scheduling are not supporting the typical rule-based reasoning that occurs in stress shops.

A case study was presented of the implementation of a software system in a stress shop. The field study illustrates how a custom system was created to address the context-sensitive issues related to reactive if-then reasoning. It illustrates the types of features necessary to support. The case study also illustrates that decision support for production scheduling tasks in stress shops should focus on rule-based human reasoning. In these shops, many unexpected events occur that cannot be solved by flexibility on the shop floor. This leads to a scheduling task that is aimed at solving a large number of problems.

Designing and implementing a custom scheduling system is not feasible for each and every stress shop. Therefore, research is necessary into how context-sensitive issues can be supported in generic and widely available standard systems. Questions that have to be answered are: how can the task structure be mapped and supported? How can the types of information needed by the scheduler be acquired and presented? How can the system support the degree of operation/task manipulation possible in practice? Future research will focus on these questions.

8.6 ACKNOWLEDGMENT

This research has been supported in part by NSERC grant OGP0121274 on Adaptive Production Control. This research has been done partly at the Institute for Business Engineering and Technology Application (BETA) at Eindhoven University of Technology.

8.7 REFERENCES

Friscia, T. (1997). ERP enters the race. *Manufacturing Systems,* 15(9), 34.

Hess, U. (1995). A plan that works. *APICS - The Performance Advantage,* 5(8), 36-40.

Higgins, P. (1996). Interaction in hybrid intelligent scheduling, *International Journal of Human Factors in Manufacuring,* 6(3), 185-203.

McKay, K. N. (1992). *Production Planning and Scheduling: A Model for Manufacturing Decisions Requiring Judgement.* Waterloo, Ontario: University of Waterloo, Ph.D. Thesis.

Melnyk, S. (1995). 1995 Finite capacity scheduling survey. *APICS - The Performance Advantage,* 5(8), 60-70.

Rasmussen, J. (1986). *Information Processing and Human-Computer Interaction: An approach to cognitive engineering.* Amsterdam: North–Holland.

Reason, J. T. (1990). *Human Error.* Cambridge: Cambridge University Press.

Sanderson, P. M. (1989) The human planning and scheduling role in advanced manufacturing systems: An emerging human factors domain, *Human Factors,* 31(6), 635-666.

Sanderson, P. M. (1991). Towards the model human scheduler. *International Journal of Human Factors in Manufacturing*, 1(3), 195–219.

Schaaf, T. W. van der (1995). Human recovery of errors in man–machine systems. In T. B. Sheridan (Ed.), *Proceedings of the 6th IFAC/IFIP/IFORS/IEA Symposium on Analysis, Design, and Evaluation of Man–Machine Systems,* USA, June 27–29, Cambridge, MA: Massachusetts Institute of Technology, pp. 91-96.

Sheridan, T. B. (1980). Theory of man–machine interaction as related to computerized automation. In E. J. Kompass and T. J. Williams (Eds.), *Man–Machine Interfaces for Industrial Control,* Barrington, IL: Control Engineering, pp. 35-50.

Weil, M. (1997). Degrees of freedom, *Manufacturing Systems*, 15(9), 72-80.

Wiers, V. C. S. (1997a) A review of the applicability of OR and AI scheduling techniques in practice. *OMEGA – The International Journal of Management Science,* 25(2), 145–153.

Wiers, V. C. S. (1997b). *Human–Computer Interaction in Production Scheduling: Analysis and design of decision support systems for production scheduling tasks.* Wageningen, The Netherlands: Ponsen & Looijen, Ph.D. Thesis, Eindhoven University of Technology.

Wiers, V. C. S. and McKay, K. N. (1996). Task allocation: human-computer interaction in intelligent scheduling. In *Proceedings of the 15th Workshop of the UK Planning and Scheduling Special Interest Group, UK,* November 21–22, Liverpool, UK: Liverpool John Moores University, pp. 333-344.

Wiers, V. C. S. and Schaaf, T. W. van der (1997). A framework for decision support in production scheduling tasks. *Production Planning and Control,* 8(6), 533–544.

CHAPTER NINE

Human Factors in the Planning and Scheduling of Flexible Manufacturing Systems

Jannes Slomp

9.1 INTRODUCTION

An FMS (Flexible Manufacturing System) is a computer-controlled system consisting of automated workstations linked by a material handling system and capable of processing different jobs simultaneously. The FMS concept evolved in the early 1970s. During the 1980s, there seemed to be general agreement that the use of FMS would spread widely. The general idea is that FMS enable firms to achieve the efficiency of automated, high-volume mass production while retaining the flexibility of low-volume job-shop production. Despite the potential advantages of FMS, their use is spreading somewhat slowly. This is mainly because flexible manufacturing systems have not been as profitable to users as some other innovations (Mansfield, 1993). Technical, organizational, and environmental (changes in the market place) problems have also had a negative impact on the profitability of FMS (Slomp, 1997). The less-than-expected profitability of FMS has negatively affected the imitation rate in industry. It is, however, expected that the majority of firms in the automobile, electric equipment, machinery and aerospace industries in Japan, the United States and Western Europe will install FMS in the course of time.

Jaikumar (1986), describing application in several countries, found that industry worldwide seems to prefer small FMS with large buffer storages for pallets/products and simple routings of workpieces. Often, the smaller systems offer the advantage of unmanned or lightly manned production for some period of time. The trend towards smaller FMS in the last decade of the 20th century was further stimulated by developments in manufacturing technology. The productivity of Computer Numerical Control (CNC) machines has been increased as a result of more revolutions per minute and improved cutting tools. Furthermore, there is a tendency to integrate more manufacturing techniques into one single machining centre. Apart from these technological changes, the move towards modular components of Computer Integrated Manufacturing (CIM) connected through intelligent (AI) systems also reduces the advantages of large FMS. There is a trend towards intelligent, open, modular, low cost systems (Kopacek, 1999).

The contribution of this chapter is that it explores the human factors which play a role in the planning and scheduling of FMS. Several authors have reviewed literature on planning, scheduling and control of FMS (Dhar, 1986; Shanker and

Agrawal, 1991; Kouvelis, 1992; Rachamadugu and Stecke, 1994). All these reviews focus on the technical aspects of planning, scheduling and control (i.e. Operations Research approaches). The importance of human factors is not recognized in this stream of research. Rachamadugu and Stecke (1994) categorize FMS from a scheduling point of view, and basically discern two types: Flexible Flow Systems (FFS) and General Flexible Machining Systems (GFMS). An FFS can either be a Flexible Assembly System (FAS) or a Flexible Transfer Line (FTL). A GFMS can be subdivided into Dedicated Flexible Manufacturing Systems (DFMS) and Nondedicated Flexible Manufacturing Systems (NFMS). These types of FMS can be described in terms of their system, environmental, and operational characteristics, (see Table 9.1). This chapter focuses on the relation between human factors and the planning, scheduling and control issues of small FMS of the NFMS type.

Section 9.2 summarizes the human factors related with automation. This can be seen as the broad context of the chapter. Section 9.3 presents a generic framework for planning, scheduling and control of FMS and indicates how responsibilities can be divided in the organization. Section 9.4 presents a longitudinal case study concerning a small FMS installed in a company in the east of The Netherlands. The case study is meant i) to illustrate the planning and scheduling complexity of a small FMS, ii) to show how the complexity may change in the course of time, iii) to illustrate the impact of these changes on the human factors related with planning, scheduling and control of the FMS, and iv) to indicate some major human factors which are only marginally dealt with in literature. Sections 9.5, 9.6 and 9.7 elaboureate on these human factors. Section 9.5 deals with the working times of operators and the impact on planning, scheduling and control. Section 9.6 concerns the division of planning, scheduling and control tasks among employees. Section 9.7 discusses the relation between operator tasks and machine tasks. Finally, Section 9.8 is a résumé section.

9.2 HUMAN FACTORS AND AUTOMATION

Automation has a huge impact on people's working situation. Conversely, the working situation of people greatly influences the success of manufacturing automation. Boyer *et al.* (1997) investigate, among other things, the effect of worker empowerment, quality leadership, and increased human co-ordination ('soft integration') on the performance of automated manufacturing technology. They show that quality leadership and human co-ordination significantly contribute to the performance (growth and profit) of automated manufacturing technology. The effect of worker empowerment is less visible in their survey, but they conclude that managers of companies with large investments in automated manufacturing technologies must not neglect this in infrastructure improvement programs. Small and Yasin (1997) show that firms that had spent greater effort on developing human factors appeared to gain more of the benefits of Automated Manufacturing Technology (AMT) than their counterparts. These human factors involve elements of worker preparation, such as the establishment of multi-skilled production workers, communication about the impact of AMT to all plant workers,

Table 9.1 Types of FMS and their characteristics

Type of FMS	System and Environmental Characteristics	Operational characteristics
Flexible Flow System (FFS)	• Is dedicated to produce a specific set of (few) part types (FTL) of assembled product (FAS) • Material flow is usually unidirectional • Part types are asked for in relatively large volumes	• Performance indicators: System outputs and machine utilization • Main operating problems: Determination of the production ratio (only if not defined externally) and determination of the input sequence
General Flexible Machining System (GFMS)		

• Dedicated Flexible Machining System (DFMS)	• Is able to produce a wide variety of part types simultaneously
	• Part routings can be different, even for parts of the same type
	• Few part types are asked for in particular ratios
		• Performance indicators: System output and machine utilization
		• Main operating problems:
		  • Determination of the input ratio
		  • Determination of the part input sequence
		  • How to keep the bottleneck machine busy

• Nondedicated Flexible Machining System (NFMS)	• Is able to produce a wide variety of part types simultaneously
	• Part routings can be different, even for parts of the same type
	• Part types and production requirements change frequently
		• Performance indicators: Throughput time, machine utilization and meeting due dates
		• Main operating problems:
		  • Dealing with routing flexibility
		  • Dealing with large number of orders with different due dates
		  • Dealing with limitations concerning fixtures, cutting tools and tool magazine capacity

emphasizing teamwork and group activities, and pre-installation training for all project participants. Also the establishment of team-based project management contributes to the success of automated manufacturing technology.

Several researchers have investigated elements of the relation between human factors and automation and human factors or ergonomics has much to offer the design of manufacturing systems. For instance, in Salvendy's Handbook (1997), the section on Job Design is probably related the most to the issues covered in focused literature on 'human factors and automated manufacturing'. This section includes chapters on function allocation, task analysis, mental workload, job and team design, participatory ergonomics, models in training and instruction, computer-based instruction, organizational design and macro-ergonomics, and socially centred design. Some of these factors will be dealt with below.

A stream of research on human factors and automation is based upon a fundamental critique of the technologically-oriented approach of the development of manufacturing systems. Brödner (1985) makes a plea in favour of skill-based design that starts from the idea that it is necessary in the development of manufacturing systems to find a division of functions between people and machines that reflects their strengths and weaknesses. This approach, which is embraced by many others (see e.g. Ravden *et al.*, 1986; Corbett *et al.*, 1990), stresses the need to integrate human factors design criteria in the process of systems design. Ravden *et al.* (1986), distinguish the following human aspects: human factor objectives, system design process, allocation of functions, job design, organizational structure, hardware ergonomics, software ergonomics, environmental ergonomics, and human factors evaluation. These aspects need to be integrated into an overall design process of flexible equipment. Siemieniuch *et al.* (1999) propose a methodology which may support ergonomics to define roles and organizational structures at the time the engineers are developing the technical view of the facility. The methodology starts from the premise that an organization can be construed as a configuration of knowledge, embodied in humans and machines. A 'set of rule-sets' and 'paste functions' are supportive to allocate functions and to link the various roles.

McLoughin and Clark (1994) have written a research book on 'technological change at work'. They distinguish three main areas in the change of work caused by new technology: i) work tasks and skills, ii) job content and work organization and iii) supervision and control. Based on case studies and literature, they argue that new technology reduces the number of complex tasks requiring manual skills and abilities and generates new complex tasks which require mental problem-solving and interpretive skills and abilities and an understanding of system interdependencies. Tacit skills and abilities associated with the performance of work with the old technology, however, are still required. Although characteristics and capabilities of technology independently influence task and skill requirements, the content of jobs and the pattern of work organization are important subjects to managerial choice in the design of work. McLoughin and Clark (1994) argue that new technology should have a complementary, supportive, role in the jobs of workers, instead of playing a replacement role only. New technology asks for a detailed consideration of work content and work organization. McLoughin and Clark (1994) illustrate two directions with respect to the organization of supervisory control tasks. In some instances line-managers have attempted to use new technology to make operations more visible and to improve the certainty and confidence of management decision-making. This has resulted in a centralization of the control of work operations. In other instances managers have sought to use new technology to lay greater emphasis on the delegation of decision-making, to points close to the production process itself. This was partly due to recognition of the fact that some new technologies eroded aspects of the role of first-line supervisors and were conducive to a degree of team autonomy among workgroups.

Benders (1993) wrote a research book on work design and manufacturing automation. Using a few case studies, he illustrates which organizational choices have to be made with respect to the division of labour around FMS. He distinguishes four types of workers the work can be allocated to i) the internal generalist, i.e. an FMS operator, ii) an internal specialist, i.e. another employee, employed within the manufacturing department, iii) an external generalist, i.e. an employee, employed in another department than manufacturing, who has a variety of tasks, and iv) an external specialist, i.e. an employee employed in a department other than manufacturing and specialized in certain tasks. The case studies presented by Benders (1993) show a variety of work allocations. A completely integrated job function, however, is not found in the cases. The complexity of some tasks (e.g. programming) makes it difficult to allocate these tasks to generalists. Furthermore, in the opinion of the management, the assignment of additional tasks (e.g. maintenance, programming, planning and scheduling) to operators may obstruct the efficient and effective use of expensive machinery.

Several authors indicate that the role of humans in modern production systems moves towards system monitoring and supervisory control. Executive tasks are to a large extent automated, or will be automated. This has an impact on the focus of some research in the area of human factors. Barfield *et al.* (1986) study human-computer supervisory performance in the operation and control of flexible manufacturing systems. They distinguish four main modes of the supervisor's function: planning (scheduling, inventory control, and capacity planning), teaching (programming, job prioritizing, dispatching), monitoring (tool life status, machine status, conveyor condition, and parts flow), and intervening. They subsequently develop a type of Fitts List (Fitts, 1951) which may be helpful for the allocation of supervisory tasks to humans or computers. With respect to the planning and scheduling task of supervisors, they mention the intuition of humans to determine algorithms and their ability to set criteria and limits. This ability of human operators is confirmed in a study of Kopardekar and Mital (1999). They show that human intuition may outperform the best possible mathematical approximation method in a cutting stock problem. This indicates the ability of humans to deal effectively with complex problems in an intuitive way. Adlemo (1999) presents a control system for FMS suitable for human operator interaction. Ammons *et al.* (1986) report that the role of people is only rarely considered while designing decision support tools, or planning and scheduling tools, for FMS. In their opinion, this has led to a recurring bad fit between human operators and the automated production control functions of an FMS. They subsequently argue that the tasks of people should be designed and described as precisely as the algorithms applied in the computer software. A nice example of such an approach is given in an article by Sharit and Elhence (1989). They discuss the computerization of tool-replacement decision-making in FMS from a human-systems perspective and conclude that a human-computer co-operative arrangement has the potential to realize the best balance between the more direct local consequences and the less direct global consequences of tool-replacement decisions on economic and throughput objectives. The supervisory role of humans in modern manufacturing systems has also stimulated research on the mental workload of operators in automated systems. Hwang and Salvendy (1988) show that task complexity is an important predictor of the mental workload of operators of advanced manufacturing systems. Lin and Hwang (1998) establish an index to measure mental workload in supervisory tasks. They demonstrate that the index can be used to predict the mental workload of FMS operators. They reveal that 'span', 'discriminate', 'predict' and 'transfer attention' are important elements relevant to the mental workload of supervisory tasks of humans in an automated manufacturing system. Their study concentrates on the control function of operators. Scheduling tasks are not taken into account. Nakamura and Salvendy (1988) focus solely on the scheduling task within FMS. They emphasize that 'predicting' is the most important mental workload element in human decision-making.

This chapter is meant to indicate the relation between human factors and the planning, scheduling and control of an FMS. A longitudinal case study presented in Section 9.4, serves as a means to uncover the relation. Only the structural and visible changes in human factors and planning, scheduling and control are documented and will be analyzed. Three factors, which have direct impact on people's working situation and which relate to planning, scheduling and control of advanced manufacturing systems, are distinguished and will be explored further in separate sections (9.5, 9.6 and 9.7). These factors concern i) the working times of employees, ii) the division of planning, scheduling and control tasks among employees, and iii) the relation between operator tasks and machine tasks. This chapter does not address issues of mental workload. The case study and the subsequent analysis, are basically meant as a useful input for human-centred research on planning, scheduling and control for advanced manufacturing systems. The next section will explore the meaning of planning, scheduling and control of FMS.

9.3 PLANNING, SCHEDULING AND CONTROL OF FMS

The planning, scheduling and control of FMS have gained significant attention in operations management literature. To cope with the complexities of an FMS, most authors propose a hierarchical framework for the planning, scheduling and control of FMS. Based upon literature (Stecke, 1985; van Looveren *et al.*, 1986) and organizational practice in many companies, a generic description can be given of the decision hierarchy of the production control for an FMS (Slomp, 1997). The hierarchy consists of three decision levels: i) the assignment level, ii) the off-line level, and iii) the on-line level (see Figure 9.1). On the assignment level, orders are distributed among the various production units, including the FMS. The assignment level is responsible for generating realizable throughput times and a realistic workload of the FMS. In many practical situations, the activities on the assignment level are executed periodically, for instance monthly, weekly, or daily. Often a Manufacturing Requirements Planning (MRP) system is used on the assignment level. Within the assignment level it may be difficult to take account of the specific characteristics (or limitations) of the FMS. In an MRP system, for instance, an FMS is merely seen as a capacity resource. There have been few attempts to integrate FMS production planning problems into a closed-loop MRP system (see, for instance, Mazzola *et al.*, 1989). The off-line level receives orders from the assignment level. The off-line level is responsible for getting a good fit between the orders received and the characteristics (limitations) of the FMS. Off-line decisions are based upon a model of the FMS, in which the most important characteristics (limitations) of the FMS are incorporated. Most off-line decisions are performed periodically, for instance monthly, weekly, or daily. The off-line decisions may concern the batching of orders to be produced in the same period and the loading (assignment) of operations and tools to the various workstations. Sometimes, scheduling is also done on the off-line level. On-line activities are based on the information from the off-line level *and* on the actual status of the FMS. On-line decisions are activated by real-time information from the FMS and can be taken at any moment. The decisions concern the release of orders to the FMS and the sequence in which competitive activities have to be performed (dispatching). The extent of detail on the off-line level determines the degree of freedom on the on-line level. An alternative for batching, loading, and release is the so-called 'flexible approach' (Stecke and Kim, 1988), where decisions about the actual order mix are based primarily on the actual status of the system with the objective of optimizing certain performance measures such as machine utilization and due date performance. Basically, the flexible approach can be seen as an intelligent form of releasing. The three-level hierarchy can be recognized in most production control systems. Differences are often due to the location of certain subproblems on another level (for instance on-line loading) or even the complete absence of the off-line level.

The decision levels are determined by the information flow required to and from the FMS (off-line versus on-line) and the knowledge needed about the specific characteristics of the FMS (global versus detailed) (see Figure 9.1). These aspects are also important for possible assignments of decision tasks to the organizational levels of the firm. The right part of Figure 9.1 illustrates how the levels of the control hierarchy may correspond with the levels in the organization hierarchy. The planning department has a global knowledge of the FMS and bases its assignment decisions on off-line, periodic, information of the FMS. The foreman has detailed knowledge of the FMS, but his production control decisions may be based on periodic, off-line, information of the FMS. The operators of the FMS have detailed knowledge of the FMS and their production control decisions are based upon real-time information of the FMS. Figure 9.1 is just an example of the division of tasks. As will be seen in the case study of Section 9.4, there are various ways to divide planning, scheduling and control tasks among employees.

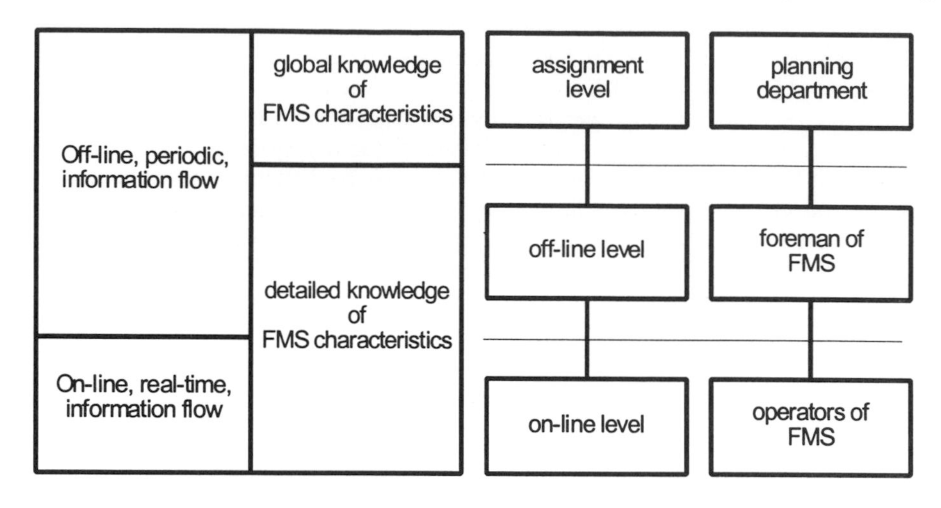

Figure 9.1 Characteristics and hierarchy in planning, scheduling and control in FMS.

9.4 A LONGITUDINAL CASE STUDY

The longitudinal case study is based on an existing firm that has installed a small flexible manufacturing system. The case study covers a period of four years, from 1988 to 1992, and offers an illustration of the impact of several human factors on the planning and scheduling of an FMS and vice versa. From 1992 to 2000, there were no significant changes in the organizational and/or planning, scheduling and control situation. The firm is now searching for a replacement of the FMS.

The firm described in this paper designs, produces, and sells pneumatic actuators for valves. In 1988, the company employed about 100 people producing 80,000 actuators, of which 80% were exported. The firm distinguishes 12 basic types (sizes) of actuators. Each basic type can be delivered in several variants. The most important parts of an actuator are its housing, two pistons, two end covers, and its drive shaft. The diversity of actuators (i.e., the variants of an actuator type) is highly determined by the actuator housings since these have to fit onto the specific equipment of the customer. In 1988, the firm faced a rapid increase in both volume and mix (diversity), resulting in, amongst other things, a dramatic increase in required metal cutting capacity. Mid-1988, the firm bought a flexible manufacturing system (FMS) for the manufacturing of the housings.

Figure 9.2 gives a schematic representation of the particular FMS involved. Identical machining centers are linked by a pallet transport vehicle and an integrated pallet buffer system with a capacity of 28 pallets. Each of the machining centres can hold a maximum of 40 cutting tools in the tool magazine. A tool robot performs the (on-line) tool changing between a central tool storage and the tool magazines. An FMS-computer takes care of the co-ordination of all activities within the FMS. The clamping and unclamping of parts on pallets is done manually on two integrated

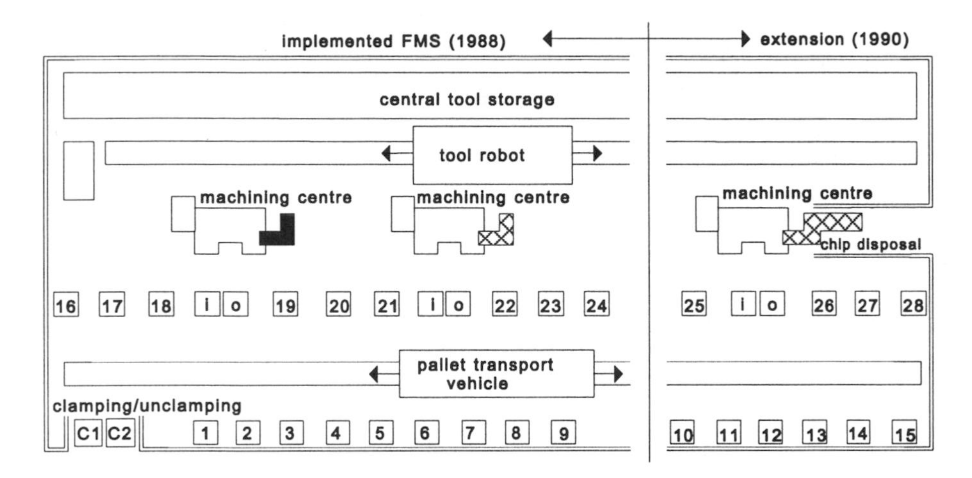

Figure 9.2. Schematic representation of the FMS.

load/unload (C1/C2) stations. Before manufacturing of an order can start on the FMS, an operator has to i) prepare the required pallet(s)/fixture(s) for the specific order, ii) build up and pre-set the required cutting tools, iii) load the central tool storage with these tools, and iv) instruct the FMS-computer concerning the order release. Next, one or more items of the order can be clamped on a pallet/fixture. This is done on a clamp/unclamp station (C1, C2). The pallet transport vehicle transports the palletized items either to the pallet pool (1-28) or directly to a machining centre (M1, M2 or M3). After machining, the pallet transport vehicle transports the palletized items either to the pallet pool or directly to a clamp/unclamp station. On this clamp/unclamp station the items are replaced by new items. The refilled pallet/fixture repeats the cycle through the system. The (unclamped) items on which an operation has been performed, may wait for their next operation.

In principle, each order (batch of identical parts) is assigned to just one pallet/fixture combination. All parts of the order need this pallet/fixture (for several part types there even exists just one unique pallet/fixture combination). Therefore, to avoid machine idle time caused by the sometimes large amount of time needed to unload and subsequently load a pallet, batches of different orders are processed in a so-called order-mix routine, e.g. A-B-C-A-B-C- etc. This aspect complicates the scheduling of orders: the time needed for an order depends on the order-mix.

The development of and around the FMS can be split up into three stages, which are briefly described as follows.

9.4.1 The 'two machining centre' stage

In the middle of 1988, the firm installed an FMS consisting of two identical machining centres. The most important advantages of installing the FMS were:

- *A more efficient manufacturing of housings.* The FMS is an efficient system. First, all the preparatory work, such as the clamping of parts on pallets/fixtures and the presetting of tools, can be done during machining time. Second, one worker is able to operate two machines. Third, the FMS is able to operate in unmanned periods.
- *An improved flexibility with respect to the housing types processed (sizes of the housings).* The FMS is able to produce the different housing types in a mix without loss of production time. Consequently, the firm can respond more rapidly to rush orders (as long as sufficient materials are available).
- *A reduction in the total lead-time and, consequently, a reduction in the amount of work-in-process.* The total lead-time is directly related to the number of processing steps; for each processing step a fixed amount of time is reserved by the production control system. Several processing steps can be performed by the FMS. This has resulted in a reduction in the total lead-time of the housings from 9 to 6 weeks.

The above advantages are directly related to the characteristics of automated manufacturing: automation/computerization, flexibility, and integration.

In order to utilize the FMS optimally the firm decided to adapt two-shift manufacturing in combination with a partly unmanned night shift. As in many automated systems, the human tasks related to the FMS show some polarity: there are simple tasks (such as clamping, unclamping, deburring, pre-setting of tools) and complex tasks (such as maintenance, NC-programming, and scheduling of jobs). The management of the firm decided to integrate the maintenance and scheduling tasks, i.e. the off-line planning task, in the job responsibilities of FMS-operator. An important argument which supported the decision was the fact that operators are aware of all the operational constraints such as the presence of an order-mix on a machine tool and the limited availability of cutting tools and fixtures (i.e., a unique cutting tool cannot be used at more than one machine tool simultaneously and a unique fixture can only be used by one production order). Also, the operators are close to the system and know the 'special sounds' of the machine tools which require maintenance action. The operators of the FMS experienced their jobs as a promotion: they became systems operator instead of machine operator. At the firm, a technical specialist was made responsible for the design of fixtures and the writing of NC-programs. This specialist works in day shift, but is on call outside working hours. The supplier of the FMS is responsible for major, urgent repairs.

9.4.2 The 'three machining center' stage

An important feature of the two-machine FMS was its modular structure. It was relatively easy to enlarge the pallet buffer and to add an extra machining centre. In 1990, the management of the firm decided to increase the capacity of the pallet buffer from 15 pallets to 28 pallets, and to implement an extra machining centre. These investments were occasioned by increased sales and the wish of production management to manufacture more part types on the FMS, such as the pistons of the actuators. The FMS was viewed as an efficient manufacturing system.

The investments appeared to have a significant impact on the operation of the FMS. The polarity of operator tasks became more apparent. The number of relatively simple manual tasks, such as clamping and unclamping, had been increased, and the scheduling and control tasks appeared to be more complex. The increase in the number of machining centres emphasized some of the most difficult scheduling constraints, such as the presence of unique cutting tools and the limited number of fixture types.

In spite of the further polarity of operator tasks, the management of the firm did not change the work content of FMS-operator. In the group of FMS-operators, however, a natural division of labour took place in which some operators only performed clamping/unclamping and burring tasks and other operators spent a significant amount of their time on controlling tasks. Because of the higher workload, more operators worked at the FMS simultaneously.

9.4.3 The 'continuous operation' stage

Early in 1991, a further increase in the FMS capacity seemed desirable. Sales were still growing. The management of the firm considered the purchase of a fourth machining centre and more buffer storage for pallets, to attain a better utilization of the unmanned night shift. However, the firm eventually did not choose to make such an additional investment. The increasing complexity of the system and the operational problems expected, were seen as major obstacles. Instead, the firm opted for an organizational solution.

Until 1991, the FMS was functioning in a two-manned shift mode and a partially unmanned third shift. The firm was operating 5 days a week. After consulting the trade unions and the operators, the management decided to change the work scheme for the FMS. Since mid-1991, the FMS has been running 24 hours a day, 7 days per week. Each day includes two short unmanned periods of three hours. Four teams each consisting of three operators are working at the FMS. Table 9.2 shows the scheduling of the teams over the weeks and the daily working scheme. A disadvantage of the daily working scheme is that the teams do not meet each other when changing shifts.

The impact of the new work scheme was quite significant. More people were needed to operate the FMS. The capacity of the FMS increased by about 70%. Several new fixtures were designed (and NC-programs written) in order to manufacture other components of the pneumatic actuators on the FMS. Also, some components of electric actuators were assigned to the FMS: in 1990, the firm took over another firm producing electric actuators. The new way of working appears to be economically advantageous. At the moment, over 10% of the work performed on the FMS is subcontracted from other firms; the firm is able to compete successfully with firms in low-cost countries.

Because of the 'continuous operation', the working situation of the operators has changed significantly. The management of the firm decided to reorganize the functions around the FMS. In the 'two and three machining centres' stages, the operators of the FMS were responsible for the clamping/unclamping activities as well as for the scheduling of operations and some technical maintenance and repair tasks.

In the 'continuous machining centres' stage, the operators remain responsible for clamping/unclamping activities and the monitoring of the system while three employees/supervisors, working in two shifts, are made responsible for the writing and testing of NC-programs, the preventive and corrective maintenance of the system, the planning and scheduling of orders on the system (including the selection of orders for the unmanned periods), and the administration of new and finished orders for the FMS. Parallel to the transition from the 'three machining centres' stage to the 'continuous operation' stage, a spreadsheet-based decision support tool was designed to facilitate the production control tasks of the three supervisors (Gruteke, 1991. i.e. the off-line planning). This tool only provides the necessary order information and enables the supervisors to deal with the information efficiently. There is no algorithm included in the spreadsheet-based decision support tool. The planning and scheduling task requires a supervisor, in spite of the availability of the decision support tool, approximately 4 hours per day.

Table 9.2 Working periods and times in the continuous operation stage

Scheduling of the teams over the weeks (in days per week)								
	Week 1		Week 2		Week 3		Week 4	
	Early	Late	Early	Late	Early	Late	Early	Late
Team 1	5	0	0	4	0	2	3	0
Team 2	0	2	3	0	5	0	0	4
Team 3	2	0	0	3	0	5	4	0
Team 4	0	5	4	0	2	0	0	3

Daily working scheme of the FMS:

Early shift: 03.00 a.m. – 12.00 a.m.
Unmanned period: 12.00 a.m. – 03.00 p.m.
Late shift: 03.00 p.m. – 12.00 p.m.
Unmanned period: 12.00 p.m. – 03.00 a.m.

Total operating time per day: 24 hours

The division of labour simplified the recruitment of new operators. The required educational level of operators is lower, which brought in operators on a lower salary scale. Another important argument in favour of reallocation of tasks was the increased complexity of the system.

The production manager indicated that the operators are highly content with the new work scheme. They work on average 29.75 hours per week, excluding a thirty-minute break per shift, and earn a full weekly salary; the shift bonus is 30%.

Furthermore, the new work scheme has changed the home situation of several operators; it offers the possibility of another division of labour at home!

9.5 WORKING TIME

The case study presented in this chapter has indicated that the working times of operators may change after the introduction of an FMS. Many firms decide to move to a multi-shift situation in order to utilize their automated systems as much as possible. This has an important impact on the planning, scheduling and control of automated systems. People in a planning department are usually working in a day shift mode. This means that they will not be able to react immediately to those changes on the work floor which occur during evening and night shifts. For this reason, operators of FMS are usually given some responsibility for the scheduling and control of orders on the FMS.

FMS offers the possibility of producing in an unattended mode. The period in which unattended manufacturing is possible is usually limited because of several constraints of the system. These constraints may be, for instance, the capacity of the cutting tool magazine and the availability of fixtures on which work pieces have to be clamped. In order to maximize the number of productive hours in an unattended period (e.g. during the night), a decision support tool may be helpful. It is conceivable to include an Integer Programming Model (Operations Research) in the decision support tool. The firm in the case study decided to maximize the productive, unattended hours in a more organizational way. Short unattended periods (3 hours per period) were created, for which it was relatively easy to select work pieces. A decision support tool with an advanced algorithm was not needed. It was sufficient to apply a spreadsheet-based information system.

The possibility of unattended productive periods increases the load of the workers at the FMS. Next to the regular clamping and unclamping activities, operators have to remove the work pieces performed in a previous unattended period and they have to prepare new loaded pallets for the coming unattended period. This extra loading of the operators, in connection with the uncoupling of the activities of operators and machines (which is inherent to new technology), stresses the need to simultaneously plan and schedule machine and operator tasks in order to utilize human and machine capacities as fully as possible. This aspect will be dealt with in Section 9.7. The extra loading of operators also supports a further division of tasks among employees. This aspect is dealt with in the next section.

9.6 DIVISION OF TASKS AMONG EMPLOYEES

The human tasks resulting from the automated equipment show some polarity. Some tasks are relatively simple (clamping, unclamping, deburring, presetting tools) while others are complex (NC-programming, testing new products, maintenance, planning and scheduling) and need training and education. This polarity complicates the design of integrated operator jobs. An employee who is highly educated is not interested in performing simple tasks for several hours per day. This fact stimulates the internal and external division of labour. Internal division refers to the existence of internal specialists and generalists. External division concerns the assignment of tasks to external specialists or generalists (Benders, 1993). The case study has shown that the complex tasks are less related to the timing of machine tasks than the simple ones are. This has made it possible to allocate the complex tasks to the three supervisors, in the 'continuous operation' stage, while leaving the simple tasks with the operators. Because the three supervisors do not work in the shift system presented in Table 9.2, the planning and scheduling tasks have to be performed in a periodical mode instead of continuous mode. This may lead to a discrepancy between plan and reality. Discrepancies are limited in the case of a reliable manufacturing system, however, in case of disturbances, for instance caused by a machine breakdown, it should be clear for operators how to reschedule activities. A certain (re)scheduling responsibility (release and dispatching) remains on the shop floor.

An example of a division of planning, scheduling and control tasks among several employees is presented in Figure 9.3. The production control hierarchy was designed and implemented for a similar type of FMS to that described in Section 9.4 (see Slomp and Gaalman, 1993). The hierarchy basically consists of four levels.

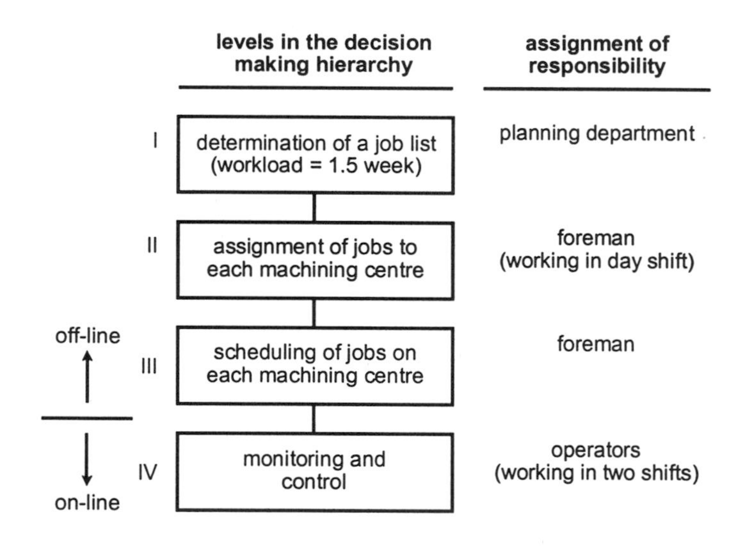

Figure 9.3 Production control hierarchy for an FMS.

Level I. At the beginning of each week, an MRP print-out is generated which contains all future orders of the FMS. The planning department selects a subset of these orders for immediate release to the FMS. This selection is based upon the due dates generated by the MRP system and the workload of the FMS. The total amount of work-in-progress is limited to 1.5 weeks. The FMS should manufacture each order within at most two weeks. In this way the FMS has a slack of 0.5 weeks to produce all orders in time.

Level II. The foreman, who works on day shift, allocates the released orders to each machining centre. Several aspects complicate this allocation activity. As far as possible, order allocation should be done such that each fixture and each unique cutting tool is needed on just one machining centre. By doing this, the machining centres can be considered independently of each other during the remainder of the procedure.

Level III. The third level in the production control hierarchy deals with the scheduling of orders on each machining centre. The human-scheduler (= foreman) has to deal with several aspects, such as the limited capacity of the tool-magazine and limited availability of pallets/fixtures. Orders which need the same pallet/fixture should not be scheduled sequentially since this would cause idle time. Finally, the throughput time of an order should not exceed two weeks. In this particular case, an interactive scheduling tool was developed to support the foreman's decision-making, (see Slomp and Gupta, 1992).

Level IV. The operators are responsible for the monitoring and control of the FMS. In case of disturbances, they are allowed to reschedule tasks on machines. The output of the interactive scheduling tool gives information required for rescheduling.

The assignment of production control tasks to the levels of the organization hierarchy plays an important role in solving production control problems. The assignment, for instance, determines the quickness of reaction on breakdowns and/or rush orders. Figure 9.3 should be seen as just one example of assigning production control tasks to the levels of the organization. The production control hierarchy has to be seen as the starting point for the design of decision support tools and an information system. The need for decision support tools depends on the complexity of the problems at each level of the production control hierarchy and the qualifications of the involved employees. Section 9.2 has indicated some human factors which need to be taken into account while developing decision support tools.

9.7 OPERATOR TASKS AND MACHINE TASKS

In modern manufacturing systems, machine tasks and operator tasks are more or less uncoupled. The availability of a pallet storage, or pallet pool in the FMS offers the possibility for unattended productive periods. It also releases the operator from the pace of the system during attended periods. Furthermore, it enables the situation in which one worker operates more than one machine. In order to reduce the number of operators and to avoid interference between the activities of machines and operators, the planning and scheduling system has to deal with machine tasks and human tasks simultaneously. There are basically three options for an FMS scheduling system to deal with operator tasks:

1. The scheduling system may schedule only machine tasks and may assume that operators have time available to do their tasks in time. Most scheduling procedures presented in the literature implicitly embrace this assumption (see Rachamadugu and Stecke, 1994).

2. The scheduling system may schedule operator tasks in the same way as machine tasks. Slomp *et al.* (1988) present a quasi on-line scheduling procedure for FMS in which an operation is defined as the combination of operator tasks, transport device tasks, and machine tasks, which are all related to the manufacturing of a work piece on a machine. Operations are designed, selected and scheduled sequentially. By doing so, temporary overload of operators, transport devices and machines is avoided.
3. The scheduling system may include the concept of time windows for operator tasks. A time window is defined as the period in which a task has to be performed. The scheduling of machine tasks fixes the earliest possible starting times and the latest needed finishing times of several operator activities (e.g. clamping, unclamping, and deburring). It determines the time windows of tasks. A scheduling system may take care of minimal time windows for operator tasks (see Slomp and Gaalman, 1998).

The above options reflect different philosophies in the design of new technology (see also the discussion in Section 9.2). The first option fits a technologically-oriented approach in which human activities follow technological developments. The second option concerns a technical approach to man-machine systems into design where human beings are seen as resources needed to perform machine operations. The third option fits in a human-centred oriented design methodology based on the needs (time windows) of workers.

9.8 RESUME

This chapter has described the relation between human factors and the planning, scheduling and control of flexible manufacturing systems. First, an overview is given of relevant literature on human factors and automation. Second, the elements of a planning, scheduling and control system of an FMS are explained briefly. Next, a longitudinal case study is presented to illustrate the complexity of the planning and scheduling of a small FMS, with illustrations of how some important human factors relate to the planning, scheduling and control of the FMS. Three major human factors have been distinguished in this chapter. They can be labelled as the *'working time aspect'*, the *'division of labour aspect'*, and the *'operator-machine relation aspect'*. These factors are dealt with in separate sections. The *working time aspect* relates to the fact that FMS are usually performing in a multi-shift system, sometimes with unattended productive periods. The fact that operators work in a multi-shift system forces a central planning department, which usually performs in the day shift, to plan in a periodical mode. Every day, or week, a production plan has to be released. The presence of unattended periods may require a sophisticated decision support tool, at the operator or at the planning level, to maximize the productive time in these periods. Another solution to the utilization problem may be the settlement of shorter unattended periods in a day period. The selection of appropriate orders for these shorter periods may be simple. Unattended periods, in general, will increase the workload for all task types (i.e. clamping, unclamping, but also scheduling and NC-programming) and lead to further polarity of complex and simple tasks. This may support a (further) division of labour. The presence of short unattended periods, furthermore, asks for careful operator planning. The *division of labour aspect* concerns the allocation of planning, scheduling and control activities to the various levels of the organization. The polarity of tasks (some tasks are simple while others are complex and need training and education) and the relation of these tasks to the timing of machine tasks have an important impact on the final allocation of tasks. Analysis of these elements may lead to a situation in which the major part of the planning and scheduling task will be performed in a central planning department. Operators, however, need to have the authority to reschedule in case of disturbances. The division of decision tasks among the levels of the organization has to be seen as the starting point for the design of decision support tools and an information system. The *operator-machine relation aspect* refers to the fact that the loose coupling between operator tasks and machine tasks in automated manufacturing systems will challenge the planning and scheduling system to optimize the use of operator and machine capacities. Three options for the planning and scheduling system are mentioned. First, the scheduling system may assume that operator capacity is not a limited resource and is always available. Second, the scheduling system may deal with operator tasks in a similar way to machine tasks. Third, the scheduling system may control the operator activities by making use of the concept of 'time-windows' which provides the operators some freedom in the timing of their activities. These options reflect different philosophies in the design of new technology. The last option fits in a human-centred design philosophy.

9.9 REFERENCES

Adlemo, A. (1999). Operator control activities in flexible manufacturing systems, part 2, *International Journal of Computer Integrated Manufacturing*, Vol.12, No.4, 325-337.

Ammons, J.C., Govandaray, J. and Mitchell, C.M. (1986). Human-aided scheduling for FMS: a paradigm for human-computer interaction in real time scheduling and control. In: K.E. Stecke and R. Suri (eds), *Proceedings of the Second ORSA/TIMS Conference on Flexible Manufacturing Systems: Operations Research Models and Applications*, Elsevier Publishers B.V., Amsterdam, pp. 443-454.

Barfield, W., Hwang, S-L. and Chang T-C. (1986). Human-computer supervisory performance in the operation and control of flexible manufacturing systems. In: A. Kusiak (ed.), *Flexible Manufacturing Systems: Methods and Studies*, pp. 377-408.

Benders, J. (1993). *Optional Options: Work Design and Manufacturing Automation,* Avebury, Aldershot.

Bi, S. and Salvendy, G. (1994). Analytical modeling and experimental study of human workload in scheduling of advanced manufacturing systems, *The International Journal of Human Factors in Manufacturing*, Vol.4, No.2, 205-234.

Boyer, K.K., Leong, G.K., Ward, P.T. and Krajewski, L.J. (1997). Unlocking the potential of advanced manufacturing technologies, *Journal of Operations Management*, Vol.15, 331-347.

Brödner, P. (1985). Skill-based production: The superior concept to the unmanned factory. In: H.J. Bullinger and H. Warnecke (eds), *Towards the Factory of the Future,* Springer-Verlag, Stuttgart, pp. 500-505.

Corbett, J.M., Rasmussen, I.B. and Rauner, F. (1990). *Crossing the Border: The Social and Engineering Design of Computer Integrated Manufacturing Systems,* Springer-Verlag, London.

Dhar, U.R. (1986). Overview of models and DSS in planning and scheduling of FMS, *International Journal of Production Economics*, Vol.25, 121-127.

Fitts, P.M. (ed) (1951). *Human Engineering for an Effective Air Navigation and Traffic-control System*. Columbus, OH. Ohio State University Research Foundation.

Gruteke, R. (1991). *Planning en besturing van het flexibele fabricagesysteem bij El-o-matic B.V. te Hengelo*. Master Thesis, Department of Mechanical Engineering, University of Twente, Enschede (in Dutch).

Hwang, S-L. and Salvendy, G. (1988). Operator performance and subjective response in control of flexible manufacturing systems, *Work and Stress*, Vol.2, No.1, 27-39.

Jaikumar, R. (1986). Postindustrial manufacturing, *Harvard Business Review*, Vol.64, No.6, 69-76.

Kopacek, P. (1999). Intelligent manufacturing: present state and future trends, *Journal of Intelligent and Robotic Systems*, Vol.26, 217-229.

Kopardekar, P. and Mital, A. (1999). Cutting stock problem: a heuristic solution based on operators' intuitive strategies, *International Journal of Computer Integrated Manufacturing*, Vol.12, No.4, 371-382.

Kouvelis, P. (1992). Design and planning problems in flexible manufacturing systems: a critical review, *Journal of Intelligent Manufacturing*, Vol.3, 75-99.

Lin, D-Y. and Hwang, S-L. (1998). The development of mental workload measurement in flexible manufacturing systems, *Human Factors and Ergonomics in Manufacturing*, Vol.8, No.1, 41-62.

LINDO Systems (1998). Inc., *LINGO Optimization Modeling Language* (User Manual) Chicago, IL.

Looveren, A.J. van, Gelders, L.F. and Wassenhove, L.N. van. (1986). A review of FMS planning models. In: A. Kusiak (ed.), *Modelling and Design of Flexible Manufacturing Systems*, Elsevier Science Publishers B.V., Amsterdam, pp. 3-31.

Mansfield, E. (1993). The diffusion of flexible manufacturing systems in Japan, Europe and the United States, *Management Science*, Vol.39, No.2, 149-159.

Mazzola, J.B., Neebe, A.W. and Dunn, C.V.R. (1989). Production planning of a flexible manufacturing system in a material requirements planning environment, *The International Journal of Flexible Manufacturing Systems*, Vol.1, 115-142.

McLoughin I. and Clark, J. (1994). *Technological Change at Work*, Open University Press, Buckingham – Philadelphia.

Nakamura, N. and Savendy, G. (1988). An experimental study of human decision-making in computer-based scheduling of flexible manufacturing system, *International Journal of Production Research,* Vol. 26, No. 4, 567-583.

Rachamadugu, R. and Stecke, K.E. (1994). Classification and review of FMS scheduling procedures, *Production Planning & Control*, Vol.5, 2-20.

Ravden, S.J., Johnson, G.I., Clegg, C.W. and Corbett, J.M. (1986). *Skill Based Automated Manufacturing,* P. Brödner (ed.), IFAC Workshop, Karlsruhe, Federal Republic of Germany, pp. 71-75.

Salvendy, G. (ed.) (1997). *Handbook of Human Factors and Ergonomics, Second Edition,* John Wiley & Sons, Inc., New York.

Shanker, K. and Agrawal, A.K. (1991). Loading problem and resource considerations in FMS: a review, *International Journal of Production Economics*, Vol.25, 111-119.

Sharit, J. and Elhence, S. (1989). Computerization of tool-replacement decision making in flexible manufacturing systems: a human-systems perspective, *International Journal of Production Research*, Vol.27, 2027-2039.

Siemieniuch, C.E., Sinclair, M.A. and Vaughan, G.M.C. (1999). A method for decision support for the allocation of functions and the design of jobs in manufacturing, based on knowledge requirements, *International Journal of Computer Integrated Manufacturing*, Vol.12, No.4, 311-324.

Slomp, J. (1997). The design and operation of flexible manufacturing shops. In: A. Artiba and S.E. Elmaghraby (eds), *The Planning and Scheduling of Production Systems*, Chapman & Hall, London, pp. 199-226.

Slomp, J. and Gaalman, G.J.C. (1993). A production control system for a flexible manufacturing system - description and evaluation, *International Journal of Flexible Automation and Integrated Manufacturing*, Vol. 1, No.2, 93-103.

Slomp, J. and Gaalman, G.J.C. (1998). Scheduling parameters in flexible manufacturing cells. In: H. Migliore, S. Randhawa, W. G. Sullivan and M. M. Ahmad (eds), *Flexible Automation and Intelligent Manufacturing 1998*, Begell House, Inc., New York, pp. 737-746.

Slomp J., Gaalman G. J. C. Nawijn, en W. M. (1988). Quasi On-Line Scheduling Procedures for Flexible Manufacturing Systems, *International Journal of Production Research*, Vol.26, No.4, 585-598.

Slomp, J. and Gupta, J. N. D. (1992). Interactive tool for scheduling jobs in a flexible manufacturing environment, *Computer-integrated Manufacturing Systems,* Vol.5 No.4, 291-299.

Small, M. H. and Yasin, M. (1997). Advanced manufacturing technology: Implementation policy and performance, *Journal of Operations Management,* Vol.15, 349-370.

Stecke, K. E. (1985). Design, planning, scheduling and control problems of flexible manufacturing systems, *Annals of Operations Research,* Vol.3, No.1, 3-12.

Stecke, K. E., and Kim, I. (1988). A study of FMS part type selection approaches for short-term production planning, *International Journal of Flexible Manufacturing Systems,* Vol.1, No.1, 7-29.

PART III

Plans, Schedules and Computer Systems

CHAPTER TEN

Design of a Knowledge-based Scheduling System for a Sheet Material Manufacturer

Vincent C.S. Wiers

10.1 INTRODUCTION

The field of production scheduling is dominated by research of a theoretical and quantitative nature (e.g., Halsall *et al.*, 1994). Although several authors have argued that there is a large gap between theory and practice (McKay, 1988; Sanderson, 1989; Wiers, 1997a), the stream of publications about production scheduling has not shown much change in focus. There is also a great need for empirical evidence regarding the applicability of scheduling techniques in practice.

Knowledge-based technology has been advocated as a promising approach in the field of production scheduling. Many researchers in scheduling believe that knowledge-based technology is able to bridge some of the gap that exists between scheduling practice and operations research.

This chapter describes the design and implementation of a knowledge-based scheduling system for a sheet material manufacturer. The design is based on a model presented by Wiers (1997b) and Wiers and van der Schaaf (1997). This model contains four aspects that drive the design scheduling decision support: autonomy, transparency, level of support, and information presentation.

The chapter is structured as follows: Section 10.2 describes the production situation and the sheet material products. In Section 10.3, the scheduling task is described and analyzed. In Section 10.4, the knowledge-based decision support system is presented. Lastly, in Section 10.5 conclusions and a discussion of the system implementation experiences are given.

10.2 SHEET MATERIAL PRODUCTION

10.2.1 Products

The company produces high quality sheet material that is used for façade cladding, interior construction and finishing, and for laboratory and project furniture. The products are made by first treating sheets of paper with resin, and then pressing a number of paper sheets together under high pressure. Figure 10.1 shows how a product is made from a number of sheets.

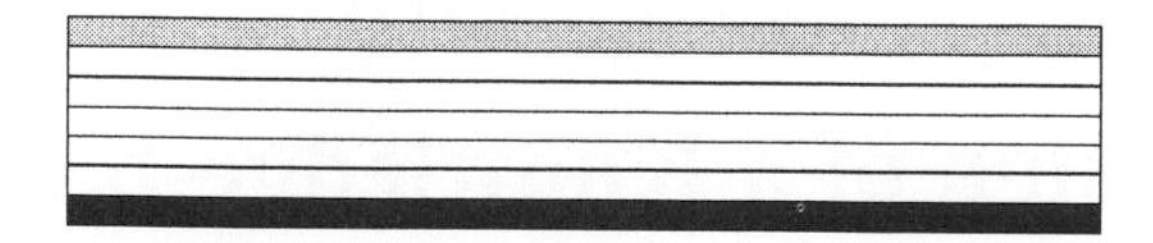

Figure 10.1 Sheet side view (actual thickness varies from 3 mm to 50 mm).

The sheets in the centre are referred to as the *core material* of the product. The number of paper sheets determines the thickness of the core material and, consequently, the sheet product. Instead of paper sheets, wood pulp is used in a minority of the products. The top and bottom sheets determine the look of the product and add much of the product variety. The sheet products have a number of predetermined sizes. The structure of the sheet product is important for its visual appearance. To summarize, the following characteristics determine the product variety:

- size
- structure
- colour
- thickness
- core material.

The sheet material is pressed in large presses with multiple levels. A high pressure (approximately 90 bar) is applied to the sheets for a fixed length of time (15-30 minutes). After pressing, the sheet products undergo a number of finishing processes. Figure 10.2 below shows a side view from sheets that are being pressed.

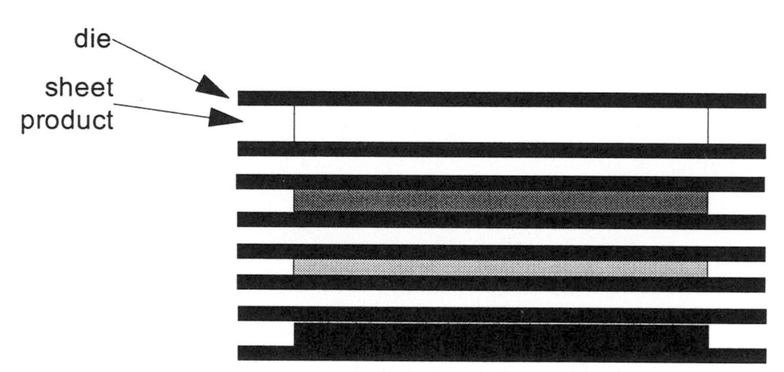

Figure 10.2 Side view of sheet products and dies.

At each level of a press, one sheet is pressed between two dies that give the sheet product its structure. This means that multiple sheets are produced in one batch. The characteristics of the sheets that are pressed in one batch are allowed to differ within strict, predefined bounds. For example, the thickness of the sheets in one batch can vary between bounds, whereas the size of the sheets and the structure are not allowed to differ. There are many constraints that have to be obeyed when combining different sheets in one batch.

10.2.2 Production process

Figure 10.3 shows the structure of the production process.

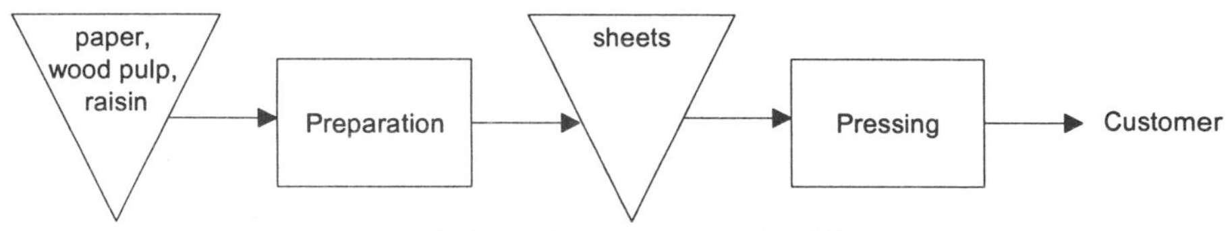

Figure 10.3 Production process (simplified).

The part of the production process until the stock point for sheets is driven by stock replenishment. The pressing process is driven by customer orders, because a large number of different products can be made by varying the types of sheets that are pressed together. The case study focuses on scheduling the pressing process.

The presses differ from each other regarding size, quality, and number of levels (15, 15, 8, 20). The factory has four presses that operate independently. Figure 10.4 below shows the layout of a press.

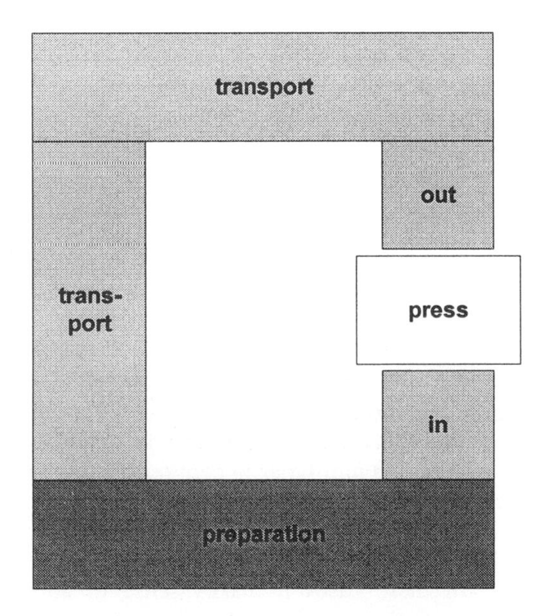

Figure 10.4 Top view of press layout.

The sheets first have to be prepared in the preparation section. At the newer presses, the preparation process is highly automated. Employees place the required number and type of sheets in the die. The die gives structure to the sheet products.

The prepared sheets are put on the transportation system, waiting for the previous batch to end. As soon as the batch is finished, the sheet products are fed out of the press on the transportation system, and the prepared sheets are fed in the press. While the press starts pressing again, the sheet products are taken out of the dies, and these dies are filled again with new sheets. As a result of this process, the planning of the presses often shows the typical A-B-A-B sequence. This means that while structure A is being pressed, structure B is prepared and waiting to be pressed, and vice versa.

The sheet manufacturer operates on a three-shift basis. This means that production continues 24 hours per day, except for weekends. However, there are not enough shifts to operate all presses continuously. The factory calendar indicates which presses can be manned at which times.

10.2.3 Scheduling situation typology

In Wiers (1997b) and Wiers and van der Schaaf (1997), a typology is presented to categorize scheduling situations based on production unit characteristics. The typology is depicted in Table 10.1.

Table 10.1 Typology of production units

		UNCERTAINTY	
		No	Yes
HUMAN RECOVERY	No	***Smooth shop***	***Stress shop***
	Yes	***Social shop***	***Sociotechnical shop***

The names of the production unit types illustrate the nature of the scheduling task, resulting from a certain division of autonomy between the scheduler and the production unit. The production unit types in counter-clockwise order are (see also McKay and Wiers, chapter 8 in this book):

- In the *smooth shop*, there is little internal or external uncertainty and as a result, there is little need for human intervention and problem solving, i.e., human recovery. In these cases it makes sense to automatically generate optimal schedules.
- In the *social shop*, the scheduler can lay out the basic schedule with sequences and timing, but allow for autonomy on the shop floor to tune the final work sequence at any resource. The scheduler may provide an optimized recommendation, but acknowledges that from a social and motivational point of view some human recovery might be preferred.
- In the *sociotechnical shop*, it is not possible to imbed the necessary flexibility into the schedule to take into account uncertainty. In these production units, human recovery should be employed to compensate for disturbances in the

production unit. Hence, schedules are only generated to provide a framework for production and are not executed exactly, optimization therefore does not make sense.

- In the *stress shop*, there is much uncertainty that cannot be compensated for by allocating autonomy to the shop floor because sufficient localized recovery is impossible. In these cases, the impact of the uncertainty affects multiple resources and tasks requiring the bigger picture to be taken into account. Therefore, disturbances have to be managed by the scheduler. To enable effective rescheduling actions, high demands are placed on the speed and accuracy of feedback from the shop floor.

The pressing process of the sheet manufacturer can be categorized as primarily a *stress shop*. The shop has to deal with a large amount of uncertainty, originating from both the internal production process (scrap, breakdowns) and from external factors (varying customer demand, rush orders). The flexibility on the shop floor to deal with these disturbances is very limited.

10.3 THE SCHEDULING TASK

10.3.1 Task description

In the sheet production process, one scheduler is responsible for scheduling the four presses. The activities in the scheduling task are depicted in Figure 10.5.

The dashed lines indicate the strong iterative nature of the scheduling task. This will be discussed in more detail later.

10.3.1.1 Visiting the shop floor

At the start of each day, the scheduler visits the shop floor to get firsthand information about the status of production. Many events have occurred throughout the night that the scheduler wishes to assess. Also, the scheduler likes to discuss forthcoming production with the operators and foremen at the presses.

10.3.1.2 Obtaining production orders and press assignment

Back at the office, the scheduler downloads a list of production orders from the ERP system. The production orders have already been assigned to a specific press; this means that the scheduler does not have to decide which press will produce which orders. Only in exceptional cases the press assignment is changed. The press assignment is done based on sheet size and product group.

10.3.1.3 Structure group sequencing

The scheduler briefly reviews the list of production orders to get an impression of the urgency of the orders. Because sheets of the same structure have to be grouped in the schedule, one of the first decisions the scheduler has to make is the sequence of the structure groups. Based on experience, the scheduler decides if a structure group will be produced 'on both sides' of the press (A-A-A-A sequence), or if two different structure groups will be produced at the same time (A-B-A-B sequence). The sequencing of structure groups also depends on the availability of dies.

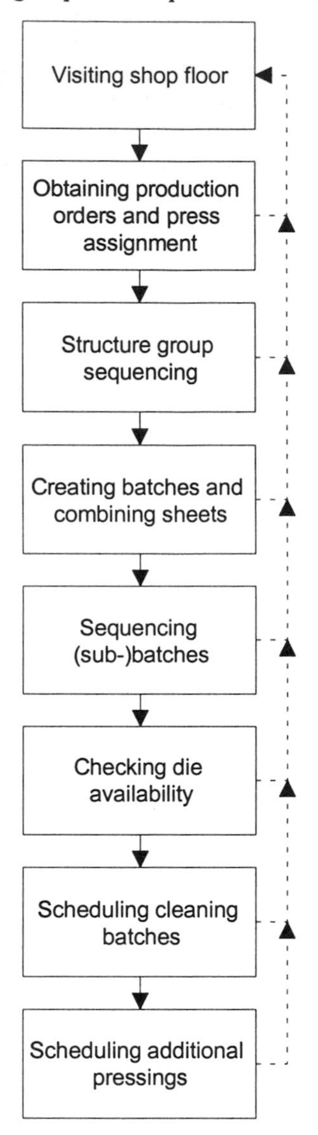

Figure 10. 5 Activities in the scheduling task.

10.3.1.4 Creating batches and combining sheets

Next, the scheduler starts generating batches within the current structure group. The size of the batches equals the number of levels of the current press. There are some exceptions to this rule: a specific type of thin sheet product is pressed two at a time on each level. Generating batches means picking up sheets from production orders and placing these sheets in the batch. A set of sheets from the same production order within a batch is called a sub-batch. The sheets within a sub-batch are identical. However, in many cases, more than one sub-batch is used to fill a batch. There are many constraints that determine if two sub-batches are allowed in one batch. Firstly, the sheets in one batch must be of the same size and structure. Secondly, the thickness of the sheets is not allowed to differ more than a predefined measure. This measure again depends on the thickness of the thinnest sheet being evaluated. Thirdly, if the batch contains a white sheet, this sheet must be the thickest one in the batch. Fourthly, a limited set of core materials is allowed to be combined in one batch.

10.3.1.5 Sequencing (sub-)batches

When sequencing sub-batches, the scheduler takes the following aspects into account:

- *Minimize 'breaking apart' production orders*. For example: when a batch is generated from a production order, the sheets in the next batch will preferably also originate from this production order, until the production order is 'empty'. Sheets in a batch that come from the same production order are referred to as *sub-batch*.
- *Sheet colour*. The preparation process prefers as few changes of colour as possible. Therefore, the scheduler tries to group similar colours. At the same time, the scheduler tries to avoid grouping white and near white sheets together to avoid confusion.
- *Thickness*. The scheduler attempts to sequence sheets from thin to thick to create a balance between preparation time and pressing time for the batches. If a batch with thick sheets must be prepared while a batch with thin sheets is pressed, there is a risk that the press will finish earlier than the preparation process, leading to idle time. Also, the preparation department prefers gradual changes of thickness.

The grouping process within structure groups is complicated by the following press characteristic: dependent on which press is used, the sheets are taken out of the press in either LIFO (last-in-first-out) or FIFO (first-in-first-out) sequence. Table 10.2 illustrates this.

Table 10.2 Grouping of sub-batches

Batch	LIFO	FIFO
1	A24.4.1	A10.3.2
	A10.3.2	A24.4.1
2	A08.2.3	A24.4.1
	A24.4.1	A08.2.3
3	A05.1.4	A08.2.3
	A04.1.7	A04.1.7
	A08.2.3	A05.1.4

The codes in the table indicate the colours of the sheets of that specific sub-batch. The LIFO-column indicates the preferred sequence of the sub-batches if the unloading sequence of that press is LIFO. Similarly, the FIFO-column indicates the preferred sequence of the sub-batches if the unloading sequence of that press is FIFO. This characteristic complicates the scheduler's task: the scheduler not only sequences structure groups and batches, but must also align the sub-batches to create a good schedule.

10.3.1.6 Checking die availability

As stated before, the scheduler guards the available number of dies. The company uses several types of die to give structure to the sheet material. Some dies are very complex to make and are very costly. This means that in producing sheets with structures that have a limited amount of dies, only a limited amount of presses can produce this structure simultaneously.

A fixed number of dies are available per structure group; however, this only applies to certain structures. The number of required dies for a structure group depends on the number of presses on which this structure group is being pressed, the number of levels of these presses, and if the sheets are structured on one side or on both sides (in the latter case, two dies are required per sheet).

10.3.1.7 Scheduling of cleaning batches

If a changeover occurs *to* a structure that uses a different set of dies, the first batch of that group must be used to clean the dies. This only applies to certain structures. No products are made in this cleaning batch. Furthermore, the second batch after the cleaning batch with that set of dies is not allowed to contain white sheet products, because there is a risk of pollution.

10.3.1.8 Scheduling additional pressings

After the pressing process, the sheet products undergo some finishing processes. Before the sheet products are packed and palletized, a quality check is performed on each sheet. If the sheet does not comply it is rejected and a new sheet has to be produced for the customer. The company explicitly chooses not to produce an excess of sheets in advance to buffer for rejections, because of past experiences with high stock levels.

This means that in the current production schedule, slack has to be inserted intelligently. The slack is put in the schedule by creating so-called *additional pressings*. A batch that contains empty space that can be utilized to re-press rejected sheets is called an additional pressing. An additional pressing is a batch that is partly empty and partly filled with sheets from planned production orders. The foreman fills the empty space when it is clear which sheets have been rejected.

The scheduler knows from experience how much scrap can be expected on average per press. Hence, the scheduler inserts additional pressings with fixed intervals in the schedule. However, the additional pressings must also comply with the following rules: if two different structure groups are being produced at the press (A-B-A-B), the additional pressing of structure group A must adjoin the additional pressing of structure group B. Moreover, the last batch of a structure group must be an additional pressing. This means that an additional pressing may need to be postponed.

10.3.2 Need for decision support

A stress shop places high demands on the scheduling task. There is little human recovery on the shop floor to relieve the scheduler of putting out fires. Fortunately, a reasonable capacity check is made before production orders are released to the press scheduler. Hence, most of the disturbances that the scheduler has to deal with are internal, such as breakdowns, rejects, etc.

The sheet manufacturing company discussed in this chapter faces a number of challenges in the near future that raise the need for decision support in the production scheduling task. Lead-times have to be shortened whereas the service level must remain constant. Moreover, the number of product variants is expected to increase. The company also expects to grow in the coming years, and plans an expansion of their production facilities.

Most schedulers only spend a small part of their time actually constructing a schedule, and use the rest of their time monitoring and problem-solving. Contrarily, in the sheet manufacturing company, the scheduler spends a major part of his time constructing a schedule. This is caused by the fact that many batches are produced daily that all have to be scheduled, and there are a large number of constraints that have to be obeyed.

10.3.3 Task analysis

The task of the scheduler has been used as foundation for the scheduling algorithm in the scheduling decision support system. To be able to translate the task to an algorithm, a suitable decomposition of the task structure had to be chosen. Figure 10. 6 shows the main activities in the scheduling task and their interrelationships.

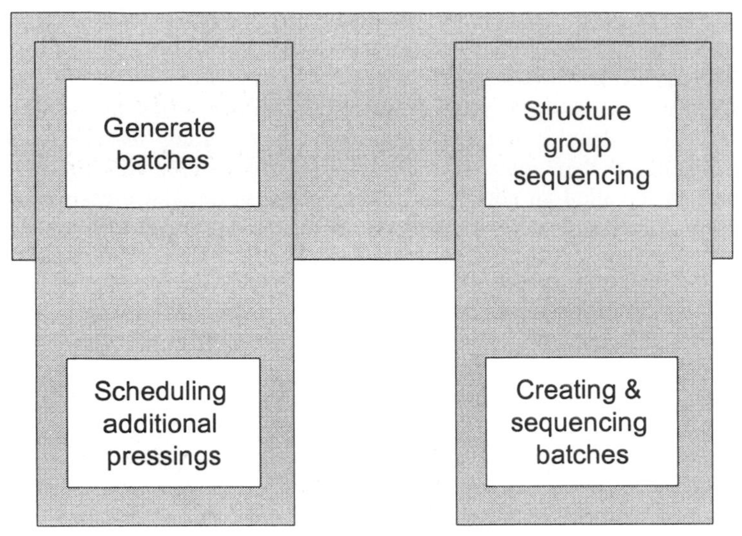

Figure 10. 6 Structure of the scheduling task.

The scheduling task consists of a number of elements that are strongly interrelated. A sequence for the structure groups is determined while the scheduler generates batches and additional pressings. The way the batches are scheduled follows directly from the sequence in which batches are generated.

To reduce the complexity of implementing the scheduling algorithm, it was felt that the task had to be split in two parts: one for dealing with batch creation, and one for scheduling batches. Such a decomposition would offer the scheduler more control on the scheduling system: after the first step, the scheduler would be able to change the intermediate results for the second step. Figure 10.7 shows the decomposed scheduling task.

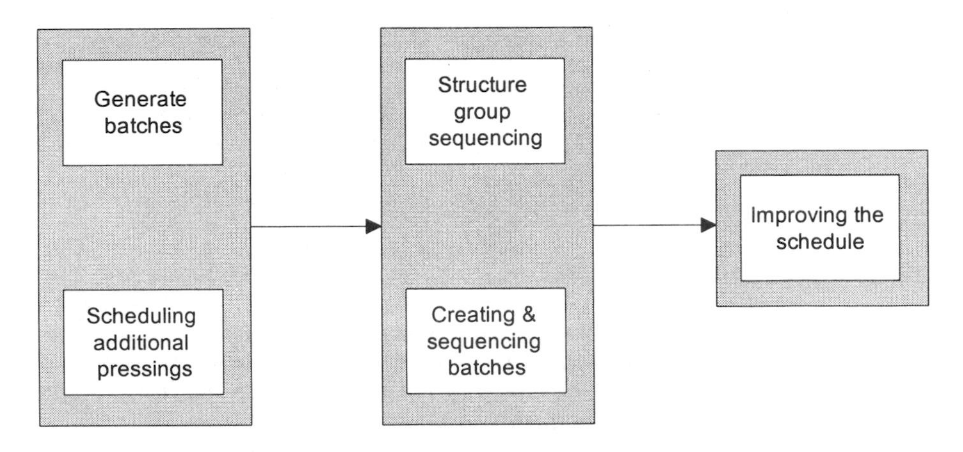

Figure 10.7 Decomposed scheduling task.

The first part of the decomposed scheduling task generates batches and additional pressings. The output therefore consists of a list of batches. The second part of the decomposed scheduling task schedules the batches, including the structure groups. Because the scheduling of batches may result in erroneously placed additional pressings, there may be a need for a third step to improve the schedule, either manually or automatically.

10.4 SCHEDULING DECISION SUPPORT SYSTEM

10.4.1 System characteristics

In Wiers (1997b) and Wiers and van der Schaaf (1997), four design aspects of scheduling decision support systems are introduced. These are: i) level of support, ii) transparency, iii) autonomy, and iv) information presentation. Below, these aspects are discussed in the context of the case study.

10.4.1.1 Level of support

The level of support for schedule generation lies in the possible variants of sharing responsibility between the human scheduler and the decision support system. At one extreme, the human acts as a principal controller, taking advice from the computer. The opposite extreme has the computer as the principal controller with the human performing corrections and adjustments. Interactive scheduling is located between these extremes (Sanderson, 1989; Higgins, 1996). The number of exceptions in the scheduling task indicates the required level of functionality.

In the case study, a medium to high level of support has been chosen. The scheduling system automatically generates schedules, and the human scheduler is enabled to change the schedule. Also, the scheduler is able to violate most constraints in the knowledge base. Because the algorithm was implemented in two steps, the scheduler has more control on the automatic generation of schedules.

As in most scheduling tasks, the human scheduler deals with many exceptions. However, the number of exceptions stands out against the large number of batches scheduled daily. In other words, the number of exceptions is not high compared to the number of task elements. This leads to the fact that the scheduler spends much time generating an initial schedule. A high level of support was therefore requested to relieve the scheduler of the time-consuming task of schedule generation. On the other hand, the scheduler is offered much freedom in adjusting the schedule.

10.4.1.2 Transparency

The transparency of a scheduling system influences the extent to which the human scheduler feels that he or she is in direct control. The need for transparency increases in situations where the scheduling task is perceived as critical, for example, when the scheduler has to deal with great uncertainty and tight delivery constraints.

The scheduling system contains a scheduling algorithm that offers a high level of support. There is a conflict between such scheduling algorithms and the transparency of the algorithm. It is difficult to design and implement sophisticated algorithms that are also transparent to the user.

To compensate for the lack of transparency of the algorithm, the schedulers participated actively in the functional specification of the scheduling system. This creates a feeling of responsibility for the system design. Moreover, attention has been given to training, coaching and other 'soft factors'. It was also felt that the individual characteristics of the scheduler are very important in the perceived transparency of the system. Fortunately, the scheduler is well-educated, computer literate and willing to adopt the new style of working.

10.4.1.3 Autonomy

Autonomy describes the degrees of freedom at a certain level in an organization. The functionality of a scheduling system should not support activities that fall outside the scheduler's autonomy. From the scheduling typology described earlier, it follows that most autonomy is in the hands of the scheduler in the case study described in this chapter. However, a number of decisions were identified that are taken on the shop floor:

- *Producing additional pressings*. The scheduler inserts slack in some batches that can be utilized by the foreman/operators on the shop floor. This means that the scheduling system must support this activity.
- *Moving batches*. The foreman/operators have limited ability to move batches in time. The scheduler notes these changes when information is automatically fed back from the shop floor to the scheduling system.

10.4.1.4 Information presentation

The aggregation level of the information presentation of a scheduling information system is a key factor for effective human-computer interaction. Textual displays are suited to display information on a low aggregation level. Graphical displays, such as the Gantt chart, are suitable to display information on a high aggregation level in a comprehensive manner (Rickel, 1988; Higgins, 1996).

The scheduling system combines a graphical representation and a textual representation in its main working screen. In the Rapid Application Development (RAD) sessions (see later), a Gantt chart was initially proposed as the main screen. However, the scheduler indicated that he needed more detailed information about batches and production orders at the same time. Therefore, a Gantt chart is combined with two lists: a batch list and a production order list. The screen shows information for one press at a time. The colours in the Gantt chart are used to indicate which structure group is being produced. Also, an extra area was defined per batch to indicate lateness and additional pressing.

10.4.2 Technology

10.4.2.1 Architecture

Figure 10.8 below shows the architecture of the scheduling system.

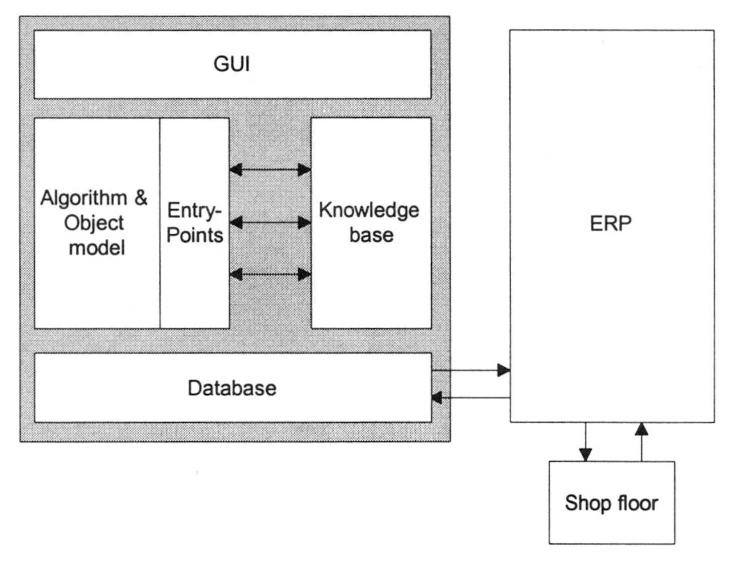

Figure 10.8 System architecture.

The system was built in Delphi/Turbo Pascal, except for the knowledge base, which was built in C^{++}. The knowledge base had been built in an earlier project and has been reused. The scheduling system runs as a stand-alone application on a Wintel platform.

10.4.2.2 Knowledge base and 'entry-points'

The scheduling system contains a knowledge base for maintaining and using scheduling knowledge. The knowledge base supports three knowledge representations as 'building blocks': rules, calculations and tables. The building blocks can be freely added to and edited. The knowledge base contains a predefined collection of domains, which link the knowledge to the planning algorithm via so-called *entry-points*. A domain contains input and output parameters: the input parameters are the 'question' and are filled in by the planning system, and the output parameters are the 'answer' and are returned by the knowledge base.

10.4.3 Project phases

10.4.3.1 Acquisition

The company evaluated a number of standard software packages for production scheduling. However, no package was found to be able to handle the situation. Many packages had difficulty in modelling the presses with multiple levels, and no evaluated package was able to offer a suitable scheduling algorithm. Therefore, the company chose to buy a customized knowledge based scheduling system, based on standard software components.

10.4.3.2 Design

At the start of the project, Rapid Application Development (RAD) sessions were used to globally specify the system. A visual prototype was built and presented to the schedulers. The comments of the schedulers were collected in each session and the prototype was adjusted accordingly. After three sessions, a global functional design was written. The RAD sessions turned out to be very valuable in specifying the graphical properties of the system. However, the scheduling algorithm could not be incorporated in the visual prototype. Moreover, it is quite difficult to formulate acceptance criteria for the scheduling algorithm, as the performance of a scheduling system is very difficult to measure objectively (Gary *et al.*, 1995).

10.4.3.3 Implementation

In the implementation phase, the components of the system were programmed, integrated and tested. During this phase, many discussions were held with the schedulers to sort out details in required functionality. In many cases, a trade off had to be made between the level of detail in the scheduling system and the length (cost) of the project. Also, it turned out to be difficult to get all details cleared out in an early phase of the project: many times a new functional design was made, new exceptions to the rules were mentioned.

In particularly, reaching consensus about the functional design of the scheduling algorithm was a laborious process. As the design of the algorithm advanced, the schedulers had more difficulty in understanding the implications for the outcome. On the other hand, reaching consensus was regarded as a necessity by the project leader to be able to make trade-offs and to manage the project.

10.5 CONCLUSIONS AND DISCUSSION

This chapter presented a case study of the design and implementation of a knowledge-based scheduling system at a sheet material manufacturer. The company went live with the system in Spring 1999. Experience with the system so far shows that the main goal of the system has been met, decreasing the time needed for generating a good schedule. The scheduler used to need 4 hours to generate a schedule and this has been decreased to 30 minutes.

The design principles as described in this chapter proved to be very useful in designing the system. However, it should be remarked that applying one's own theories blurs the effectiveness of the theory itself to an uncertain extent. Also, whereas the design principles offer a suitable way of setting up the stage for the scheduling system, many non-trivial details still have to be sorted out during the implementation.

The case study produced a number of lessons learned that are difficult to tackle with scientific theories: instead, pragmatism turned out to be an essential survival trait. Firstly, it turned out to be very difficult to make a good task allocation in automatic scheduling. When details in functionality have to be sorted out, it is hard to find clear arguments why certain activities have to be automated and others not. The effects of some decisions in design only show much later, if they show at all in an isolated manner.

Secondly, specifying the automated scheduling algorithm was a laborious process. It requires much abstraction and conceptualization, which sometimes even dazzled the knowledge engineers. At some points in time, the process was only kept going thanks to the sheer stamina of the participants.

Thirdly, the project benefited greatly from 'soft' factors such as a helpful IT department that provided quick response when necessary, a company culture that provided a positive attitude towards effective planning and IT, a manager that understood the importance of scheduling and had a similar background as the knowledge consultant.

Fourthly, in designing knowledge-based systems it is always difficult to decide what knowledge to hardcode and what knowledge to model in knowledge base. Nevertheless, the knowledge-based approach offered suitable mechanics to capture the scheduling task, although a controversy existed throughout the project between the quality of the model and the budget of the project.

10.6 REFERENCES

Gary, K., Uzsoy, R., Smith, S. P. and Kempf, K.G. (1995). Measuring the quality of manufacturing schedules. In D.E. Brown and W.T. Scherer (Eds), *Intelligent Scheduling Systems*, Boston, Kluwer Academic Publishers, pp. 129-154.

Halsall, D., Muhlemann, A. and Price, D. (1994). A review of production planning and scheduling in smaller manufacturing companies in the UK, *Production Planning and Control,* 5(5), 485–493.

Higgins, P. (1996). Interaction in hybrid intelligent scheduling, *International Journal of Human Factors in Manufacturing*, 6(3), 185–203.

McKay, K. N., Safayeni, F. R. and Buzacott, J. A. (1988). Job–shop scheduling theory: What is relevant?, *Interfaces*, 18(4), 84–90.

Rickel, J. (1988). Issues in the design of scheduling systems. In M. Oliff (Ed.), *Expert Systems and Intelligent Manufacturing*, New York: Elsevier Science Publishers, pp. 70–89.

Sanderson, P. M. (1989). The human planning and scheduling role in advanced manufacturing systems: An emerging human factors domain, *Human Factors*, 31(6), 635–666.

Wiers, V. C. S. (1997a). A review of the applicability of OR and AI scheduling techniques in practice, *OMEGA – The International Journal of Management Science,* 25(2), 145–153.

Wiers, V. C. S. (1997b). *Human–computer Interaction in Production Scheduling: Analysis and Design of Decision Support Systems for Production Scheduling tasks,* Wageningen, The Netherlands: Ponsen & Looijen, Ph.D. Thesis, Eindhoven University of Technology.

Wiers, V. C. S. and Schaaf, T. W. van de, (1997). A framework for decision support in production scheduling tasks, *Production Planning and Control,* 8(6), 533–544.

CHAPTER ELEVEN

Design and Implementation of an Effective Decision Support System: A Case Study in Steel Hot Rolling Mill Scheduling

Peter Cowling

11.1 THE PRODUCTION ENVIRONMENT

The work described here was largely performed at a medium-sized European steel mill, with an unusually diverse product portfolio, and hence a difficult scheduling problem to solve. The decision support system developed as a result has been successfully used, with minor modifications, at many steel mills worldwide. The work was carried out while the author was employed at A. I. Systems, Belgium, a software house building software for planning and scheduling decision support.

A steel hot rolling mill transforms steel slabs into steel coils. First, slabs must be moved from a storage area to the furnace area using manual or semi-automatic cranes. After heating, the slabs are subjected to very high pressures in a series of rolls to produce a steel coil of a few millimetres thickness. The rolls which are in contact with the hot steel band quickly become worn, so coils are milled in programmes of a few hours. Between programmes some or all rolls must be replaced. Several different types of programmes may be chosen, where the programme type chosen determines the types of coil which may be rolled in the programme and the preparation which must be carried out prior to executing the programme. The correct choice of programme type is an important planning decision. A rolling schedule is a sequence of 100 to 300 coils to be milled in a programme lasting several hours, which satisfies complex technical, commercial and logistical constraints. Producing such a rolling schedule is a difficult task, which is carried out by a small group of experienced schedulers. A schematic diagram of the production processes surrounding steel hot rolling is given in Figure 11.1.

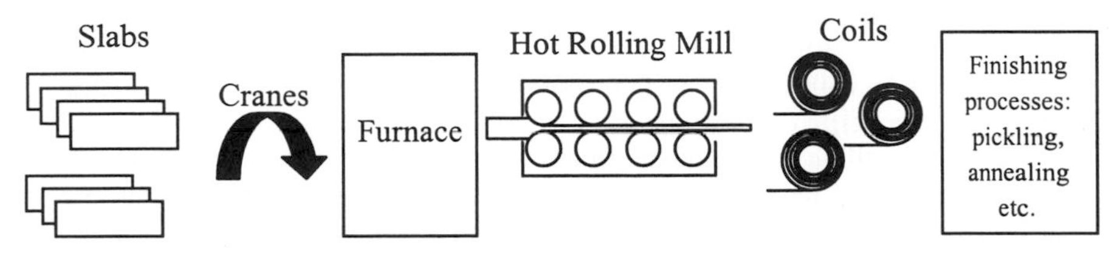

Fig. 11.1 Schematic of the hot rolling mill production processes.

Technical and logistical constraints arise from the wide range of machinery involved in steel hot rolling. It is possible that each coil could be manufactured from one of a number of different slabs that are available, so that slabs may be assigned to coils in such a way as to facilitate the rest of the production process. The cranes, which transport steel slabs from the slab yard storage area, may present a production bottleneck, particularly when the rolling schedule calls for slabs which are not at the top of a stack and crane-intensive unpiling operations must be carried out. Short slabs should be present in consecutive pairs in the sequence, otherwise there will be energy-wasting gaps in the furnace. Steel with different mechanical and chemical properties will need to attain different temperatures before milling. It is important that the slabs in a furnace at any given time require similar temperatures. Since there may be several furnaces, each having different properties, the schedule for each furnace should group coils having similar milling temperatures. Furthermore, the flow of slabs leaving each furnace needs to be balanced since each slab will have a minimum and maximum heating time. Changes in the dimensions or hardness of consecutive coils in the sequence may require changes in the pressure exerted by each rolling stand. Making changes in the pressures exerted is likely to result in reduced coil quality. Large changes give the risk of machine failure, expensive downtime and unplanned maintenance. Each programme should have a particular form, starting with easy-to-roll narrow, thick, soft, high tolerance coils, moving through a series of smooth changes in dimensions and hardness to difficult-to-roll wide, thin, hard, low tolerance coils. From these difficult, wide coils we then finish with a long sequence of coils having gradually reducing width in order to stop grooves which are made in work rolls by the edges of each steel coil rolled from marking future coils. Hence the width profile of each programme has a 'coffin' shape. The status and type of the work rolls used determine the coffin types and hence the coils which can be rolled. The choice of coffin type, taking into account technical and commercial constraints, is an important medium term planning problem. In particular, the population of coils/slabs which remains after rolling the scheduled coffins, has to be sufficiently heterogeneous to allow future coffins to be rolled without large jumps in dimensions or hardness.

Commercial constraints arise since, in this make-to-order environment, each coil has a due date and selling price and is destined for a customer who has certain tolerances for non-adherence to due date and quality targets. These tolerances will usually be known for internal customers and must be guessed for external ones. In an environment where the order book is full then throughput of appropriate coils must be maximised, so that careful selection must be made of coils to be rolled, and hence the coffin types to be made. Throughput must be maximised by ensuring that each programme is as long as possible between downtimes caused by work roll changes. When the order book is not so full, emphasis can be placed on coil quality, or on rolling coils with unusual dimensions or physical properties that require highly specific coffin types.

Whilst hot rolling is one of the most reliable processes in a steel mill, rolling mill schedules must be able to react quickly to unforeseen events, including the failure of cranes or furnaces, non-availability of certain coffin types due to non-availability of work rolls, rush and cancelled customer orders and problems with manufacturing processes upstream and downstream. Schedule generation must be able to react to these unforeseen events.

Different performance measures exist to measure the effectiveness of the hot rolling mill. Production personnel often use measures associated with the throughput and quality of steel coils. Commercial staff have measures associated with sales volume and delivery performance. Senior managers have measures that attempt to aggregate and measure overall customer satisfaction and production costs. Each of these groups may receive bonus payments in accordance with their performance measures. It is often the case that these measures will conflict with each other.

The information that is used to make schedules for the hot mill comes from several sources. Data concerning the slabs currently in the slabyard and client orders are usually held in the plant computer systems. Caster scheduling systems provide information as to which slabs will arrive in the slabyard in the immediate future. However, it is not always possible for caster schedules to be met and this information may be unreliable. Other rules and procedures governing the technical capabilities of the mill are usually known by a small group of engineers and schedulers. These rules change rapidly in response to conditions in the plant and external forces. Other inputs to the scheduling process come from sales and marketing divisions and from strategic directives emanating from higher management.

11.2 CAPTURING SCHEDULER EXPERTISE

Typically the manual planning of an eight-hour shift would take the scheduler a couple of hours. This long planning time meant that it was difficult to react to unforeseen production events, particularly on night and weekend shifts when a scheduler might not be available. When several programme schedules had to be generated at the same time, sequential generation of the schedules in advance meant that the quality of the manually generated schedules deteriorates with time. Too much scheduler time was spent in the mundane activity of coil sequencing, when this time could be better spent investigating higher-level issues of mill scheduling. Both schedulers and production managers recognised that the manual scheduling procedures in place were incapable of dealing with all of the technical, commercial and logistical constraints of the production process. Moreover, the performance measures considered were simple ones concerned with steel throughput, coil quality and customer due dates, which lacked the ability to discriminate at a sufficiently high level of detail.

The procedure used by the hot rolling mill scheduler was to print off a large mainframe listing of the orders whose due dates fell in the current week, sorted by coil width and then cut and paste slabs from this list by hand until a sufficiently good sequence was found. The resulting sequence was then keyed into the mainframe systems for dissemination to production and commercial departments. The schedulers work office hours, but the production process is continuous. Hence some schedules would be rolled shortly after being entered into the mainframe, whereas others would use data of the forecast production over the next 24 hours. Forecast data are often unreliable due to failures to attain quality targets in the steel-making processes. The forecast data would usually be ignored in practice, with only steel slabs currently in the slab yard being scheduled, so that some schedules (rolled on Sunday evenings) used data which are up to 60 hours out of date. We have now seen the manual scheduling activity at several steel mills, and the procedure appears to be similar at most of them. Some steel mills provide a simple computer tool for editing schedules. This does not greatly speed up schedule generation, but it does reduce problems due to errors of transcription. Whilst the scheduler generated schedules, he absorbed a good deal of information via telephonic and verbal communications from colleagues. This information concerned particularly errors in the data held, and commercial and production considerations. The programmes found by this manual method were of very variable quality, in terms of throughput, coil quality, technical characteristics and customer satisfaction. One systematic effect that was introduced by the manual scheduling system was that each scheduler did not wish to include difficult material in his schedules, since he received no extra reward for this material and this material may have resulted in reduced coil quality and throughput and increased production risk.

A model for the hot rolling mill scheduling problem was constructed by a team consisting of a specialist in Operational Research, a software engineer and a production engineer with many years' experience in the steel industry. This small team had regular meetings with a group of schedulers, managers and engineers from a given steel plant. The steel plant contribution to the project was championed by a senior manager with a good knowledge of both the engineering process and the commercial requirements of the mill. His continued enthusiasm for the project was critical for success. Meetings with engineers and schedulers from other steel plants ensured general applicability of results.

One of the most important activities of the modelling process was to identify which features of schedule generation were mundane, where computer-generated information would be most useful, and which corresponded to the interesting tasks which required the skills, insight and experience of a human scheduler. Presentation of the project to schedulers, engineers and managers as removing some of the mundane part of their jobs aided the process of gathering data. Here again the input of experienced managers, themselves ex-schedulers, was of more direct use than information provided by schedulers. We found that schedulers were defensive in regarding *all* aspects of their work as requiring a high degree of creativity, unsuitable for computer assistance. The scheduling managers were more willing to listen to new ideas and propose those areas amenable to computerised decision support. Table 11.1 briefly summarises the results obtained.

Table 11.1 Division of tasks into those suitable and those unsuitable for computer assistance

Tasks suitable for computerisation	Tasks requiring human flexibility
1. Coil sequencing, applying 'rules of thumb' and engineering principles.	1. Deciding which programme types should be milled.
2. Assigning slabs to coils.	2. Assigning priorities to customer orders.
	3. Assigning priorities to production goals.
	4. Rescheduling in response to production events.
	5. Revising 'rules of thumb' in response to information received.
	6. Obtaining information concerning likely future events.
	7. Giving out information.

Most of the schedulers' time was spent in sequencing the coils to be rolled in each shift. However the choice of sequence to be rolled was largely governed by well understood engineering principles. Even so, the task of assigning slabs to coils and sequencing the coils is a generalisation of the *Asymmetric Travelling Salesman* Problem (Lawler *et al.*, 1985) which is a member of the NP-hard class of difficult combinatorial optimisation problems (Garey and Johnson, 1979). The complexity of these principles had caused other attempts at creating a useful decision aid for hot rolling mill planning to fail. If the schedulers' time were freed from the mundane task of sequencing coils then he would have more time for the interesting tasks associated with the generation of good schedules. These interesting tasks were concerned with dissemination of information and reflection upon planning practices and external information received. In particular he could better decide the priority of coils to be sequenced and the types of programme to be investigated and be more proactive in response to information received, for example concerning machine failure or rush orders. This would allow the scheduler to build schedules which better balanced the requirements of production staff, commercial staff and senior management, whilst the majority of the technical constraints could be handled by the computerised coil sequencing process. Information as to the tasks performed by schedulers and scheduling managers was obtained both through observing these personnel at work and through discussions. It was felt that the information obtained from the meetings and from middle-level scheduling managers was more useful than that obtained from meetings with, and observation of, schedulers. Most of the middle-level scheduling managers were themselves former schedulers and a view was formed as to what constituted 'ideal' practice rather than simply current practice. In the process of identifying the mundane and the interesting tasks that the scheduler carries out the managers in charge of the scheduling and production process were able to consider the impact of the personality and mood of human schedulers in schedule generation. Whilst it was felt by the scheduling managers that personality and mood could have a highly positive, or negative impact on how well the interesting tasks of the schedulers were performed, it was regarded as unlikely that personality and mood would have a particularly positive effect on the tasks of coil sequencing and the assignment of slabs to coils. We illustrate this later with the cases of 'Risky Ron' and 'Safe Sam'.

It quickly became apparent that the constraints and objectives of the hot rolling mill schedule were not at all stable, so that any successful computerised scheduling aid would have to be easily configurable. In particular the coil sequencing problem, although mundane, required input from many other information sources, particularly the engineering and scheduling expertise of plant employees. Easy configuration would enable the system to react to a wide range of circumstances, in particular real-time events and changes in plant machinery or working practices. In addition methods were required to enable the scheduler to quickly assimilate, communicate and act upon a large amount of information, particularly on the population of slabs currently held in the slab yard and the production possibilities offered by that population.

In our long series of meetings and observations with schedulers and managers, it became evident that these two groups of people would have different expectations and require decision support in different ways. Engineers and managers needed to be able to change most aspects of the system, most notably they needed to be able to change aspects of the model which arose due to changes in plant configuration or working practices. They would also need to be able to set the performance measures, which would determine the nature of the schedules produced. Whilst engineers/managers had faith in the technical abilities of their scheduling staff, they felt it was rather dangerous to give access to the performance measures to them. This might reintroduce behaviour which was perceived as undesirable, such as the tendency for schedulers not to include difficult coils in their schedules, in order to maintain popularity with those responsible for milling the sequence. The schedulers would need to select coffin types to be milled and those slabs that were available for milling, but this could be done whilst hiding much of the detailed information and parameter-setting possibilities. There was a feeling from the schedulers that they would benefit by having clearer goalposts, inasmuch as they would simply have to maximise their score, in order to satisfy the purely technical task of sequence generation. The engineers/managers perceived benefit, since they would be determining the weighting of factors, which went to make up the score, which would give them more control over the detailed scheduling process. The performance measures which this put in place could be more detailed and less prone to abuse than simple ones used previously. It is interesting to note that the measures, which were arrived at for the final system, varied a great deal from those that were already in place. The 'maximise throughput while maintaining acceptable quality' measure which was predominant was felt to be prone to abuse, for example since it discouraged the rolling of difficult coils. New measures would consider performance from both commercial and technical points of view, using a combination of five different measures considering the:

- commercial value of coils, including aspects such as timeliness and profitability;
- coil quality;
- work roll usage (throughput);
- slab yard management;
- adherence to ideal programme shape.

Much of the work done to capture the knowledge of schedulers and managers was done 'from first principles', with the key being a very high level of user involvement in all stages of development of the model and decision support system. The schedulers and managers spent much of their time fire fighting and had little time to reflect on what their jobs actually involved, prior to the BetaPlanner project. There was no body of knowledge then available to the developers (who did not include any human factors specialists) as to how schedulers work in practice. (The current volume should be a valuable aid for future practitioners who attempt to implement scheduling decision support systems in areas where the scheduling process is currently manually intensive.)

11.3 TECHNIQUES FOR HOT ROLLING MILL SCHEDULING

There have been several papers written about decision support for hot rolling mill planning. The literature in this area may be divided into two groups. Jacobs *et al.* (1988), Balas and Martin (1991), Sasidhar and Achary (1991), Petersen *et al.* (1992) and Assaf *et al.* (1997) all use mathematical programming techniques to solve simplified models of the hot rolling mill scheduling problem. Cowling (1995), Stauffer and Liebling (1997) and Lopez *et al.* (1998) propose more generally applicable models which are solved using heuristic optimisation techniques. Using a simple model produces nearly optimal solutions more often, which may, however, be harder to implement in practice. Using a more complex model, solved using heuristic approaches, presents no absolute guarantee of distance from optimality in the model, but gives results that are usually easier to apply in practice. Each of these approaches has its merits and deficiencies. None of these papers have addressed the issues of the capture of scheduler expertise or the effects of implementation upon the working practices of a steel mill. It was clear throughout our study of the problem, that only a complex model solved using fast heuristics would provide a mechanism for capturing scheduler expertise and production complexities for adequate decision support within the client steel plants.

11.4 THE BETAPLANNER DECISION SUPPORT SYSTEM

The decision support system for the hot rolling mill planning problem would be used by a range of users with different levels of expertise. Functionality must be hidden from lower-level users both to give rise to a system that is sufficiently easy to use and for reasons of security. In order to cope with the different levels of expertise, the scheduling decision support system was made configurable on three levels: OR analyst, engineer/expert and scheduler/mill foreman.

The OR analyst is needed only on initial set-up to tune parameters of the heuristic which were hidden from other users. Due to the wide variation between the specifications of different steel plants, the heuristics used to solve the problems modelled need tweaking for best possible performance. The users of the system so far have not been able to do this tuning themselves, and so this user level contains parameters that would not be changed following initial implementation. An early prototype of BetaPlanner did not hide these parameters sufficiently well and curious users induced undesirable behaviour by changing the parameters for the OR/analyst user level!

The engineer/expert is able to alter all technical aspects of the system. In particular he can reflect changes in operating procedures and machinery through an intuitive graphical user interface. He can define the categories of slabs and orders which the scheduler uses to define the population to be scheduled, model the commercial value of coils to reflect the desires of sales and marketing departments and define new coffin types to reflect changes in the technological and commercial environment. The system allows him to balance all of the technical, commercial and strategic performance goals to arrive at a best compromise for a given situation. He can save configuration data, using descriptive names, so historical data are built up over time, which captures some elements of the knowledge and expertise of the engineer/expert, and the system becomes easier to use over time. For example, configuration data might handle the two situations when the order book is full and throughput considerations dominate and the situation where the order book is nearly empty and quality considerations dominate. They could also handle a variety of strategic directives, for example concerning a marketing push for specific types of steel or customer.

The scheduler possesses the tools required to analyse the slabyard and generate and analyse a number of schedules to the engineers' specification. He can also analyse the impact of a particular schedule on the remaining slab population to ensure that the population remains sufficiently heterogeneous for future schedule generation. He can choose schedule types and has some control over coil priorities. The scheduler's interface is sufficiently easy to use that it may be, and has been, used by the mill foreman to generate a schedule in response to unforeseen events when there is no scheduler present at the steel plant. This very simple operation is aided by the engineers' interface described above. Typically the engineer will have defined programmes with names such as 'Night shift emergency coffin – full slab yard' and schedules for these coffins can be generated with a couple of mouse clicks. As maintained above, it is very important that certain tools and parameters are hidden from certain users. An early prototype of the system did not hide this information and the schedulers demonstrated considerable ingenuity in locating configuration files and trying new values, with the result that schedule generation slowed by a factor of several thousand times.

Initial attempts to solve the model by mimicking the schedulers' approaches were unsuccessful, due to the difficulty of capturing the approaches in sufficient detail so as to make computer implementation possible. As we have noted above, managers/engineers within the steel plant wanted to introduce a new system of performance measurement. However, since a very detailed level of modelling, starting from basic engineering principles, had been carried out it was possible to approach the problem using heuristic techniques from Combinatorial Optimisation, particularly local and Tabu search. The results were highly satisfactory. For further details see Cowling (1995) and Baccus, *et al.,* (1995).

A high-level view of the schedule generation procedure used is given in Figure 11.2. In the *slab selection* stage first the slabs which are released for sequencing are selected. Certain slabs may be unavailable, for example due to the breakdown of one of the slab yard cranes. Then commercial considerations are used to determine the priority of each coil. This step is usually carried out every few days by the engineer/expert in response to commercial considerations and strategic goals. The *coffin selection* is carried out, usually by the scheduler/mill foreman, to determine which programme types are to be rolled to produce coffin programmes containing slabs of the highest possible value. Slab and coffin selection are both supported by powerful and easy to use graphical user interfaces that allow the user great flexibility in doing 'what-if' analyses. The interfaces provide statistics concerning the value of rolled coils and the production possibilities of the remaining population, suitable for both the scheduler and the engineer/expert. The slab and coffin selections, together with the technical rules and weights for each performance measure are then used by the *sequencer* to heuristically generate a high quality schedule, within a time frame of a couple of minutes. Since the time frame to change the mill specification and reschedule is short, new schedules can be generated quickly in response to unexpected events such as machine breakdowns and rush orders. When more time is available the sequencer can use this time to generate schedules of slightly higher quality. Finally, the scheduler is able to use the *programme editor* to carry out detailed editing of the sequence, if necessary. Experience shows that this is almost never necessary in practice, but it is important that the scheduler has this final element of control. We will return to this point later. After this editing process the schedules are sent to mill mainframe systems for dissemination throughout the steel mill.

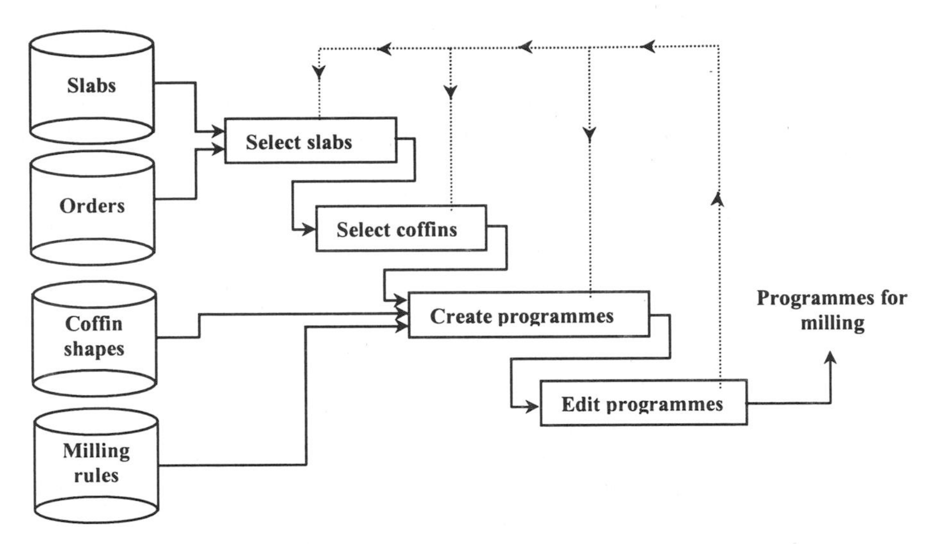

Fig. 11.2 BetaPlanner sequence generation procedure.

11.5 BENEFITS OF THE BETAPLANNER SYSTEM

The most obvious benefits of the BetaPlanner system over a manual planning system are the improved technical and commercial characteristics of the schedules generated, due to the system's ability to consider a wider range of factors than could be considered by a human planner. Of course, the BetaPlanner system is not able to bring to bear reasoning and other faculties, unlike a human scheduler. Since the tasks which are undertaken by the computer system have been chosen as those where the human qualities of reasoning are less important, and since the computer system can make simple decisions many millions of times faster than a human planner, the result is far better schedules overall. Faster sequence generation allows a quick response to disturbances and frees up the schedulers so that they can spend more time on the more interesting tasks, such as data-gathering, which will also improve sequence quality, although the benefits from this are harder to quantify. The fast scheduling time also allows the decision support tool to be used to consider 'what-if' questions. For example, one BetaPlanner user has used the system to appraise the possibility of investment in a new type of furnace. Since BetaPlanner can plan several shifts within a reasonable time (a few minutes) and each shift has equal priority, the planning horizon can be increased to improve the characteristics of sequences in the medium term, without the decrease in quality over time which manual schedules demonstrate.

Personality, mood and opinion influence the schedulers' work. Whilst this is critical to scheduling flexibility, scheduling managers may prefer that the mundane activity of sequencing is not influenced by these factors. An interesting example occurred at a particular steel plant where there were two schedulers, who we shall name 'Risky Ron' and 'Safe Sam'. Ron believed that it was his duty to stretch the technical capacity of the mill to its limits. His programmes were very long and often contained dangerously large changes in coil dimensions and hardness. Sam's sequences were often significantly shorter since he was unwilling to approach the limits of the mill. Under normal conditions, the company would prefer Sam's conservative approach, leading to a higher quality of steel produced. Under unusual conditions, for example when slab stocks were low, then Ron's approach was necessary to make commercially viable sequences. The hours worked by Ron and Sam were governed by shift patterns and not mill conditions. By building configurations corresponding to the best sequencing characteristics of Ron and Sam, management have been able to have a greater influence on schedule characteristics. These configurations contain information of the attitudes of Ron and Sam concerning the ideal 'shape' of a schedule, and the penalties associated with violations from this shape. Since both Ron and Sam had significant input into the scheduling configuration to be used under normal conditions, they were happy to use the standard configuration (corresponding closely to Sam's conservative style). However, when the number of slabs in the slab yard was reduced, an appropriately titled configuration was used by both Ron and Sam with similar results, corresponding more closely to Ron's more vigorous scheduling style. Acceptance of a computerised decision support system by the schedulers was improved since the system was presented, truthfully, as a system that would make their jobs more interesting and not replace them. There was also some enthusiasm to learn computer skills, which was boosted by the introduction of simple computer games into the workplace.

11. 6 USER INGENUITY IN APPLYING BETAPLANNER

BetaPlanner is currently in use at several steel mills worldwide and possesses a small club of 'power users'. The open-ended nature of the planning tool has allowed it to be used in several ingenious ways that its creators had not originally foreseen.

A pickling line passes hot rolled steel coils through a bath of hot acid in order to remove surface oxides before further processing or shipping. The coils must be welded together in a continuous band prior to passing through the acid bath. As may be imagined, the pickling line is governed by quite different constraints and objectives than the hot rolling mill. One user noted particular similarities in that there were significant set-up costs in passing from one coil to another, and overall sequence shape was governed by complex rules. He designed a series of configurations for BetaPlanner in order to address the pickling line scheduling problem and has now used it successfully for pickling line scheduling for some time.

Another user considered simultaneous implementation of BetaPlanner, together with an advanced fürnace. The ability of BetaPlanner to handle the resulting complex furnace arrangement allowed the user to carry out extensive 'what-if' analysis to aid the decision whether to invest in new furnace technology. In this particular case the steel manufacturer was at some considerable distance from the BetaPlanner system developers and nearly all of the modelling and 'what-if' testing was done over the Internet. The result was that both the furnace and the decision support system were purchased and used successfully.

Further users have used BetaPlanner extensively to carry out 'hot charging' where slabs pass directly from the caster to the hot rolling mill, with a large consequent energy saving. Hot charged slabs have a relatively short time window in which they must be placed in a hot rolling mill sequence, requiring detailed timing information from schedule generation. In this case the existence of accurate sequence data greatly facilitated the incorporation of timing information and this feature was successfully incorporated into BetaPlanner with only a couple of weeks of development effort.

The application of our decision support system in areas not originally foreseen provides evidence that including a far greater amount of flexibility into the model than is necessary for providing scheduling decision support in the short-term, has two significant benefits. Not only does it provide a good deal of 'future-proofing', but it can also aid users in applying the decision support tool in novel ways and in seeing new areas of application of other computerised decision support tools. We believe that the extra effort required for this 'over-modelling' may be worthwhile for any decision support system that will be used over a long period. Providing a mechanism whereby experienced users can continually update, improve and extend their knowledge of the production process and the decision support system using an open-ended model and interfaces which support 'what-if?' questions has increased acceptance of the decision support system at all levels. It should be noted in this case that the cost of development of the system is increased enormously by the incorporation of this flexible approach, which may, therefore, be most appropriate for large systems, or those which have a potentially very large user base.

11.7 LOOKING BACK AT THE APPROACHES TAKEN

The BetaPlanner system has now been installed and used successfully at a range of steel mills throughout the world. In this section we reflect on the methods for capturing the expertise of human schedulers and development and implementation of the BetaPlanner decision support system.

We believe that there were two keys to the development of a successful prototype of the decision support system. First, a champion within the main pilot site at senior management level ensured a very high level of interest and that resources were devoted to the project. Second, a wide range of personnel were involved and *felt* involved in the project, so that ownership of the project was clearly joint from the beginning. However, the great deal of information and a great variety of opinions available made available gave rise to a good deal of 'nervousness' in the model. The model of the scheduling process included, at some stage in the modelling process, many features that are either no longer present in the current system, or present, but not used to the best of our knowledge. Initially this nervousness resulted in frequent changes to models and software prototypes. Once a reasonably stable model and prototype were in place, a waiting time was instituted for all user requests that were felt to be transitory. Essentially this meant that user requests would have to be made more than once before they were implemented. Much time was saved using this technique, at only a small loss in user satisfaction levels.

Spending a good deal of time talking to and observing schedulers at work provided much information as to the overall nature of the scheduling task. It was our experience that the schedulers themselves found it difficult to step back and separate those parts of their work which could usefully receive computer support, from those which could not. In particular most schedulers felt that the number and instability of the rules, which they had to deal with when sequencing coils, would be prohibitive to computer implementation. Talking with engineers and managers at a higher level and an analysis of the sequences actually produced, allowed us to capture some important aspects of the scheduling process using a mathematical model. This technique was alien to both the schedulers and the engineers/managers, who had to be kept on board through regular dissemination of the results produced. The steel engineer who was part of the development team was constantly appraised of the development of the model and heuristics for solving it. He was able to successfully communicate details of the system in a language which the engineers and managers could understand. The schedulers were kept on board through constant reassurance that a decision support system would make their jobs more interesting, rather than replace them. This constant explanation and reassurance was a time-consuming activity that was critical to overall project success.

Considerable development time was spent in developing a flexible and powerful sequence-editing capability for the system. This is a capability which is seldom used now. However, this capability was important in prototyping, since observation and automatic capture of the changes which scheduling staff made, allowed the models and heuristics to be improved. In the final system, it is almost never the case that a human operator can find changes to a schedule that would improve the overall score. However, being able to make these changes is important in the early stages of implementation since it both exposes situations where the performance measures should be changed, and once the performance measures are trusted, demonstrates that the solutions produced are not subject to easy improvement on a technical level. This encourages schedulers to spend far less time on the mundane task of coil sequencing and more time looking at wider opportunities for schedule improvement through better reaction to production events and exchange of information with personnel involved in scheduling or production at upstream and downstream processes.

11.8 CONCLUSIONS

We believe that BetaPlanner provides an example of a system for scheduling support that has enjoyed considerable success due mainly to the following factors:

- only those aspects which may genuinely be eased by computer support are tackled;
- extensive configurability allows the user to express his preferences in a way which can influence detailed scheduling behaviour;
- three different levels to suit the level of user expertise, and for security;
- hiding information and features of the system to suit user level and requirements;
- allowing users to maintain historical configuration information in a way which enables and encourages its reuse;
- results are provided sufficiently quickly for experimentation to be possible;
- an easy to use graphical user interface.

We have described the approaches used to develop the scheduling decision support system and discussed the issues that arise when the system is used in practice, particularly the impact upon performance measures and the effect of user ingenuity in using the system. On reflection it is clear that in this case too much of the time spent in the analysis of the system requirements was spent with schedulers working at a low level. Whilst these schedulers were good sources of information concerning the details of the scheduling process, they found it very hard to think 'outside the box'. More senior managers and engineers, particularly those who had formerly been schedulers, provided more useful insight as to ideal future practice, as opposed to the current practice.

The ability of the system to evolve naturally to capture the knowledge of the schedulers, engineers and managers is very important to acceptance of the system. Whilst the hardware platform upon which the decision support system runs is likely to date quickly, the ability of the system to 'learn' and 'evolve' through the capture of changing schedule specifications and performance measures, should greatly increase its practical lifespan.

Had there existed a structured framework of knowledge concerning the abilities and capacity for training of scheduling personnel, this would have provided a useful start to the process of systems requirements capture. In particular, it would have aided the division of the scheduling problem into aspects where computer automation is useful and those requiring human creativity, flexibility and communication skills. However, it is likely that the problems within the steel industry alone demand sufficiently different skills from schedulers that only general conclusions would be possible. For example, continuous caster planning, where liquid steel is turned to steel slabs, is a highly flexible production process requiring good skills of order aggregation, very different to those sequencing and assignment skills required by hot rolling mill planning. Scheduling problems across different industries might be categorised as to type of skill employed, so that information from other manufacturing and service sector applications may bc uscd in requirements specification. This book will add useful insight for commercial software developers in specifying requirements for these future scheduling decision support systems.

11.9 REFERENCES

A. I. Systems (1994) BetaPlanner Product Description, A. I. Systems, J. Wybran Ave. 40, B-1070 Brussels, Belgium.

Assaf, I., Chen, M. and Katzberg, J. (1997) Steel production schedule generation, *International Journal of Production Research*, 35(2), 467-477.

Baccus, F., Cowling, P., Vaessen, N. and Van Nerom, L. (1995) Optimal rolling mill planning at Usines Gustave Boël, *Steel Times International – Rolling Supplement*, March 1995.

Balas, E. and Martin, C.H. (1991) Combinatorial Optimization in Steel Rolling, *DIMACS/ RUTCOR Workshop Proceedings*, April 1991.

Cowling, P. (1995) Optimization in steel hot rolling, in *Optimization in Industry 3,* (ed.) Anna Sciomachen, John Wiley & Sons, Chichester, 55-66.

Garey, M.R. and Johnson, D.S. (1979) *Computers and Intractability: A Guide to the Theory of NP-completeness*, W.H. Freeman, New York.

Jacobs, T.L., Wright, J.R. and Cobbs, A.E. (1988) Optimal inter-process steel production scheduling, *Computers and Operations Research*, 15(6), 497-507.

Lawler, E.L., Lenstra, J.K., Rinooy Kan, A.H.G. and Schmoys, D.B. (1985) *The Travelling Salesman Problem: A Guided Tour of Combinatorial Optimization*, John Wiley & Sons, Chichester.

Lopez, L., Carter, M.W. and Gendreau, M. (1998) The hot strip mill production scheduling problem: A tabu search approach, *European Journal of Operational Research,* 106, 317-335.

Petersen, C.M., Sørensen, K.L. and Vidal, R.V.V. (1992) Inter-process synchronization in steel production, *International Journal of Production Research,* 30(1), 1415-1425.

Sasidhar, B. and Achary, K.K. (1991) A multiple arc network model of production planning in a steel mill, *International Journal of Production Economics,* 22, 195-202.

Stauffer, L. and Liebling, Th.M. (1997) Rolling horizon scheduling in a rolling mill, *Annals of Operations Research,* 69, 323-349.

CHAPTER TWELVE

A Field Test of a Prototype Scheduling System

Scott Webster

12.1 INTRODUCTION

This chapter documents the scheduling function at a machine tool plant and reports the development and testing of a prototype scheduling system at this facility. The objective is to add to the descriptive literature on how scheduling is done in practice with particular attention to the challenges and opportunities of computer-intensive approaches to production scheduling.

Descriptive work helps clarify the complexities and critical issues arising in practice. While individual studies suffer from questionable generalizability, when taken as a whole this body of literature plays a useful role in directing research to areas of high potential impact. There is a need for additional descriptive studies of scheduling practice for three reasons. First, there remains a significant gap between the scheduling literature and scheduling practice. It is generally acknowledged that the impact on scheduling practice of over 40 years of research, manifested in over 20,000 journal articles (Dessouky *et al.,* 1995) and 20 books, has been minimal. Second, the descriptive scheduling literature is sparse. Early work appearing in the 1960s and early 1970s was followed by a lull that has only begun to turn around in recent years (see MacCarthy *et al.,* 1997 or Wiers, 1997 for a review of the literature). Third, the rapid advancement in information infrastructure (e.g. through SAP and other ERP systems) and the increasing practitioner interest in computer-based scheduling systems suggest that the opportunity to reduce the gap between theory and practice has never been greater.

The field test project began with a proposal to the company in July 1995. Intensive analysis, design, and development efforts took place over the ensuing 11 months. The prototype scheduling system was implemented in June 1996. The field test lasted for approximately one year, during which time the scheduler experimented with the system and offered suggestions. We continued to develop the prototype in response to feedback from the scheduler and, during the latter part of the test, we focused on assessing the strengths and weaknesses of computer-based scheduling at the company.

The chapter makes two main contributions. First, as noted above, it adds to the case study literature of scheduling practice. Second, it documents our experience with one possible field-based approach for enriching a research program. We describe the design, development, implementation, and project management of a prototype scheduling system in a plant. Researchers considering similar field-based activities may benefit from our experience.

The remainder of the chapter is organized into four sections. The next section provides background on plant operations. Section 12.3 describes how scheduling is traditionally done at the plant. Section 12.4 documents the project and Section 12.5 provides conclusions.

12.2 BACKGROUND

The plant produces inserts that attach to steel cutter bodies (see Chapter 4 in this book for a case study of the plant that produces the cutter bodies). The inserts, which are prismatic-shaped pieces of tungsten carbide about 2cms. across, serve as the blade of a cutting tool (see Figure 4.1 in Chapter 4). The plant manufactures approximately 2,000 basic types of inserts. Some finishing operations, such as honing and coating, may be required for some inserts, yielding a total of over 4,000 varieties of finished products. These 4,000 varieties of inserts are referred to as *standard inserts*. Standard inserts make up a little over 95% of the volume. The remaining volume is comprised of customized inserts that are engineered to customer requirements.

There are about 1,000 to 1,200 shop orders in process at all times. Of these, approximately 20% are customer orders and 80% are orders to replenish safety stock levels. The number of inserts per order averages 600 and can range up to 6,000 pieces. Approximately 50% of customer orders were shipped on time in 1996.

The plant, which received ISO 9001 certification in 1996, has 20,000 square feet of manufacturing space and employs about 120 people. The plant is normally running in some capacity around the clock and seven days per week. There are 31 work-centres of six different types that perform various grinding, honing, cleansing, and coating operations. An insert requires anywhere from two to eight machining and coating operations. The coating operation is outsourced on about 60% of the shop orders.

12.3 THE SCHEDULING FUNCTION

A number of researchers have stressed the importance of the linkage between intermediate-term planning systems and short-term scheduling, as well as the need for more direct consideration of this linkage in the literature (Buxey, 1989; MacCarthy and Liu, 1993; Conway, 1994; Kanet and Sridharan, 1998). Material requirements planning and the interface between MRP and shop floor scheduling are interesting at the plant because of the special character of the product. The product requires no assembly - just machining and coating operations - and finished inserts are produced using one of about a dozen possible insert bases. These characteristics, combined with the fact that a given insert base is used for both make-to-stock (MTS) and make-to-order (MTO) finished inserts, are why the company uses MRP to help plan shop floor releases, while a reorder point approach is primarily used to plan purchase orders.

Material requirements planning for MTS inserts works as follows. Weekly demand forecasts for each MTS insert are periodically updated and used by the

MRP system to compute finished goods weeks-of-supply (WOS). Each MTS insert has a pre-specified safety stock level stated in terms of WOS. An MRP report is generated weekly. The report lists recommended shop floor releases for MTS inserts based on a pre-specified lot size and a pre-specified lead-time offset from when insert inventory is projected to dip into safety stock. The safety stock level, lot size, and lead-time for an insert can only be changed with agreement among the inventory control, purchasing, and manufacturing groups. The report also lists projected inventory at the end of each week for each insert base (i.e. raw material for finished insert). These projections, however, ignore possible demand for MTO inserts. Furthermore, the usage rate for each insert base tends to be fairly steady (due to the small number of different bases used for the 4,000 possible finished inserts). Consequently, the purchasing and inventory control groups manage raw material inventory largely by comparing current inventories with reorder point targets. Purchase orders are released as inventory levels approach or dip below these targets.

Shop floor releases are jointly determined by the master scheduler, who we will refer to as Jim, and the manufacturing manager, using two key reports - the MRP report with recommended releases for MTS inserts and a booked order report. The booked order report lists orders for MTO inserts that have not yet been released to the shop floor. The information in these reports combined with Jim's sense of shop status (e.g. amount of WIP, loads at various work-centers, etc.) are the main information sources used when deciding which orders to release in the upcoming week. The number of shop floor releases in a week usually ranges between 100 and 500. While the order release decision is normally a once per week activity, there are exceptions (e.g. customer calls in mid-week to request an earlier ship date).

Before a shop order for some quantity of an insert can be released to the shop it must be assigned a route. There are over 5,600 routes defined in the system. Manufacturing engineering either selects a previously defined route or creates a new one. A shop order number is automatically generated when the information is entered.

Order booking at the company is straightforward. If a customer places an order for MTS inserts, the inside sales group first checks if there is sufficient inventory. If so, then nothing needs to be manufactured and the order is shipped from inventory. Usually inventory is sufficient to cover the order, but if not,then inside sales quotes a shipment date based on the standard manufacturing lead-time in the MRP system. If a customer places an order for MTO inserts, then the inside sales group checks with Jim to determine when the order can be shipped. Jim estimates lead-time as one week per remaining operation (e.g. an insert requiring four operations is projected to be available in four weeks). Due to lot sizing considerations, the shop order quantity for a MTO insert will sometimes be larger than the quantity on the particular customer order(s) that generated the release. The company customarily defers processing final operations on these excess inserts (i.e. a tactic of delayed differentiation). These are referred to as *semi-finished* inserts and, depending on the choice of the final operations, these semi-finished inserts can be transformed into a number of different finished inserts. Jim uses his knowledge of the types and quantities of semi-finished inserts when determining the number of remaining operations for lead-time estimation. He provides lead-time estimates on an average of about 50 orders per week. There is no standard rule used for estimating lead-times for customized inserts. Customized inserts are engineered-to-order and make up less than 5% of total volume.

Jim's main responsibilities begin once an order has been released to the shop floor. The schedule at the plant is formally communicated to operators through dispatch lists that are posted at each work-center. As the master scheduler, Jim is responsible for updating dispatch lists and for moving orders between work-centers. His approach to scheduling, which draws on over ten years of experience, is detailed below.

A key source of information for the scheduling function is a database that contains real-time information on shop status. Each operator at a workcenter is assigned a unique employee number. Prior to processing a shop order, the operator logs on to the database and enters his or her employee number along with information on the shop order about to be processed. When the procedure is finished, the status of the order is changed from *awaiting processing* to *in-process* at the workcenter. A similar procedure is followed when an operator stops working on an order. If an operator stops working on an order prior to completion (e.g. operator leaves at the end of a shift), information on the degree to which the order is complete is entered in the database. The database not only helps Jim understand what is going on the shop floor, but contains information that is used for planning purposes (e.g. develop estimates of operation processing times) and to track improvement efforts.

Jim normally arrives at the plant at 6:30 a.m. One of the first things he does is review a report of information in the shop floor database. The report is sorted by workcenter type and contains data on shop orders due within four weeks that are waiting to be processed. After scanning the report, he meets with the manufacturing manager, shop floor supervisors, and maintenance personnel. The purpose of the meeting is to report and discuss any problems. Problems are documented, corrective actions are initiated, and follow-up activities on previous activities are performed.

After the meeting, Jim walks through the shop floor, checks all workcenters, and browses through the dispatch lists at every resource in order to find out what is going on and what is left from the previous shift. His objective is to get an accurate picture on what is done, what is in process, and what needs to be done next. He looks for discrepancies between what is actually happening and the account in the shop floor database (e.g. operator forgetting to sign-off on the completion of an order) and he identifies workcenters that are getting low on work. As a general rule, Jim maintains a maximum of three days of work in the dispatch list of each workcenter, and every workcenter should always have at least 24 hours of work at the end of a workday.

Each workcenter has an input bin and an output bin for storing shop orders. During the initial walk-through, and periodically throughout the day, Jim moves shop orders from output bins of workcenters to the input bins of the next workcenters. As he does this, he looks for opportunities to group orders together to save set-up time and he updates the dispatch lists (e.g. adds new orders to the dispatch list and/or changes the processing sequence).

Jim is interested in completing orders on time, freeing up capacity through set-up savings, and keeping work-centers occupied. He attempts to make the most of existing capacity because work-centers periodically break down and the plant has been running at or near full capacity. His approach to achieving these objectives through scheduling a group of similar work-centers encompasses two steps. He first identifies those unscheduled orders with similar due dates (e.g. due within a two-week window) that are due soonest, and then concentrates on assigning and sequencing these orders at the work-centers. The two steps are repeated until all orders are scheduled.

Jim balances three main considerations when assigning orders with similar due dates to a specific work-center of a given type. He strives to i) save set-ups by grouping similar jobs together, ii) maximize future assignment flexibility by filling the queue of the least flexible work-center first, and iii) maintain a similar amount of work in the work-center queues. He tries to keep workload relatively balanced across the work-centers because he would like completion times at the stage to be approximately proportional to due dates (e.g. orders due within the next two weeks are completed before orders due within the follwing two weeks, etc.). The rule of thumb for sequencing the orders with a similar due date at a particular work-center is: outside customer orders first, reservation and inside customer orders second, and stock replenishment orders last. Reservation orders are linked to a blanket purchase order. A blanket purchase order refers to a commitment by a customer to order a total quantity of product over some time frame (e.g. six months), but the timing and quantities of individual shipments have yet to be determined. Inside customer orders refer to orders from another plant within the company. Exceptions to this general sequencing rule occur in order to take advantage of set-up savings.

With ten years of experience, Jim feels that it is relatively easy for him to assign orders to work-centers and update the dispatch lists from observing what is happening on the shop floor and analyzing information in the shop floor database. He feels comfortable with the logic he has developed and refined over the years. A bigger problem is a lack of time. It is a challenge to keep ahead when moving orders between work-centers, keeping dispatch lists at 31 work-centers up-to-date, maintaining an accurate picture of ever changing shop status (both in the present with some 1,000 orders in process and projected status over the near future), and resolving problems.

12.4 FIELD TEST PROJECT

12.4.1 Project background

The plant manager (hereafter, Mike) felt that a computer-based scheduling system could be a means to improve scheduling efficiency, thereby giving Jim more time to anticipate and solve problems. An interest in improving schedule quality (e.g. better on-time delivery and more efficient use of resources) was not a major motivating factor, but instead was viewed as a desirable consequence of improved efficiency. Mike and Jim became aware of various scheduling systems through in-house studies and from a vendor presentation series organized at the University of

Wisconsin-Madison. Mike saw some potential in these systems, but was not ready to make an investment because of the high cost with uncertain benefit. This is the background that led to a joint academic/industry project to develop and test a prototype scheduling system.

The project attempted to combine academic and industry interests for mutual benefit. The academic objectives were to i) gain insight into scheduling practice, ii) develop a software platform that could be used as a link to industry and serve as a test-bed for scheduling research, and iii) provide valuable experience and training for students. The company's objective was to gain deeper insight into the benefits and risks of computer-based scheduling systems. Three possible outcomes for the company were envisioned: i) conclude that scheduling software is not worthwhile at the present time, ii) purchase commercial scheduling software, and iii) continue as a research-focused test site (i.e. test-bed for scheduling methodologies with no commitment to ongoing maintenance or enhancements).

The project was approved in July 1995 with company and university funding. In addition to the author, the project team included three students, Mike, Jim, and a person from the information systems group at the company. Due to differing backgrounds and terminology, one of the biggest challenges with any project of this nature is clear communication. The centerpiece of the data-gathering and communication process was what we called the *project document*. Its purpose was to provide a clear ongoing record of the project; it contained all pertinent information and was regularly updated and expanded. The first draft, which was completed in June 1995, contained the proposal, the budget, and timing of project milestones. Over the two-year period the document continually expanded in both scope and detail to include memos (e.g. progress reports) and descriptions of information systems, data file layouts, resources, markets, manufacturing process, relevant planning and control processes, prototype system design, system documentation, and lessons learned from the project. We primarily collected information through interviews and analysis of written documentation. It was not unusual for our initial interpretation of some point to be incorrect, and we found that answers to the same questions were not always consistent among different people or even the same person at different times. A given section of the document would typically go through several iterations before it became stable. In addition, one student volunteered a week of his time at the plant working as a machine operator and inspector. We found this informal mode of data collection to be especially valuable for both validating our understanding and for exposing inconsistencies and gaps in our knowledge.

To summarize the evolution of the project, intensive data-gathering and systems analysis began in July 1995 with the project approval. During our weekly visits, we focused on gaining a basic understanding of how the scheduling function was performed and a detailed understanding of the information systems that supported the scheduling function. By the end of August 1995 this information in the project document had stabilized. Between August and October, we prepared the system design, coded the primary user interface, and developed a detailed system development plan. Coding and testing took place from October until a workable version of the system was completed in March 1996. The system was then tested and refined until June 1996 when it was implemented at the plant. For the next 14 months, Jim experimented with the system and offered suggestions for improvement. The following two subsections outline the design of the prototype and our experiences with the field test.

12.4.2 System design

The prototype was developed and implemented on a Windows platform (originally Windows 3.11 and later upgraded with Windows 95) in APL★Plus III version 1.2 from Manugistics. APL★Plus III is a general array processor computer programming language that is well suited to manipulating schedules as matrices; it also includes features for programming Windows' graphical user interface objects.

A computer-based scheduling system assumes an underlying model of a manufacturing operation. Our goal was to develop a model general enough to be applicable to a range of discrete manufacturers. The model for the prototype is defined through four data objects - *calendar*, *resource*, *product*, *order* - each of which is a large nested array. The *calendar* data object contains as many shop calendars as a user wishes to define. A shop calendar specifies downtime periods. For example, there is typically a shop calendar that specifies when the plant is shut down as well as several shop calendars that define periods of down-time for planned maintenance on various resources. The *resource* data object defines shop resources. These resources may be work-centers, tooling, labor, or whatever is relevant for scheduling at the plant. When defining a resource, it is necessary to specify the number of identical copies of the resource and the set of shop calendars that apply to each copy. A resource may also be designated as an infinite capacity resource, in which case the number of copies is assumed not to be a binding constraint (e.g. a subcontractor would typically be designated as an infinite capacity resource; processing time would be set to the estimated lead-time). The *product* data object defines the products that can be produced. The data object contains one or more routes for each product. A route defines all the necessary operations (including subcontract work if appropriate) and precedence relationships among operations. An operation is defined by a set of resources linked by and/or relationships ('and' to require multiple resources simultaneously and 'or' to allow a choice among alternative resources). A variety of data may be specified for each operation-resource pair, including set-up time/cost, set-down time/cost, processing time/cost per unit, an in-to-out ratio (used to account for yield rates and assembly/disassembly operations), and a quantity independent loss due to scrap. Each product includes a code that is used to signal whether a set-down and set-up is required when changing from one product to another. The last data object contains information on shop orders. A shop order defines one or more jobs, where a job is a required quantity of a specific product. Each job is defined by a product, order quantity, due date, early cost rate, tardy cost rate, a route code, and information on when raw material will be available and the degree to which various operations are complete.

The order in which the data objects were presented reflects the natural hierarchy of relationships; the *resource* data object points to calendars in the *calendar* data object, the *product* data object points to resources in the *resource* data object, and the *order* data object points to products in the *product* data object. The data objects may be filled through an extract feature that reads data from files in a prespecified format. Alternatively, calendars, resources, products, and orders can be added, copied, deleted, edited, and saved through a series of dialog boxes.

The scheduling engine uses the four data objects as input and either schedules selected orders around an existing schedule or generates a new schedule from scratch. Before a schedule is constructed, the data objects are checked for consistency (e.g. a resource referenced in the *product* data object is defined in the *resource* data object, etc.).

Our top priority was to design an elegant underlying model and a rich user interface in order to provide a stable foundation for possible future algorithmic experimentation and enhancement. We chose a relatively simple and easy to implement scheduling algorithm, namely *priority-based forward simulation*. The algorithm works as follows. Given an initial schedule (e.g. scheduled downtime on the various resources and any jobs that have already been scheduled) and a start time specified by the user, the algorithm first identifies those operations that can be scheduled to begin processing the start time. This set includes only those operations with sufficient raw materials (as indicated by an operation early start time), no unfinished precedent operations, and at least one of each type of resource required for its processing is available for the amount of time necessary to complete the operation. If this is nonempty, then the top priority operation is selected. Operation priority is determined according to an *earliest filtered due date* priority rule with ties broken according to the *weighted shortest processing time* rule. The earliest filtered due date rule is controlled by a user-specified filter width during which shop order due dates are viewed as equivalent. For example, a filter width of one week means that all shop orders due within the same week are viewed as having the same due date. Among all schedulable operations with the same earliest filtered due date, the operation with the smallest processing time to tardiness weight ratio is the highest priority. Operation processing time, in this case, is based upon the maximum processing time required among those resources that are available. The rationale for the maximum operator, as opposed to say average or minimum, stems from the possibility that some resources such as a set-up crew may only be needed for a short time compared to workcenter or tooling resources. Once the highest priority operation is identified it must be scheduled, but the routing may allow for choices among the resources. The logic for resource selection makes use of a distinction between the primary resource (e.g. workcenter) and secondary resources (tooling, labor) as identified in the routing. The primary resource that will complete the operation soonest is selected. In other words, the algorithm favours low set-up time and low unit processing time when choosing from multiple primary resources that are available. Secondary resources are selected arbitrarily if there is a choice. At this point, the top priority schedulable operation has been identified and the resources that will process the operation have been identified, so the next step is scheduling the operation on the resources. The set of schedulable operations is updated to account for the fact that some resources are no longer available, and the logic for operation selection and resource identification is repeated. Once there are no operations that can be scheduled to begin at the current time, time is advanced to the point when a resource becomes available or an operation becomes available for scheduling (e.g. raw material available and precedent operations complete), whichever occurs first.

Schedulable operations, if any, are identified and scheduled. The logic continues until all operations of all shop orders have been scheduled.

A schedule can be viewed in either text or Gantt chart format (see Figure 12.1). There are a variety of text formats for viewing a schedule and schedule performance statistics. One may select specific orders or resources, data can be sorted in several ways (e.g. by order code, product code, or tardiness), and a schedule of operations can be viewed by resource or by order code. Similarly, orders or resources can be selected and sorted in several ways when viewing a Gantt chart. The Gantt chart includes a zoom feature to control the amount of data displayed.

Text reports and the Gantt chart include editing capability. Schedule changes may be saved to a file or undone. One text report displays the operations scheduled at a selected resource. The mouse can be used to select an operation and drag it to a new position in the sequence; job completion times and schedule performance statistics are updated accordingly. The editing features incorporated into the Gantt chart are richer. Operations can be deleted, operation processing times can be changed, and operations can be dragged and dropped to a new position on the same resource or to a different resource. A right mouse click on an operation exposes detailed data (i.e. resource, order code, product code, operation code, operation start and completion time, order start and completion time). The prototype contains context-sensitive online help that explains the available menu options and features throughout the system. Figure 12.1 illustrates a few sample views of the user interface.

12.4.3 Field test experience

The information systems group at the company wrote a series of programs that write information from company databases to three files - one file each for the *resource*, *product*, and *order* data objects. Shop calendars were not stored in the company databases so this information was maintained manually (i.e. calendars created, copied, and edited by pointing and clicking a mouse). These programs were set to run automatically at 5:00 a.m. every workday morning. After the programs were run, the three files were automatically sent via email as file attachments to Jim and the author. When opening the prototype in the morning, the first step is to update the *resource*, *product*, and *order* data objects with information in the files. The extract program could be run any time during the day in order to update the system according to current shop status, though this was not normally done.

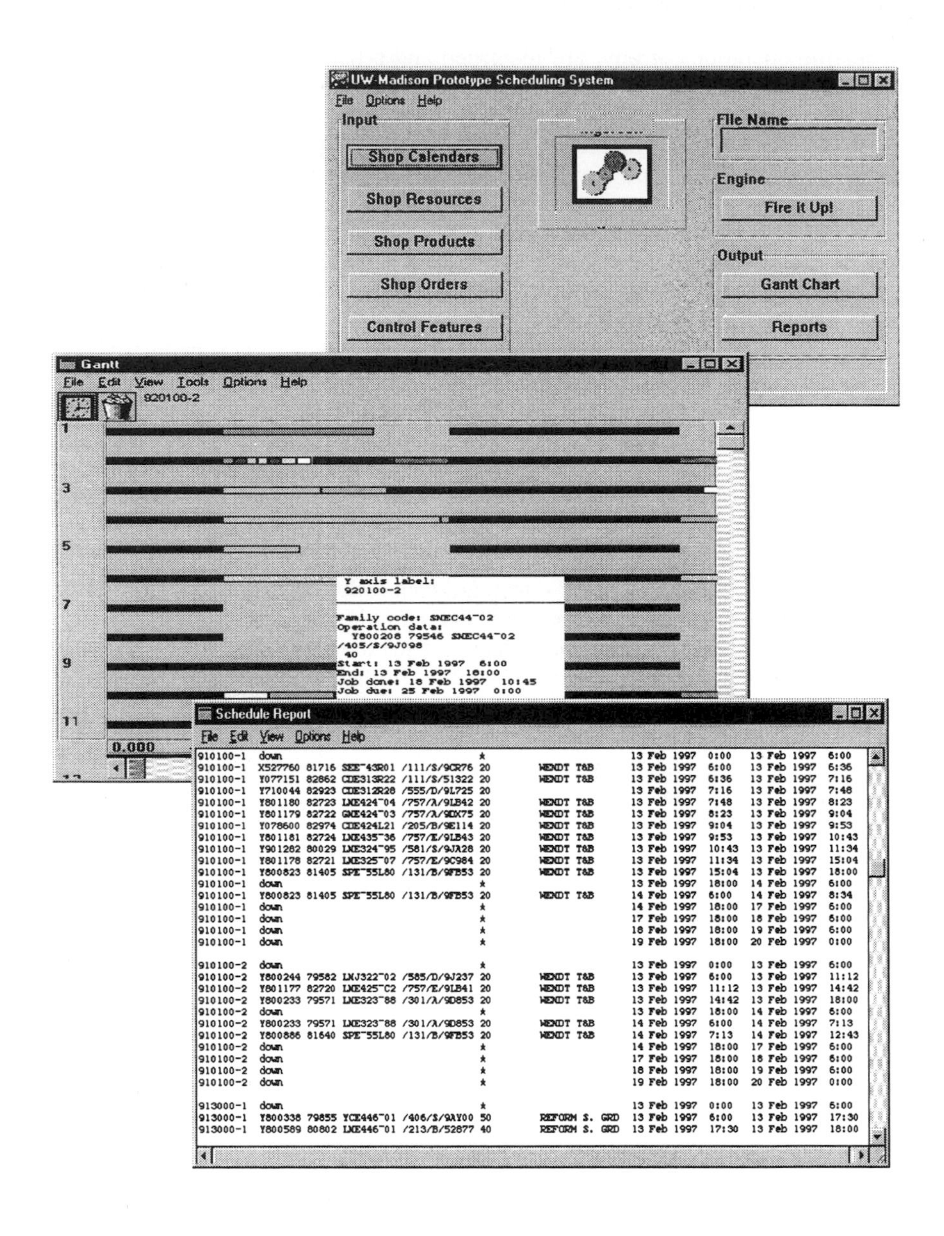

Figure 12.1 Example forms from the system. The text window in the Gantt chart displays information on the scheduled operation (appears when the right mouse button is clicked).

We tested and debugged the system during the three months prior to implementation at the plant. In addition, the first few weeks after implementation were devoted to identifying and correcting errors. By the end of June, Jim began experimenting with the system.

We continued to develop the system in response to feedback from Jim. Jim helped us identify weaknesses in the system design. We found, for example, that we incorrectly assumed an operation should not begin processing unless it could be run continuously until completion. In reality, operations are routinely scheduled around planned maintenance and weekend breaks (e.g. an operation will start Monday morning at the point where it left off on Friday evening), and we changed the scheduling engine accordingly. We also found that Jim was not comfortable scheduling beyond what was waiting to be processed at a particular work-center. Due to the dynamic nature of the plant (e.g. machine breakdowns, absenteeism), he felt it was not worthwhile to schedule an order at a work-center on the assumption that the order would be complete at a previous work-center by a particular time. As a result, the information systems group at the company wrote a program that only used information on the impending operation when defining product routes. One other difficulty we encountered related to the limits of technology, an issue likely to be less significant today. Our approach was to store the entire model in core memory in order to avoid time spent reading and writing to a database. We started with a 16 megabyte RAM PC and found this to be insufficient (usable space was closer to 8 megabytes due to space consumed by other open applications). The prototype would make use of 'virtual RAM,' which meant that it would read and write to Windows system files when there was insufficient space. This slowed down the engine considerably. We doubled the RAM to 32 megabytes. This helped, but did not eliminate the problem (e.g. 4 minutes to schedule 230 single operation jobs on 30 resources using a 75 MHz Pentium processor).

Jim also suggested changes that improved the reporting and editing features of the system. Some of these changes dealt with different ways of sorting, selecting, and displaying information on a text report. We also added an editing feature that allowed a user to easily rearrange the sequence of jobs in a dispatch list.

By February 1997, key features identified during field testing had been implemented. While we continued to develop and implement minor refinements after this point in time, the focus of our academic/industry group shifted to assessing what we had learned from the field test. Ideally, scheduling systems support three basic functions: i) development and repair (i.e. making adjustments in response to change) of a schedule, ii) generation of lead-time estimates for due date negotiation, and iii) data collection and assimilation of shop status (i.e. how work is progressing and availability of resources). While the prototype addressed all three functions, it was by far the weakest on supporting data collection and assimilation for two reasons. First, user intervention was required to access current information on shop status. Jim would have to run a computer program to retrieve the latest shop floor information, then click buttons in the prototype to load it into the system. Second, the virtual shop as represented in the shop floor database was incomplete and inaccurate. For example, the database does not identify what jobs are stored in the input and output bins of each work-center. In addition, shop floor personnel do not always update job status correctly or in a timely manner. We did not anticipate this second factor during the early stages of the project because we viewed the plant's information systems as one of their strengths. We now believe it is difficult to overestimate the sensitivity of effective computer-based scheduling to the quality and scope of the underlying information systems.

From our experience over the two-year period our team concluded that three issues should be addressed in the near future. First, information systems supporting data collection and assimilation were most in need of development. A scheduling package and supporting systems would need to be especially strong in this area in order to be viable at the plant. Second, additional discipline in procedures affecting the information systems was required, as operator errors in logging orders were prevalent. This underscores the importance of the human element in effective information systems. Third, it had become apparent that management did not have a good understanding of the causes of relatively poor on-time performance, and that this needed to be resolved.

The project officially ended in July 1997. In the meantime the plant has been concentrating on improving current operations with the anticipation of possible implementation of commercial scheduling software within two years. The decision will not be made in isolation. The company operates three plants and will look for a single system that can be applied in all three settings. In fact, a sister plant is currently testing an ERP package. If the test proves successful, then management will take a close look at the scheduling application that is offered by the ERP system vendor.

12.5 CONCLUSIONS

We have described how scheduling is done at one plant in the machine tool industry and we reported our experience with a project to develop and test a prototype scheduling system at the plant. The project combined industry and academic interests. Plant management was considering an investment in commercial software and wanted to first gain deeper insight into computer-based scheduling systems through hands-on experience. They also held out the possibility of continuing to operate as a test site using the prototype system. From an academic perspective, we wanted to gain insight into scheduling practice, provide experiential learning for students, and develop a software platform suitable for testing research results in an industry setting. While we realized our objectives, we regret not having the opportunity to use the system as a test-bed for research on scheduling algorithms.

In addition to hard work by all involved, we think there are two factors that contributed most to the effectiveness of the project. First, the jointly developed statement of objectives and possible outcomes meant that we had a common set of expectations going into the project (nontrivial because of different interests). Second, the evolving project document forced misunderstandings to become apparent early on and provided a formal mechanism for resolving these differences.

This chapter is motivated by the gap between scheduling research and scheduling practice. It adds to the descriptive literature on how scheduling is done in practice - literature that when taken as a whole is useful for guiding research activity to areas of high impact. In addition, the chapter may prove a useful reference for researchers considering similar field-based approaches as part of their research program.

12.6 ACKNOWLEDGEMENTS

I would like to acknowledge the hard work of three students on this project - Suwito Hadiprayitno, Ta-Yao Wei, Jinhwa Kim - and the significant time and effort of the plant manager and the master scheduler at the plant. I also appreciate the perceptive comments by Professors MacCarthy and Wilson on an early draft of this manuscript.

12.7 REFERENCES

Buxey, G. (1989). Production scheduling: practice and theory. *European Journal of Operational Research*, **39**, pp. 17-31.

Conway, R. W. (1996). Linking MRP-II and FCS. *APICS – The Performance Advantage*, June, pp. 40-44.

Dessouky, M. I., Moray, N. and Kijowski, B. (1995). Strategic behavior and scheduling theory, *Human Factors*, **37**, pp. 443-472.

Kanet, J. J. and Sridharan, S. V. (1998). The value of using scheduling information in planning material requirements. *Decision Sciences*, **29**, pp. 479-497.

MacCarthy, B. L., Crawford, S., Vernon, C. F. and Wilson, J. R. (1997). How do humans plan and schedule? *The Third International Workshop on Models and Algorithms for Planning and Scheduling Problems,* Queens College, Cambridge.

MacCarthy, B. L. and Liu, J. (1993). A recent survey of production scheduling. *International Journal of Production Research*, **31**, pp. 59-79.

Wiers, V. C. S. (1997). *Human–Computer Interaction in Production Scheduling: Analysis and design of decision support systems for production scheduling tasks.* Wageningen, The Netherlands: Ponsen & Looijen, Ph.D. Thesis, Eindhoven University of Technology.

CHAPTER THIRTEEN

Architecture and Interface Aspects of Scheduling Decision Support

Peter G. Higgins

13.1 INTRODUCTION

The gap between classical scheduling theory and actual industrial practice has been discussed throughout this book. When it comes to the complexity and uncertainty of many manufacturing environments, the scheduling activity addressed by classical researchers and the activities followed by industrial practitioners are disparate. Both constitute goal-directed allocation of resources over time to perform a collection of tasks: however, their specific goals and decision-making activities differ markedly. In classical theory, goals are simple quantitative measures (e.g., resource utilisation and tardiness) and scheduling activities are algorithmic procedures (e.g., heuristics such as Shortest Processing Time (SPT) or sub-optimising methods, for instance, branch-and-bound).

In real plants, scheduling is a social activity; schedulers are subjected to the social dynamics operating within the organisation (see also Crawford in Chapter 5). Rich and assorted, intrinsic and extrinsic values and goals of the system of people of which they are part, affect them (Gault, 1984). Their behaviour is situated activity and is therefore embedded in the particular work environment (Bødker, 1991 and see Wäfler in Chapter 19). They have to make effective decisions in circumstances where the state of the system cannot be clearly predicted, information regarding jobs, materials and resources are ill-defined and scheduling goals are diverse. Scheduling under 'perplexity' rather than combinatorial complexity characterises many schedulers' worlds, especially those working in small job shops. Perplexity, derived from the Latin word *perplexus* meaning involved, is an apt descriptor:

> Perplexity: inability to determine what to think, or how to act, owing to the involved, intricate, or complicated conditions of circumstances, or of the matter to be dealt with, generally also involving mental perturbation and anxiety. (Oxford English Dictionary).

Confusion and uncertainty are the hallmarks of perplexity (McKay *et al.*, 1989). Schedulers operating in the 'real world' of manufacturing must continually surmount environmental perplexity. As McKay discusses in another chapter, decision strategies are often complex and hinge on the schedulers' awareness of the subtle relationships between factors that comes from an intimate knowledge of the plant, products, and processes. Indeed, McKay expresses amazement that schedulers make so few mistakes in coping with the amount of data and degree of complexity usually present.

In managing perplexity, schedulers do not use clearly defined, easily articulated, models of the scheduling process. Their knowledge develops over time as they continuously gather information as social beings (McKay, 1987). It does not stem from consciously elaborated principles or procedures, but is a matter of habit, experience and practice, which matures over years of experience in the work environment (Polanyi, 1966). This presents a basic problem for software development, as the automated construction of schedules requires all the constraints and rules to be prescribed. Instead of trying to elicit schedulers' knowledge to place in the software, developers of computer applications should give schedulers a leading role in the decision-making process. Then they can take advantage of the special abilities of humans (Sheridan, 1976);

- their ability to recognise patterns in the data, identifying what is, and what is not, essential from the current context (Sharit, 1984; Papantonopoulos, 1990);
- their ability to handle unexpected events and information that may be diverse, inexact, or conflicting (Higgins, 1996a);
- their ability to formulate general rules from specific cases (Sharit, 1984; Meister, 1966).

To tease out the form of a suitable methodology for designing tools that suit perplex environments, the chapter opens with a theoretical discussion on scheduling activity as purposive-rational action. It is argued that a methodology for developing scheduling tools must produce a decision architecture that encompasses the manufacturing system and the problem-solving operations of human decision-makers, without requiring complete knowledge of the scheduling domain. Constraint management is then used as a unifying concept for discussing the efficacy of prevailing scheduling tools and for expounding a new decision architecture and a new methodology for their development. Using data collected from a field study, the elements of a hybrid intelligent production scheduling system (HIPSS) based on the new architecture and methodology are then presented.

13.2 THEORETICAL FOUNDATIONS

While the consciousness of schedulers may form through shared activity with other people, their behaviour is purposeful as it is directed towards the accomplishment of specific tasks. The focus is a functional psychology relating perception and action. Schedulers are usually practical persons who have a keen understanding of their resources, the capabilities of machines and work practices. Their competence comes from a deep understanding of the work domain, evolved through experience gained under a variety of circumstances (Dutton and Starbuck, 1971; Brödner, 1990; Higgins, 1999a, 1999b). They acquire much of their knowledge by solving pressing problems (De Montmollin and De Keyser, 1986). They learn to deal with a vast array of factors arising in the working environment, which may be unpredictable and are not easily placed into a theoretical scheduling context.

Gault (1984) puts forward a theory of Social OR Action that places purposeful action into the context of operations research. He focuses on the actions of people in organisations inhabited by purposeful beings. Schedulers are subjected to the social dynamics operating within the organisation. Individual goals and relationships with other persons are subject to change. That is, schedulers in practice cannot isolate themselves from other activities within the organisation. They have to cope with complex systems of changing problems that interact with each other. Ackoff (1979a, 1979b) sees planning under these conditions as a process of the management of messes, not the solving of problems. He therefore sees it as pointless to seek optimal schedules. Scheduling becomes a process of synthesis more than analysis. Gault argues that scheduling support should help actors improve the quality of their actions as the complex array of consequences associated with any action includes many that schedulers experience subjectively. Merely focusing on improving the performance of a few quantifiable measures may be detrimental to the performance of the subjective dimensions, which may, in turn, be more important for organisational viability. For instance, the mix of challenging but stressful work and simple but undemanding work affects a skilled operator's performance.

Having decided upon a goal to meet, schedulers can be understood to form intentions, make plans and carry out actions (Norman, 1986). Their intentions derive from their value structures and internal goals (Rasmussen, 1985). To know how to shape a scheduling system, in which humans actively partake in the making of decisions, we need to know how humans perceive the scheduling process. How do they interpret signals from the environment and work out appropriate actions (Green, 1990)? What mechanisms do they apply in generating descriptions of system purpose and form, explanations of system functioning and observed system states, and predictions of future system states (Rouse and Morris, 1986)? Obtaining answers to these questions depends on using formal methods for describing human decision-making processes in scheduling. A systems-oriented method of analysis is therefore required that encompasses both the manufacturing system and the problem-solving operations of the human decision-maker.

As it is difficult to observe in short field studies the full complexity and subtlety of the decision strategies of schedulers operating in dynamic environments, scheduling tools have to be able to be developed without thorough knowledge of all scheduling behaviours. A methodology is needed for developing tools that assist schedulers to follow strategies they find effective, without having to encompass the actual scheduling practices in the software. While the methodology must be capable of handling incomplete domain knowledge, the decision architecture of a scheduling tool should permit refinement as knowledge of the domain improves when the tool is used in the field. That is, the design of the decision architecture should support iterative refinement of domain knowledge.

13.3 DESIGNING THE HUMAN-COMPUTER INTERACTION PROCESS

Schedulers in practice require tools that help them to do their work more effectively. Primarily, a computer aid must support schedulers to apply accustomed methods developed through years of experience. The interaction process must cope with the ways they tackle scheduling problems. Using their intuition and knowledge, schedulers should be able to guide searches for suitable schedules in directions they would like to follow. While not interfering with their perception of scheduling problems, the computer system should over time help them extend their knowledge of scheduling. These are all attributes that the 1991 SIGMAN (Special Interest Group in Manufacturing of the American Association for Artificial Intelligence) workshop on interactive scheduling identified as desirable (Kempf *et al.*, 1991).

As scheduling is event-driven, the interaction process between human and computer must allow schedulers to experiment flexibly and opportunistically with different strategies (Woods, 1988; Woods and Roth, 1988; Sanderson, 1989). For example, the software must include functions for undoing and redoing all actions so users can manoeuvre freely in the decision space. Software development has to shift from a focus on OR models and solutions to the provision of decision support for domain experts, who are at the core of the decision-making process. The decision architecture is therefore designed so schedulers are active participants in schedule construction. Unlike conventional, interactive scheduling systems that present completed schedules to users, there is constant interactivity between human and computer as they share the decision activities, though under the control of the user.

When making decisions, schedulers usually go beyond the restricted set of information displayed by Gantt charts. Gantt charts by themselves are inadequate vehicles for displaying all the information that schedulers may want to use. A Gantt chart should be seen as the product of a decision-making process; it is the plan for running the shop. While its form may well suit the exposition of a plan, it may not be suitable as the primary display for decision-making. The difference is communication as against discovery. A Gantt chart's purpose is to *communicate* the plan of work to those people who have to use it. The imperative for communication is simplicity (Bertin, 1981). However, a display for decision-making must show comprehensively all data a scheduler may use to *discover* relationships, in a form that helps them to visualise abstract relationships between data. That is, it reveals relationships formed by the interplay of the data.

Realising the inadequacies of a standard Gantt chart as an interface for decision-making, designers of interactive scheduling systems populate their screens with extra objects, either added to the Gantt chart or placed in additional windows or pop-up boxes. For example, users of LEKIN[1] can see more details of a job by double clicking any of its operations. The operation becomes encircled, the due date and expected completion date of the job are marked on the time line, and a box appears showing the routing of all the operations for the job and its contribution to the performance measure.

A 'hybrid intelligent' production scheduling system (HIPSS) that brings human and computer 'intelligence' together in the process of decision-making is

---

[1] A flexible job-shop scheduling system developed for educational purposes by Pinedo and Chao (1999).

distinctly different to other interactive systems. Unlike other interactive systems, a HIPSS allows users a broad degree of action. It preserves the schedulers' initiative to evaluate situations and to make decisions. The focus of the development of a HIPSS is on what the scheduler is doing with the computer and not what the computer is doing. In using a HIPSS, schedulers should feel engaged in scheduling activity, and not engaged in managing a computer.

To develop a 'hybrid intelligent' system requires a systematic approach to software development. The methodology for developing a HIPSS addresses decision-making in complex systems in which there are many competing and conflicting goals. It:

1. has schedulers at the centre of the decision-making architecture;
2. uses Cognitive Work Analysis (CWA) to identify the information a HIPSS needs to display;
3. uses configural display elements that replace schedulers' memory storage with *external memory* and supports schedulers to make visual inferences thereby reducing the need for any mediating inferential process.

13.3.1 Human-centred architecture

By locating humans centrally in the decision architecture, software can be designed that is not tightly bound to an overly restrictive perspective of the problem. The design emphasis moves away from including all the scheduling constraints and conditions within the software. The focus moves to the development of software that supports expert schedulers in applying their vast knowledge of the physical domain and its operational constraints to the decision-making process. Figure 1 gives one example of a human-centred architecture that allows schedulers to freely make decisions. They have full control over the application of the scheduling rules and can accept or reject advice at will from the knowledge-based adviser. Just as skilled artisans can freely wield their tools to accomplish purposeful actions, schedulers can employ a HIPSS as a cognitive tool. Schedulers can move freely within the decision space, seeking patterns within data, recognising familiar work situations, and exploring decision-making strategies under novel circumstances (Higgins, 1996a).

A human-centred architecture does not presume abandonment of classical OR methods, but instead accepts Morton and Pentico's (1993) proposition that 'All useful approaches should be pursued'. The human can act as an intermediary between the real-world manufacturing environment and the abstract world of operations research, by:

- dealing with the stated and unstated conflicting goals;
- resolving how to use information that is incomplete, ambiguous, biased, outdated, or erroneous;
- grouping jobs to meet the specific criteria for applying selected heuristics.

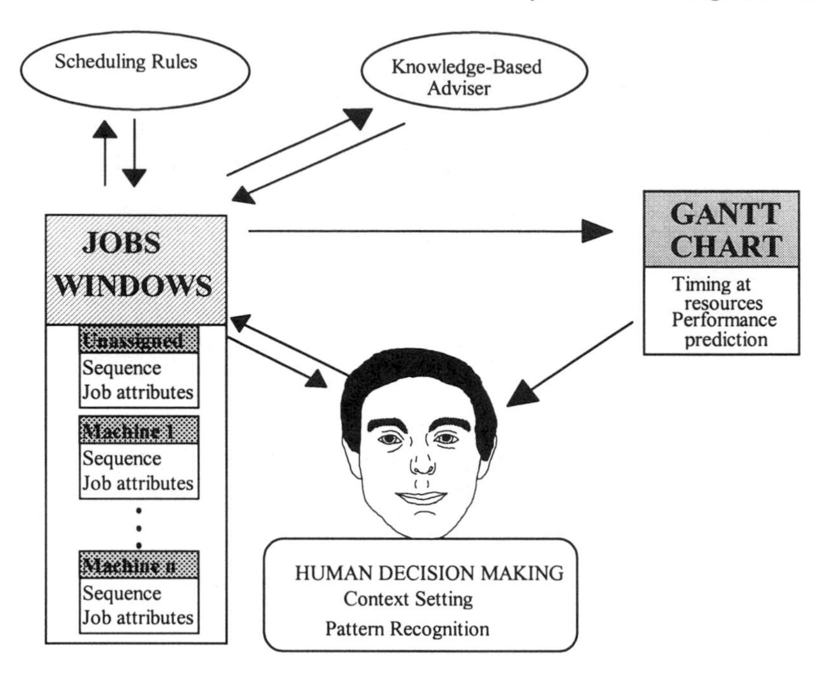

Figure 13.1 Interface elements and their location in a hybrid intelligent production scheduling system (HIPSS).

In identifying human capabilities, Nakamura and Salvendy (1994) stress the proficiency of schedulers to adapt to changing production goals and priorities, to make decisions on realistic criteria and to consider multiple goals. Schedulers can also recognise conflicts between goals and decide how to resolve them. However, to be proficient users of a HIPSS as a cognitive tool, schedulers have to completely understand its functions and behaviour (Woods and Roth, 1988). In using a HIPSS as a tool, schedulers must be able to see the connection between their own intentions, or actions, and the effects produced by them.

13.3.2 Cognitive work analysis

The development of a HIPSS for a particular industrial setting depends upon an analysis of the sociotechnical system: the manufacturing system and the problem-solving operations of human decision-makers. Cognitive Work Analysis (CWA) provides the requisite methodology (Rasmussen, 1986; Sanderson, 1998; Vicente, 1999). It incorporates two different types of analysis: Work Domain Analysis (WDA) and Control Task Analysis (CTA). WDA uses an abstraction hierarchy (AH) - a generic framework for describing goal-oriented systems - to describe a system in a way that distinguishes its purposive and physical aspects. WDA is event independent and is quite separate from Control Task Analysis (CTA), which is a subsequent event *dependent* analysis of the activity that takes place within a work domain.

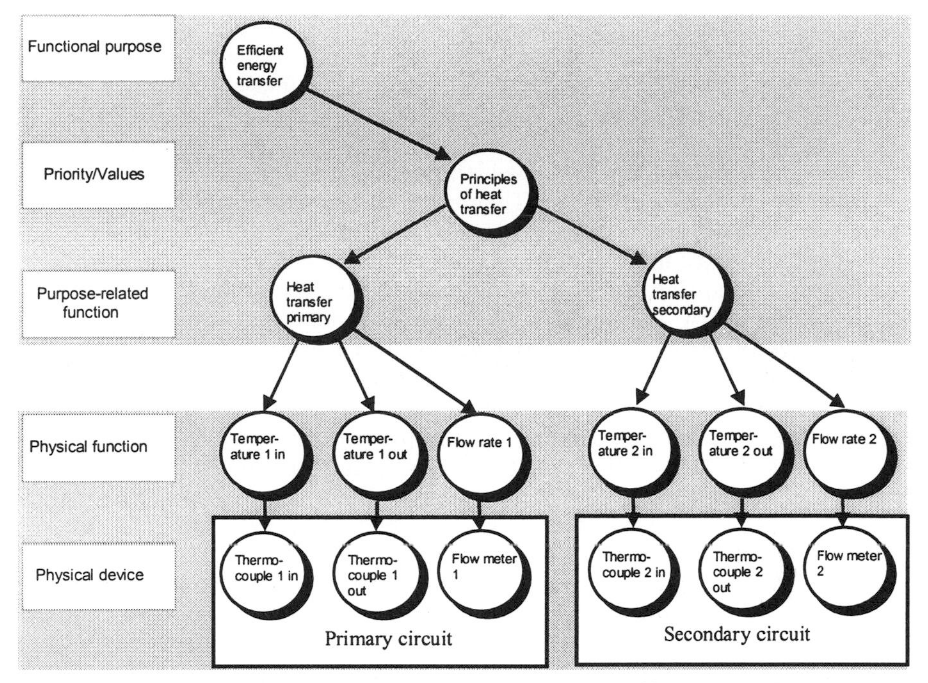

Figure 13.2 State variables associated with a heat exchanger at different levels of abstraction.

The basic features of CWA are most clearly demonstrated using a supervisory control example. While it is not a scheduling example, it is easy to explain the meaning of the AH using a system bound by physical laws. The AH in Figure 13.2 is a representation of the work domain for a heat exchanger, a component that is common in chemical processing and thermal power plants. The links show how to instantiate the design from the level of functional purpose to physical devices. The *functional purpose* of the heat exchanger is the efficient transfer of heat from one fluid to another. At the level of *priorities or values* are the intentional constraints: through the application of the physical principles of heat transfer, the designer places constraints on the behaviour of the system. The *purpose-relate function* is the transference of heat between the primary and secondary circuits. Below this are the *physical functions* and *devices*. At the physical level, there are tubes, fluids and sensors. The sensors measure the appropriate state variables. The nodes and links are clear and unambiguous.

Control-room operators (i.e. the supervisory controllers) monitor the operation of the heat exchanger, troubleshoot, and intervene where necessary to modify its performance. Their activities are the subject of CTA. Just as the AH is a formal descriptor for WDA, Rasmussen's (1986) *decision ladder* provides the rigour for representing each set of activities in CTA (Vicente and Rasmussen, 1990; Vicente, 1999). As a template of *generic* activities on which to overlay supervisory control activities, it acts as a prompt for identifying the information processing activities in the CTA (see Figure 13.2). It consists of a series of nodes, alternating between data processing activity and states of knowledge. States of knowledge are the outputs or products of the information processing activities engaged in by a decision-maker.

The decision ladder is not used to describe the decision process - the actual cognitive processes of the decision-maker - but to describe the products of information processing. For Vicente, 'The goal is to find out what needs to be done, not how it can be done'.

In monitoring the operation of the heat exchanger, the operators' activities may start at the bottom left of the ladder. Some event alerts them to monitor the condition of the heat exchanger. They scan all the instruments, observing the various temperature and flow rate values. Their 'reading' of the temperatures and flows depends on the instrument panel transforming the electrical signals from the thermocouples and flowmeters into information displayed at the level of physical function in the WDA. To identify the state of the system, they need to perceive the energy flows in and out of the heat exchanger. To do this they must know the thermodynamic relationships. The interpretation of imbalance of energy flows may be that the system is undergoing a thermal perturbation: it is not in equilibrium. If the goal state is 'maintain equilibrium', the operators may need to define a task to bring the system into equilibrium: for instance, decrease the temperature of the primary circuit fluid leaving the heat exchanger. The procedural steps may include increasing the flowrate of the fluid in the secondary circuit and concomitantly decreasing its entry temperature.

As the decision ladder provides a systematic means for analysis of the operators' decision activities, the chances of oversight are lessened. Even from this brief explanation of the decision ladder, it is clear that operators may benefit from a display that allows them to directly perceive energy flows. The procedural steps for the defined task may be automated if they are routine.

The decision activities do not have to use every node, climbing the ladder on the left and then descending on the right. Arcs can shunt across the ladder, reflecting short-cuts in the information processing activities. They are not in set positions but vary with the specific application. Consider the case of operators trouble-shooting a problem with the heat exchanger. They see an unusually low reading of the 'temperature 2 out' (see left side of Figure 13.2), while the other instruments are within their normal range. An operator who associates this pattern with past experience right away goes to the task 'to check the thermocouple electrical circuit (see right side of Figure 13.3)'. Before the task can be executed, the procedural steps must be drawn up. This stage may be bypassed, if the problem arises so frequently that the trouble-shooter can readily recall the procedure. The decision to check the electrical circuit depends upon previous experience of this type of problem. If the operators have not come across this situation before (or if on checking the electrical circuit it had been found to be functioning normally), their decision activities would return to the left side of the ladder and move upwards to more fully identify the state of the system. That is, decision-making moves to observation at a higher level of abstraction: the flow of energy in the primary and secondary circuits.

A well-designed computer interface, therefore, must depict the different levels of abstraction shown in the WDA in a way that supports all the activities shown in the decision ladder.

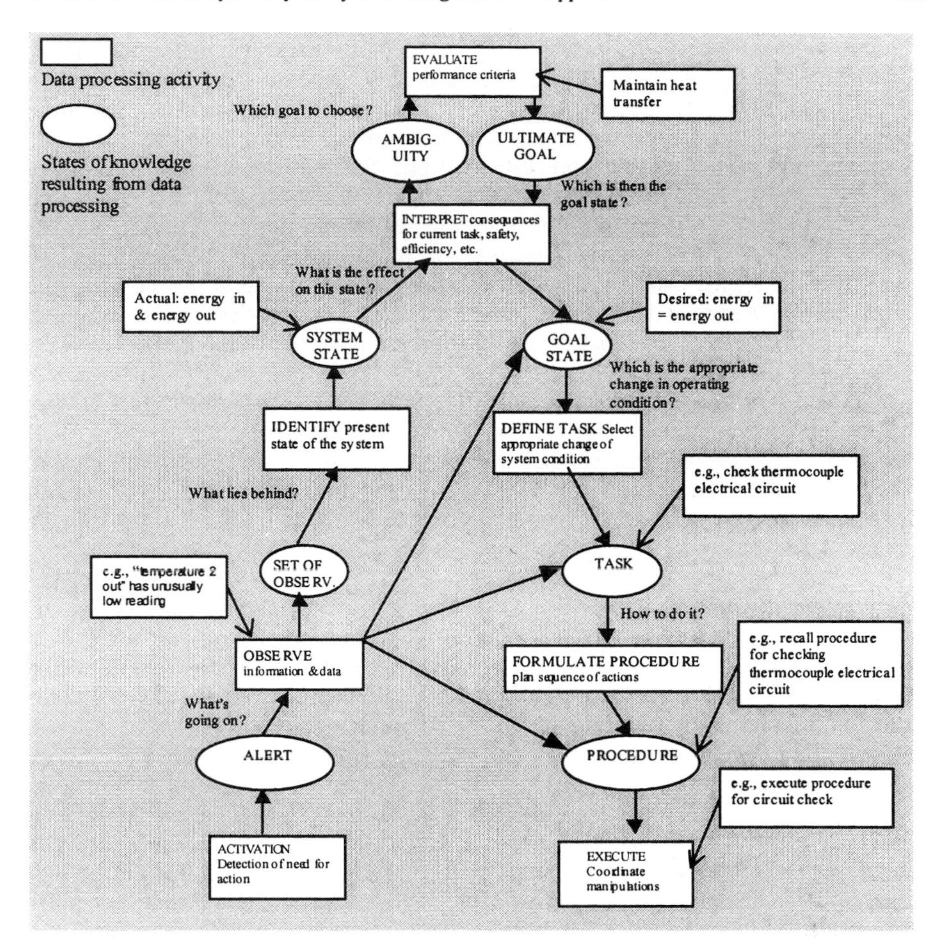

Figure 13.3 The decision-making activity of the supervisory controller in monitoring the heat exchanger.

In the process control industry, the WDA is founded on the physical laws that constrain the system: for the heat exchanger, these are the principles of fluid flow and heat transfer. In production scheduling, the WDA describes the properties of the system that are associated with the scheduling activities portrayed by the CTA. However, in describing a system from a scheduling perspective, at the level of priorities or values there are no physical laws to call upon.

13.4 APPLICATION OF COGNITIVE WORK ANALYSIS TO SCHEDULING

In job-shop scheduling, the WDA is based on a means-ends analysis of constraints in regard to meeting desired outcomes. The control activities of schedulers relate to the management of constraints. By choosing to place a particular job on a machine, a scheduler places a temporal constraint on the machine. While the machine deals with the job, it is unavailable for other jobs. Some constraints are physical, set by physical laws, while others are intentional, set to meet the intentions of the designers and users. From the schedulers' perspective, a machine's physical constraints are rigid and they define its capabilities. Allocation of a job to one machine in preference to another, where there are several similar machines, is an act of intentionally setting the variable physical constraints on one machine at a chosen time to the informational constraints defining the job. The setting of constraints is the consequence of the scheduling *activity* of an agent, either human or computer. The scheduling process is therefore the subject of CTA and not of WDA.

To demonstrate the application of CWA to scheduling, Higgins (1992, 1999a) used information he had collected from a field study of a printing company that mainly produces continuous stationery: typical products are invoice and cheque forms. In using his example, we will only discuss it here in sufficient detail to elucidate the technique of CWA.

13.4.1 Work domain analysis

The printing company has two types of offset press - The Akira presses and the Trident press, which are web and sheet-fed, respectively. Only the Akiras could print fan-fold jobs, although any press could run the sheeted jobs. The Akiras, however, being web presses must be fed paper in continuous form. After printing, the paper is cut into sheets. Cutting may take place on the press itself or as a separate operation on the Bowe cutter. The input to the Trident is stacked sheet, which the shop obtains in sheet form or converts from continuous paper. The Akiras vary in the number of colours they can print. Web presses operate at high speed and can print on both sides of a continuous roll of paper. They generally have in-line binding and cutting features. After impressing the image on the page, a web press may either place perforations across the sheet to produce fan-fold forms or cut the paper into individual sheets. On fan-fold forms made for track-fed computer printers, the press also places perforations on each side of the sheet and makes holes required by the sprockets. The final product is either fan-fold paper or stacked sheet. Sheet-fed presses operate more slowly than web presses, but are faster to set up (known as 'make ready' within the trade): accordingly, they suit short production runs.

The four Akiras are alike, except for the number of colours - one, two, four and six — that they can print and for the ancillary attachments that provide extra functionality. Each colour requires a set of three cylinders: a plate cylinder, a blanket or offset cylinder and an impression cylinder. A principal constraint in offset printing is the cylinder size. The size of the impression cylinder depends upon the required depth of paper for the job, as it must be an exact multiple of the sheet's depth. The ancillaries place additional constraints on the allocation of jobs to presses. For instance, all jobs using ink that needs ultraviolet (UV) fixing must go onto Akira 3, the six-colour press. For machine parts that are swapped between machines, the configuration of one machine affects other machines. For example, across the four Akira presses, only six colours can be printed concurrently on 297mm-depth sheet, as each colour requires a cylinder set and the cylinder size is an exact multiple of the sheet's depth.

Where a job has multiple parts, they can run either consecutively or separately on one machine. Equally, the split can be between machines. A collator then joins the parts. Which collator is used depends upon the specifications for the job, as they differed in their capabilities.

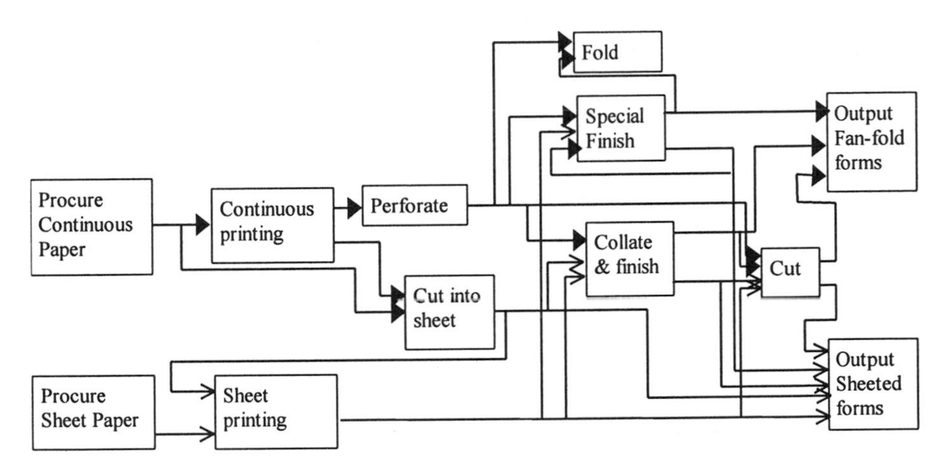

Figure 13.4 Major physical functions in printing.

The production process consists of the major physical functions shown in Figure 13.4. Finishing and special finishing can be broken down into separate procedures. The arcs show the order of operations.

Figure 13.5 shows the major resources used to perform these functions. The closed and open arrows on the arcs show the routing for fan-fold and sheeted paper, respectively. The dashed arcs from Akiras 1, 2 and 4 signify that their links will be the same as those shown for Akira 3.

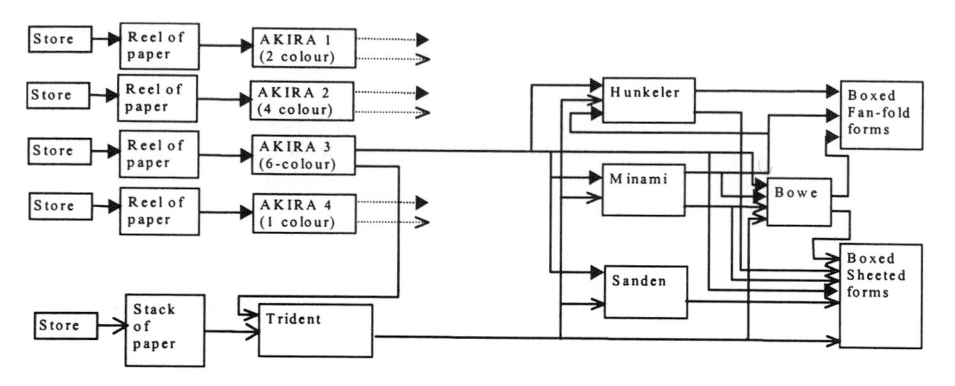

Figure 13.5 Major physical resources in printing.

Arcs in Figure 13.6 show the relations, from the ends to means, between the physical functions and physical resources. For simplicity, the four Akiras are lumped into a single generalised representation. Note that for some operations different resources can be used (e.g., folding, collating and finishing). In some cases, they are alternatives (e.g., in many cases it is immaterial which collator is used). For others it depends upon the specific operation, for example, the Hunkeler can die-cut the paper and glue transparent window on envelopes and Akira 3 is the only resource that can apply ultraviolet light to cure special inks. Arcs from the 'Special Finish' to the Hunkeler and the generalised Akira indicate these ends-means relations.

The ends-means relationships between physical functions and physical resources in Figure 13.6 form the levels of 'physical function' and 'physical device' in an AH for the WDA. For any particular job, the purpose-related level of the AH consists of the specification for the manufactured components. In other words, the job specifies the purpose of the system. The attributes of a job form the intentional constraints on the physical parameters (material, geometry, batch size, etc.) for the manufacture of the final product. The purpose-related function of the manufacturing system changes with the change in jobs. Hence, in production scheduling the WDA focuses on the configuration of the manufacturing system to meet the purpose-related function, that is, the production of a specified job.

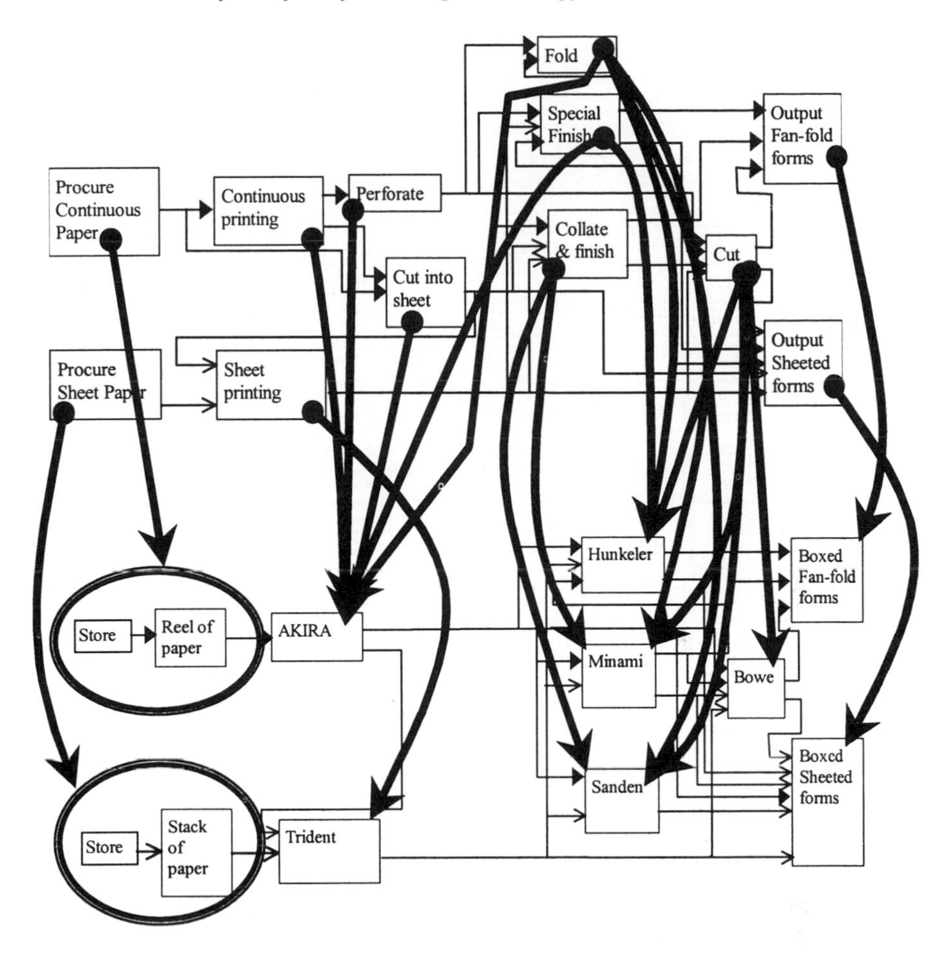

Figure 13.6 Ends-Means relationships between physical functions and physical resources.

13.4.2 Focusing on the presses

To reduce the complexity of the scheduling problem for the sake of a manageable discussion, only the formation of an AH for the Akira presses is examined. While Figure 13.7 only shows the Akira presses, the AH for a full WDA for the manufacturing environment includes all the other physical resources as physical devices.

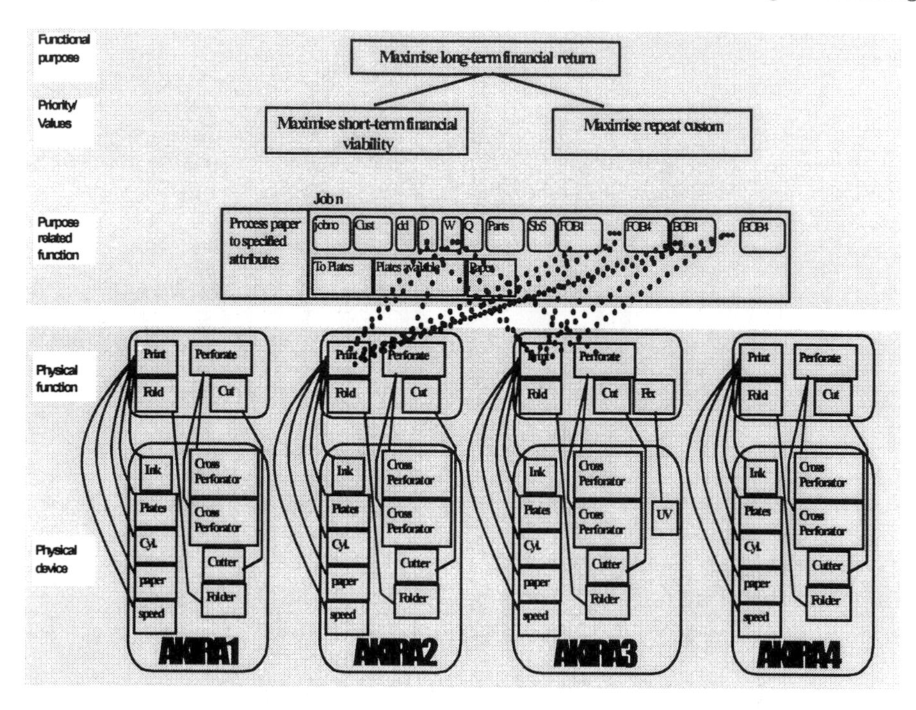

Figure 13.7 Work Domain Analysis for scheduling the Akira presses showing feasible means-ends links for a particular job specification.

For each job, links from the set of constraints, that is, the specified attributes, to particular physical functions represent the feasible alternatives. All constraints on the purpose-related function that may map to constraints on the physical function must map to a single aggregated node for a link to exist. For example, if a job requires four colours, there must be four links from the front-of-bill (FOB) and back-of-bill (BOB) nodes at the purpose-related level to four-colour nodes within the aggregated physical function 'print,' which in turn map to four separate applicators within the 'ink' device as shown in Figure 13.7. To produce a four-colour job, there are only links to Akira 2 and Akira 3, as Akira 1 and Akira 2 do not have enough colour applicators. A job allocated to a machine has an ends-means chain instantiated between the purpose-related function and the machine as shown in Figure 13.8.

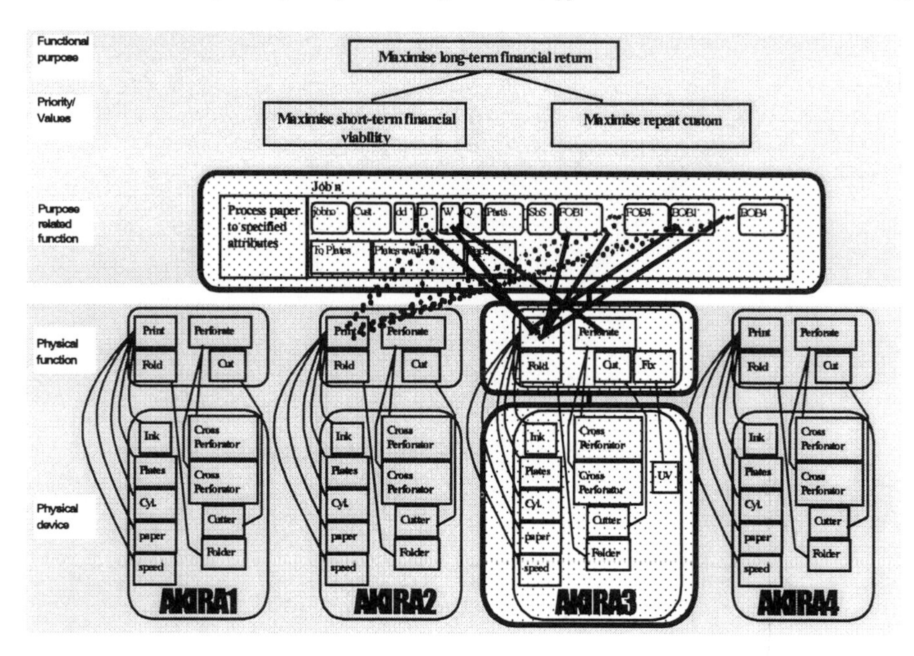

Figure 13.8 The means-ends chain that is instantiated is shown by solid links.

The hard technical constraints of the machines, which are causal constraints, project up to the purpose-related function level. Means-ends links only form where the causal constraints match the requirements of the purpose-related intentional constraints. For instance, the number of colours required for the job must be within the constraint boundary of the number of colours that the press is capable of producing. As the physical system can be 'redesigned' by changing the set up of the machines, intentional constraints at the purpose-related level set constraints on the physical device. For instance, a press's cylinder size is changed to meet the constraints set by the depth of paper required by the job.

The functional purpose of production control at the company was to 'maximise the long-term financial return' of the company. To meet this purpose, manufacture was organised to 'maximise short-term financial viability' and 'maximise repeat custom.' These are shown at the priority/values level of abstraction. How links form between the nodes the priority/values level and the lower levels will be shown next.

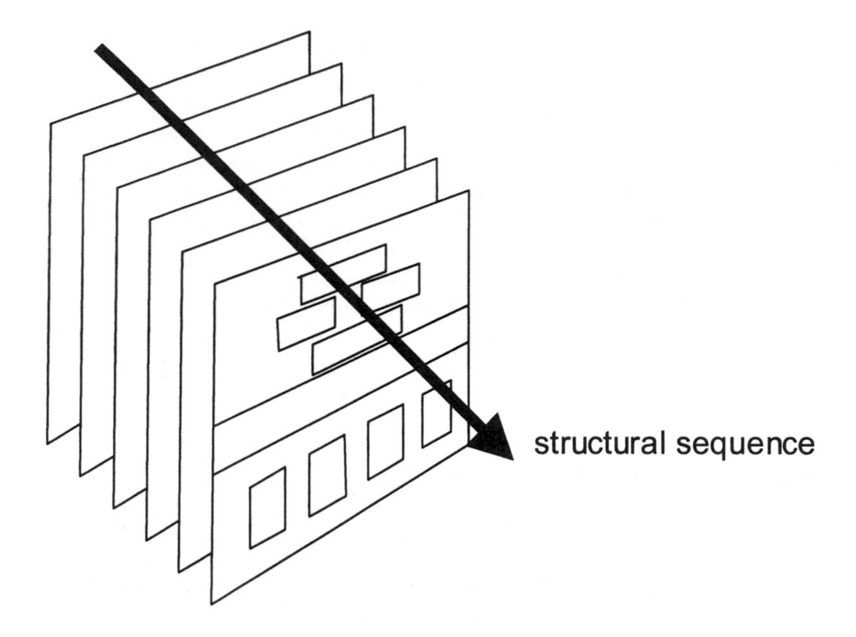

Figure 13.9 Structural sequence of Abstraction Hierarchies.

As the configuration of the physical device must meet the purpose-related end, the AH is redesigned for each job as the purpose-related function changes. That is, each job has a different AH. The AHs for different jobs can therefore be considered metaphorically as a batch of cards. Each card shows the potential configurations for the processing of the batch defined by the job. Moving from one card to another denotes the changeover of jobs. If in changing from one job to another, the machine's configuration must change, then the values of the constraints change between cards. The order of the cards sets the structural sequence of jobs (Figure 13.9): the temporal order of processing. Scheduling activity is therefore a process of ordering the cards and is therefore the subject of control task analysis (CTA).

13.4.3 Control task analysis

Control task analysis of scheduling behaviour identifies the processes in setting the intentional constraints identified by WDA. Sanderson (1991) was first to use the decision ladder for the control task analysis of the control structure for scheduling. She emphasises that the decision ladder provides a template for 'a rational and complete analysis of the type of information in which humans must engage to solve scheduling problems', and as such it does not reflect the way that humans internally represent scheduling (if at all!).

Figure 13.10 shows a decision ladder for the decision-making activity associated with developing and maintaining a production schedule. The type of information processing varies with the position on the decision ladder. The arcs link nodes that cycle between the recognition of a state of knowledge and information processing activities. Progressing up the ladder on the left, the application of knowledge characterises the decision behaviour in analysing the state of the manufacturing system. Having identified the consequences of the system state, decision behaviour moves down the right side of the ladder. Schedulers use their knowledge to plan scheduling actions, that is, actions affecting the intentional constraints. In contrast, arcs that shunt from left to right, near the bottom of the ladder, are associated with rule-based behaviour; that is, the recognition of a familiar pattern in the data triggers the application of an accustomed routine. Which arcs form depends upon the domain, the scheduler's expertise and the presence of 'stimuli' for invoking particular behavioural responses.

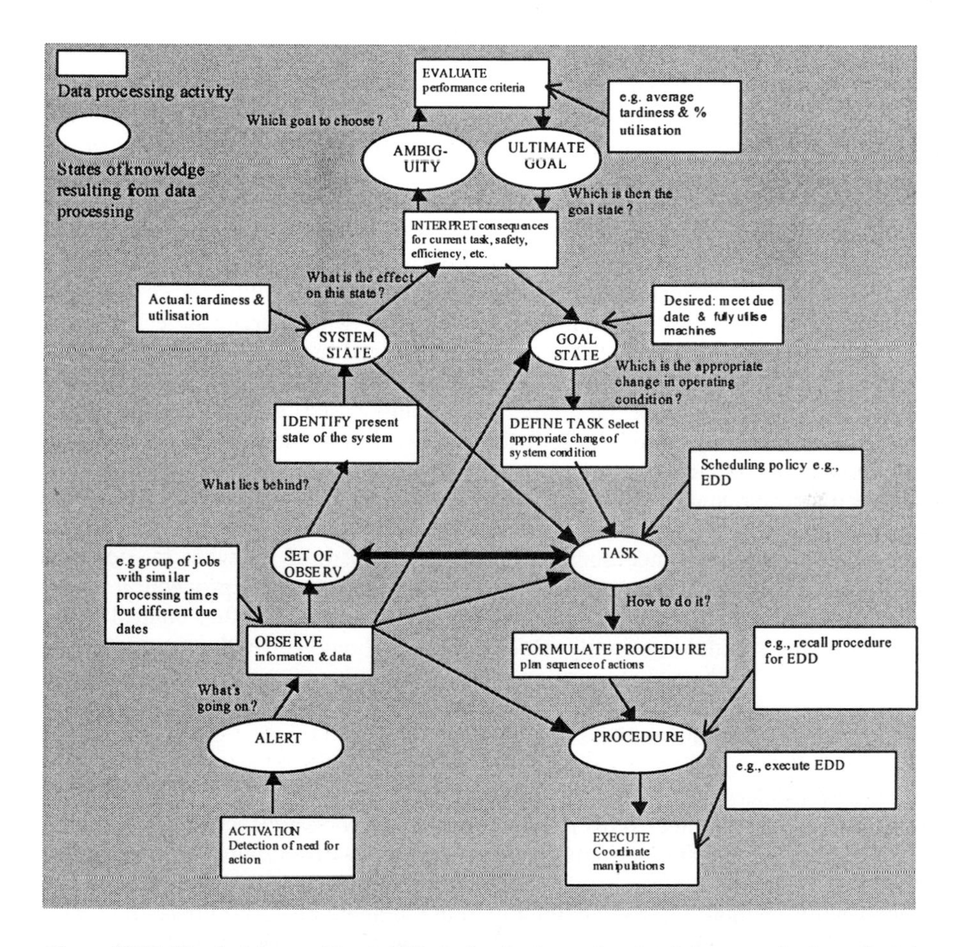

Figure 13.10 The decision-making activity in developing and maintaining a production schedule.

In executing a procedure, schedulers search for key attributes among the data. For instance, to compose a schedule for a single machine using an EDD policy, they are likely to search all available jobs for the one that has the earliest due date. They would then most likely place it at the head of the queue. Again they would search the remaining jobs for the one that has the earliest due date. They would then place it next in the queue. They would repeat this procedure for the remaining jobs. The information extracted by schedulers from the data during the execution of the policy differs from the information used to select a suitable policy. A crude example demonstrates this. Schedulers are more likely to select EDD over SPT where the processing times for all jobs are similar but with the due dates broadly spread, than for the converse situation. In making this decision they link a 'set of observations' to a control 'task' as shown in Figure 13.10. The data display and representational form for supporting this activity may be different to that required for carrying out the procedural steps.

A greater challenge for schedulers is making decisions in circumstances where no known heuristics seem to apply. With no policies clearly pertinent, they try drawing some meaning from the 'set of observations.' Their identification of the present state of the system depends upon them making a connection between the data presented and their perception. By changing the presentation - by modifying, adapting or transforming the form in which data is presented - a mapping may emerge. If schedulers find a pattern in the data for which they know a policy that is efficacious, then they may enact this policy; on the decision ladder there would be an arc between the 'system state' and 'task'. If a scheduling policy is still indeterminate, schedulers may have to contemplate what 'target state' they are trying to satisfy (e.g., minimise average tardiness), and then define a heuristic that meets it ('task'). Where they do not know what performance criterion to follow, decision-making moves to the higher knowledge-based level of performance criteria. If their behaviour were describable as pure rational action, then they would trial various functional performance criteria until they found one that suitably matches the performance goals.

It is clear from the decision ladder that decision-making by schedulers extends far beyond a focus on scheduling heuristics. Developers of software for decision support need to widen their purview from the automation of procedural steps, which concerns only the bottom right side of the ladder, to the other control tasks, such as helping schedulers identify which goal they should currently pursue.

13.4.4 Multiple decision ladders

Because schedulers seek different goals at different times, CTA of scheduling requires a series of connected decision ladders. Let us look how activities map onto multiple decision ladders. Assume that a scheduler, in observing insufficient jobs queued at a particular press, associates this state with a goal state, 'extend the queue length', without referring explicitly to the ultimate goal, 'no press idle time'. The scheduler's task is to add jobs to the queue by executing a procedure that minimises changes to the press set-up. The elements of the decision are mapped on *activity 1*, the leftmost decision ladder shown in Figure 13.11. While executing this procedure, the scheduler is alert (*activity 2*) for adverse consequences. The observation process is that of scanning patterns in job attributes. As long as the scheduler does not find any distinctive patterns in the data not associated with the current goal, the procedure for *activity 1* continues. If, however, the scheduler identifies a pattern associated with a significant change to the system state, then *activity 2* progresses and *activity 1* halts. If the new system state includes tardy jobs for which meeting due date is critical, then decision activity leaps to the ultimate goal 'meet the due dates of selected jobs'. The goal state is redefined and the task becomes meeting major set-up constraints, but relaxing minor set-up constraints where necessary to allow inclusion of jobs that must meet their due date. If a known procedure exists, then activity leaps to its execution. During the execution of the new procedure, the scheduler again is on the alert for adverse effects. In scanning jobs allocated to other machines, the scheduler may, for argument's sake, identify that the current scheduling procedure violates the set-up configuration across all parallel machines for the immediate processing of unannounced premium jobs. This triggers *activity 3*. To resolve the relative importance of competing goals, decision activity then moves to the top of the ladder. On figuring out appropriate performance criteria, the scheduler may then define a target state, associate a task to meet the target, and then carry out a procedure.

In Figure 13.11 switching between the different activities seems to be serial. But, scanning for a pattern that triggers a new activity occurs concurrently with the search for jobs that meet the constraints of the current procedure. Therefore, the left side of the decision ladder (activities dealing with analysis) along with the associative leap to the 'ultimate goal' (e.g., *activity 2* in Figure 13.11) runs concurrently with the extant activity (e.g., *activity 1*). The right side of the ladder (planning and execution) is activated only when the scheduler decides to direct activity towards the new goal.

In the above example, the focus is on the utilisation of a particular machine. Schedulers also focus on jobs. For instance, as new jobs arrive, they may consider where to place them in the current schedule. In considering possible scheduling choices, their attention would be on the requirements of the jobs (job attributes) and relating them to the operational constraints (machine capabilities and current set-up configuration). This requires an understanding of the structural means-ends relations between the job attributes and constraints associated with the different physical devices.

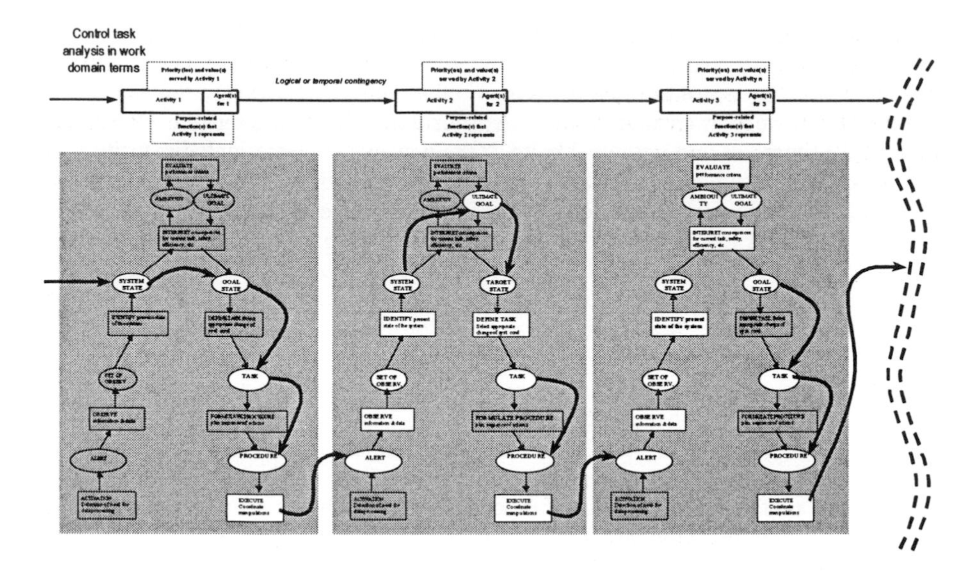

Figure 13.11 Scheduling control tasks as a series of decision ladders.

13.4.5 Goal structure links Control Task Analysis to Work Domain Analysis

The many goals that schedulers seek may seem unrelated. They are, however, bounded by rational action. A structural relationship between goals can be found by mapping the actual operational objectives to goals at higher levels of abstraction. In developing the structure, cognitive engineers need to find the principal goal of the company. The principal goal and the schedulers' operational objectives are linked. Without any links there would be no effect on company performance by scheduling behaviour. This does not imply all operational objectives are the most effective means for attaining the principal goal. Between the operational objectives and the principal goal are intermediate goals. Some intermediate goals may clearly be substantiated from field data, whereas others may have to be inferred. Instead of having to extract all the rules of the domain, cognitive engineers only have to find a logical basis for structurally linking the goals.

The goal structure for the printing company is shown in Figure 13.12. At the lowest level of the hierarchy are goals *A* - the immediate operational objectives that directly guide the building of the schedule. These include low set-up time (2A) for each machine, particular jobs meeting their due dates (13A), priority given to favoured customers (15A), little change to a machine's set-up between discounted jobs (9A), matching an operator's ability to the work task (14A), and invoicing all jobs in the month the required raw materials are purchased (8A). Above this are goals *B*, to which schedulers may also directly attend; however, the extent of the focus varies between goals and schedulers. They are more abstract than operational objectives: for instance, 'fully utilise all machines' (1B) and 'maximise return on discounted jobs' (5B). At the higher levels are goals that indirectly relate to shopfloor parameters. The focus is on short-term financial viability and customer patronage. At the top of the goal structure is the *raison d'être* of the company, the maximisation of long-term financial return. It depends upon short-term financial viability (1D) and customer patronage (2D). Short-term financial viability, in turn, depends upon productivity (1C) and cash flow (2C). By aiming to maximise customer satisfaction (3C), the goal of maximum repeat custom is achieved.

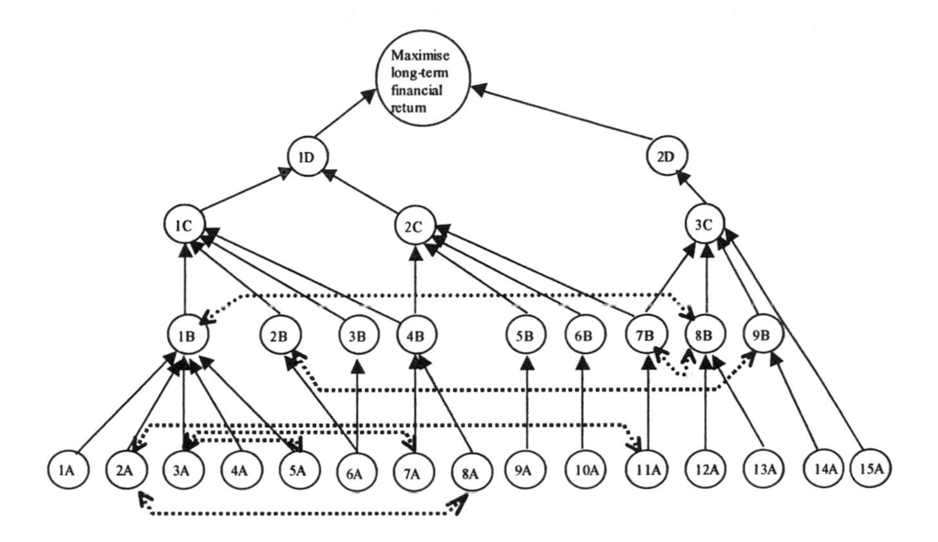

Figure 13.12 The scheduling goal structure.

The higher a goal is up the hierarchy, the less directly it relates to immediate shop floor activity. High-level goals tend to be attained through satisfaction of low-level goals, rather than by the direct attention of the scheduler. Nonetheless, schedulers sometimes directly considered them when making scheduling decisions. Directed arcs into a goal do not depict direct causation. Instead, they indicate a tendency of a goal to move with its underlying subgoals towards satisfaction. Underlying goals at times may move away from satisfaction, to improve performance at the higher level. For example, changing a machine's configuration is an activity that is directed away from 'low machine set-up time' (goal 2A). This seemingly reduces 'machine utilisation' (goal 1B). Yet, if no jobs are available for the current configuration the machine becomes idle and consequently utilisation reduces. A dotted arc within a level of the hierarchy denotes that the linked goals are constrained by the relationship that ties them. For example 'the quality of jobs is to the standard the customer expects' (goal 9B) sets the upper bound on 'maximise processing speed' (goal 2B). Therefore, in pursuing goals that constrain other goals, experienced schedulers consider adverse effects on other parts of the goal structure.

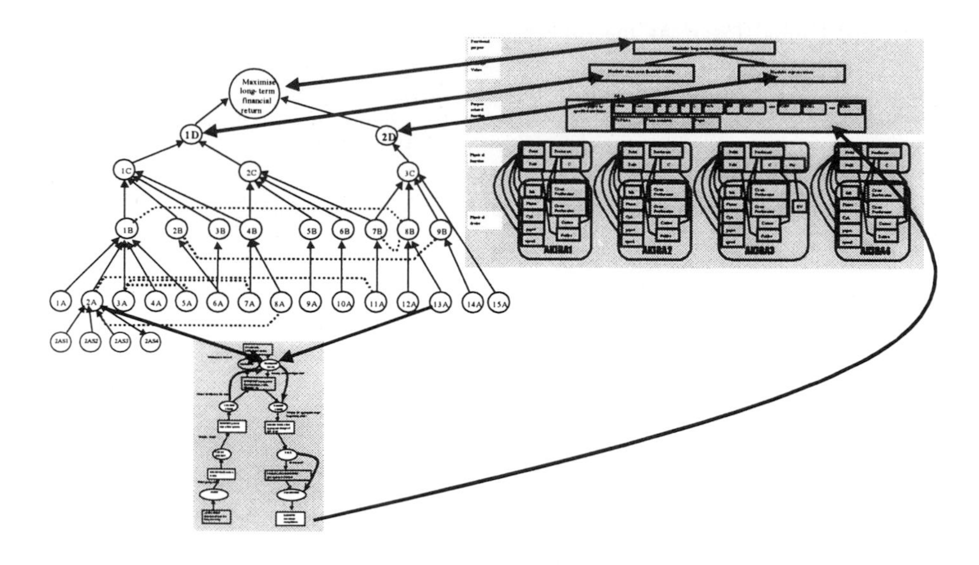

Figure 13.13 The relationship between the goal structure, decision ladder and abstraction hierarchy.

Normally the ultimate goal in the decision ladders comes from the bottom level (i.e., goals A) of the goal structure in Figure 13.12. To realise a goal, an objective is set, a scheduling policy is defined and operational steps are executed; these are equivalent to the ultimate goal, target state, task and procedure, respectively, in the decision ladder. The relationship between a decision ladder, the goal structure and the AH is shown in Figure 13.13. In the relationship between the goal structure, decision ladder and abstraction hierarchy, the apex of the goal structure coincides with the functional purpose level of the AH, and the level immediately below coincides with the highest-level priorities in the AH.

The ultimate goal may often consist of a combination of goals shown in the goal structure. For instance, to satisfy combined goals of utilisation and tardiness, the procedure schedulers would follow would be distinctly different to the procedure for either goal by itself. As schedulers step through the procedure they scan the job tags for those that meet the desired constraints. For each job schedulers scan, they observe whether the feasible means-ends links (Figure 13.7) in the AH meet the current procedural requirements. From the subset that meets the requirements they select a job, that is, they instantiate a particular means-ends chain (Figure 13.8). During the scanning process they see patterns among job attributes that may trigger the consideration of other goals. In effect, they compare the AHs for many jobs, grouping them in a logical structural sequence (Figure 13.9)

13.5 DEVELOPING A HIPSS FROM CWA

The information on the scheduling constraints and decision activities, identified using CWA, forms the basis for developing a HIPSS (hybrid intelligent production scheduling system). A minimal form of a HIPSS merely displays the constraints identified in the AH: the constraints at the purpose-related level (job attributes) and the constraints at the physical function and device level (machine constraints). All control tasks in developing a schedule would be completely manual. By including feasible means-ends links shown in the AH, the computer shows which machines can process it. That is, the computer helps schedulers match the intentional constraints (job attributes) to the physical constraints (machine constraints).

The appropriate form of the data display varies with the type of decision being made. The presentation and control of information between the computer and human may vary with the recognition-action cycles in a decision ladder. On the left of the ladder, the decision behaviour is that of seeking dominant patterns in the data. The computer should support schedulers recognising current goals and procedures from patterns in the data. Observation and identification activities in the decision ladder are linked to the abstraction hierarchy of the work domain. Goal-relevant constraints from the work domain are mapped onto salient perceptual properties of the display: principles relevant to their design are covered by *Ecological Interface Design* (Vicente and Rasmussen 1990, 1992; Vicente, 1999), *representation aiding* (Woods, 1991) and *configural displays* (Bennett and Flach, 1992). Effortless extraction of information from the data depends upon the form of its representation. Each representation has its own set of constraints and intrinsic and extrinsic properties that emphasises some relationships and properties at the expense of others. Therefore, the form of data representation for showing the state variables at various levels of abstraction must be well grounded in theory.

Table 13.1 Legend for the labels is Figure 13.14.

Label	Description
1	Horizontal line links all the elements of a JSO into a unified image.
2	Horizontal position of the vertical bar signifies cylinder size. The height of the bar signifies paper depth.
3	The cylinder size is the same as label 2, but the depth of paper is a third.
4	Signifies an increase in width between jobs.
5 & 6	The bars (labelled 5) map to the press icons labelled 6, with the current choice indicated by the tallest bar.
7	The length of the thick bar depicts the processing time.
8	The length of the thin bar depicts the set-up time.
9 & 10	The horizontal position of the vertical line indicates which colour is to be printed.
11	The bar touching the baseline signifies that the colour is to be printed on the front of the bill, whereas the position shown for label 9 indicates the colour is to be printed on the back of the bill.

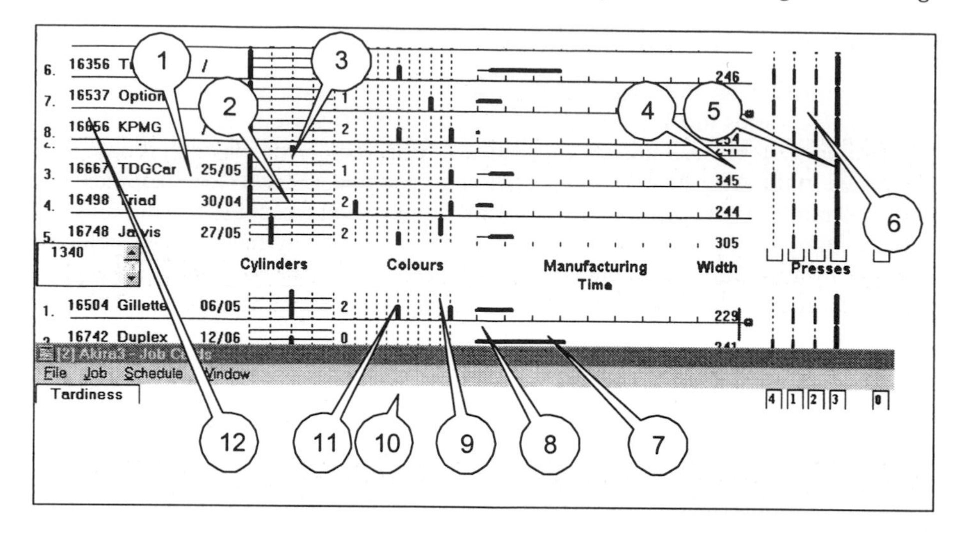

Figure 13.14 Jobs Window for Akira 3: Jobs and their characteristics. A partial display of objects representing job attributes.

An example of a display for the printing case is the Jobs Window in Figure 13.14. It shows the detailed information for jobs queued at a particular machine as symbolic objects; Table 13.1 provides an explanation of the different elements. From the order of jobs in the Jobs Window the computer produces a Gantt chart (Figure 13.15). In the display each job is represented by a job specification object (JSO), which denotes the characteristics of the final product. It shows the intentional constraints associated with the purpose-related function, 'process paper,' (i.e., the job's attributes). The features of a JSO are such that it:

- Clearly and distinctively shows the job attributes;
- Clearly displays, unambiguously, the values of the attributes;
- Supports the scanning of jobs to locate those jobs having a particular attribute value; and
- Clearly displays patterns in attributes across jobs.

Each element in the JSO signifies a particular job attribute and is visually quite distinctive, thereby enabling schedulers to clearly distinguish between elements depicting different attributes. Their design is guided by theories relating substitutive and additive scales and global and local features (Higgins, 1996a; 1996b). For each job, the element that refers to the same attribute has some features in common with the JSOs for all other jobs. These 'global' features allow identification of all elements denoting the same attribute. As the global features for each element are distinctly different to the surrounding elements, the scheduler can easily identify the element that signifies a particular attribute. For each element, the scheduler can easily recognise the value of the referent attribute using 'local' features within the element that changes with the value. By scanning job attributes, at the purpose-related level of the AH, schedulers can observe both the patterns across jobs and also the system constraints (the current machine configurations).

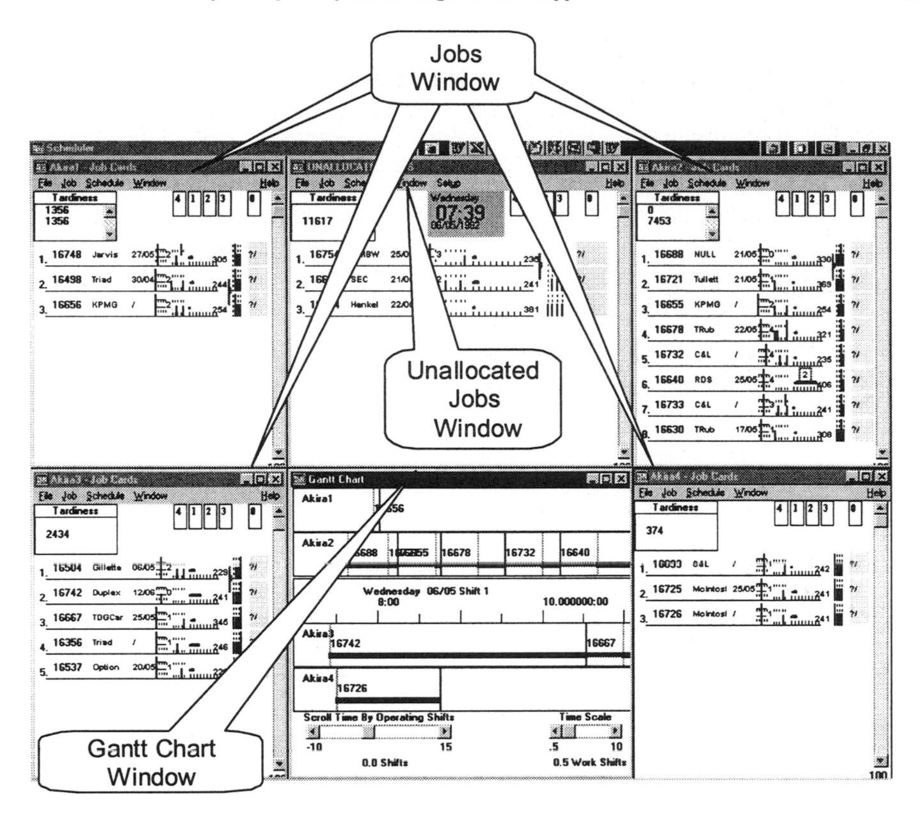

Figure 13.15 IPSS overview of the printing shop during schedule construction, which shows the four presses and the unallocated jobs.

To the right of each JSO is a sign (labelled 5) that shows possible means-ends links. The short, vertical lines indicate alternative machines that can produce the number of colours required for the job, whereas the large line shows the machine to which the job has been allocated (the means-ends link currently instantiated). Schedulers can perceive the work domain at different levels of abstraction. At the purpose-related function level, the HIPSS shows the constraints that specify the function, 'process paper', in the abstraction hierarchy (AH) in Figure 13.7. The list of objects, in effect, forms a sequence of abstraction hierarchies in Figure 13.9. The Jobs-Window display in the HIPSS is equivalent to a plan view of a pack of abstraction hierarchies. Peering downwards, metaphorically, the user can see the work domain constraints for each card in the pack. For each job, the JSO provides affordances to the different levels of the associated AH. Figure 13.16 shows some of the links between the screen objects and the AH. Curved broken lines depict the association between the constraints in the JSO and the AH. Users can also observe how the intentional constraints at the purpose-related level place constraints on the physical devices. The cylinder element of the JSO shows the depth - an intentional constraint set by the job specification - and the cylinder size, which is a configuration requirement for the press. Solid arcs having arrow ends show the link between the depiction in the Jobs Window and the constraint on the physical device.

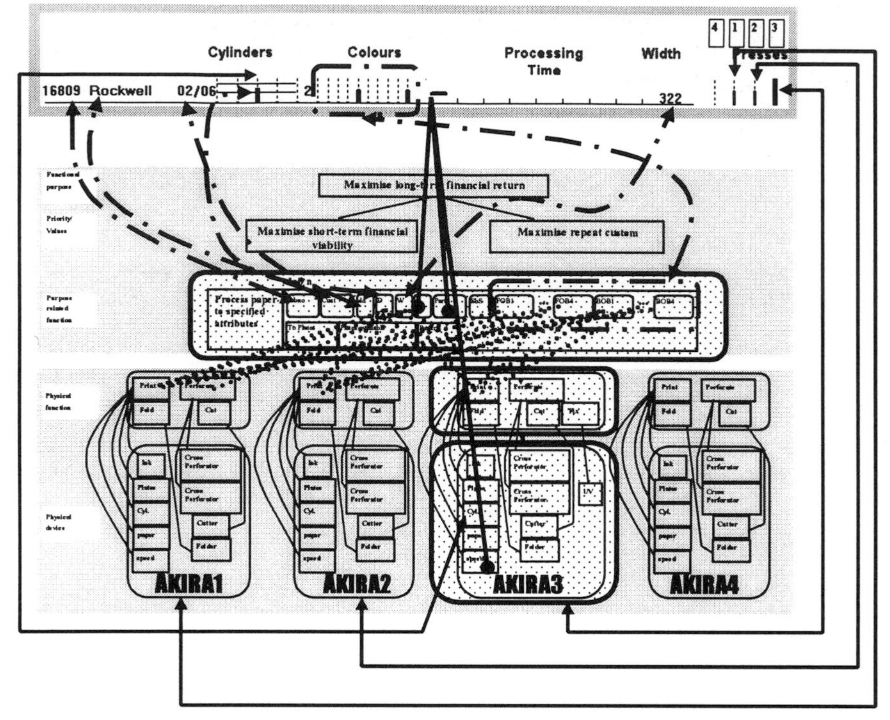

Figure 13.16 The relationship between the abstraction hierarchy and the signs in the Jobs Windows in the HIPSS for printing.

Set-up time is not derived from the mapping of the intentional constraints at the purpose-related function level to the physical-device level of the AH. Instead, it is a product of the juxtaposition of AHs. In juxtaposing JSOs, differences in the local features of the elements signify a configuration change in moving from one AH to another. A change from one cylinder size to another is a major set-up, which is clearly observable, as labels 2 and 3 in Figure 13.17 show. A graphic object (labelled 1) signifies a minor set-up occurring when the width of paper decreases from one job to another.

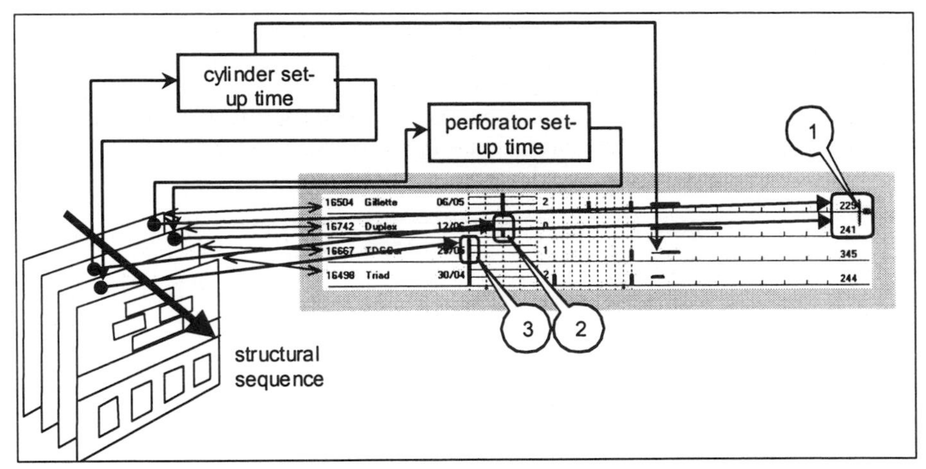

Figure 13.17 Set-up time is a factor that arises when the value of particular constraints differs between abstraction hierarchies.

To move the problem solution closer to meeting a set of goals, schedulers shuffle the job objects in the Job Window until a satisfactory pattern is obtained. For example, to maximise the number of jobs before a cylinder change, jobs would be rearranged so that the vertical line through the cylinder element of the JSOs would be unbroken.

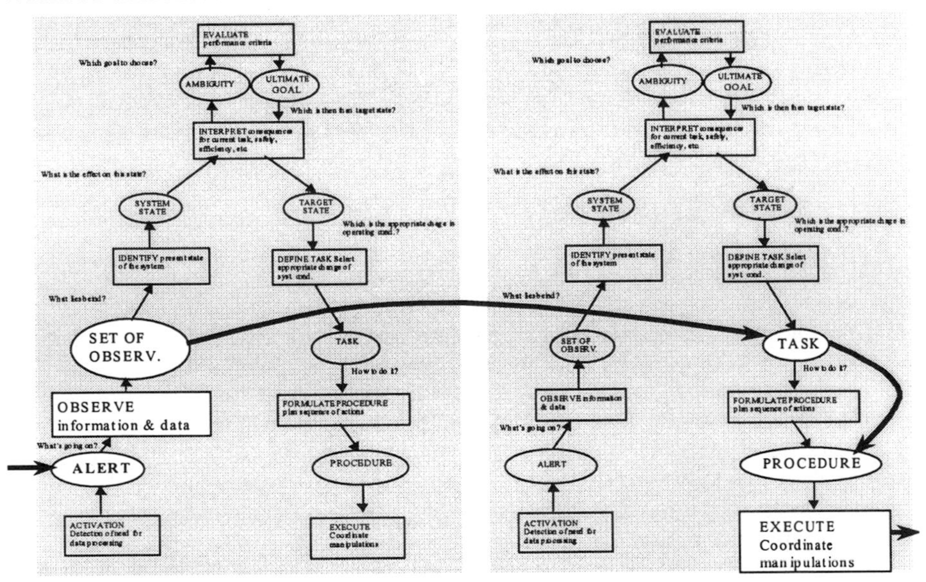

Human Scheduler Computer

Figure 13.18 Human scheduler recognising relevant policy and computer carries out the procedure.

In familiar situations, behaviour is rule-based as schedulers apply familiar scheduling heuristics. The HIPSS can help schedulers execute rule-based procedures, shown on the right side of the ladder. As they group jobs, the computer shows which machines the string can be allocated (Figure 13.19). Where possible the steps of procedures can be automated (Figure 13.18). The scheduler first selects a string of jobs and then chooses a heuristic to apply.

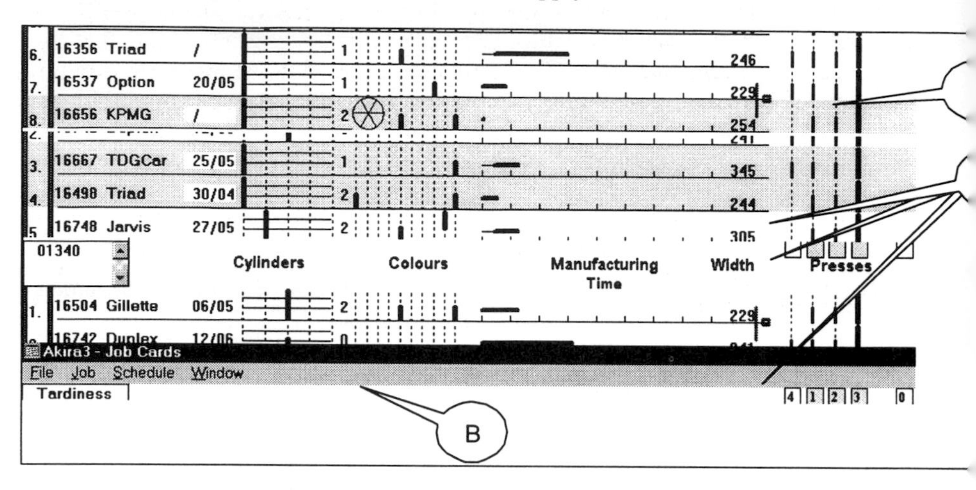

Figure 13.19 As a string of jobs is collected, the collected objects become shaded (labelled A) and the permissible presses become shaded (labelled C). The number of spokes on the blue 'collector' (labelled B) shows how many have been collected. When the 'collector' is right clicked, the job numbers are displayed in a pop-up box, listed in the sequential order of collection.

13.5.1 Schedulers at the centre

By designing a HIPSS using CWA, it is possible to position schedulers centrally in the decision-making architecture as depicted in Figure 13.1. In the case of the HIPSS for printing, schedulers use the Jobs Windows (see Figure 13.14 for a detailed view of a window) to choose classification strategies for forming groups and then place them into machine queues. They can reflect upon the characteristics, that is, the attributes, of jobs and the calls that these make upon the shop. The job attributes, and patterns among attributes across jobs, act as stimuli. Using information that is displayed and other domain knowledge, they can seek patterns in the data on which to draw inferences about possible scheduling strategies. In an opportunistic way, they can try various groupings, make amendments and backtrack on previous decisions. They can select sets of jobs to move as one within a queue or to another machine. In ordering the jobs within a set, they can apply various OR heuristics.

The interface is designed so users can attend to constraints at times appropriate to them. The HIPSS is designed so that it does not needlessly get in the way of decision-making. Users can explore potential decisions in their own way, guided by their understanding of the scheduling process. The HIPSS passes messages unobtrusively to users. For example, the black bars denoting constraints on permissible presses (label 5, Figure 13.12) are always visible and can be considered at any time. For soft constraints graphical signs have particular efficacy. The graphic object (labelled 1 and 4 in Figure 13.17 and Figure 13.14, respectively) denotes a soft constraint that occurs when a job has a lower value of width than its immediate predecessor and the predecessor uses the same cylinder. The message draws attention while not being disruptive. While the knowledge-based adviser warns the user when soft constraints are infringed, it disallows violation of hard constraints (see label 4 in Figure 13.14). Of course, if a user attempts to violate hard constraints the activities of the HIPSS will not remain in the background. It then sends an intrusive warning through a pop-up message box. For example, as the string of jobs shown in Figure 13.19 is collected, permissible presses are shaded. This message is non disruptive. However, if the scheduler attempts to allocate the string to an inappropriate press (the unshaded press 4), then the computer pops up a warning in a standard message box, to which the scheduler is compelled to attend.

13.5.2 Improving scheduling performance

Although schedulers using a HIPSS may follow familiar practices, the HIPSS should be designed to encourage schedulers to improve scheduling behaviour. Often they are preoccupied with immediate operational objectives (e.g., low press set-up time across the shift), which are at the lowest level of the goal structure (Figure 13.20). Occasionally, their attention may be drawn to high-level goals. If their focus could be generally raised to higher-level goals, then the scheduling performance relating to these goals may improve. Goals at the level above the immediate operational objectives tend to be directed towards functional goals. Some goals at this level are commensurate with traditional OR goals. For the printing example, percentage utilisation is a suitable measure of performance for the goal 'fully utilise all machines' (1B). Similarly, average tardiness is a suitable measure for the goal 'all jobs delivered on their due date' (8B).

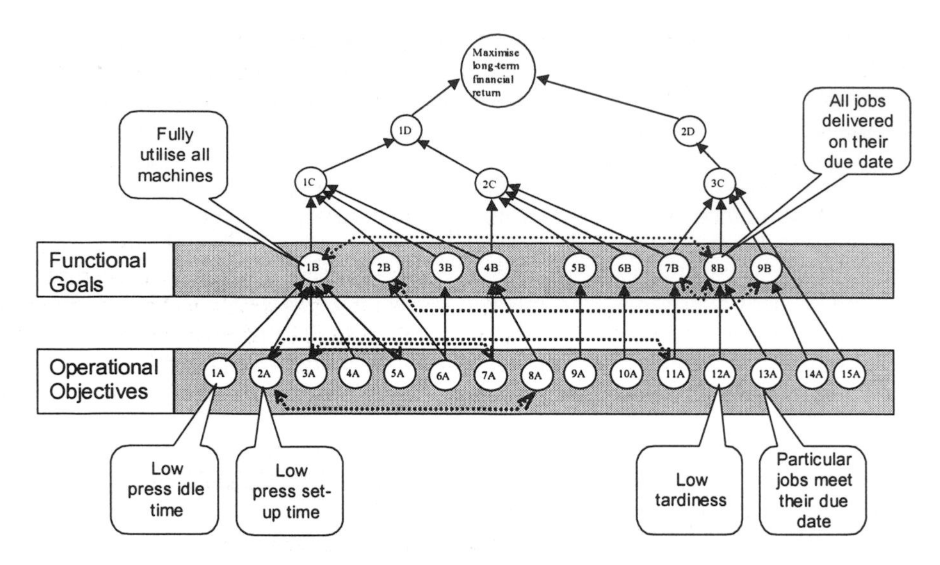

Figure 13.20 Schedulers' practices commonly address operational objectives. A HIPSS should extend their interest to functional goals.

Schedulers using the HIPSS in Figure 13.15 can track the average tardiness at each press (labelled 12 in Figure 13.14) and across all presses (in the window for unallocated jobs in Figure 13.15). Inclusion of measures for other functional goals would further improve the HIPSS. The HIPSS gives schedulers the opportunity to seek ultimate goals that encompass those functional goals for which performance is displayed. In formulating a procedure (the lower right side of the decision ladder) to move a schedule towards an ultimate goal, schedulers must be able to experiment with various strategies. This HIPSS does not show which jobs to move, and where they should move, for performance to improve. The Jobs Window in Figure 13.21 is for a HIPSS that addresses this issue. This HIPSS was designed for a system having only a few attributes (release date, due date and processing time). Each job's tardiness, priority weight and weighted tardiness are shown graphically. (Refer to Table 13.2 for a description of the labels). Schedulers can see where jobs need to move in the queue for due date to be met. Placing the mouse cursor on job 03 causes a bar to appear between jobs 11 and 06. This bar shows where the job has to move for it to be finished on time. A similar bar shows the location on the Gantt chart.

There may be various arrangements that would give the same value of overall tardiness. By observing the distribution of tardiness schedulers can adjudge other aspects of performance qualitatively. From experience, they may, for example, prefer a balanced distribution of tardiness. Or perhaps they may prefer to concentrate the tardiness in only a few jobs. Having established the immediate objective, they can try out various strategies for meeting it.

Table 13.2 Legend for the labels in Figure 13.21

Label	Description
1	Transparent rectangle signifies a non-tardy job
2	The length of the horizontal line shows a non-tardy job's earliness
3	The width of the rectangle shows the job's contribution to the performance measure.
4	The height of the rectangle shows the weight used for the job.
5	The horizontal line shows the unweighted measure: tardiness for this case.
6	The bar shows where a job indicated by the cursor for the mouse (labelled 7) has to be placed so that the job will not be tardy.
7	The cursor for the mouse. Note that job 03 is tardy.
8	The performance measure for the machine

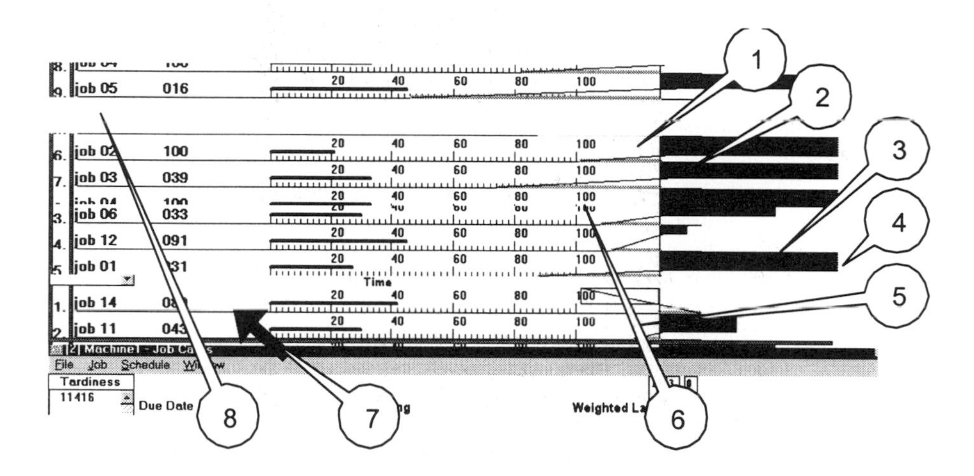

Figure 13.21 The HIPSS Jobs Window for a simple scheduling model.

13.6 CONCLUSION

For those production environments that must rely on the knowledge and skills of human schedulers, scheduling software should support their decision-making. A hybrid intelligent human-computer production scheduling (HIPSS) paradigm that supports human schedulers operating in environments characterised by uncertainty and instability has been advanced. Its contribution to scheduling practice is a methodology for addressing decision making in complex systems in which there are many competing and conflicting goals.

The decision architecture of a HIPSS locates schedulers at the centre of the decision-making process, where they can act as intermediaries between the real-world manufacturing environment and the abstract world of operations research, by:

- dealing with the stated and unstated conflicting goals;

- resolving how to use information that is incomplete, ambiguous, biased, outdated, or erroneous;
- grouping jobs to meet the specific criteria for applying selected heuristics.

Just as in the process industry, where the operator in the control room has instruments for displaying the state variables and performance measures of the plant and has alarms to warn of critical constraint violations, a HIPSS provides an environment for schedule control, with features for showing the state of the schedule, for indicating performance and for warning the violation of constraints. As the understanding of scheduling factors in the domain is improved, new indicators and automated procedures, which are under the command of the schedulers, may be incorporated in a hybrid intelligent production scheduling system. By being actively involved in decision-making, the human can deal with contingencies and other aspects of scheduling jobs that are difficult to vest in a computer decision-maker. The use of intelligent human decision-makers with vast local knowledge also obviates the need for an exhaustive knowledge base.

For humans to play a coherent and active role in schedule construction, they must have ready access to all the information they use to make decisions. Cognitive Work Analysis (CWA) provides a useful formalism for designing a HIPSS. Work Domain Analysis (WDA) identifies the system constraints that should, where possible, be displayed by the computer. Using Control Task Analysis (CTA) cognitive engineers can construct decision ladders associated with the different scheduling goals. The details in the goal structure and the decision ladders vary between schedulers as the particular problem-solving technique a person applies depends upon on experiential familiarity with the task. Therefore, it is important to design a HIPSS that has many degrees of freedom to allow for wide-ranging problem-solving strategies. Where cognitive engineers can construct decision ladders for prototypical behaviours, then there is the potential for determining the states of knowledge and control tasks that schedulers may use in making decisions. Cognitive engineers can then consider how best to represent the states of knowledge (e.g., patterns in data, system constraints, goal states and measurement of goal performance) and to automate control tasks (e.g., procedures for ordering jobs). Where possible, information is displayed using graphical elements that are configured by the value of the parameters they represent. The graphics are designed so there are visual cues to significant factors and visual pattern recognition replaces pattern-matching activity in the schedulers' memory.

The extent of the analysis and the scope of the design for a HIPSS may vary with the domain and the degree of support sought. A minimal HIPSS replicates the constraints in the WDA. The computer displays the job attributes and machine attributes and warns when there is violation of technical constraints. All the control tasks associated with the decision ladders are undertaken by the schedulers. For relatively simple and fixed manufacturing systems, the path between subgoals and the control tasks for each subgoal may be well understood. Under these circumstances, it may be easy to automate decision processes. More complex environments may depend upon human schedulers identifying the relevant goals from patterns in the data and then recognising which scheduling procedure is applicable: frequent procedures may be automated.

As HIPSS may be developed in stages, 'hasten slowly' may be the appropriate motto!

13. 7 REFERENCES

Ackoff, R. L. (1979a). The future of operational research is past. *Journal of the Operational Research Society*, 30, 93-104.

Ackoff, R. L. (1979b). Resurrecting the future of operational research. *Journal of the Operational Research Society*, 30, 189-199.

Bennett, K. and Flach, J. M. (1992). Graphical displays: Implications for divided attention, focused attention, and problem-solving. *Human Factors,* 34(5), 923-935.

Bertin, J. (1981). *Graphics and Graphic Information-Processing*. De Gruyter.

Bødker, S. (1991). *Through the Interface: A Human Activity Approach to User Interface Design*, Hillsdale, NJ: Lawrence Erlbaum Associates.

Brödner, P. (1990). Technocentric-anthropocentric approaches: towards skill-based manufacturing. In M. Warner, W. Wobbe and P. Brödner (Eds.), *New Technology and Manufacturing Management: Strategic Choices for Flexible Production Systems*, John Wiley & Sons, pp. 101-112.

De Montmollin, M. and De Keyser, V. (1986). Expert logic versus operator logic. In G. Mancini, G. Johannsen and L. Martensson (Eds.), *Analysis, Design, and Evaluation of Man-Machine Systems*, New York: Pergamon, pp. 43-49.

Dutton, J. M. and Starbuck, W. (1971). Finding Charlie's run-time estimator. In J. M. Dutton and W. Starbuck (Eds.) *Computer Simulation of Human Behaviour*, New York: John Wiley & Sons, pp. 218-242.

Gault, R. (1984). OR as education. *European Journal of Operational Research*, 16, 293-307.

Higgins, P. G. (1992). Human-computer production scheduling: contribution to thc hybrid automation paradigm. In P. Brödner and W. Karwowski (Eds.), *Ergonomics Of Hybrid Automated Systems - III: Proceedings of the Third International Conference on Human Aspects of Advanced Manufacturing and Hybrid Automation*, Gelsenkirchen, August 26-28 1992, Germany, Elsevier, pp. 211-216.

Higgins, P. G. (1996a). Interaction in hybrid intelligent scheduling. *The International Journal of Human Factors in Manufacturing*, 6(3), 185-203.

Higgins, P. G. (1996b). Using graphics to display messages in an intelligent decision support system. In F. Burstein, H. Linger and H. Smith (Eds.) *Proceedings of the Second Melbourne Workshop on Intelligent Decision Support Systems, IDS'96*, Monash University, September 9th 1996, Monash University, pp. 32-38.

Higgins, P. G. (1999a). Production scheduling: some issues relating to the location of the user in the decision-making architecture. In J. Scott and B. Dalgarno (Eds.) *Proceedings of the 1999 Conference of the Computer Human Interaction Special Group of the Ergonomics Society of Australia. OzCHI'99,* November 28-30, Wagga Wagga, Charles Sturt University, pp. 44-51.

Higgins, P. G. (1999b). *Job Shop Scheduling: Hybrid Intelligent Human-Computer Paradigm.* Ph.D. Thesis, The University of Melbourne, Australia.

McKay K.N. (1987). *Conceptual Framework for Job Shop Scheduling*, MASc Dissertation, Department of Management Science, University of Waterloo, Ontario, Canada.

McKay, K.N. (1997). Scheduler adaptation in reactive settings - design issues for context-sensitive scheduling tools. *Proceedings of IE Conference Practice and Applications*, November 1997, San Diego.

McKay, K. N., Buzacott, J. A. and Safayeni, F. R. (1989). The schedulers desk - can it be automated? Decisional structure in automated manufacturing. *Proceedings of the IFAC/CIRP/IFIP/IFORS Workshop*, Genoa, Italy, September 18-21, pp. 57-61.

Meister, D. (1966). Human factors in reliability. In W. G. Ireson (Ed.) *Reliability Handbook*, McGraw-Hill.

Morton, T. E. and Pentico, D. W. (1993). *Heuristic Scheduling Systems: With Applications to Production Systems and Project Management*, New York: John Wiley & Sons.

Nakamura, N. and Salvendy, G. (1994). Human planner and scheduler. In G. Salvendy and W. Karwowski (Eds.), *Design of Work and Development of Personnel in Advanced Manufacturing*, New York: John Wiley and Sons, 331-354.

Norman, D. A. (1986). Cognitive engineering. In D. A. Norman and S. Draper (Eds.) *User Centered System Design: New perspectives in human-computer interaction*, Hillsdale, NJ: Lawrence Erlbaum Associates, pp. 31-61.

Papantonopoulos, S. (1990). *A Decision Model for Cognitive Task Allocation*, Doctoral Thesis, Purdue University.

Pinedo, M. and Chao, X. (1999). LEKIN® - Flexible Job-Shop Scheduling System, Stern School of Business. URL (http://www.ieor.columbia.edu/~andrew/scheduling/Lekin.html). 9th August 1999.

Polanyi, M. (1966). *The Tacit Dimension*, London: Routledge & Kegan Paul.

Rasmussen, J. (1985). *Human Error Data: Fact or Fiction*, Risø Report M-2499, Roskilde, Denmark: Risø National Laboratory.

Rasmussen, J. (1986). *Information Processing and Human Machine Interaction: An Approach to Cognitive Engineering*, New York: North-Holland.

Rouse, W. B. and Morris, N. M. (1986). On looking into the black box: Prospects and limits in the search for mental models. *Psychological Bulletin*, 100, 349-363.

Sanderson, P. M. (1988). Human supervisory control in discrete manufacturing: translating the paradigm. In W. Karwowski (Ed.) *Ergonomics of Hybrid Automated Systems I: Proceedings of the First International Conference on Ergonomics of Advanced Manufacturing and Hybrid Systems,* Louisville, Kentucky, August 15-18, 1988, Elsevier, pp. 15-22.

Sanderson, P. (1989). The human planning and scheduling roles in advanced manufacturing systems: an emerging human factors domain. *Human Factors,* 31 (6), 635-666.

Sanderson, P. M. (1991). Towards the model human scheduler. *International Journal of Human Factors in Manufacturing*, 1(3), 195-219.

Sanderson, P. M. (1998). Cognitive work analysis and the analysis, design, and evaluation of human-computer interactive systems. In P. Calder and B. Thomas (Eds.) *Proceedings 1998 Australian Computer-Human Interaction Conference, OzCHI'98,* November 30 –December 4, Adelaide, IEEE, pp. 220-227.

Sharit, J. (1984). *Human Supervisory Control Of A Flexible Manufacturing System: An Exploratory Investigation*, Doctoral Dissertation, Purdue University.

Sheridan, T. B. (1976). Toward a general model of supervisory control. In T. B. Sheridan and G. Johannsen (Eds.) *Monitoring Behavior and Supervisory Control*, New York: Plenum, pp. 271-281.

Vicente, K. J. (1999). *Cognitive Work Analysis: Towards Safe, Productive, and Healthy Computer-based Work*, Hillsdale, NJ: Lawrence Erlbaum Associates.

Vicente, K. J. and Rasmussen, J. (1990). The ecology of human-machine systems ii: mediating 'direct perception' in complex work domains. *EPRL-90-91*, University of Illinois at Urbana-Champaign, USA.

Vicente, K. J. and Rasmussen, J. (1992). Ecological interface design: theoretical foundations. *IEEE Transactions on Systems, Man and Cybernetics*, SMC-22, 589-606.

Woods, D. D. (1988). Coping with complexity: the psychology of human behaviour in complex systems. In L. P. Goodstein, H. B. Andersen and S. E. Olsen (Eds.) *Tasks, Errors, and Mental Models: a festschrift to celebrate the 60th birthday of Professor Jens Rasmussen*, London: Taylor & Francis, pp. 128-148.

Woods, D. D. (1991). The cognitive engineering of problem representations. In J. Alty and G. Weir (Eds.), *Human-Computer Interaction in Complex Systems,* London: Academic prcss, pp 169-188.

Woods, D. D. and Roth, E. M. (1988). Cognitive systems engineering. In M. Helander (Ed.) *Handbook of Human-Computer Interaction*, Amsterdam: Elsevier Science Publishers, pp. 3-43.

CHAPTER FOURTEEN

Designing and Using an Interactive MRP-CRP System Based on Human Responsibility

Nobuto Nakamura

14.1 INTRODUCTION

Generally, a production plan will be divided into long-range, medium-range, and short-range (Hax and Candea, 1984). The long-range (years) plan is necessary to develop facilities and equipment, suppliers, and production processes and becomes, in turn, the constraints on the medium- and short-range plans. The long-range capacity planning problem is not handled in this chapter.

Aggregate planning develops medium-range (6-18 months) production plans concerning employment, inventories, utilities, facility modifications, and material-supply contracts. A master production schedule, which is made from data determined by the aggregate plan, is a short-range (several weeks to a few months) production plan for producing finished goods or end-items. This plan develops production schedules of parts and assemblies to be manufactured, schedules of purchased materials, shop floor schedules, and work force schedules (Gaither, 1996).

On the other hand, the capacity planning which plans the balance of demand and supply for manufacturing capacity is divided into resource planning, rough-cut capacity planning, and capacity requirements planning corresponding to each level of the production planning mentioned above. Of course, these levels are mutually linked, although the content of the capacity planning at each level is different, making it possible to say that human judgement is necessary for final decision-making.

In a short-range production plan, one of most important developments has been that of material requirements planning (MRP). MRP is a computer-based information system for planning production and purchases of dependent demand items. It uses information about end product demands, product structure and component requirements, production and purchase lead-times, and current inventory level to develop cost-effective production and purchasing schedules.

Why have so many production companies today adopted MRP systems? The reasons result mainly from the philosophy of MRP systems, that is, 'each raw material, part, subassembly and assembly needed in production should arrive simultaneously at the right time to produce the end items in the master production schedule (MPS)' (Gaither, 1996).

However, since it has become evident that MRP alone is not sufficient to cope adequately with balancing capacity with workload, further development has produced the system which consists of both material requirements planning and capacity requirements planning (CRP).

The main purpose of CRP is to test that sufficient capacity exists to accomplish the detailed work orders planned by MRP, which needs prompt decision-making using appropriate information in order to link to the executive plan (Orlicky, 1975; Vollman *et al.,* 1988). In particular, the CRP process contains complex decisions such as the adjustments of the resource and scheduling of the load. Since this cannot be easily processed with the computer, it is necessary for the human (called planner, manager, supervisor, and so on) to judge on final decisions.

In this chapter we will begin to address the different levels of capacity planning corresponding to production planning levels. We will then examine in greater detail the human role and responsibility at the level of capacity requirements planning. We will develop this together with a detailed explanation of the CRP process itself. We will then go on to show the design and implementation of an interactive MRP-CRP system which helps the human to balance workload and capacity, or in other words, to balance the demand for manufacturing capacity and the supply of that capacity. Finally, we will illustrate a simple example which tests the validity and the effectiveness of the system.

14.2 CAPACITY PLANNING HIERARCHY

In general, the capacity planning process is expressed by a layered structure which consists of three hierarchies corresponding to each of the three hierarchies of production planning, as shown in Figure 14.1. The decision-making functions and content of each stage are discussed below.

14.2.1 Capacity Requirements Planning

Capacity Requirements Planning (CRP) located at the lowest level, affects decision-making at the most detailed level of capacity planning, and then is processed concretely at each work-center. It's purpose is to check whether the operation schedule of each order, obtained from Material Requirements Planning (MRP), is executable.

MRP on its own assumes that capacity is available when needed unless told otherwise. CRP, on the other hand, takes material requirements from MRP and converts them into standard hours of load of labor and machine on the various work-centers. By utilizing the management information system, CRP attempts to develop loading plans that are in good balance with capacities. This loads backward from the due date using operational lead-times from production routings, assuming there is no resource capacity limitation. That is, CRP is founded on backward loading to infinite capacity. There are alternative loading methods: backward loading to finite capacity and forward loading to finite capacity. It is important to understand the difference between infinite capacity loading and finite capacity loading because it affects the roles of the human and the computer.

Infinite capacity loading is based on the notion that it is possible to adjust the available resources to match the workload. Therefore, infinite capacity loading

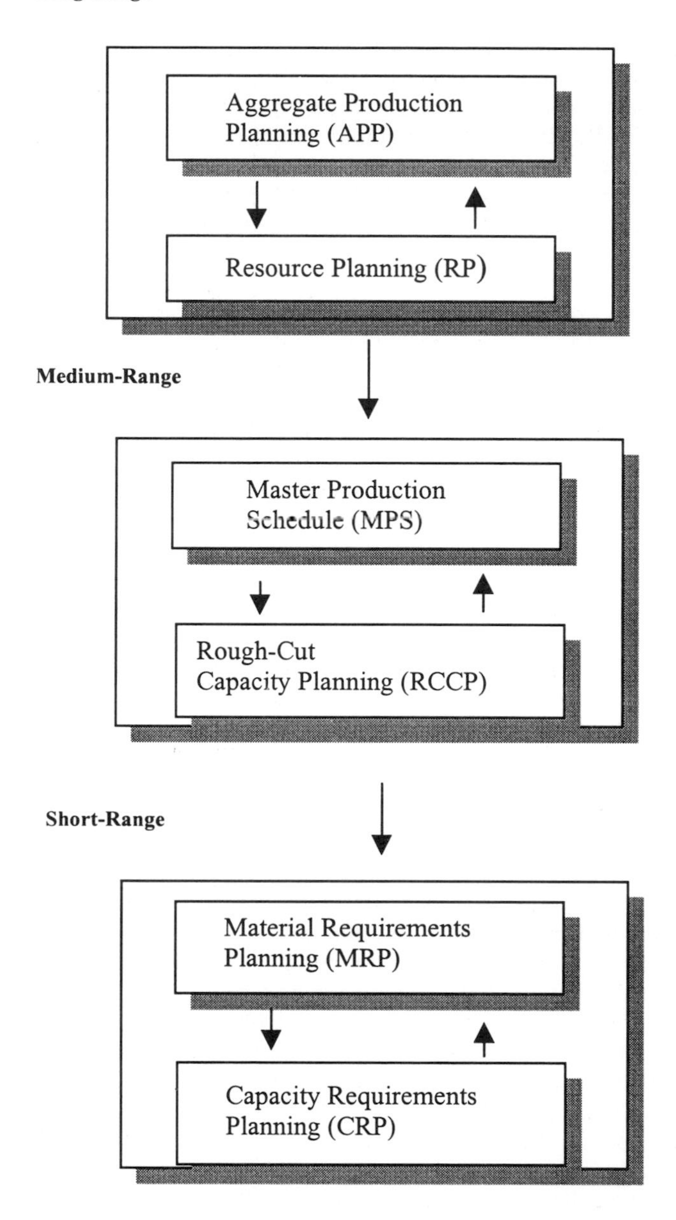

Figure 14.1 Hierarchy of production planning and capacity planning.

works well when the capacities are flexible and can be expanded or contracted when resources can be reallocated as needed. Conversely, infinite capacity loading cannot be used in fixed capacity situations.

Finite capacity loading is based on the notion that we should put no more into the work-center than it can be expected to produce. The concept of finite capacity loading is a generally appropriate method, but the following problems exist concerning this approach:

1. The capacity limitation is usually flexible within a certain range, and can be achieved by doing overtime, increasing the number of workers, increasing the knowledge and experience of workers. However, a decision on limitations is a difficult one.
2. When the load exceeds the limit, the computer automatically decides and schedules whether to make the operation in or to make it out. If it is made out, it is necessary to reschedule all the operations in the downstream. As a result, the work-centers of the downstream might become overloaded, necessitating rescheduled operations. This would also influence the master production plan and customer delivery dates may not be satisfied. This could become a serious problem for the company, as it may inspire a loss of customer confidence. When operations are rescheduled ahead of time, the work-centers of the upstream are influenced, and finally the material purchase schedule has to be changed. Even if this succeeds, extra labor and costs are generated.
3. A large problem of finite capacity loading is the apparent lack of transparent results. Hence, the planner cannot see which work-center may be bottlenecked in the near future, nor be able to take suitable action to resolve the bottleneck.
4. The main problem of finite capacity loading is at what point to entrust the computer with the load adjustment and schedule decision-making. The computer cannot take responsibility for the result, that must be taken by the planner.

The alternative to finite capacity loading is Capacity Requirements Planning (CRP). The concept behind CRP, like finite capacity loading, is to put no more into the work-center than it can be expected to produce. However, the difference with finite capacity loading is that the computer does not adjust the load of the work order but the human does. The first step is to load the work order in priority sequence. The second step, rather than the computer automatically rescheduling overloads, is for the planner to balance the planned capacity by changing the work-force, overtime, and so on. Usually, when the planner wishes to adjust the workload, he/she takes the moving of the workload to right/left period on the load report to cope with overload/underload (see later) (Render and Heizer, 1997). Moreover, the required capacity is changed by alternative routings, subcontracting, and make or buy decisions. Subsequent to these changes, rescheduling is done only as a last resort as the existing schedule reflects what is needed to meet the master production plan. Usually, the capacity problem of a large-scale change is not handled in CRP, because its inherent assumption is that the load never exceeds capacity in the global situation.

Therefore, a technique is needed to assure that the capacity is in the ballpark prior to running CRP (Correll and Edson, 1999). This technique is called Rough-Cut Capacity Planning (RCCP).

14.2.2 Rough-Cut Capacity Planning

Rough-Cut Capacity Planning (RCCP) is used to test the feasibility of the Master Production Schedule (MPS) from the perspective of capacity requirements prior to committing to the changes and running MRP and CRP.

Although both RCCP and CRP are very similar, there are a number of differences. RCCP differs from CRP in the following aspects (Scott, 1994):

1. The capacity profiles for RCCP are developed for each master scheduled item rather than materials at every work-center.
2. RCCP loads only on selected critical resources rather than at every work-center.
3. RCCP uses a bill of resources instead of the production routings.
4. RCCP may have interactive facilities which help the planner in the task of judging production plans and capacity in order to achieve a balance.

As mentioned above, the purpose of RCCP is to examine whether MPS can be guaranteed from the viewpoint of capacity. As a result, the planner of MPS can adjust capacity in a medium-range plan at as early a stage as possible. Moreover, receiving excessive orders in production capacity can be avoided. It is better to know that MPS is at least feasible before running MRP and CRP. A little effort up front with RCCP can alleviate much of the uncertainty in MRP.

14.2.3 Resource Planning

Resource Planning (RP) is capacity planning located in the highest level of the capacity planning hierarchy. Its main purpose is to provide a statement of resources needed for achievement at product family level, which is determined by Aggregate Production Planning (APP).

Varying levels of computer support are used for RP. Some companies have complete modules as part of their Manufacturing Resource Planning (MRP) software. Others use standard spreadsheet programs on a personal computer, still others execute the process manually (Correll and Edson, 1999).

14.3 HUMAN RESPONSIBILITY IN CRP

As previously described, the purpose of CRP is to test whether the capacity of the work-center is balanced or not and whether the due date of the work order is satisfied or not, when each operation of the work order is allocated to the work-center. The CRP process is as follows (Figure 14.2): first, it needs data from MRP - the planned order release for all materials. The second factor is the routing, which defines the steps necessary to produce the product, the operation sequences, the standard times, and the work-centers where the operations are to be performed. The third factor is the work-center information, which includes planned lead-times and scheduling rules. Finally, these data are processed as the results dispatch list (for example, Figure 14.3) and capacity plan (for example. Figure 14.4) are primarily output. While the dispatch list is used for making the work order schedule in each work-center, the capacity plan is used for making a balance between what is available and what is required in each work-center.

The feature of the CRP process is that the final decision-making depends on the human rather than the computer (see Figure 14.5).

14.3.1 The supervisor's responsibility in dispatching and scheduling

Dispatch List output from MRP is only a priority list; it is not a job sequence mandate. The supervisor must decide this job sequence. Thus, the following information necessary to determine priority is contained on the dispatch list: the order number, the operation start date, the operation due date, and order due date. In addition to this information, the dispatch list must differentiate between what is currently at the work-center and what has not yet arrived. From this information, the supervisor can see what work will be available for assignment. Once the priority has been established, the supervisor has to take into consideration the capability of the machine. Also, the supervisor could run into problems with an operator who has never before performed a specific job. The supervisor must take such matters into consideration and would most likely decide to run a different sequence from that on the dispatch list. Regardless of these potential problems it is important to recognize that the dispatch list helps the supervisor to decide the work-centre schedule.

If the supervisor realizes that the due date of a work order can not be satisfied, he might utilize one or more of these five fundamental techniques, each of which can compress the full lead-times, in order to meet the due date.

1. Compressing queue and move times
2. Operation overlapping
3. Order splitting
4. Operation splitting
5. Additional shift.

Which technique should be used depends on the individual circumstances. Usually, the supervisor uses a combination of these techniques depending on the material and status of the machines and/or workers.

Although priority is the most significant factor in work allocation, a good supervisor knows there are a lot of matters that should be considered over and above this. A good supervisor knows how to achieve the highest levels of productivity better than the computer does. Moreover, a good supervisor knows whether the worker and equipment can work in harmony on any particular job. These rules give the supervisor the flexibility to improve productivity at the same time as preserving the integrity of the priority process. However, when the rule is established as normal practice change may become difficult.

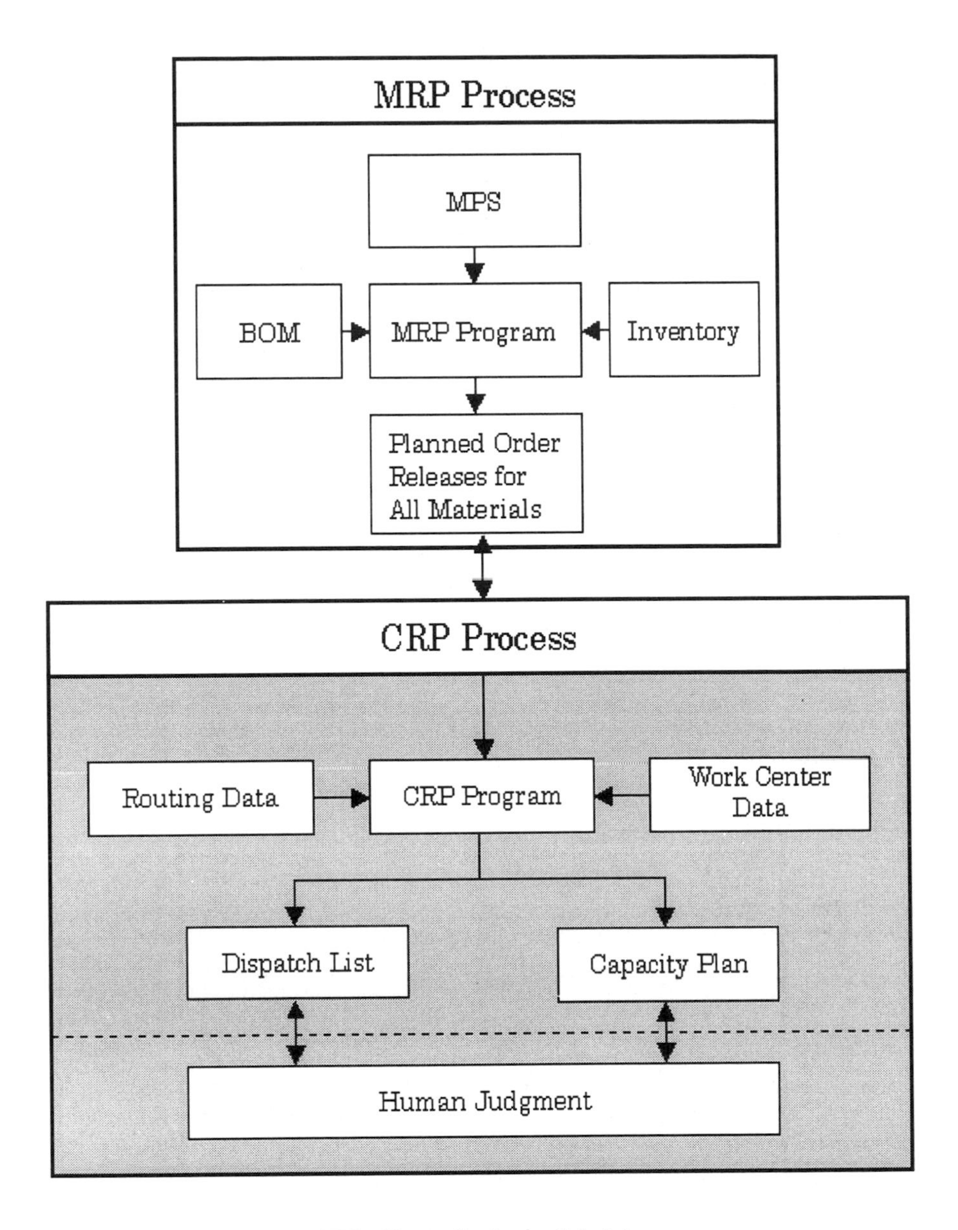

Note: MPS: Master Production Schedule

BOM: Bill of Materials

MRP: Material Requirements Planning

CRP: Capacity Requirements Planning

Figure 14.2 CRP process.

WORK CENTER NO.:24			DESCRIPTION: LATHES							DATE: 7/2		
ORDER NO.	PART NO.	PART DESCRIPTION	QTY.	OPN. NO.	OPN. START DATE	OPN. DUE DATE	ORDER DUE DATE	SET UP HRS.	RUN HRS.	NEXT/ PREV. W C	OPN. STATUS
JOBS CURRENTLY AT THIS WORK CENTER											
W123	144398	SHAFT	200	20	6/26	6/27	7/19	0	7.0	04	R
W123	144398	SHAFT	200	30	6/29	7/2	7/19	1.0	10.0	07	Q
W124	428876	BOLT	3000	30	6/29	7/3	7/24	2.0	30.0	07	Q
W110	330246	GEAR	500	20	6/28	7/3	7/18	1.0	15.0	07	Q
W120	407211	BOLT	4000	30	7/2	7/5	7/20	2.0	50.0	07	Q
W112	163726	HUB	40	20	7/3	7/5	7/23	3.0	8.0	07	Q
W128	118132	GEAR	400	20	7/5	7/10	7/24	2.0	40.0	07	Q
							TOTAL	11.0	160.0		
JOB COMING TO THIS WORK CENTER											
W129	186846	SHAFT	20	20	7/3	7/5	7/12	1.0	4.5	01	R
W138	258721	SPACER	2000	20	7/9	7/11	7/25	0.5	20.0	05	S
W140	321406	HUB	50	30	7/10	7/12	7/27	1.0	30.0	01	H
							TOTAL	2.5	54.5		

Figure 14.3 Dispatch list for work-center 24

(Source: *Gaining Control: Capacity Management and Scheduling*, Correll and Edson, © 1999. Reprinted by permission of John Wiley and Sons, Inc.)

DATE	**: 8/6**	**NO. MACH.**	**: 5**	**HOURS/SHIFT**	**: 8**
WORK CENTER	**: 24**	**NO. OPER.**	**: 2**	**SHIFTS/DAY**	**: 1**
DESCRIPTION	**: LATHES**	**MACH./OPER.**	**: 2**	**DAYS/WEEK**	**: 5**
DEMO. CAP'Y.	**: 120**	**MAX. CAP'Y.**	**: 336**	**LOAD FACTOR**	**: 75%**

WEEK	MACHINE CAPACITY			LABOR CAPACITY		
	REQ'D. CAP'Y. (HRS.)	PLAN CAP'Y. (HRS.)	LOAD VS. CAPACITY (%) 50 100 150	REQ'D. CAP'Y. (HRS.)	PLAN CAP'Y. (HRS.)	LOAD VS. CAPACITY (%) 50 100 150
8/6	128	150	XXXXXXX	64	60	XXXXXXXXXXXX
8/13	118	150	XXXXXX	59	60	XXXXXXXXXXX
8/20	116	150	XXXXX	58	60	XXXXXXXXXX
8/27	126	150	XXXXXXX	63	60	XXXXXXXXXXXX
9/3	130	150	XXXXXXX	65	60	XXXXXXXXXXXXX
9/10	144	150	XXXXXXXXX	72	60	XXXXXXXXXXXXXXX
9/17	140	150	XXXXXXXXX	70	60	XXXXXXXXXXXXXX
9/24	150	150	XXXXXXXXXX	75	90	XXXXXXX
10/1	166	150	XXXXXXXXXXXX	83	90	XXXXXXXX
10/8	160	150	XXXXXXXXXXX	80	90	XXXXXXXX
↓ 7/28						

Figure 14.4 Capacity plan for work-center 24

(Source: *Gaining Control: Capacity Management and Scheduling*, Correll and Edson, © 1999. Reprinted by permission of John Wiley and Sons, Inc.)

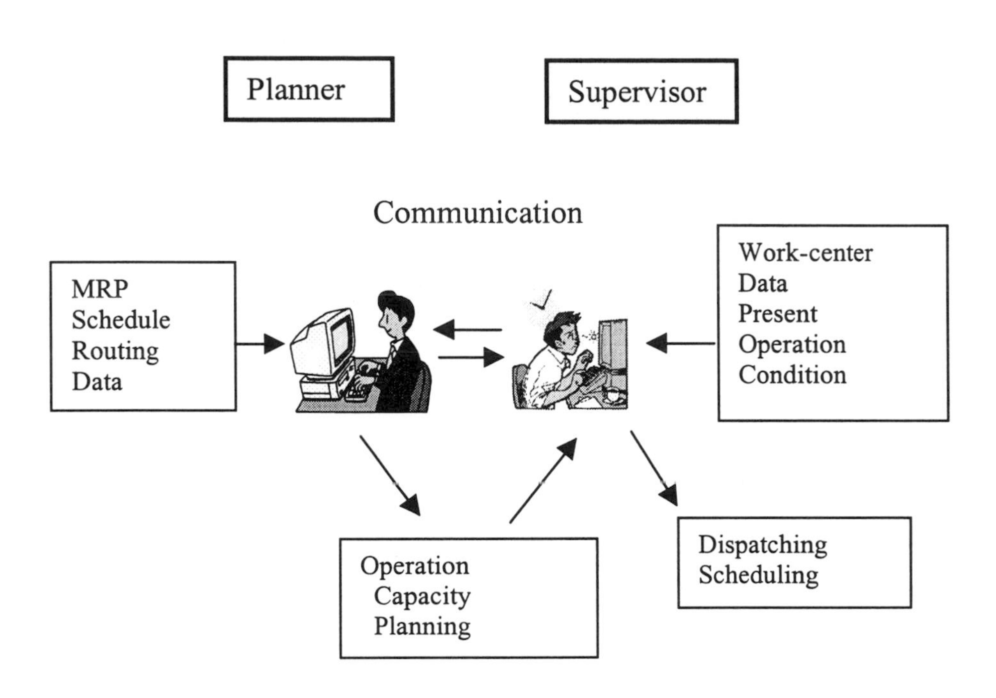

Figure 14.5 Roles of planner and supervisor.

It is clear that a schedule that does not achieve due dates costs much. The added cost or inconvenience needs to be assessed against missing the due date. Cost is generated as soon as overtime work begins or money is spent on alternative methods. Efforts should be made to meet the scheduled work days initially rather than acting after the work is delayed. It is more important to pay attention to the schedule of each work-center rather than to worry about the performance of individual work-centers. The productivity of each work-center increases when the schedule is consistently defended by all work-centers.

It is important that the supervisor can always obtain correct and accurate information from the shop floor, as decisions must be based on that information. If the information is not accurate, the supervisor cannot produce a good schedule.

14.3.2 The planner's responsibility in operation capacity planning

The planner is responsible for solving inconsistencies between load and capacity. If the capacity plan which is obtained from MRP development is not executable, the solution does not make sense. The planner has a number of courses of action he can take, but this action is constrained by two variables: capacity and lead-time, as shown in Figure 14.6.

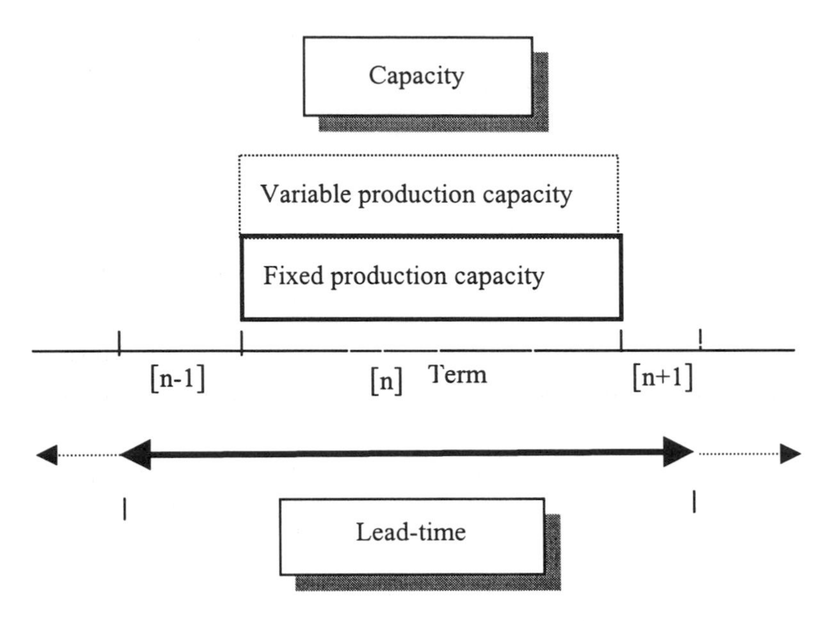

Figure 14.6 Two variables in capacity decision-making

A schedule that overloads the resources available may only be solved by modification of one of the two variables in Figure 14.6.

1. Capacity is increased, or
2. Lead-time is extended.

Which of the two methods is adopted, or a mixed plan by which they are both combined, is the planner's decision.

The generalized decision-making framework within which capacity planning by the planner takes place is indicated in Table 14.1.

When a load in the future is uncertain, the problem of balancing capacity and load is very difficult, especially because the load in the future cannot be forecasted for the job shop and the product mix cannot be fixed. That is to say, MPS is not planned easily. Very often the only feasible solution is a short-range modification to the capacity, by the use of overtime, standby equipment, and subcontracting. The key to the balance of capacity and load is whether or not capacity is flexible.

If the production system has greater flexibility, the load adjustment does not become a problem because of the introduction of work cells operated by multi-skilled workers. Provided there is not an entire plant overload, balancing capacity may normally move operators of the work cell to other work cells in each product group. Also, balance of the load in a plant where the production system has excessive equipment beforehand does not become too much of a problem. However, it will give rise to an economic problem. The option of maintaining excess capacity is economically feasible only where longer-range demand is relatively stable. As for the role of the planner in this case, effective use the excess capacity becomes a problem.

Table 14.1 Action courses of capacity planning decisions

Key variables	Action policy	Action
Capacity	• Building in excess capacity • Temporarily increase capacity (Only flexible) • Allow capacity to be varied to match demand	• Operate standby machines • Use overtime, subcontractors idle machines (Only flexible) • Use multi-skilled workers • Change work cell layout
Lead-time	• Extend lead-time to fit capacity	• Move jobs (Compression of lead-time)

The other variable is lead-time. The lead-time may be extended or compressed so as to move the load to an alternative time period where capacity is available. The planner should decide which load to be moved at which time period. When the moved load and the moved period are decided the planner must check whether the due date of the work order is satisfied or not. If not satisfied, the planner must compress the lead-time by using one or more of the five techniques mentioned in Section 14.3.1.

The planner's decision may affect many work-centre loads. Therefore, commitment and teamwork between the planner and the supervisors is necessary. A change of MPS might also be necessary, according to the planner's decision. In this case, the consent of the MPS planner will be needed. If a major change of current schedule is forced by changing MPS, it is necessary to refer to the material purchase report, the product sales report, the level of worker's skill, running of the machines, and the quality of the material, etc.

However, the key to success is for both planner and supervisor to work together and to make the best use of each person's expertise and experience. Problems which require human judgement cannot be solved by computer no matter how it is served. Therefore, it seems that an interactive simulation process in which the computer offers appropriate information to the planner and the supervisor and they make final decisions based on that information is most suitable. This idea is illustrated in Figure 14.7.

14.4 AN INTERACTIVE MRP-CRP SYSTEM

The effectiveness of the human-computer interaction system is described in several papers (for example, Bauer *et al.*, 1994; Nakamura and Salvendy, 1994). A Leitstand is very notable, a computer graphics-based decision support system for interactive production planning and control. The word 'Leitstand' is German for command centre or a type of directing stand which is developed in Germany (Kanet and Sridharan, 1990).

The underlying idea of this approach is that good planning requires a human planner in making, evaluating and adjusting plans. An interactive system gives him/her all possible support, by combining human knowledge and experience and computerized information. Thus, the responsibility for planning is placed on the human planner.

We have designed and built an interactive MRP-CRP System which the planner uses as a tool to carry out the decision-making of CRP.

14.4.1 System configuration

This system adopts the composition as shown in Figure 14.8. The system divides into 'MRP system', 'CRP system', 'Load adjustment', 'Report generation system', and 'Interface'.

MRP system: MRP is developed from the Job data, MPS, and BOM, and the reports which are output.

CRP system: The load at each work-center is calculated from the report generation system and the MRP system.

Report generation system: From the data collected by the MRP system, various reports are generated, which are shown to the user through the interface. Two primary outputs result: the dispatch list and the capacity plan.

Load adjustment: The load at the work-center adjusted by the CRP system is shown to the user through the interface.

The user also sees this load schedule, and decides the amount of load to be moved as well as the movement of the time period. The system specifies changes from the information it has gathered, and outputs the instructions which determine the dispatching list and the capacity plan which are, in turn, passed on to the MRP and CRP systems.

Interface: Information passed from the report generation system and the load adjustment is shown to the user, and 'Undo' (cancellation of processing immediately before), 'Save' (preservation of specific information) and 'Load' (reading of specific information) are processed. Moreover, the change in each time period at each work-center caused by having moved the load is displayed.

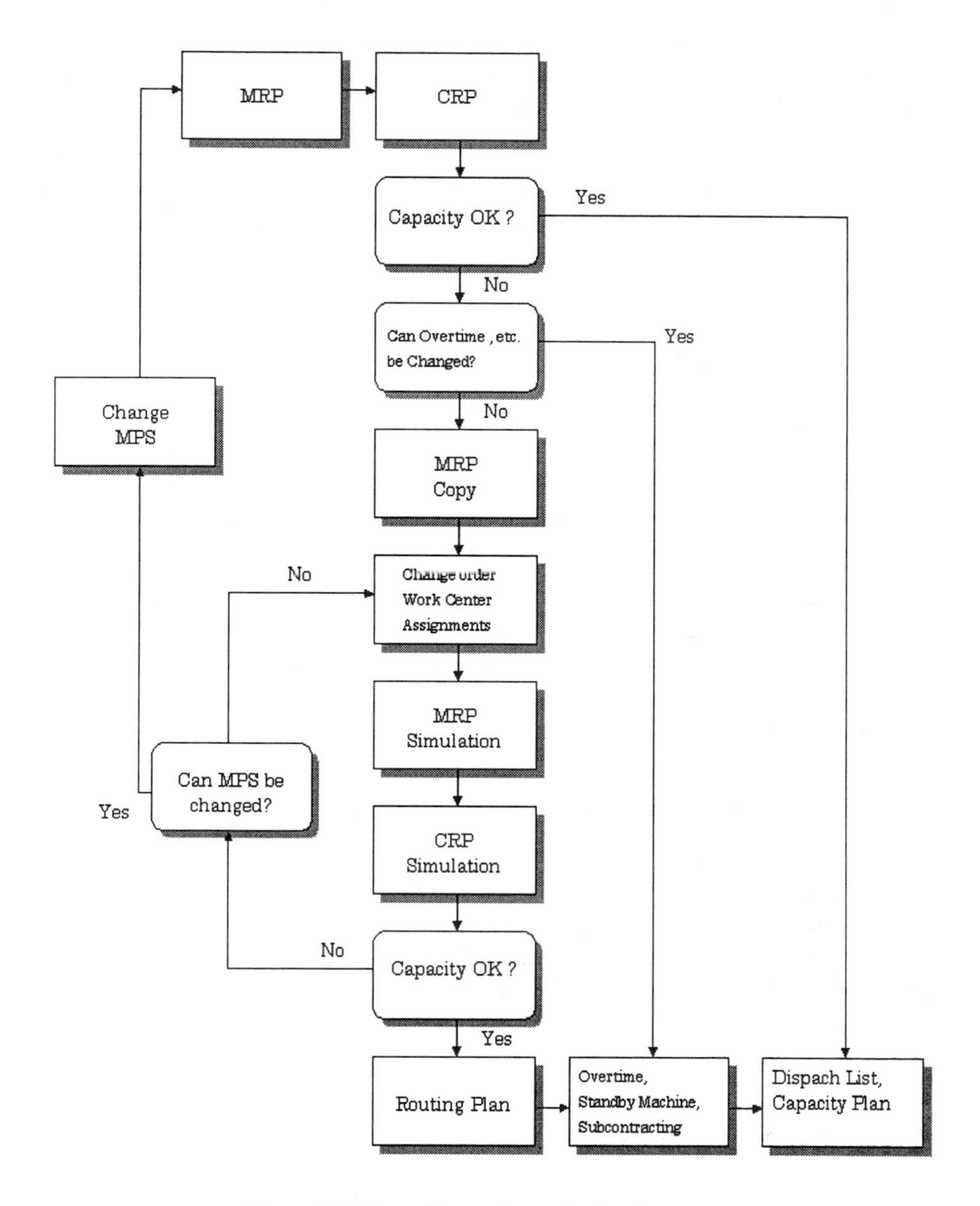

Figure 14.7 Capacity requirements planning process.

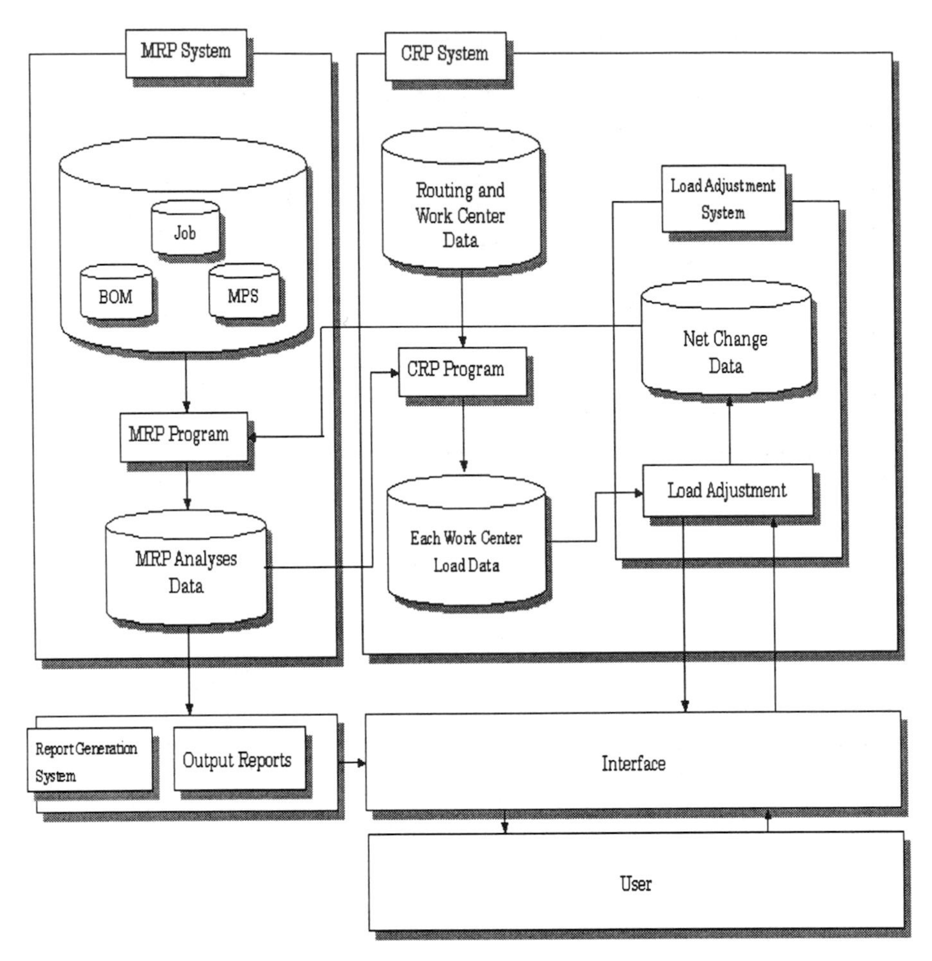

Figure 14.8 System structure.

14.4.2 Methods of interaction

The interaction between the user and the system is shown in Figure 14.9. The system develops MRP from the given data (MPS, BOM, and inventory information), and makes the load schedule first. Next, the user evaluates the information through the interface, sets the target (resolving overloading), and gives appropriate instructions (order movement, overtime, etc.) to achieve the target. The system processes the instruction from the user, and feeds back the resultant information. The user repeats this process until the target is achieved.

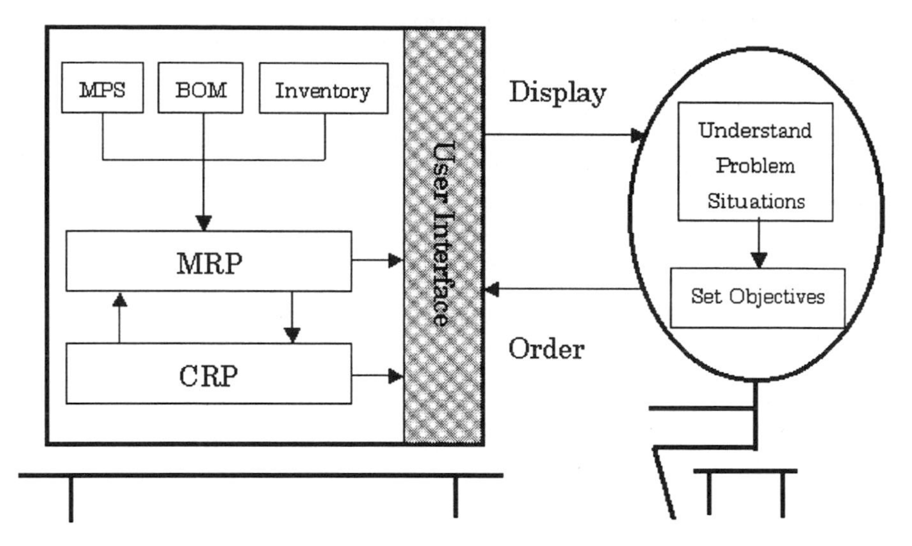

Figure 14.9 Interaction in the MRP – CRP system.

14.4.2.1 Overload adjustment method

Let us assume that certain parts are overloaded in a certain period as shown in Figure 14.10. In this research, the load is adjusted by moving the processing of the overload elements back and forth at either period. For instance, when assuming that the user wants to move the overloaded part to the left, the system side should think it is one of the left side of the load which is the objective in Figure 14.10(a). (Processing following overload is done from the period that it has belonged to back at the planning stage. Therefore, because it is necessary to complete processing following an overload, it is not necessary to be concerned when the load is moved left.)

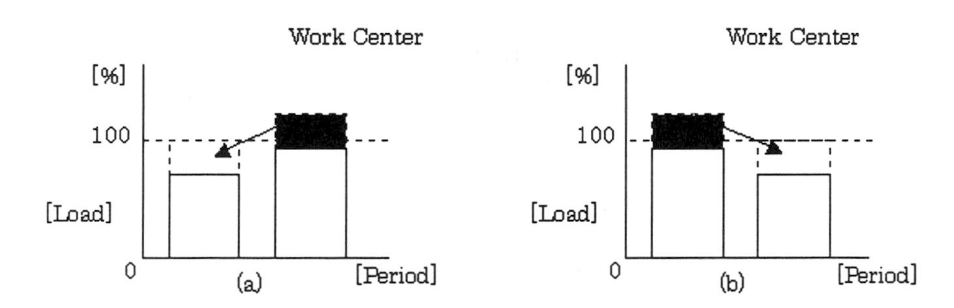

Figure 14.10 Movement.

Thus, it is necessary to consider what situation may occur when the load is moved left. For instance, it is assumed that the load schedule like Figure 14.11 is given. At this time, when the user moves the load in a state of overload left, the system should move up the related load. At this time, executing the production plan might become impossible because there is a chance that the system will allocate the load allocated at work-center 1 by moving the load earlier. Therefore, the system should check out this possibility. (If it is so, the system should tell the user he is not able to move the load in a state of overload left.)

If this does not happen, all loads are moved up. However, another overload might be generated afterwards. It is necessary to inform the user of that if it happens.

On the other hand, when assuming that the user wants to move the load in the state of the overload to the right, it is necessary to think only about the right side of the load which is the objective of the system side as shown in Figure 14.10(b). We must consider what situation may arise well ahead. For instance, when the amount of the load like Figure 14.12 is given, it is the same as a previous example. However, the possibility of making a late plan by moving the load will not be output but the possibility of delivery date delay is output. In this system, the user is informed if there will be a delivery date delay, and the decision about whether or not to allow this is left to the user.

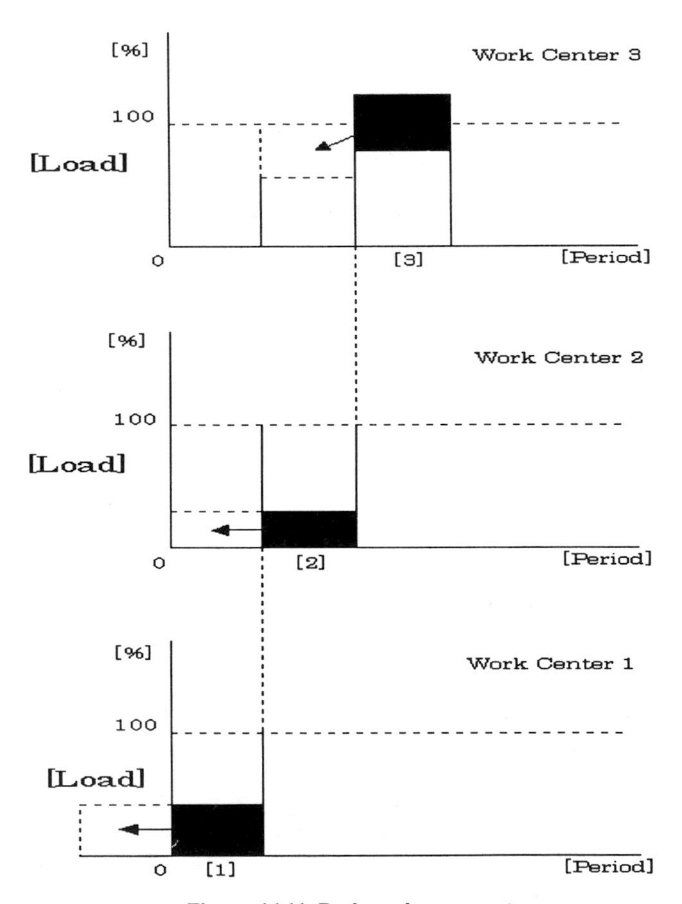

Figure 14.11 Backward movement.

Moreover, delivery date delays will occur when they have not been decided upon at the time of processing the overload because parts are shared in end-products. Therefore, it is necessary for the system to show this.

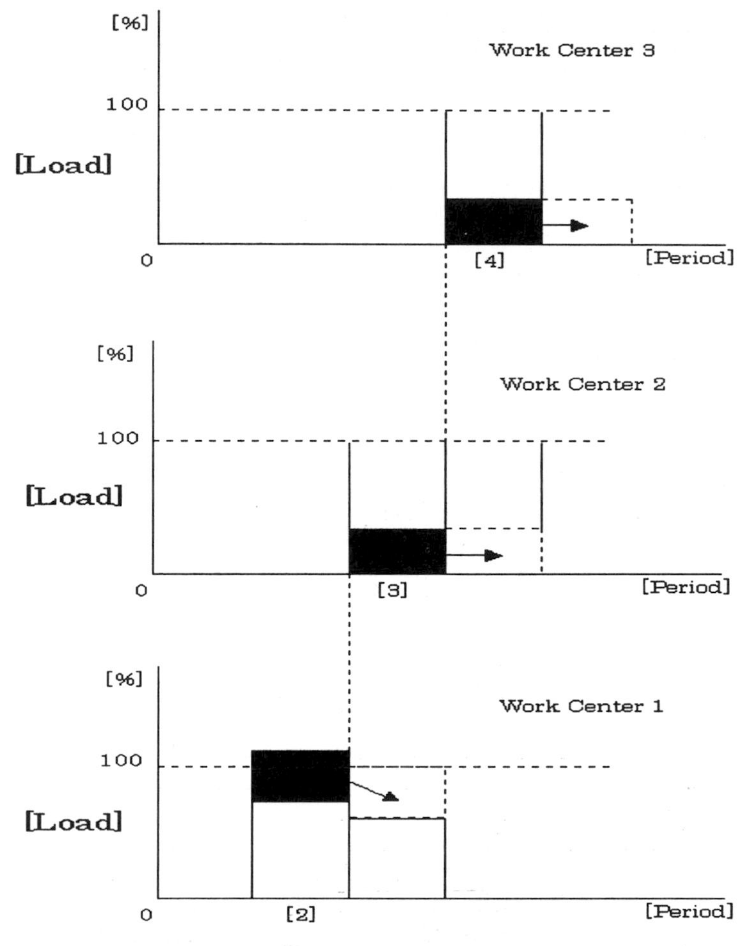

Figure 14.12 Forward movement.

14.4.2.2 Overload selection method

In the previous paragraph, we explained how the system should deal with moving the load in the selected state of overload. Which load the user ought to select is the problem. The targeted load is all loads from the period which caused the overload. The user should select one from among those, and process it. When the user needs to select one load in the state of overload, a checklist is required. Because the amount of inventory, the lead-time, and the process route are thought of as good decision aiding material, this system can properly offer the user this information. The system's constituent parts are also important, as the load table changes according to both the movement of these parts and to which product was targeted.

Therefore, because information for that is also necessary, it can be presented to the user.

14.5 AN EXAMPLE OF A CASE STUDY

14.5.1 Setting problem

The sample problem is a company which manufactures birdhouses, taken up in Scott (1994)[1] as a practice problem. This company builds six kinds of birdhouses. The number of parts in the birdhouses is 19, as shown in Table 14.2. Birdhouses are made by combining these parts.

Table 14.2 Parts description

Part no.	Description
A123	Wall mounting, red roof
A124	Wall mounting, green roof
A125	Freestanding, red roof, short stalk
A126	Freestanding, red roof, long stalk
A127	Freestanding, green roof, short stalk
A128	Freestanding, green roof, long stalk
B234	House subassembly
B235	Red roof
B236	Green roof
B237	Short stalk subassembly
B238	Slong stalk subassembly
C345	Roof
C346	Side
C347	Front
C348	House base
C349	Short stalk subassembly
C350	Long stalk
C351	Ground base
C352	Tube

(Source: Scott, 1994, p. 58)

The Bill of Materials (BOM) is necessary information for MRP, shown in Table 14.3, and MPS is given as Table 14.4. This shows how many products are produced in a week. Also, inventory information is given in Table 14.5. In this example there are four processes, each of which constitutes a work-center.

- Assembly process (A)
- Painting process (P)
- Welding process (W)
- Cutting process (C)

[1]We used Scott's data without modification for the example problem.

Production routings by which each part is processed are shown in Table 14.6.

Table 14.3 BOM

Parent part no.	Child part no.	Quantity per unit of parent
A123	B234	1
A123	B235	1
A124	B234	1
A124	B236	1
A125	A123	1
A125	B237	1
A126	A123	1
A126	B238	1
A127	A124	1
A127	B237	1
A128	A124	1
A128	B238	1
B234	C346	3
B234	C347	1
B234	C348	1
B235	C345	1
B236	C345	1
B237	C349	1
B237	C351	1
B238	C350	1
B238	C351	1
C249	C352	0.25
C350	C352	0.5

(Source: Scott, 1994, p. 59)

Table 14.4 MPS

	Week no.						
Part no.	13	14	15	16	17	18	19
A123	0	22	0	46	0	18	0
A124	0	0	37	5	0	0	0
A125	0	28	20	0	44	0	0
A126	0	0	0	60	0	0	0
A127	0	0	36	0	0	12	0
A128	0	54	0	26	48	26	0

(Source: Scott, 1994, p. 317)

Table 14.5. Actual inventory

Part no.	Inventory
A123	40
B234	0
C345	37
C346	350
C347	175
C348	145
C349	20

(Source: Scott, 1994, p. 318)

Table 14.6 Routing list

Part no.	Work Center Center	Setup (mins)	Run (mins)	Production operation
A123	A	15	4	Assemble red roof to house s/assy
A124	A	15	4	Assemble green roof to house s/assy
A125	A	0	2	Assemble short stalk to red house unit
A126	A	0	2	Assemble long stalk to red house unit
A127	A	0	2	Assemble short stalk to green house unit
A128	A	0	2	Assemble long stalk to green house unit
B234	A	45	12	Glue sides and front to house base
B235	P	60	6	Paint roof red
B236	P	60	6	Paint roof green
B237	W	25	4	Weld short stalk to ground base
B238	W	25	4	Weld long stalk to ground base
B249	C	10	2	Cut short stalk from 4 m tube
B350	C	10	2	Cut long stalk from 4 m tube

(Source: Scott, 1994, p. 332)

14.5.2 Implementation

The CRP process is implemented as follows. First, MRP is executed, and the MRP schedule as shown in Figure 14.13 is output.

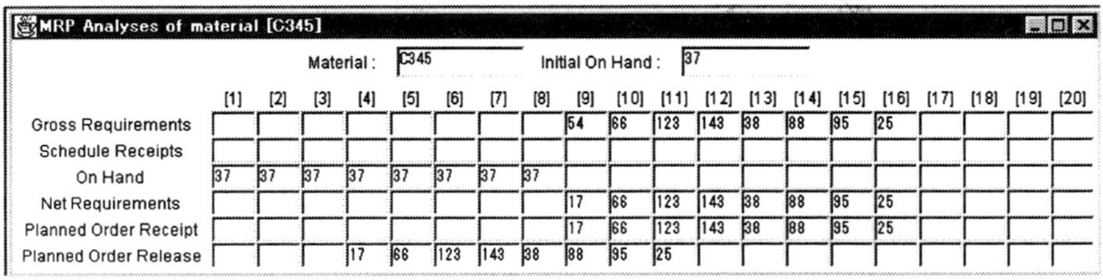

MRP Analyses of material [C345]

Material : C345 Initial On Hand : 37

	[1]	[2]	[3]	[4]	[5]	[6]	[7]	[8]	[9]	[10]	[11]	[12]	[13]	[14]	[15]	[16]	[17]	[18]	[19]	[20]
Gross Requirements									54	66	123	143	38	88	95	25				
Schedule Receipts																				
On Hand	37	37	37	37	37	37	37	37												
Net Requirements									17	66	123	143	38	88	95	25				
Planned Order Receipt									17	66	123	143	38	88	95	25				
Planned Order Release				17	66	123	143	38	88	95	25									

Figure 14.13 MRP schedule: #c345.

When 'work-center' is selected from the 'window' menu after the MRP is executed, Figure 14.14 is displayed. The horizontal axis of this figure shows the period, and the vertical axis shows the capacity. '100%' shows the limit of the capacity at the work-center. It is the user's objective to allocate the load below this limit.

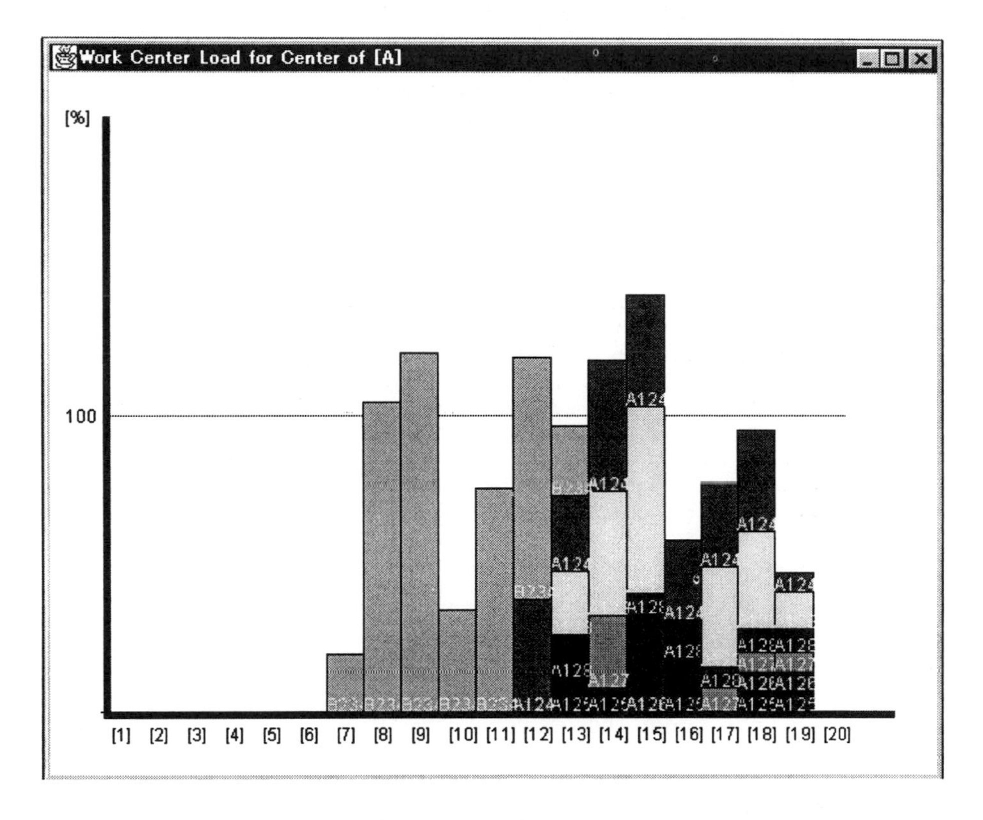

Figure 14.14 Capacity loading on work-center A.

The load is moved so that the user may click any load on the screen. A window as in Figure 14.15 appears when clicked. This window is sequentially displayed.

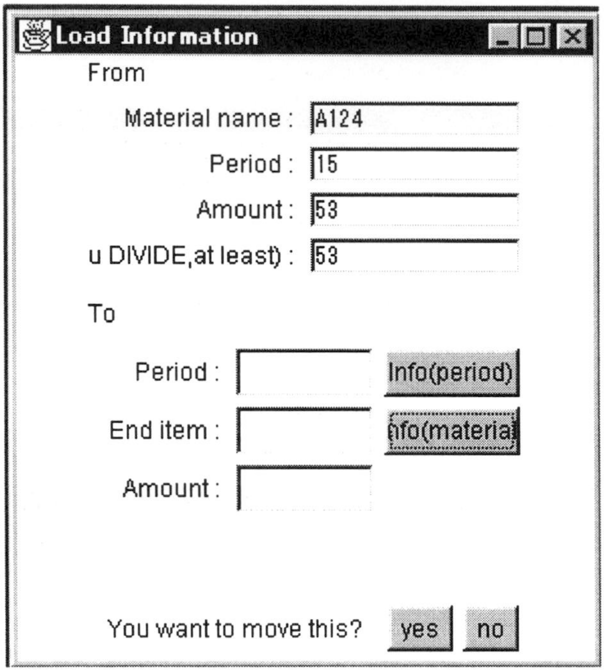

Figure 14.15 After click.

The lines under 'From' show Material name, Period, and Amount of the load which the user selected on the load schedule, and the amount that needs to be moved when the user is about to meet the capacity limit.

Three lines under 'To' are for the user to input the values. Period, End item, and Amount which need to be moved in parts shown in 'From' are input. Moreover, it is possible to refer to information like Figure 14.16 by pushing the button once the period and end item have been input.

After a necessary input is clarified, the user can move parts by pushing the 'yes' button and can cancel by pushing 'no'.

The user learns what change took place because of the movement of the load in the window as shown in Figure 14.17. Colored circles are used.

The following five kinds of colored circles are displayed in this window.

- *Red circle*: Period which has newly exceeded the capacity restriction by the moved load.
- *Small blue circle*: Period which has not satisfied the capacity limit yet, though the load decreased by moving the load.
- *Blue circle*: Period without change even if load is moved.
- *Large blue circle*: Period that load has increased though the capacity limit has been exceeded by the moved load.
- *White circle*: Period for load to satisfy the capacity limit with the moved load.

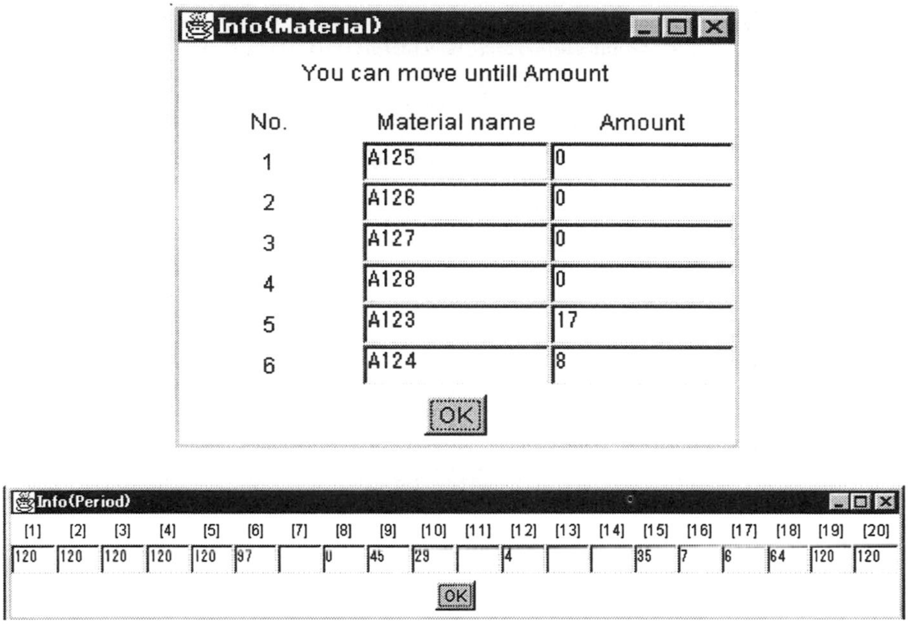

Info(Material)

You can move untill Amount

No.	Material name	Amount
1	A125	0
2	A126	0
3	A127	0
4	A128	0
5	A123	17
6	A124	8

OK

Info(Period)

[1]	[2]	[3]	[4]	[5]	[6]	[7]	[8]	[9]	[10]	[11]	[12]	[13]	[14]	[15]	[16]	[17]	[18]	[19]	[20]
120	120	120	120	120	97		0	45	29		4			35	7	6	64	120	120

OK

Figure 14.16 Check data.

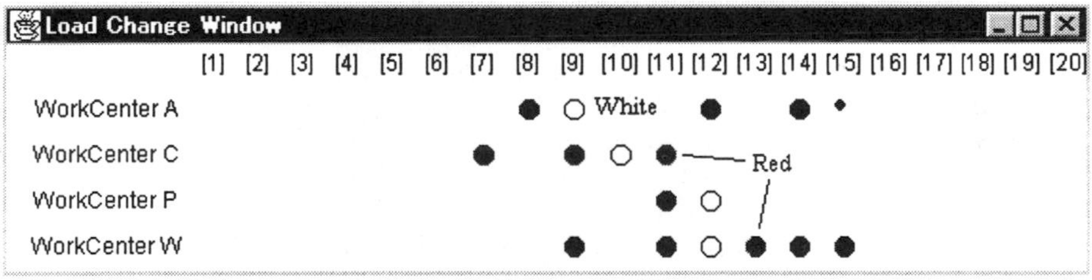

Load Change Window

[1] [2] [3] [4] [5] [6] [7] [8] [9] [10] [11] [12] [13] [14] [15] [16] [17] [18] [19] [20]

WorkCenter A

WorkCenter C

WorkCenter P

WorkCenter W

White

Red

Figure 14.17 Load change indicators (note: Except for red and white circles all others are blue).

It may appear that this result is in a good direction. At this time, the user can preserve the state by selecting 'save' from the 'File' menu. Follow up directions are given by the computer.

It is possible to return to the previous state by selecting 'Undo' from the 'Edit' menu and 'Load' from the 'File' menu as shown in Figure 14.18, when the result of a movement is not as desired.

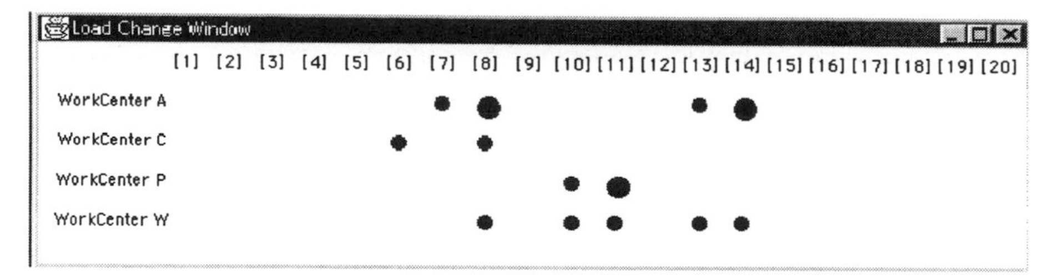

Figure 14.18 Unsuccessful move (The color of all circles is blue).

The user finally obtains a result like that in Figure 14.19 by repeating these steps. However, only the load of work-center A has been adjusted up to now. Therefore, it is necessary to adjust the load by repeating similar steps with other work-centers. Final results in this example are shown in Figures 14.20 to 14.22.

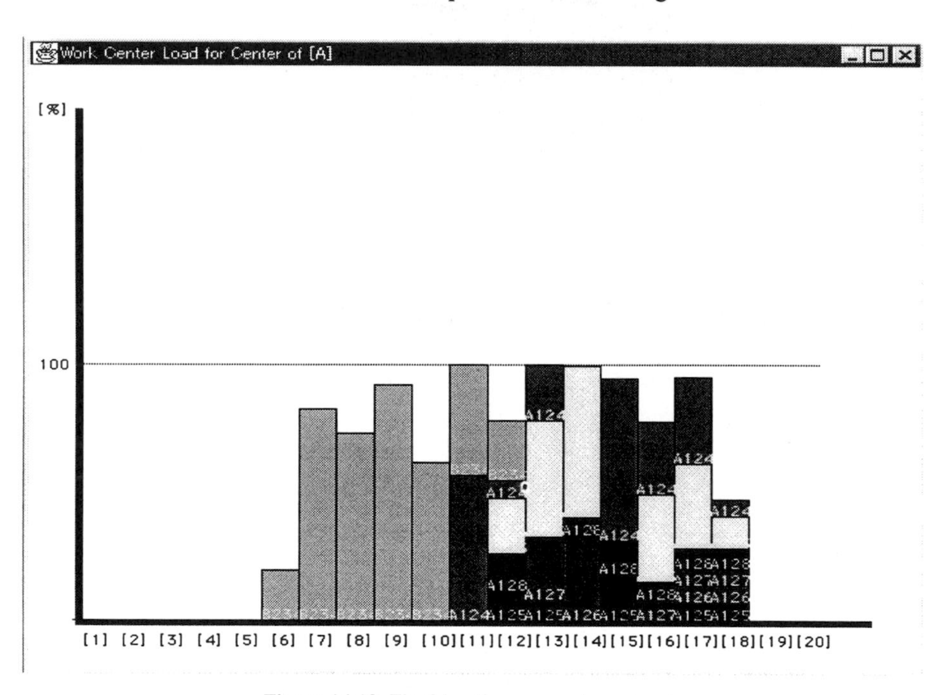

Figure 14.19 Final loading on work-center A.

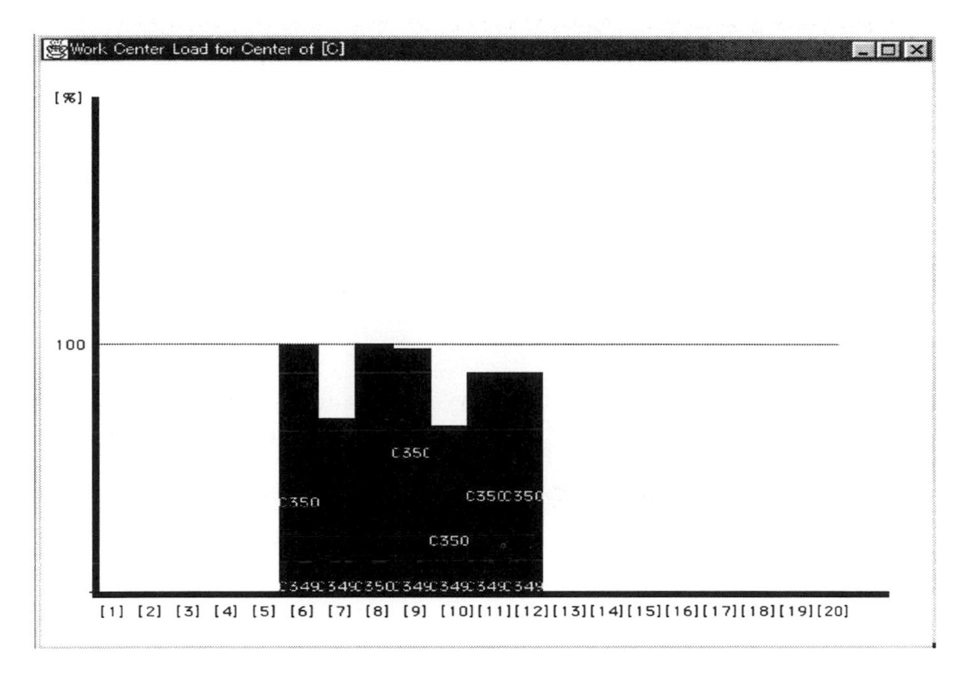

Figure 14.20 Final loading on work-center C.

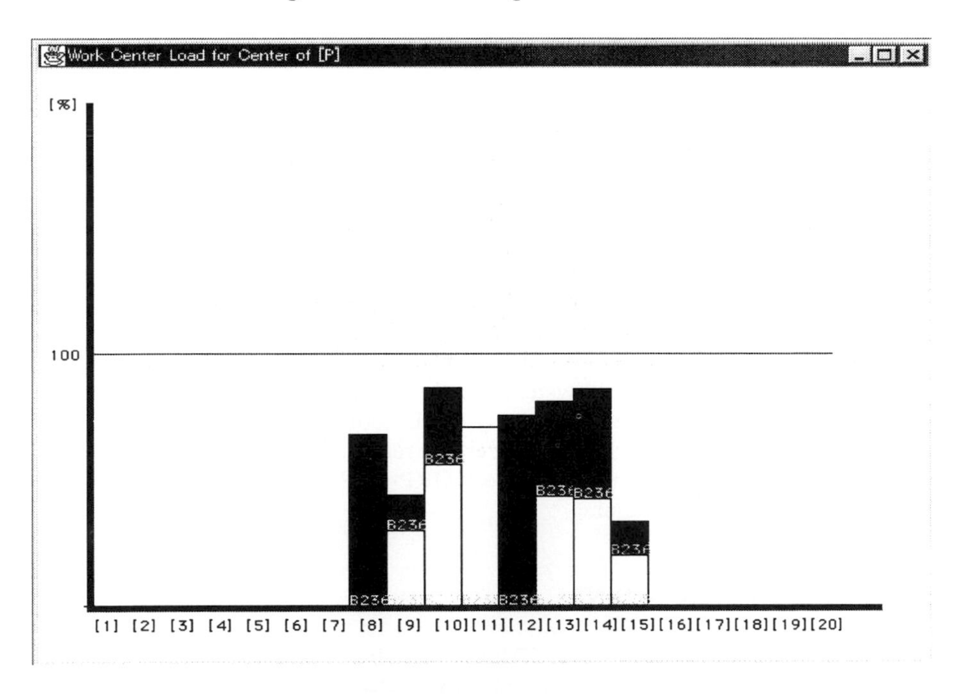

Figure 14.21 Final loading on work-center P.

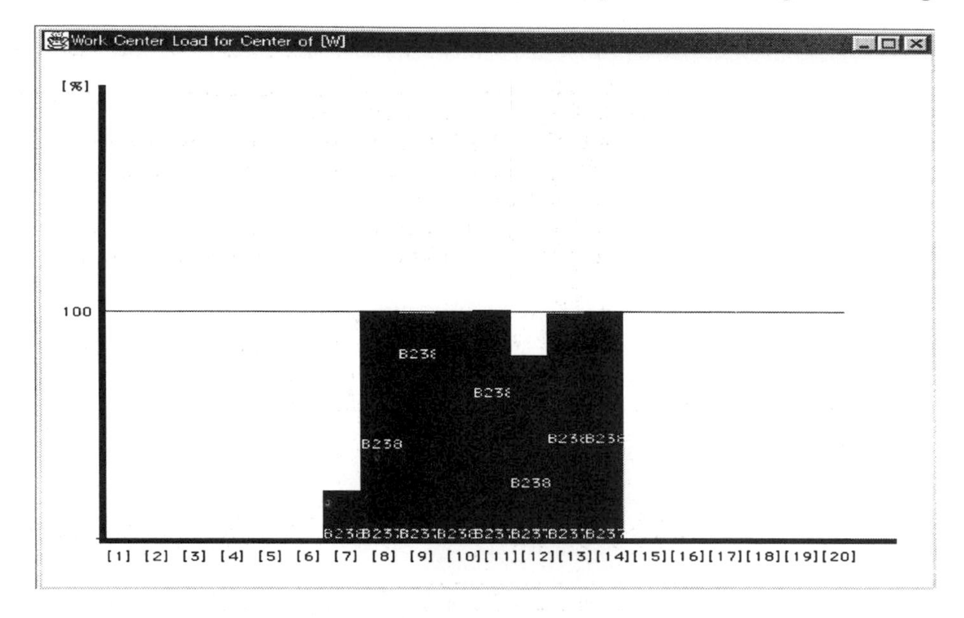

Figure 14.22 Final loading on work-center W.

14.6 CONCLUSION

In this chapter, we described Capacity Requirements Planning (CRP) which checked the feasibility of outputs obtained from the development of Material Requirements Planning (MRP). In particular, it focused on human responsibility in CRP.

Since there are many options in capacity planning, it is best to have a human make these decisions rather than the computer. It is thought that this situation will not change in the future. Therefore, we recognized that the computer is by far the best tool in supporting the decision-making of the human element who may be a planner, supervisor, or manager.

From such recognition, an interactive MRP-CRP system was designed and implemented as a useful tool when MRP-CRP was developed for the user as a final decision-maker. However, this system is only a prototype. More improvement is needed for this system to become a truly useful tool. A first task is to investigate and assess the impact of the system concept on real large-scale problems. A second task is to develop an interactive intelligent MRP-CRP system with knowledge and information for optimal planning.

14.7 ACKNOWLEDGEMENT

The author wishes to thank Akinori Takeda of Nishikawa Rubber Industry Co. Ltd. for his contributions to the section titled ‘An Interactive MRP-CRP System’.

14.8 REFERENCES

Bauer, A., Bowden, R., Browne, J., Duggan, J. and Lyons, G. (1994). *Shop Floor Control Systems : From design to implementation*, London: Chapman & Hall.

Correll, J. G. and Edson, N. W. (1999). *Gaining Control: Capacity management and scheduling*, New York: John Wiley & Sons, Inc.

Gaither, N. (1996). *Production Operations and Management*, Belmont, California: Wadsworth Publishing Company.

Hax, A. C. and Candea, D. (1984). *Production and Inventory Management*, Englewood Cliffs, California: Prentice-Hall.

Kanet, J. J. and Sridharan, V. (1990). 'The Electronic Leitstand: A new tool for shop scheduling', *Manufacturing Review*, Vol.3, No.3, 161-170.

Nakamura, N. and Salvendy, G. (1994). 'Human Planner and Scheduler', in G. Salvendy and W. Karwowski (eds.), *Design of Work and Development of Personnel in Advanced Manufacturing*, New York: John Wiley & Sons, Inc., pp. 331-354.

Orlicky, J. A. (1975). *Material Requirements Planning: The new way of life in production and inventory management*, New York: McGraw-Hill Book Company.

Render, B. and Heizer, J. (1997). *Principles of Operations Management*, Second Edition, Upper Saddle River, New Jersey: Prentice-Hall.

Scott, B. (1994). *Manufacturing Planning Systems*, London: McGraw-Hill Book Company.

Vollman, T. E., Berry, W. L. and Whybark, D.C. (1988). *Manufacturing Planning and Control Systems*, Second Edition, Homewood: Richard D. Irwin, Inc.

PART IV

Context and Environment for Planning, Scheduling and Control

CHAPTER FIFTEEN

Assessing the Effectiveness of Manufacturing Information Systems

Janet Efstathiou, Anisoara Calinescu, John Schirn, Lars Fjeldsøe-Nielsen, Suja Sivadasan, Julita Bermejo-Alonso and Colin J. Neill

15.1 INTRODUCTION

The increasing competitiveness of the global marketplace is forcing manufacturers to become more customer-focused, with mass production being superseded by 'flexible specialisation' or mass customisation. Competitiveness is no longer dependent on price alone, but delivery times and quality are becoming as, if not more, important (Wiendahl and Scholtissek, 1994; Schonberger, 1996). Increased customer expectations lead to higher product variety, reduced product life cycles, shorter customer lead-times and unpredictable demand patterns (Higgins *et al.*, 1996). The reaction has been to develop strategies that call for increased flexibility and responsiveness of manufacturing operations.

These demands create a constant pressure for change in the way the facilities are operated, perhaps exceeding the intentions of the original planners. To cope with changes in product volume, variety etc, the schedulers, who are responsible for the day to day operation of the facility, may adopt practices that solve short-term product problems, but which may become standard practice thereafter. Together with differing practices, shifting market and supply environment, the changes in product range, volume and mix all combine to affect the complexity of the facility.

If the complexity of the facility increases beyond control, the facility will become unpredictable and will not be able to deliver to schedule. In attempting to regain control of the facility, the planners and managers may respond in a number of ways, including:

- reduce the product range;
- modularise the products;
- change the manufacturing layout to cells;
- install a computer-based manufacturing information system;
- add extra manufacturing capacity.

The problem for the planner or manager is to choose amongst these and other options. Which option will do most to improve the reliability and flexibility of the facility? What will be the effect on lead-time? What other strategies are available?

This chapter will present an integrated methodology that will attempt to answer these questions.

15.1.1 The role of scheduling and planning

In a manufacturing facility, we understand the word planning as referring to the allocation of fixed resources to a particular group of tasks or products. Thus, the planning function would decide the number and arrangement of workstations, tools and labour to produce a particular set of combinations of products. This might also include, for example, choosing the routes for particular jobs, the lot size and the set-up times. The planning task is performed infrequently, with an initial planning phase at the establishment of a facility, and perhaps reviews at annual intervals.

Scheduling is the day-to-day activity of allocating the fixed resources to the current set of orders. This is a much more dynamic activity, carried out in response to the market. Scheduling may be done daily or weekly, depending on the complexity and dynamic behaviour of the facility.

In addition to the planning and scheduling activities, the facility must also be monitored. This involves checking the progress of work through the facility, making sure that everything is proceeding according to the schedule. This also involves monitoring the status of the workcentres, verifying that they are in an operational condition.

Let us consider now the flow of material and documents between these three levels of the facility. We shall use the term 'documents' to refer to the computer- and paper-based records, charts, tables, instructions and reports that are circulated within a manufacturing facility. The term 'information' will be reserved for use in a more formal way, as defined in Section 15.2.4 below.

The scheduling function receives plans from the planning level of the organisation, defining the layout, routing, production parameters, product range, volume and mix for the facility. The scheduling function also receives documents from the sales function specifying current demand for the products. The dynamic data are likely to be updated daily or weekly, with the addition of more frequent requests, caused by machine or material unavailability or changes required by customers. The scheduling function feeds performance and status reports outwards to the sales function, and may provide more long-term performance reports to the planning function. Furthermore, the scheduling function may generate and retain more operational data for its own internal use.

Operational scheduling forms the link between the control system and production, and will therefore have to deal with all the problems of inadequate planning at previous levels, variability and unreliability of the production environment and uncertainty of the external environment. Being at the lowest level of planning and control, this is the area where most problems are made visible and have to be dealt with. Hence, the quality (as defined in Section 15.2.3 below) of the documents received by the scheduling function is critically important in making sure that the scheduling function can do a rapid and effective job.

The ability of the scheduling function to do its task of adjusting and repairing schedules in response to dynamic behaviour of the facility depends too on how well the production environment was designed and its suitability of production planning (Roll, Karni and Arzi, 1992; Bauer *et al.*, 1994). However, it is difficult to assess the quality of the planning output, since this is affected by how well the scheduling function performs.

For flexibility of manufacturing material flow to be realised into competitive advantage, the controlling system has to have adequate capacity and be supported by information of adequate quality. Failing this, potential flexibility acts only to increase the complexity of the manufacturing environment, leading to short-term 'fire-fighting' problem-solving. Furthermore, an unsuitable information system can also increase the complexity of the manufacturing environment, making it harder to predict and control.

15.1.2 Problems in controlling the scheduling task

Manufacturing companies face many problems in controlling the scheduling task, as implied by the extensive use of performance measures for schedule adherence and stability, and the volume of research that has concentrated on schedule repair. Far from being fully automated, the day-to-day schedule often requires (human) negotiation with internal and external suppliers and customers. Problems in controlling the scheduling task include:

- uncertain and variable demand, including customer changes to order quantities;
- unreliable supply of raw materials;
- an unreliable and variable production environment due to, for example, machine breakdowns, operator absenteeism, scrap and rework;
- unavailable, out-of-date and/or inaccurate information required to make effective scheduling decisions;
- unclear and/or conflicting 'goals' for driving scheduling decisions;
- interdependence with other departmental and functional schedules.

There may be a number of features of the organisation's custom and practice that may cause these problems to persist, including:

- poor communication between planning levels;
- production managers' obsession with short term and localised performance issues that do not necessarily support the organisational strategic goals (Hill, 1995);
- manufacturing operations in traditional mass production industries (such as the automobile industry) have opposing objectives to marketing pressures (Womack *et al.*, 1990);
- a lack of structured methods for assessing the cost effectiveness of information systems and the effect of 'information quality' which can lead to unfocused data-gathering and extensive end-user effort to obtain the required information in the right format at the right time;
- unclear responsibilities and authority regarding the maintenance of information accuracy;
- a tendency to work-around problems, fire-fighting, rather than determining causes and fixing them;
- difficulties of introducing changes to organisational culture.

The problems and causes above illustrate that the scheduling function acts as a focal point for many organisational and technical aspects of the manufacturing system at the point where value is to be added by the manufacturing system. How well the scheduling function is being controlled will therefore give a good indicator of the effectiveness of the manufacturing system organisation.

15.1.3 Motivation for a new methodology

This chapter presents an approach to measuring static and dynamic information – thematic complexity in a manufacturing facility and discusses its application in two case studies. In order to investigate the issues set out in the subsection above, a combination of research techniques was adopted and adapted throughout the life of the case studies. The research methodology followed was more than just measuring the measurable, and collecting and packaging data. In each case, appropriate research tools were carefully selected to assist with analysing and understanding activities and processes as well as being able to explain why things occur in one manner as opposed to another. Several methods were used to accumulate knowledge on the entire system including the scheduling process, information flow, internal and external relationships, and shop floor functionality. This approach of selecting existing methods revealed gaps within the research agenda, particularly within the manufacturing, information and scheduling context. To conduct a thorough investigation there was a need for the research team to go beyond existing investigative techniques. The research team was driven to address this matter by the development of new techniques (Section 15.3) to provide more variety and balance in assessing manufacturing, information and scheduling.

15.1.4 Outline of chapter

The outline of the chapter is as follows. Section 15.2 provides a review of the literature on scheduling, flexibility, information quality and complexity, as applied to manufacturing systems. Section 15.3 provides a description of the methodology that we are developing and applying to manufacturing systems. The methodology is illustrated briefly with references to current case studies, but a more complete description of the case study findings is given in Section 15.4. Sections 15.5 and 15.6 conclude the chapter with a brief discussion on human issues and a summary.

15.2 RELATED LITERATURE

15.2.1 Scheduling

A number of researchers have highlighted a gap between scheduling theory and practice, including McKay *et al.* (1988, 1995), MacCarthy and Liu (1993), Bauer *et al.* (1994) and Stoop and Wiers (1996). Classical scheduling theory makes many simplifying assumptions that do not apply to real factories but that are essential to produce workable schedules in the literature. Some examples of actual performance requirements that are often neglected in the literature are:

- *Many scheduling goals and constraints*, including capacity planning and load balancing, labour availability, material availability and tool availability (MacCarthy and Liu, 1993; Stoop and Wiers, 1996).
- *The variability of the manufacturing environment.* Actual performance will usually deviate from planned performance, often due to the stochastic nature of many manufacturing processes (Wiendahl, 1995).
- *The uncertainty of the manufacturing environment.* Disturbances occur unexpectedly, such as rework, rush orders, extra orders caused by scrap, machine breakdowns and operator absenteeism (MacCarthy and Liu, 1993). Much of the data used by schedulers may be incomplete, ambiguous, biased, outdated or erroneous (MacKay *et al.* 1988). For example, many processing times may be estimates, so that even if a schedule is optimal with respect to a deterministic or stochastic model, quite poor performance could result when evaluated relative to actual processing times (Stoop and Wiers, 1996).
- *Dynamics of the manufacturing environment.* The interaction between scheduling and higher level decision-making and the time framework of the planning and control system will determine the period over which inputs to the scheduling function will remain static (MacCarthy and Liu, 1993). Along with the factors mentioned above this leads to a need for schedules to be adjusted frequently. In this dynamic environment schedule repair becomes an important activity that needs to be supported by the scheduling system (Efstathiou, 1996).
- *Assessing schedule performance.* In practice schedule performance is not assessed by a single criterion, but as a compromise between multiple goals. Stoops and Wiers (1996) highlight three factors that complicate the assessment of a schedule's performance further: the time horizon that the schedule is judged over, different and possibly conflicting performance goals, and the fact that the schedule of one production unit may affect how well other production units can operate.

McKay *et al.* (1998) reported that there is little evidence that the first generation of AI scheduling systems produced results better than any traditional heuristic approach. The success of knowledge-based systems depends on the representation of the problem domain, which depends on the knowledge elicitation process. However, Charalambous and Hindi (1991) state that most documented AI approaches tend to use many of the rules and assumptions of OR and only the shop floor features and organisational constraints are extracted by interviews with managers. More advanced AI approaches (such as distributed systems, genetic algorithms, stochastic and heuristic repair techniques), which attempt to overcome some of the shortcomings mentioned above, are still a long way from being practical operational tools (Prosser and Buchanan, 1994).

Recent research into the scheduling problem has highlighted the importance of the role that human schedulers play. McKay *et al.* (1995), in studies of scheduler's decisions in an unstable manufacturing domain, found that the schedulers' decision processes were not based on the traditional models of scheduling, but that by using enriched information and predictive anticipation the scheduler improved performance. Stoop and Wiers (1996) consider that scheduling tasks which require flexibility, communication, negotiation and intuition benefit from the cognitive strengths of human schedulers, whereas handling many homogeneous jobs, procedures, capacity constraints etc., to optimise specified performance criteria is best supported by techniques. Stoop and Wiers (1996) also state that the success of techniques in practice is likely to depend on how the advantages of both computer-based and human-based scheduling skills can be integrated effectively.

The view of scheduling as an optimisation problem is being increasingly challenged. Scheduling is a dynamic activity, and should aim to produce schedules which are robust and flexible in adapting to the inevitable environmental changes (Bauer *et al.*, 1994). Amongst future requirements for scheduling systems prescribed in the literature are: better heuristics, decentralised control structures, real-time resource negotiation, provision of decision support and interactive scheduling techniques (MacCarthy and Liu, 1993; Bauer *et al.*, 1994).

15.2.2 Flexibility

The general definition of flexibility in the manufacturing literature is widely understood as *the ability of a manufacturing system to respond to change* (Kumar, 1987; Roll *et al.*, 1992; Das, 1996), often extended with the proviso that response should be *rapid* and *cost effective* (Gerwin, 1987; Benjafaar and Ramakrishnan, 1996; Chen and Chung, 1996; Corrêa and Slack, 1996). Research directed at increasing the understanding of flexibility has suggested a number of classifications of flexibility. However, the terms used often conflict: manufacturing flexibility has been described as a fuzzy, multi-dimensional phenomenon (Sethi and Sethi, 1990; McIntosh *et al.*, 1994). Many authors have developed a flexibility measure or set of measures, aimed at relating system performance to flexibility level, to provide decision support for factors such as: manufacturing system design/configuration, product mix changes, introducing new products, investment in new equipment and control/scheduling strategies (Brill and Mandelbaum, 1989; Chen and Chung, 1996).

Many of the studies on flexibility have focused on Flexible Manufacturing Systems (FMS) – see also chapter 9 in this book by Slomp. However, the factors considered are also relevant to more traditional manufacturing facilities, especially job shops that are, in their very nature, flexible. Due to the high automation of FMS a more formalised understanding has been required of the subsystems which make up the manufacturing system, prompting much of the research into flexibility. Indeed, many of the benefits associated with FMS have come about due to the improved 'managerial' understanding gained during the planning process (Slack, 1988).

Although flexibility as a concept is widely agreed as being an important competitive strength, the general consensus in the literature is that there is still insufficient understanding of the nature of flexibility to assess its effect on performance and implications for management (Gerwin, 1987; Sethi and Sethi, 1990; Benjafaar and Ramakrishnan, 1996; Chen and Chung, 1996; Das, 1996). Benjaafar and Ramakrishnan (1996) state that environmental conditions under which flexibility may be required, and/or justified, and operating requirements under which the associated benefits can be realised, are not yet clearly identified.

Until these factors are understood the economic benefits which result from flexibility will be difficult to measure (Azzone and Bertelè, 1989). The majority of research into flexibility has concentrated on these factors that provide the potential of flexibility. There is very little work that addresses the implications for management and the control system (scheduling in particular) to ensure that potential flexibility is realised in practice (Sethi and Sethi, 1990; Benjafaar and Ramakrishnan, 1996).

15.2.3 Information quality

With the increased responsiveness demanded in the current market environment more demands are made on information flow. How information is used and transmitted within and between companies has become a critical factor for the success of a company. The value of information depends on the usefulness of information and as such must be related to the improvement it can bring to a decision or action. Kehoe *et al.* (1992) define the effectiveness of information in fulfilling its required needs as 'information quality', which they characterise by six essential, qualitative criteria:

- *Relevance*: Information is relevant if a better decision can be made because of it. Information that is not relevant may result in increased decision-making time or mean that not all the available relevant information is utilised.
- *Timeliness*: The degree to which information is available in time for a decision. In practice this may be dependent on the frequency with which information is updated, or how out-of-date the available information is when a decision needs to be made.
- *Accuracy*: How near to the truth, or free from error, and with what precision the information is known.
- *Accessibility*: The degree to which information is available, or how easily information can be obtained by the decision-maker.
- *Comprehensiveness*: The degree to which the information is complete enough, concise enough and at the right level of detail so that the decision-maker understands the information correctly rather than having to make some interpretation or judgement on the information content.
- *Format*: The effectiveness with which information is perceived by the decision-maker due to the manner in which the information is presented.

In order to achieve high quality information a number of organisational factors need to be addressed, particularly where people interact with the information system. Gavin and Little (1994) assert that personnel have to understand the necessity of providing data and be made aware of the importance of the accuracy and consistency of these data. From an operational point of view, these characteristics of information quality could be assessed to determine the additional complexity in decision-making due to shortcomings of the data flows. Research into manufacturing information systems reported in the literature has mostly concentrated on the quantity of information required for different strategies, such as Ronen and Karp (Karp and Ronen, 1992; Ronen and Karp, 1994), who developed entropy-based measures to evaluate the relationship between information requirements and lot sizes.

15.2.4 Complexity

Like flexibility, complexity is a term often used when discussing manufacturing operations. Several definitions of the term exist in the literature, and will be reviewed below. However, we shall define and discuss further below an entropic measure of complexity, based on the work of Frizelle.

The complexity of a system is usually seen as a factor of the following elements (McCarthy and Ridgway, 1996; Mitleton-Kelly, 1997; Edmonds, 1998):

- number and variety of component parts;
- the connectivity or interactions between the component parts;
- the behaviour of a system, in terms of the difficulty of predicting or understanding the observed evolution.

No common framework is used for describing and/or comparing the various sources or causes of manufacturing complexity, although a large number of features of the manufacturing environment have been identified as adding to complexity (Wiendahl and Scholtissek, 1994; Efstathiou *et al.*, 1996; Bermejo *et al.*, 1997; Calinescu *et al.*, 1997; Schirn, 1997), as illustrated in Figure 15.1.

The volume of literature on manufacturing complexity measures is not as large as that for flexibility. Measures of manufacturing complexity proposed in the literature mostly use the concept of entropy taken from information theory (Shannon, 1949), e.g. (Deshmukh *et al.*, 1992, 1998; Frizelle, 1995; Frizelle and Woodcock, 1995). The majority of these measures are concerned with combinations of parts, machines and operations, and as such they are similar to some measures of flexibility (e.g. routing flexibility) and are restricted to a static perspective of material flow. The method proposed by Frizelle and Woodcock (1995) also considers dynamic aspects, such as machine breakdowns and queue behaviour, by direct observation of the operating environment (see Section 15.3).

The complexity measure that we adopt is based on the work of Frizelle (1995), although this chapter will also present our alternative derivation. Frizelle proposes two measures of the complexity of a manufacturing system - static and dynamic. The static complexity measures the complexity of the system's schedule, and the dynamic measure assesses how well the system obeys the schedule. Both measures indicate the expected amount of information needed to describe the state of the system.

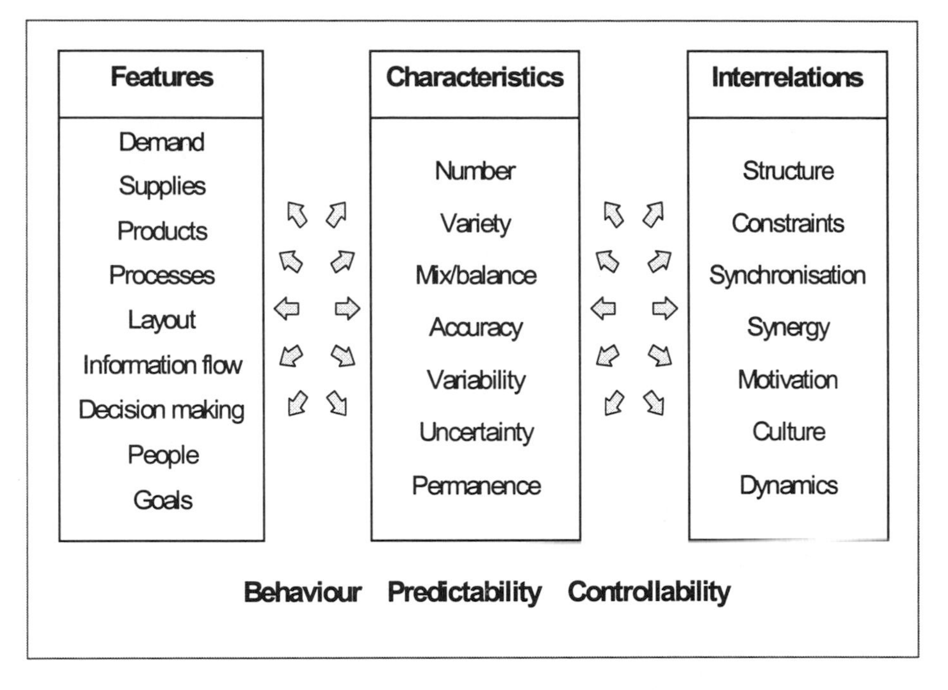

Figure 15.1 Factors of complexity in manufacturing.

The static complexity of a manufacturing system is defined as the expected amount of information necessary to describe the state of the system.

Suppose a manufacturing system consists of M machines or work centres. Let each machine have S_i possible states of interest. The states may include, for example:

- product it is working on;
- maintenance;
- idle, awaiting set-up;
- idle, awaiting operator;
- inoperable.

The probability of machine i being in state j is given by p_{ij}. The amount of information needed to describe the state of one machine at any point in time is:

$$I = -\log_2 p_{ij} \tag{15.1}$$

Hence, the expected amount of information needed to describe the state of machine i is given by:

$$I_{\exp} = -\sum_{j=1}^{S_i} p_{ij} \log_2 p_{ij} \tag{15.2}$$

Summing this over all M machines in the facility, the expected amount of information needed to describe the state of the facility is given by:

$$I_{total} = -\sum_{i=1}^{M}\sum_{j=1}^{S_i} p_{ij} \log_2 p_{ij} \tag{15.3}$$

This quantity is known as the *static information-theoretic complexity* of a manufacturing facility. This quantity can be measured for many manufacturing facilities, either through direct observation or analysis of computer records or manual records.

In practice, it is a fairly straightforward task to identify the state of a machine or workcentre, but the full picture on the information content of a facility can only be seen by including the likely duration of the states. This means that the measurement of the states of the machines must be done at least as often as the state of the resource changes. In other words, the state of the system must be checked at least once during the duration of a state, and the interval between checking the state of a resource must be at least as long as the duration of the shortest state.

This information-theoretic model may be extended slightly to take account of the duration of the states. Let the state with the shortest duration on each machine have duration Δ_i. The time-averaged expected amount of information to describe the state of machine i is therefore given by:

$$I_{average} = -\frac{1}{\Delta_i}\sum_{j=1}^{S_i} p_{ij} \log_2 p_{ij} \tag{15.4}$$

For the whole facility, this becomes:

$$I_{Totalaverage} = -\sum_{i=1}^{M}\frac{1}{\Delta_i}\sum_{j=1}^{S_i} p_{ij} \log_2 p_{ij} \tag{15.5}$$

This quantity may be called *the time-averaged static complexity* of a facility.

Although the quantity of time-averaged static complexity of a facility may seem very abstract, it is in fact very familiar to the people who run a manufacturing facility. They see a document every day that contains exactly this information – it is the schedule. The schedule specifies the planned state of each resource and the duration of that state. What the schedule cannot specify is the unplanned states, i.e. the breakdowns and unplanned idle spells. This will be covered by the second measure, i.e. the dynamic complexity.

The static measure of complexity measures the complexity of the facility as planned, i.e. according to the schedule. However, intuition and experience would suggest that a facility that deviates from the schedule is more complex. This is in the sense that it involves more effort to control and manage it, and it has a greater impact upon the business partners with which it trades, whether they are external or internal to the business. We will see now how this informal notion can be captured and formalized by the notion of dynamic information-theoretic complexity.

Suppose a facility is operating in accordance with the schedule, to some specified level of precision. This may mean that the start or end times of jobs are accurate to some precision as understood by the personnel who manage and operate the facility. (This may be different from the level of precision as specified on the formal, circulated document.) Further, the number of components of the correct quality is also within some tolerance of the job size, as specified on the schedule. When the facility is performing to the standard understood by the users of the schedule, the facility may be said to be under control.

The proportion of time the facility is under control, or the probability of sampling the system and finding it to be operating under control, we shall denote by P. Hence, neglecting time-averaging effects, the expected amount of information required to report whether a facility is under control is given by:

$$D_{control} = -P\log_2 P - (1-P)\log_2(1-P) \tag{15.6}$$

This is the amount of information needed for the facility as a whole, but in practice the facility manager would probably require a report from each of the individual workcentres. Assuming that the workcentres are independent, this would yield the following expression for dynamic complexity:

$$\overline{i=1}$$

$$D' = -\sum_{i=1}^{M}\{P_i \log P_i + (1-P_i)\log(1-P_i)\} \tag{15.7}$$

In the circumstances that the facility is not under control, the manager would require a report on the actual status of the facility. We have already written an expression for the amount of information needed to describe the state of a system, so we need to augment the expression for D with an extra term to denote the extra information needed to describe the state of a system when it is not under control. Note that in this case, the probabilities are measured over the unscheduled or uncontrolled states, NS_i. Hence, the expression for the dynamic complexity (not time-averaged) is:

$$D'_{Total} = -\sum_{i=1}^{M}\left\{P_i \log_2 P_i + (1-P)\log_2(1-P_i) + (1-P)\sum_{j\in NS_i} p_j \log_2 p_j\right\} \tag{15.8}$$

This expression tells us straight away that the dynamic complexity of a facility is reduced as P is increased. In other words, as the facility is more likely to be under control, the less information we need to report.

The next section describes the methodology that we have developed to measure the information-theoretic complexity, both static and dynamic, of manufacturing facilities.

15.3 DESCRIPTION OF METHODOLOGY

The research methodology that we have developed for the assessment of manufacturing information systems integrates several methods so as to build up a complete picture of the flow of data and material through a manufacturing facility. In this section, we shall describe the methods used and some of the obstacles that need to be overcome in their application. Our methodology integrates three key scientific approaches:

- fieldwork;
- computer modelling and simulation;
- mathematical modelling.

This chapter will concentrate on the fieldwork aspects of the methodology.

The key feature of fieldwork is the opportunity to observe real manufacturing processes. This is an effective way of taking researchers beyond the theory and into practical applications. In this area of manufacturing research, real-life industrial exposure develops an understanding of the links between manufacturing systems, information and human systems.

Access to industrial partners is often the prime barrier to many academics wishing to conduct case study research. The chances of industrial partnership depend largely on the strength of the project proposal in terms of benefits (*'what's in it for us?'*) for the company. The other influential factor is the *project champion,* the right person within the company who can take industrial ownership of the project and drive it forward. Access to the right people, at all necessary levels, is critical.

Selecting the right combination of research tools can be critical to the quality, quantity and ease of data collection and evaluation. Case study researchers have an extensive range of approaches to choose from (Yin, 1994), depending on the particular type of investigation and organisation targeted. A toolbox of methods can be adopted and arranged to suit the investigation at hand. Once the appropriate techniques have been selected, meticulous planning of the research process is required on both a macro and micro level. The on-site time available to researchers for access to personnel and measurements is often limited due to the constraints of the industrial partners. This fact, combined with the relatively high research costs (time, personnel and money) of conducting fieldwork, emphasises the need to take a strategic approach in all details of the research: from planning, designing and execution through to dissemination.

The three phases of the fieldwork are as follows:

1. mapping of document flow and analysis of scheduling practices;
2. observation of states; and
3. computation of complexity indices.

Phases 1 and 2 are carried out on site in the manufacturing organisations. Phase 3 is usually performed back at the researchers' base, since the preparation and analysis of the data can be quite time-consuming.

15.3.1 Duration of the phases

Phase 1 should take no more than one week to carry out. At the end of that phase, a presentation is made to the organisation on the document flow and scheduling practices. It is important to complete this phase quickly so as to provide some feedback to the organisation and demonstrate progress on, and commitment to, the project. The presentation also gives an important opportunity for members of the organisation to verify the findings and correct any misconceptions and misconstructions. Where the findings of the project team differ from the expectations of the members of the organisation, these areas should be highlighted for more detailed investigation in the future.

Phase 2 needs to be long enough to make reasonably sure that all the relevant states have been observed often enough to give representative estimates of the probability of each one occurring. The amount of time required to achieve this goal depends on the problem being solved, but it is not a good idea to allow this phase to last longer than about two weeks. This might mean compromising on the states that are to be observed, possibly amalgamating several infrequently occurring states into one, so that the amalgamated states occur often enough.

Phase 3 can occur in two stages. The first stage produces a preliminary analysis of the complexity indices, which can be presented to the organisation, no longer than about two weeks from the end of the fieldwork. This gives another opportunity to update the findings and obtain further feedback. The investigated organisation can also suggest other areas of interest that can be investigated during the second stage of the analysis of results. The second stage of the analysis requires thorough analysis of the results, computing indices for each document type, workstation, queue or buffer etc. A full write-up of the findings should be delivered to the organisation within about three months of the end of the fieldwork, with another opportunity to discuss further outcomes of the analysis. The methodology for each phase is described in more detail below.

15.3.2 Phase 1: Mapping the document flow

The key deliverables of this phase are:

- description of the layout of the facility;
- description of the documents which are used to monitor and control the production process; and
- identification of the issues which are perceived as affecting the efficiency of operation.

The first task is to identify all the documents that are used to communicate to, from and within the scheduling function. The data items that occur in each document are logged, together with those that are or may be changed. The frequency of generation of the documents is determined, as well as the lifetime of the document and its frequency of update. Subsequent analysis of the documents compares the actual contents of the documents as they progress through the manufacturing process. This is used to calculate complexity indices in Phase 3.

The main technique that we have used during the mapping of the document flow is the IDEF approach. In the remainder of this section, we shall outline the

IDEF approach, and illustrate its application to one of our industrial partners. The section concludes with an assessment of the IDEF approach as a tool for mapping document flow.

IDEF was developed by the U.S. Airforce in the early 1980s (U.S. Airforce, 1981). An IDEF model consists of a series of linked diagrams. Each diagram must contain at least three boxes (to ensure enough level of detail) and not more than six boxes (more would be difficult to understand in a unique diagram). The basic element consists of a box and the interface arrows as shows in Figure 15.2. This starting diagram is called A0 by convention.

The box represents a function in the system, described by an active verb phrase inside it, e.g. 'create schedule'. The arrows represent the data involved in that function and are described by a noun phrase label, e.g. 'technical information'. The position of an arrow with respect to the box specifies its role in the function as it is outlined below.

- An input arrow represents an object or information needed for the function.
- An output arrow represents the result of applying that function to the input.
- A mechanism arrow represents the method of applying that function to the input.
- A control arrow represents the information that governs the function.

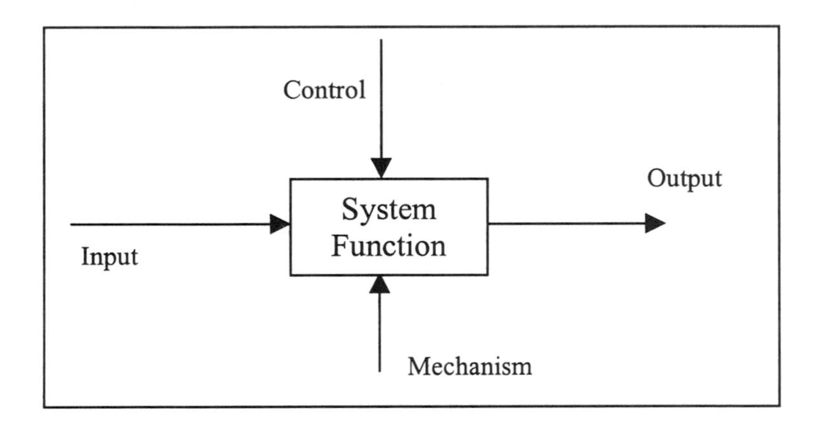

Figure 15.2 The A0 function box and interface arrows.

For a particular box, there may exist more than one input/output mechanism. Each box must have at least one control arrow.

The system is modelled in a structured and hierarchical way, from upper, less detailed diagrams down to lower, more detailed diagrams. The A0 diagram specifies the scope of the model. The next diagram, A1, represents the first level of detail of representing the system. This is further elaborated in the succeeding diagram A11.

Prior to the graphical modelling of the shop floor activities, data had been gathered about the scheduling process, the documents involved and the exchange of information. The next step was the creation of the function model using IDEF.

The first stage in creating an activity model is to define the Context (subject of the model), Viewpoint (what can be seen in the model) and the Purpose (intent of the model). For this application, these elements were defined as:

- *Context*: the scheduling function within the shop floor.
- *Viewpoint*: Activities, people, decision-making processes and document flow related to the scheduling AS it IS on the shop floor.
- *Purpose*: to specify current activities and information flow regarding scheduling on the shop floor (AS IS model).

In Figures 15.3 and 15.4, the A0 diagram reflects the main activities involved in scheduling on the shop floor. The input for the process is the MRP document. The MRP contains the weekly demand requirements. The outputs of the process are the schedule, in the form of the priority sheet containing the list of urgent jobs, and the status sheet containing the jobs to run for a particular day and shift.

The scheduling activities can be carried out both by the planner and the planning team. Usually the planner is in charge of the creation of the priority and status sheet. The planning team sometimes helps with additional activities. The status of the finished parts and WIP, the changes in demand from customers and the availability of resources (labour, machines and raw material) are the elements controlling the scheduling process.

The A1 diagram reflects the process of creating the MRPD (MRP Download) document for the next week. Firstly the MRPD is created using as input the MRP (II) whose information is transformed, according to the rules defined by the planner, into the final format. The result is an A3 sized document created manually by the planner. This document is sent to the manufacturing manager.

The final model of the scheduling function at Industrial Partner A (described in the case study in Section 15.4) contained 11 diagrams. To give an idea of the work involved in this task, the final model contained 34 inputs, 29 outputs, 51 mechanisms and 53 control arrows.

15.3 3 Findings from the case study

The IDEF methodology identified the main issues listed below:

- loops present in the job procedures;
- unclear defined responsibilities;
- unnecessary informal activities;
- repeated checking cycles;
- unnecessary documents; and
- excess data quantities.

In this particular shop floor case, a combination of manual processes, updated scheduling-related information arriving at any time and the existence of several documents dramatically increased the complexity of the scheduling function. Also, the literature review of scheduling (Section 15.2.1) showed that this proliferation of documents associated with scheduling is not uncommon.

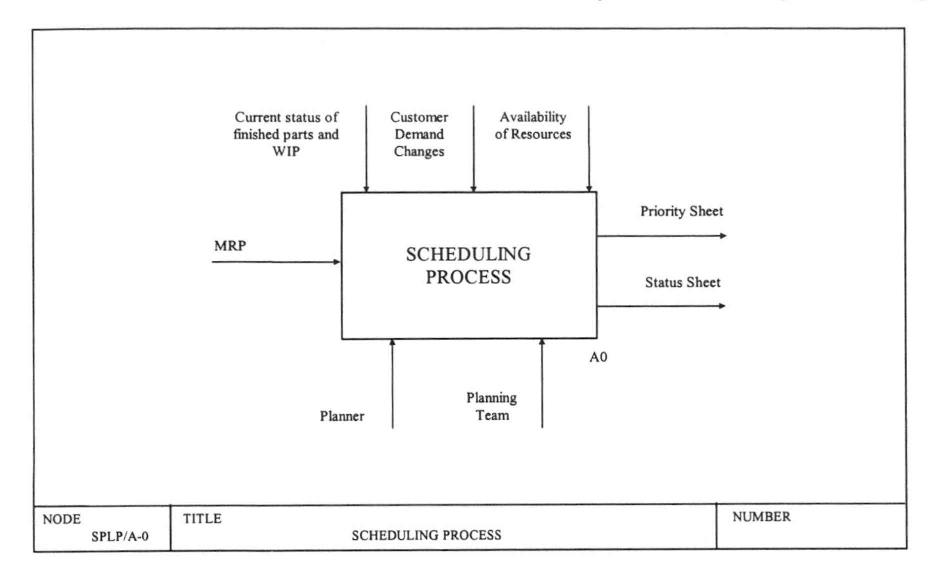

Figure 15.3 The A0 model.

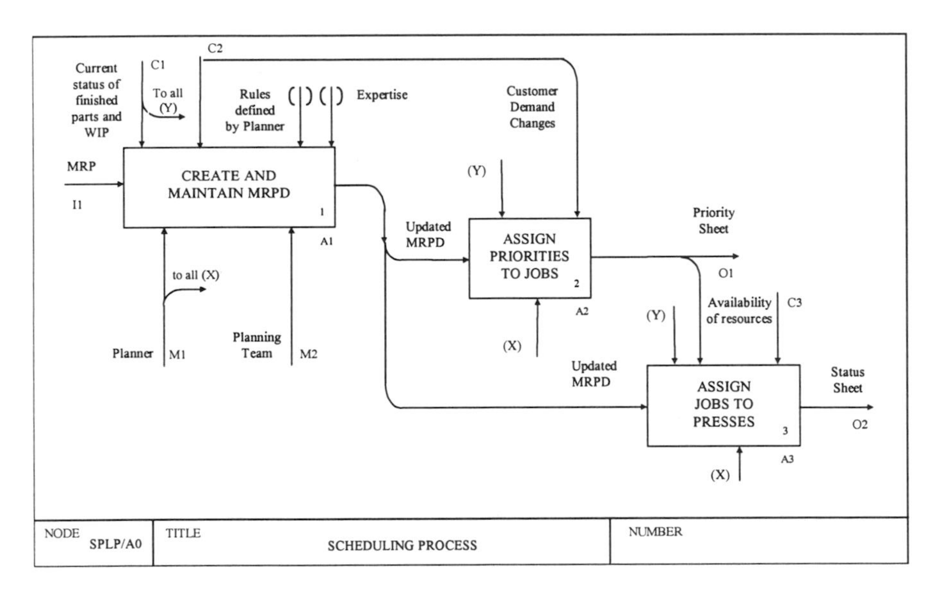

Figure 15.4 The A1 diagram.

Complexity in the documents leads to complexity in the scheduling function. This complexity does not refer to complicated numerical or graphical data being used. It must be understood as the existence of several sources of information, the different formats of the sources, the unavailable, inaccurate or ambiguous data, the exchange of the data, and the interpretation of the raw information. The multiplicity of documents used in the scheduling function has proved to be one of the main elements affecting the scheduling function and, consequently, the level of production achieved.

15.3.4 Role of IDEF

The IDEF model was chosen due to its capability of modelling different aspects of the process. Moreover, being a methodology it draws upon several data acquisition processes such as interviews, analysis of the data, construction of the model and review. Throughout all the stages of creating the model, these different procedures were applied. Also, its extended use in different applications made IDEF the most suitable model. However, some shortcomings and drawbacks of the IDEF modelling techniques were detected during the modelling process. These are summarised as follows:

- time-consuming;
- impossible to represent time in the model;
- difficulty in distinguishing between input and control data;
- inability to model organisational features;
- inability to model people and teams;
- not modelling the idea of uncertainty;
- difficulty in modelling constraints;
- necessity of structured information.

The application of the IDEF methodology highlighted some critical factors. These were reported to management through the feedback presentations, enabling them to make decisions to address these problems. Hence, despite its shortcomings, the IDEF method seemed to be a useful approach to map documents and information flow.

15.3.5 Phases 2 and 3: Observation of states and calculation of complexity indices

After obtaining feedback on the first phase of mapping the document flow, the next phase is to observe the states that occur in the facility. It is possible to choose any arbitrary number and set of states, but it is important in practice to choose a set of states that are:

- pertinent to an agreed problem;
- all likely to be seen during the allocated observation period.

In practice, this means that the number of states is not likely to exceed about a dozen for a two-week observation period. The problem to be addressed, however, is the most important decision, and this must be one of the decisions agreed after the presentation at the end of Phase 1. The problem should be one that is of current relevance to management, but one which is likely to be a source of disagreement, since the exact nature of the problem and the responsibility for its solution, and possibly definition, cannot be agreed. Sample problems are listed in Table 15.1. These were obtained from two industrial collaborators, one a press shop (Industrial Partner A) and the other a PCB assembly shop (Industrial Partner

B) in two completely separate organisations. The degree of similarity in the problems they faced is striking.

Table 15.1 Common issues in two case study organisations

Issue		Industrial Partner A	Industrial Partner B
1	Changes in customer requirements (in terms of new order, order size, specification or due date) taken into consideration; internal customers are the main sources of changes	√	√
2	Reactive short-term scheduling	√	√
3	Poor plan stability and schedule adherence	√	√
4	Poor computer support for schedulers	√	√
5	Large amounts of WIP, poorly monitored and controlled; the WIP area contains jobs which are on hold or static for long periods, or jobs which were pulled short	√	√
6	The shop floor management considers the workforce is the limiting resource	√	√
7	High absenteeism	√	√
8	Information problems – in terms of information flow, quality, timeliness and accuracy of data, and non-value-adding information-processing operations	√	√
9	Poor inter-departmental communication; decisions made at management level or higher not always discussed and/or explained	√	√
10	Lack of consensus on the business objectives between shop floor managers and top management	√	√
11	Frequently pulling jobs short once enough has been produced to satisfy the short-term demand, in order to run urgent jobs	√	
12	Quality problems, due to the technically advanced nature of the products		√
13	Quality problems, due to machine or die breakdowns	√	
14	The performance measures in use do not capture the problems and do not reflect their causes	√	
15	Highly variable work rate		√
16	Use of the WIP area to buffer work		√
17	Low product yield		√

The states that are chosen depend on the problem, but typically involve issues around the flow of material. Hence, it is usual for the states to cover the combination of workstations and the products that they operate upon. Other 'programmed' or 'scheduled' states include planned maintenance and scheduled idleness. Unscheduled states include broken down, awaiting supplied material, awaiting operator etc.

Once the states have been decided it is useful to draw up a form listing codes for all the relevant states. The form should tabulate the following data:

- time of observation
- resource observed
- state
- if an unscheduled state, the reason for its deviation from schedule.

These observations may cover resources such as machines on a shop floor, and also desks in a design or scheduling office, or the contents of a queue.

The third and final phase of the fieldwork is the calculation of the complexity indices. These are calculated according to the formulae given in Section 15.2.4. The static index is calculated by estimating the probability of each resource occupying each identified state, and calculating the static complexity for that resource. The dynamic complexity is calculated with respect to the schedule for the facility for the period under investigation. This enables the identification of those periods when parts of the facility were not under control, i.e. not in the state indicated on the schedule. From this, a value for P and hence the dynamic complexity can be calculated.

The findings of the calculations should be fed back to the collaborating organisation within about two weeks of the end of the fieldwork. This will allow further verification and discussion of the findings.

Subsequent to this phase, the academic study is likely to proceed with yet more calculation of indices for individual documents, queues and decision points. Computer simulations would be constructed to model and investigate further the observations made, and to check the completeness and veracity of the data.

15.3.6 Further developments in the methodology: expert systems and workbooks

A number of problems were identified in practice with the application of the Frizelle complexity measure, such as the long time required to gather the data necessary for the calculations and the difficulties of applying the method to job shops. The derivation of the information theoretic basis for the measure (Section 15.2.4) suggests that there may be a more straightforward way to gather or estimate the data necessary to make an estimate of the complexity of the facility. In order to structure and simplify the process of gathering the data, a workbook was constructed, which will be described further below. After completing the workbook, the answers to the questions were analysed to produce an estimate of the complexity of the facility. The profile of the facility was compared with a small number of possible alternative layouts, so that a recommendation could be made on how the facility may be reorganised in order to reduce the complexity and could improve predictability.

The workbook was a development of a simple quiz, which was devised to guide facilities on whether they were 'too flexible' (Efstathiou *et al.* 1998). This was in the format of a simple multiple-choice questionnaire, with numerical scores assigned to each optional answer. By adding up the scores, the respondent could obtain a rough indication of the flexibility of their performance, and whether it was increasing internal costs or providing flexibility to customers. This was an amusing demonstration, but the technique could not permit the interaction between some of the answers, and as a format it was inappropriate for the analysis of numerical data.

Compared to the quiz, the workbook provides a more structured and detailed analysis of the facility. It does require detailed knowledge of product ranges, downtime etc, but the questions could be answered based on the sort of knowledge that a production manager ought to be able to estimate. This includes estimates of the number of products produced, the number of machines, the use of cells and the amount of downtime.

15.4 CASE STUDY EXAMPLES

The first facility investigated (Industrial Partner A) is a low-technology process-based job-shop within a major UK automotive manufacturer, which accommodates a high number and type of tool-based machines. A high number of products is manufactured in a wide variety of batch sizes, requiring a varying number of operations, often on machines of different types. This requires frequent machine changeovers, the effect of which is amplified on the system performance by the significant changeover times when compared to processing times. Jobs are often pulled short of the planned production quantity once enough has been produced to satisfy the short-term demand, in order to set and run urgent jobs.

Labour is required in all the production stages (machine set-up, run and repair, and part movement), and there are less operators than the number of machines. The shop planning is MRP–based and done on a weekly basis. The scheduling is manual and shift-based, and consists in assigning the jobs to specific machines, so that they are completed before the due date. The schedule therefore contains only sequencing and routing information. No home lines are defined and used for the parts/operations produced within this facility. The information on a job is manually updated on the computer at the end of each shift only if a change in its status took place.

These characteristics give an insight into the high level of flexibility and complexity associated with the scheduling process. Once a job is set, however, running it is a simple and reliable process. The faulty parts are mainly due to machine or tool breakdown.

The second facility investigated (Industrial Partner B) is a high-technology and high-accuracy multi–cell shop within a major UK manufacturer, and produces printed circuit boards (PCB). It is characterised by multi-level technically advanced products on automatic and semi-automatic highly reliable machines. The shop manufactures a high number of types of technically advanced products in low batch sizes. The changeover times are significantly higher than processing times.

The workforce is highly qualified and flexible, and is required with a different degree at different operational levels. Home lines exist for some classes of operations. Computer-based weekly MRP planning is used, and the scheduling is partly done using computers, and partly manually. A bar code scanner exists in the facility, with the aim of updating the computer system with the progress of each job. However, this was only used occasionally.

This company is therefore characterised by a lower degree of scheduling flexibility than the one at Industrial Partner A, but encompasses higher process complexity. The labour work rate and quality significantly influence the process and product quality.

Although the nature of the products and processes in the two organisations are different, they confront similar major problems (issues 1–9 in Table 15.1). Furthermore, this list of problems shows that in both facilities the high levels of flexibility are not entirely controlled. These problems were persistent and in general did not depend on the nature of the products manufactured, but on the manner in which the organisations perceived and approached major issues. The company–specific issues (10–16 in Table 15.1) are generated by the procedures adopted to address the problems, which are either global or locally perceived (such as customer changes), or by systemic (e.g. process, machine or workforce) characteristics.

15.4.1. Findings from application of methodology to case study organisations

15.4.1.1 Static complexity analysis

The static complexity was calculated for a three-month period from the MRP master schedule. With the majority of parts being produced in a four or six week cycle this period was considered appropriate. The basic static measure was applied in a number of ways to reveal the relationship between part-operation complexity and the flexibility due to routing options.

An *'upper bound'* for the static complexity has been calculated by considering total flexibility of assigning part operations to presses of the relevant type. Each press is considered equally likely to process each part/operation requiring that press type (the potential flexibility due to the resources).

A *'lower bound'* for the static complexity has been calculated to represent a situation where each part is run on a home line, i.e. no routing flexibility. This will give an indication of the complexity due to the part/operation requirements only.

A *'sequence constrained'* calculation has been made to represent the routing flexibility available to the schedulers, due to the layout constraints and their preferred routing conventions. Constraints are placed on the routes available due to running sequential operations that require the same press type on adjacent presses and always in the same direction (ascending press ID numbers).

The values obtained are shown in Figure 15.5. The *lower bound* static complexity is significantly less than the *upper bound*, which indicates that a large proportion of the static complexity is due to the routing flexibility. The *sequence constrained* index is only marginally less than the *upper bound*, indicating that even with the constraints imposed a large proportion of the routing flexibility remains.

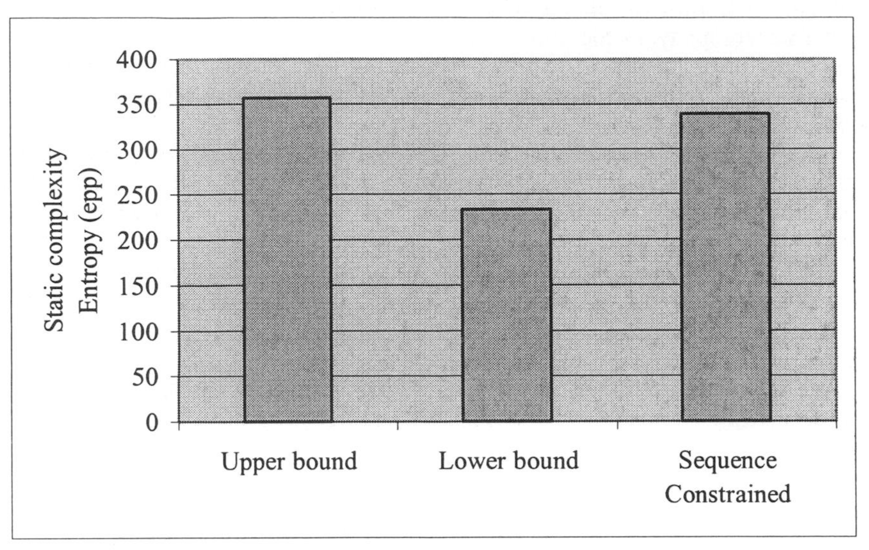

Figure 15.5 Static complexity analysis of industrial partner A.

15.4.1.2 Dynamic complexity analysis

Once the data had been collected the dynamic complexity measure was applied in a number of ways. Two are described here.

- *Multi-server model:* This method allowed us to combine the effects of machine states and queues to a group of machines. Due to the large number of machines in each group the 'state' complexity dominated the queue complexity, but the effect of individual machines could not be ascertained in detail. This may be illustrative of the way in which the schedulers handle the complexity of high routing flexibility – queues being combined and decisions delayed.
- *Machines as individual resources*: The queue variability could not be included, however the contribution of non-programmed events could be established. The operational complexity was found to be low in comparison with the static contribution. The non-programmed complexity was found to be highest for the most heavily loaded group of machines.

At the site of Industrial Partner A, the job priorities in creating the schedule were often influenced by the performance measures in place: since an important performance indicator encouraged the maximisation of the number of jobs completed, preference was usually given to processing short–sequence jobs. This measure was reported by the shop manager on a daily basis, and was a crucial criterion for calculating the shop budget.

Furthermore, the top management considered that the plans were correct and emphasised the importance of processing the jobs in the order they appeared in the plan. On the other hand, the shop managers and schedulers felt that the information in the plan was inaccurate (in terms of jobs to be processed, lot sizes and due dates), and stated that their policy is to satisfy the customer. This was often achieved at a price of pulling other jobs short and creating WIP. A fire–fighting strategy was thus adopted.

Unfortunately, the information system in place did not help to improve this situation: several loops were detected between the human issues, information and systemic characteristics and system performance (Figure 15.6). For example, the poor assessment of the number of parts produced, poor timing in recording important data on the computer, the lack of synchronisation between the available computer systems, correlated with their misuse, bring more complexity in the planning and scheduling. Furthermore, they encourage the establishment of personal practices for scheduling, based on the schedulers' expertise.

In terms of reporting performance measures, the indicators for schedule adherence, process stability and machine utilisation were regularly close to the targets and relatively stable. The fact that they were not affected by the system uncertainty and variability (machine breakdowns, absenteeism, etc.) raised a question of how correctly they were defined and/or calculated. Furthermore, due to

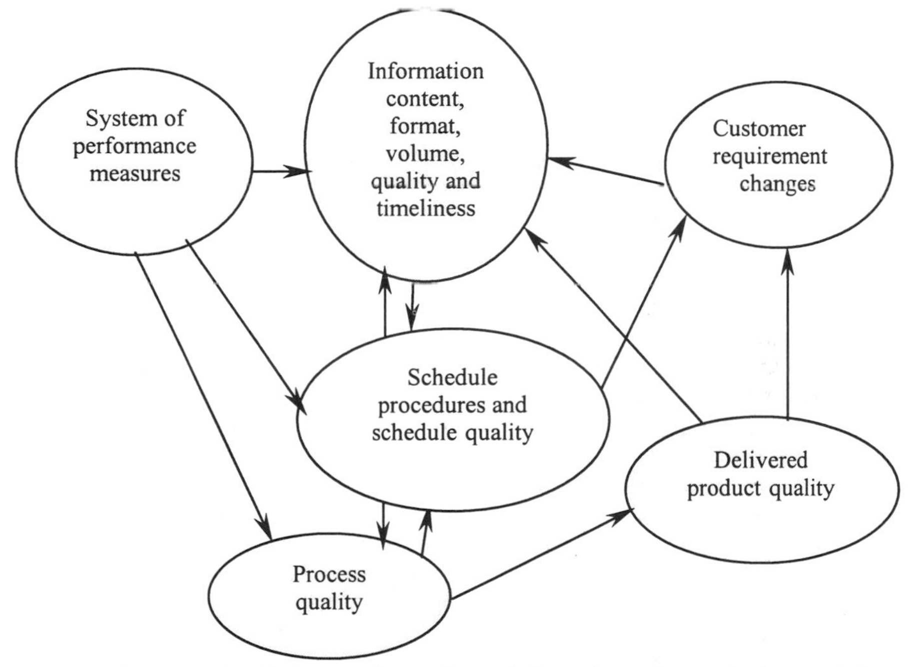

Figure 15.6 The relationship between human issues, information and systemic characteristics and system performance.

the above-mentioned behaviour, they did not help the shop to detect nor to predict the problems. On the other hand, the costs and benefits of recording information, calculating and reporting performance, were not taken into account and not mentioned by any of the interviewees. All these issues indicate that the performance measures should be dynamic and matched with the organisational context.

15.5 HUMAN ISSUES IN MANUFACTURING RESEARCH

Within the manufacturing environment, it is difficult to separate the human factors from the information systems content, as the two are often entwined. An early assessment and understanding of the social context and structure of the organisation (e.g. hierarchy, information flow, knowledge transfer, motivation, training and education) are valuable to form a foundation for measurements, observations and data collection. To understand the scheduling process it is essential to have some experience of working closely with schedulers: this method of obtaining first hand information is starting to become more recognised (McKay *et al.*, 1988, 1995).

Responsibilities lie with the researcher for ethical conduct and critical evaluation of all information. Researchers may be expected to take on many perceived roles such as acting as a channel for communication, as messenger or negotiator. Researchers also need to be continually aware of emotional issues or hidden agendas and be ready to act rationally in validating opinions and identifying key issues. The role of the researcher as an independent body from outside the organisation, should be clarified to all parties involved in the project, to emphasise neutrality.

Due to the human aspects in manufacturing there is a need to focus on the qualitative as well as the quantitative methods of research (Frizelle and Woodcock, 1995). Field data is often a mixture of formal, informal, qualitative and quantitative information, requiring a framework for analysis based on the conceptual model. Existing research techniques used are summarised in Yin (1994) in view of assessing manufacturing, information and scheduling systems.

Researchers conducting case study work in manufacturing industry need to be fluent with designing measures, recording quantitative data, and collecting qualitative data through questionnaires, interviews, focus groups and other sociologically grounded research techniques. There is also a need to develop new techniques (the techniques being developed are discussed in Section 15.4) that are more appropriate to helping answer the core research questions at hand, as has been the case in this project. Because of the often-entangled nature of complexity, qualitative information can often be the basis of forming a starting platform for research, extracting essential data that would otherwise have been impossible to collect quantitatively.

15.6 CONCLUDING REMARKS

The chapter presents the motivation for a formal methodology for measuring the information-theoretic complexity of a manufacturing facility. This involves measuring the flow of material and supporting documents through the facility. The quality of the documents is assessed according to simple measures of relevance, completeness etc. The complexity of the facility is estimated through direct observation of the states and used to indicate key features of the facility that contribute to uncertainty and unpredictability in the operation and management of the facility. Combined with the use of computer simulations and mathematical analysis, the measurements can be used to compare alternative strategies for improving the performance of the facility. The deeper knowledge and understanding of the facility and its operating practices obtained by applying this methodology can be very helpful to production managers and planners in identifying the root causes of persistent problems, and taking appropriate steps to relieve them.

15.7 REFERENCES

Azzone, G. and Bertelè, U., 1989, 'Measuring the economic effectiveness of flexible automation: a new approach', *International Journal of Production Research,* Vol. 27, No. 5, pp. 735–746.

Bauer, A., Bowden, R., Browne, J., Duggan, J. and Lyons, G., 1994, *Shop Floor Control Systems: From Design to Implementation*, Chapman & Hall, London, UK.

Benjafaar, S. and Ramakrishnan, R., 1996, 'Modelling, measurement and evaluation of sequencing flexibility in manufacturing systems', *International Journal of Production Research*, Vol. 34, No. 5, pp. 1195-1220.

Bermejo-Alonso, J., 1997, *Analysis of the Role of Information in Scheduling: A Case Study*, MSc Thesis, University of Oxford.

Bermejo, J., Calinescu A., Efstathiou, H.J. and Schirn, J., 1997, 'Dealing with Uncertainty in Manufacturing: The Impact on Scheduling', *Proceedings of the 32nd International MATADOR Conference,* A.K. Kochhar (ed.), Manchester, UK, 10-11 July, Macmillan, London, pp. 149-154.

Brill, P.H. and Mandelbaum, M., 1989, 'On measures in manufacturing systems', *International Journal of Production Research*, Vol. 27, No. 5, pp. 747-756.

Calinescu, A., Efstathiou, H.J., Bermejo, J. and Schirn, J., 1997, 'Modelling and Simulation of a Real Complex Process-based Manufacturing System', *Proceedings of the 32nd International MATADOR Conference*, A.K. Kochhar (ed.), Manchester, UK, 10-11 July, Macmillan, London, pp. 137-142.

Cassell C. and Symon G., 1997, *Quantitative Methods in Organisational Research*, Sage Publications, London.

Charalambous, O. and Hindi, K.S., 1991, 'A review of artificial intelligence-based job-shop scheduling systems', *Information and Decision Technologies*, Vol. 17, No. 3, pp. 189–202.

Chen, I.J. and Chung, C.-H., 1996, 'An examination of flexibility measurements and performance of flexible manufacturing systems', *International Journal of Production Research*, Vol. 34, No. 2, pp. 379–394.

Corrêa, H.L. and Slack, N., 1996, 'Framework to analyse flexibility and unplanned change in manufacturing systems', *Computer Integrated Manufacturing Systems*, Vol. 9, No. 1, pp. 57–64.

Das, S.K. 1996, 'The measurement of flexibility in manufacturing systems', *The International Journal of Flexible Manufacturing Systems*, Vol. 8, pp. 67–93.

Deshmukh, A.V., Talavage, J.J. and Barash, M.M., 1992, 'Characteristics of part mix complexity for manufacturing systems', *IEEE International Conference on Systems, Man and Cybernetics*, Vol. 2, New York, USA, IEEE, New York, pp. 1384-1389.

Deshmukh, A.V., Talavage, J.J. and Barash, M.M., 1998, 'Complexity in manufacturing systems, Part 1: analysis of static complexity', *IEE Transactions*, Vol. 30, pp. 645-655.

Edmonds, B. 1998, 'What is complexity? - The philosophy of complexity per se with application to some examples in evolution', in *The Evolution of Complexity*, Heylighen, F. and Aerts, D. (eds.), Kluwer, Dordrecht.

Efstathiou, H.J., 1996, 'Formalising the repair of schedules through knowledge acquisition', Advances in knowledge acquisition - Proceedings of the 9th European knowledge acquisition workshop, Nottingham, UK, May, Proceedings, in *Lecture Notes in Artificial Intelligence 1076*, Nigel Shadbolt, Kieron O'Hara and Guus Schreiber, (eds.), (subseries of Lecture Notes in Computer Science), Springer-Verlag, Berlin; Heidelberg.

Efstathiou, H.J., Calinescu A. and Bermejo, J., 1996, 'Modelling the Complexity of Production Planning and Control', *Proceedings of the 2nd International Conference on Production Planning and Control in the Metals Industry,* London, November 12-14, pp. 60-66.

Efstathiou J., Schirn J. and Calinescu A., 1998, 'The too flexible factory', *Manufacturing Engineer*, April 1998,Vol. 77, No. 2, pp. 70-73.

Frizelle, G.D.M., 1995, 'An entropic measurement of complexity in Jackson networks', *Working Paper No. 13,* Manufacturing Engineering Group, Department of Engineering, University of Cambridge,.

Frizelle, G. And Woodcock, E., 1995, 'Measuring complexity as an aid to developing operational strategy', *International Journal of Operations & Production Management*, Vol. 15, No. 5, pp. 26-39.

Gavin, C.J. and Little, D., 1994, 'Application of CASE within manufacturing industry', *Software Engineering Journal*, Vol. 9 No. 4, pp. 140-152.

Gerwin, D., 1987, 'An agenda for research on the flexibility of manufacturing processes', *International Journal of Operations & Production Management*, Vol. 7, No. 1, pp. 38–49.

Higgins, P., Le Roy, P. and Tierney, L., 1996, *Manufacturing Planning and Control: Beyond MRP II*, Chapman & Hall, London, UK.

Hill, T., 1995, *Manufacturing Strategy: Text and Cases*, MacMillan Press Ltd, London.

Karp, A. and Ronen, B., 1992, 'Improving shop floor control: an entropy model approach', *International Journal of Production Research*, Vol. 30, No. 4, pp. 923-938.

Kehoe, D.F., Little, D. and Lyons, A.C., 1992, 'Measuring a Company IQ', 3rd International Conference on Factory 2000: Competitive Performance Through Advanced Technology, *Conference Publication No. 359*, IEE, London, pp. 173–178.

Kumar, V., 1987, 'Entropic measures of manufacturing flexibility', *International Journal of Production Research*, Vol. 25, No. 7, pp. 957–966.

MacCarthy, B.L. and Liu, J., 1993, 'Addressing the gap in scheduling research: a review of optimization and heuristic methods in production scheduling', *International Journal of Production Research*, Vol. 31, No. 1, pp. 59–79.

McCarthy, I.P. and Ridgway, K., 1996, *A Portrait of Manufacturing Complexity,* Complexity Workshop, Department of Engineering Science, University of Oxford, 24 September.

McIntosh, R.I., Culley, S.J., Gest, G.B. and Mileham, A.R., 1994, 'Achieving and Optimising Flexible Manufacturing Performance on Existing Manufacturing Systems', Factory 2000 – Advanced Factory Automation, 3–5 October, *Conference Publication No. 398*, pp. 491–495.

McKay, K.N., Safayeni, F.R. and Buzacott, J.A., 1988, 'Job-shop scheduling theory: what is relevant?', *Interfaces*, Vol. 18, No. 4, pp. 84–90.

McKay, K.N., Safayeni, F.R. and Buzacott, J.A., 1995, ' "Common sense" realities of planning and scheduling in printed circuit board production', *International Journal of Production Research*, Vol. 33, No. 6, pp. 1587–1603.

Mitleton-Kelly, E., 1997, *Complexity and Organisations as Complex Social Systems,* Complexity in Manufacturing Workshop (II), Churchill College, University of Cambridge, 20 May.

Oppenheim, A.N. 1992, *Questionnaire Design, Interviewing and Attitude Measurement,* Pinter Publications, London.

Prosser, P. and Buchanan, I., 1994, 'Intelligent scheduling: past, present and future', *Intelligent Systems Engineering*, Vol. 3, No. 2, pp. 67–78.

Roll, Y., Karni, R. and Arzi, Y., 1992 'Measurement of processing flexibility in flexible manufacturing cells', *Journal of Manufacturing Systems*, Vol. 11, No. 4, pp. 258–268.

Ronen, B. and Karp, A., 1994, 'An information entropy approach to the small-lot concept', *IEEE Transactions on Engineering Management*, Vol. 41, No.1, pp. 89-92.

Schirn, J.,1997, *Manufacturing Complexity - A Case Study*, Complexity in Manufacturing Workshop (II), Churchill College, University of Cambridge, 20 May.

Schonberger, R.J., 1996, *World Class Manufacturing: The Next Decade*, The Free Press, New York.

Sethi, A.K. and Sethi, S.P., 1990, 'Flexibility in manufacturing: a survey', *The International Journal of Flexible Manufacturing Systems*, Vol. 2, pp. 289-328.

Shannon, C.E.,1949, *The Mathematical Theory of Communication*, University of Illinois Press.

Slack, N., 1988, 'Manufacturing systems flexibility - an assessment procedure', *Computer-Integrated Manufacturing Systems*, Vol. 1, No. 1, pp. 25–31.

Stoop, P.P.M. and Wiers, V.C.S., 1996, 'The complexity of scheduling in practice', *International Journal of Operations & Production Management*, Vol. 16, No. 10, pp. 37–53.

Strauss, A.L. and Corbin, J.M., 1990, *Basics of Qualitative Research: Grounded Theory, Procedures and Techniques*, Sage Publications, London.

Thorpe, R., 1999, *Introduction to Research Methodology*, Seventh Research Methodology Workshop, Cambridge University, 4 March.

U.S. Airforce, 1981, *Integrated Computer-Aided Manufacturing (ICAM) Architecture Part II, V-Information Modelling Manual IDEF1X-extended*, Materials Laboratory, Wright Patterson AFB, OH45433.

Wiendahl, H. P., 1995, *Load-orientated Manufacturing Control*, Springer-Verlag, Berlin.

Wiendahl, H. P. and Scholtissek, P., 1994, 'Management and control of complexity in manufacturing', *Annals of the CIRP*, Vol. 43, No. 2, pp. 1-8.

Womack, J., Jone, D. and Roos, D., 1990, *The Machine that Changed the World*, Rawson Associates, NewYork.

Yin R.K., 1994, *Case Study Research: Design and Methods,* 2nd edition, Applied Social Research Methods Series, Volume 5, Sage Publications, London.

CHAPTER SIXTEEN

Planning and Scheduling in the Batch Chemical Industry

Jan C. Fransoo and Wenny H.M. Raaymakers

16.1 INTRODUCTION

The process industry can be considered as a specific domain within planning and scheduling. This is due to a number of specific characteristics, which have been well documented and will be discussed in this chapter. The subject of study in this chapter is the batch chemical industry, being a subset within the process industry domain. The batch chemical industry is characterised by very complex and long processes and process structures. Process structures are complex due to the large variety in processes and consequently the large variety in resources that are used in that industry. In addition, many sequencing constraints dictate the operational scheduling of the jobs, such as sequences with no-wait restrictions on them. The processes themselves are long, with days more typical than hours needed to characterise their length. In addition, a customer order consists of a fairly large number of jobs, thus resulting in lead times typically measured in months rather than weeks.

These complex and time-spanning characteristics lead to a necessity to decompose the overall planning and scheduling problem. In such a division of tasks, the planners are responsible for long-term co-ordination of the various jobs out of which a customer order is built. The planners make commitments to customers regarding quantities and due dates. They also plan the various jobs over time and load them into time buckets and onto processing departments. The processing departments usually have a departmental scheduler who constructs the detailed job schedule. The main task of the departmental scheduler is to schedule the set of jobs allocated to the department in a specific period such that this set of jobs can be completed within the specified time frame. In this chapter, the process will be described in more detail, based on actual observations at a batch chemical manufacturer.

The main dilemma in such a hierarchical structure is the content and extent of communication between the planners and schedulers. On the one hand, the planners are expected to make an adequate assessment of the feasibility of the job set, requiring a fairly detailed insight into the constraints on the shop floor in the departments. On the other hand, they are expected to make a long-term plan, covering the entire horizon of the customer order, estimating lead times and variability in lead times. This enables them to commit to the customers and to make sure that the various departments in the factory are loaded as evenly as possible over time. The schedulers, in their turn, are expected to make a schedule for a job set which they basically cannot influence, such that the job set is completed within the given time frame, which again allows the planner to complete their long-term commitments to the customers. Further, the schedulers are expected to use their resources as efficiently as possible.

We propose linear regression as a means to adequately represent the capabilities of the shop floor when estimating capacity, which is a difficult problem. We give some theoretical results for this, and discuss why and how that can be applied in business practice. Based on the situation found in the company, a series of theoretical experiments have been conducted. The main purpose of the experiments was to answer the question of whether or not the expected and actual performance of the processing departments could be predicted fairly accurately, based on a limited number of characteristics of a job set loaded onto that department. The idea is that the complexity of the scheduling task can be grasped through a limited number of characteristics of the job set to be scheduled and the available resource set. Based on this, a regression model can then be built to estimate the makespan of a job set, based on job set and resource set characteristics.

The experiments were completed in a theoretical setting to be able to create a large number of instances and settings that increased the diversity of the experiment far beyond the practical setting of this particular company. The main conclusion of these experiments was that it is possible to construct a regression model with a small number of factors that makes a very accurate prediction of the makespan of a job set. In this chapter, the experimental setting and results will be outlined in further detail. The research questions that are still unanswered are noted.

We also discuss the application of the regression approach in an industrial setting. Based on earlier research, we have indications that our approach is intuitively appealing to production planners. Furthermore, we argue that we can also use this approach to estimate the performance of operations completed by people, e.g., estimating the schedule quality output of the operational human scheduling function.

16.2 BATCH PROCESS INDUSTRIES

The planning and scheduling domain has traditionally focused most of its research and development efforts in the discrete manufacturing industry. Only since the mid 1980s can a consistent research line in production control in process industries be seen. The American Production and Inventory Control Society (APICS) defines process industries as ‘businesses that add value to materials by mixing, separating, forming, or chemical reactions, [while] processes may be either continuous or batch and generally require rigid process control and high capital investment’ (Wallace, 1984).

Process industries are characterised by a number of very specific and typical characteristics. Although these characteristics are predominantly found in process industries, they are definitely not found in all process industries and thus are not general in that sense. These typical characteristics include variable process yields, possible use of alternative raw materials, variable product yields (potency), and a divergent product structure (co- and by-products). Extensive overviews can be found in Taylor *et al.* (1981) and Fransoo and Rutten (1994).

16.2.1 Flow and batch process industries

For the understanding of batch process industries, it is important to understand what distinguishes these types of process industries from flow process industries. This distinction is captured in a typology presented by Fransoo and Rutten (1994), based on definitions by APICS (Connor, 1986). A batch process industry is defined as a process business that primarily schedules short production runs of products. A flow process industry is defined as a manufacturer that produces, with minimal interruptions in any one production run or between production runs, products that exhibit process characteristics. Figure 16.1 displays their typology. The typology is one-dimensional and shows different industries positioned between the two extremes. The oil and steel industry are prime examples of flow process industries, whereas the pharmaceutical industry is a typical example of a batch process industry.

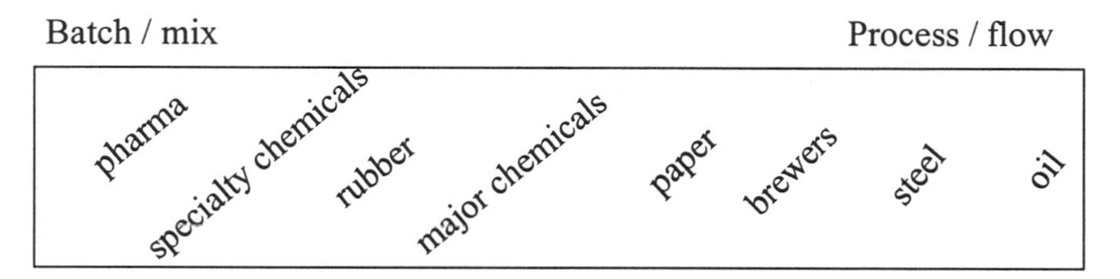

Figure 16.1 Process industry typology (Fransoo and Rutten, 1994).

The discriminating characteristics of each type are presented in Table 16.1. This shows that batch process industries are characterised by a large number of

Table 16.1 Characteristics of flow and batch process industries (Fransoo and Rutten, 1994)

Flow process industries	Batch process industries
High speed, short throughput time	Low speed, high throughput time
Clear capacity, single routing, no volume flexibility	Unclear capacity, complex routings
Low product complexity	More complex products
Low added value	High added value
Strong impact of set-up times	Less impact of set-up times
Small number of operations	Large number of operations and process steps
Limited number of products	Larger number of products

operations and a high level of product complexity. In fine chemicals production, more than ten separate operations is not uncommon. The variety in products requires the use of general multi-purpose equipment, allowing multiple different operations to be completed consecutively on one resource. Together, this leads to a complex routing structure. All this causes long lead times and high levels of work-in-process.

16.2.2 Pharmcom

The setting in this chapter is based on the Pharmcom company, a manufacturer of active ingredients for the pharmaceutical industry. Pharmcom produces about 140 products in three processing departments. Routings are divergent, and consist of 4 to 25 operations, with a total throughput time of 3 to 9 months. Each operation results in a stable intermediate product that can be stored. Subsequent operations in a routing are often performed at different processing departments.

Each operation consists of one or more processing steps, each of which requires a single resource. Processing steps may be overlapping in time. During an operation, the intermediate product is generally not stable, resulting in no-wait restrictions between processing steps. Figure 16.2 presents an example of an operation consisting of three processing steps, for which two vessels and a filter are required. The bold lines show when the resources are needed. The time at which vessel 2 and filter 1 are required in relation to the start time of the processing step on vessel 1 is given by a fixed time delay (∂_i).

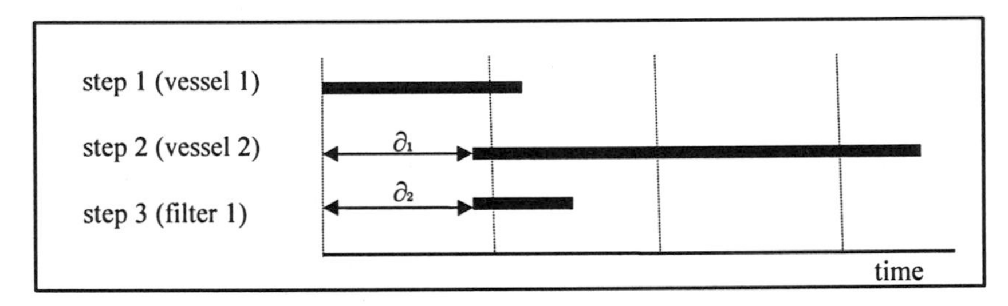

Figure 16.2 Example of an operation.

The three processing departments (PDs) are organised functionally. The PDs contain, on average, 30 vessels including some highly specialised vessels and several general, multi-purpose vessels. For the specialised vessels, no alternatives are available at the production site. General-purpose vessels are grouped into work centres within the PDs. Each processing department can perform a specific set of operation types. When an operation type can be performed in more than one PD, the operation is allocated to one of these PDs. That PD then completes all production orders for the operation type. The typical utilisation level in the PDs is between 50 and 60%.

16.3 HIERARCHICAL PLANNING

The production planning process at Pharmcom is organised in a hierarchical manner (Meal, 1984; Bertrand *et al.*, 1990; Schneeweiss, 1999), as is common in many industrial companies. The complex and time-spanning characteristics of the production process leads to a necessity to decompose the overall planning and scheduling problem. Not only is the planning and scheduling problem computationally intractable due to its complex detail and long horizon, but also the responsibilities at Pharmcom are divided amongst many people. Long-term and short-term responsibilities are split between a central planning department and the decentralised processing departments. In this division of tasks, the planners are responsible for long-term planning of the subsequent jobs out of which a customer order is built. The planners make commitments to customers regarding quantities and due dates. They also plan the various jobs over time and load them into time buckets and onto processing departments. The processing departments all have a departmental scheduler. The main task of the departmental scheduler is to schedule the set of jobs allocated to the department in a specific period such that this set of jobs can be completed within the specified time frame. The following description is slightly simplified for briefness and clarity of exposition. A full description can be found in Raaymakers *et al.* (2000).

16.3.1 Roles and dependencies at the two levels of control

Essentially, two levels of control can be distinguished, namely aggregate customer order planning and detailed production order (i.e., job) scheduling. The complex goods flow between the PDs requires close co-ordination by the customer order planning function. In addition, the goods flow needs to be co-ordinated with the Sales Department regarding the acceptance of customer orders.

At the aggregate planning level, the planning horizon is one year, divided into monthly periods. Given the customer orders, their due dates, and the estimated total lead time, the period in which the first operation of an order needs to be started is determined. The subsequent operations in the routing are then planned allowing some time-slack between successive operations. To estimate whether sufficient capacity is available to realise the aggregate production plan, workload rules are used. For each work centre, the maximum workload is specified as a percentage of the available capacity of that work centre. The workload limits have been determined empirically, based on evaluations of the realised production in previous periods. The limits provide the capacity slack that is needed to account for the capacity losses that will be incurred by interaction between production orders and by the disturbances on the shop floor. Interactions between production orders result from the fact that for each operation several resources are required, and that no-wait restrictions apply to the processing steps of an operation. The maximum workload allowed determines the maximum utilisation of a resource. Using the product and resource structures, while applying the workload rules, leads to a production plan for each PD for the coming monthly periods. This production plan contains the production orders that have to be carried out, and the release and due dates of these production orders. Each month a new production plan is provided to the PDs.

The schedulers in the PDs then construct a detailed schedule with a one-month horizon for the production orders released to them. At this point, the no-wait restrictions between processing steps have to be taken into account. If a feasible schedule cannot be constructed that completes all planned production orders in time, notice is given to the aggregate planners. It is negotiated what can be removed from the production plan in order to be able to realise a feasible schedule. Small disturbances that occur during execution of the schedule are resolved by the PD. For major disturbances, the aggregate planners are consulted to set priorities.

16.3.2 Problems experienced at Pharmcom

The main dilemma in such a hierarchical structure is the content and extent of communication between the (aggregate) planners and (detailed) schedulers. On the one hand, the planners are expected to make an adequate assessment of the feasibility of a particular set of jobs to be completed in a monthly time period. This requires a fairly detailed insight into the constraints on the shop floor in the processing departments. On the other hand, they are also expected to make a long-term plan, covering the entire horizon of the customer order, estimating lead-times and variability in lead times, to make reliable customer commitments and realise a high factory utilisation. The detailed schedulers, in their turn, are expected to make a schedule for a job set which they basically cannot influence, such that the job set is completed within the given time frame, which again allows the planner to complete their long-term commitments to the customers. Furthermore, the schedulers are expected to use their resources as efficiently as possible. As one can imagine, this may lead to friction between the planners and schedulers. The planners tend to plan in detail as much as possible, thereby taking flexibility from the schedulers and leaving them with job sets that are not feasible. The schedulers like to have job sets that fit their departments' capabilities, allowing them to schedule with high utilisation, in (locally) optimal sequence, with buckets as long as possible. Longer time buckets provide them with a greater solution space to the scheduling problem and thus make the scheduling problem easier.

Table 16.2 illustrates the resulting problem at Pharmcom. It lists the percentage of released jobs accepted by the departmental schedulers over a period of three months, and the percentage of the released jobs included in the final schedule constructed by the departmental scheduler. It is reasonable to assume that the final schedule was also completed within the available period, as uncertainty in processing times and availability of resources is small. The same data are given expressed in man and machine hours. We can observe that a substantial portion of the job set has not been accepted by the schedulers. Over the months August and September, the actual production in the department came fairly close to what was accepted by the detailed schedulers. In July, the estimate of the departmental scheduler was not very good.

Table 16.2 Actually accepted (acc) and scheduled (sch) jobs at Pharmcom (Raaymakers *et al.*, 2000).

	Number of jobs		Man hours		Machine hours	
	Acc	sch	acc	sch	Acc	sch
July	78%	67 %	90 %	68 %	95 %	66 %
August	86%	84 %	82 %	85 %	81 %	84 %
September	41%	45 %	69 %	70 %	72 %	73 %

It is thought desirable that these percentages are increased. A low percentage means that a lot of negotiation is going on between the centralised aggregate planners and the decentralised detailed schedulers. Furthermore, since the aggregate planners have experienced a considerable part of their production orders not being accepted, they tend to include more slack in the overall planning of the customer orders, leading to excess work in process and even longer lead times for the customers.

16.4 TACKLING THE COMPLEXITY USING LINEAR REGRESSION

There are various options for addressing the capacity estimation problem of the aggregate planners. A common approach would be to use queuing models to describe the capacity (Buzacott and Shantikumar, 1993). There are two problems associated with the use of queuing models to address this problem. First, the complex no-wait restrictions prevent reliable aggregate modelling of the PDs. Second, the situation to be modelled is non-stationary; this means that the model has to support short-term decision making. Queuing models are usually not transient, so that they are more suitable to support long-term and stationary decisions such as layout design.

In this chapter, we apply linear regression (see e.g. Montgomery and Peck, 1992 for an extensive explanation) to construct an aggregate model. We will first review some experimental findings on using linear regression for aggregate capacity modelling in batch chemical industries, and will then discuss some issues related to practical implementation.

16.4.1 Factors determining job interaction

At Pharmcom, essentially the workload of a job set is used to estimate the time span needed to complete that job set. In the scheduling literature, this time span is usually denoted as the makespan. The workload of a job is the quantity of capacity needed to complete the job and is usually expressed in hours. The workload on the bottleneck resource puts a lower bound on the makespan of a set of production orders (the job set). The bottleneck resource is the resource that has the highest utilisation. If there were no relation between the operations of a job (due to the δ_i's defined in Section 16.2.2) and between the jobs (due to shared resources), the workload of all jobs on the bottleneck resource would be the actual makespan. Job interaction occurs because each job requires several resources at the same time or consecutively without waiting time. Furthermore, in batch chemical industries each job may require a different combination of resources. To meet all no-wait restrictions, idle time generally needs to be included on the resources. Consequently, job interactions result in a makespan that is longer than the lower bound on the makespan, which is based on the workload of the job set.

Different job sets have different levels of interaction. Raaymakers and Fransoo (2000) have researched the type and number of characteristics of a job set that influence this interaction in multipurpose batch process industries. Their conclusion is that only a small number of factors actually influence this interaction, and thus the makespan. In a controlled simulation experiment, based on the typical characteristics of the setting at Pharmcom, the following factors explained more than 90% of the variance in the interaction across a large number of different job set and resource structures:

- average number of parallel resources;
- average number of processing steps per job;
- average overlap of processing steps within a job;
- standard deviation of processing times;
- workload balance over the resources.

The performance of the regression model turned out to be rather insensitive to variations in the characteristics of the job sets and resource structure. Since a regression model is easy to apply, it makes sense to investigate the performance that can be obtained by using such a model for aggregate planning purposes in situations such as the one described above at Pharmcom.

16.4.2 Simulation experiments

We conducted a series of controlled computer simulation experiments. These experiments simulated a hierarchical planning situation as outlined above, with the upper control level deciding about the composition of job sets, and the lower control level attempting to schedule these job sets within the available time frame. One notable exception to the situation at Pharmcom was that in our experiments each customer order consisted of exactly one production order or job. At the upper level, two policies were compared. The first policy was a workload-based policy. In this policy, similar to what happened at Pharmcom, a production order (or *job*) was allocated to a planning period as long as the total workload does not exceed a specified maximum workload and the workload per resource does not exceed the available capacity per resource. The second policy, called the makespan estimation policy, used a regression model for estimating the makespan of a set of jobs. Under this policy, a production order may be allocated to a planning period as long as the (regression-based) estimated makespan does not exceed the period length.

In the experiments, orders arrive randomly with the same lead time for all orders. A decision on accepting the order is based on either one of the two policies described above. The decision on acceptance is taken immediately upon arrival of the order. When a period starts, the allocated orders to that period are released for scheduling and execution. The schedule is constructed using the simulated annealing algorithm developed by Raaymakers and Hoogeveen (2000). If orders cannot be completed, they are postponed to the next period. If all orders are completed before the end of the period, orders from the next period are shifted forward.

Since the makespan estimation policy includes more characteristics of the job set than just the workload, we may expect this policy to perform better than the workload policy. We are interested in investigating under which conditions this difference is largest. The performance is measured by the realised capacity utilisation and service level. We chose the capacity utilisation as a performance indicator because in an over-demand situation the utilisation that can be realised is directly related to the number of orders that can be accepted. In turn, capacity utilisation influences the revenues of a company. The service level is used as the second performance indicator because it indicates the reliability of the due dates agreed with the customers. The service level is defined as the percentage of orders that are completed before their due dates. As a third, internal, performance measure, we measured the fraction of jobs that needs to be rescheduled in the decentralised scheduling and execution step.

A result of using policies for order acceptance and capacity loading that are based on aggregate information is that the job sets will not always be achievable. Achievable here means that it may not always (and in most cases it will not) be possible to construct a detailed schedule with a makespan shorter than the provided time bucket length. Hence, some replanning is always required because order acceptance and capacity loading decisions are based on an aggregate model of what can be realised by the processing departments. The amount of replanning is determined by how close this aggregate model is to the actual situation in the processing departments. Many jobs need to be shifted backwards if the aggregate model makes a too optimistic estimate of what can be realised by the processing department. On the other hand, many jobs can be shifted forward if the aggregate model makes a too pessimistic estimate of what can be realised by the processing department. In either case, replanning jobs requires time and effort of the planner in a company. In industrial practice, a minimum of replanning is therefore preferred, as has been discussed above.

In Table 16.3, the demand/capacity ratio and lead-time parameters used in the simulation experiments are given. With respect to the demand/capacity ratio, we consider two levels. At the high level, the average demand requirements for

Table 16.3 Parameters settings for the simulation experiments, levels 0 and 1

	0	1
Demand/capacity ratio (β)	0.7	1.0
Job mix variety (γ)	4-7 processing steps 20-30 processing time	1-10 processing steps 1-49 processing time
Workload balance (δ)	30, 25, 20, 15 and 10% of demand requirements for resource type 1 to 5	20% of demand requirements for each resource type
Standard lead time (ε)	2 periods	4 periods

capacity are equal to the total available capacity per planning period. At the low level, the average demand requirements for capacity are equal to 70% of the total available capacity per planning period. As has been described above, due to the no-wait restrictions on the processing steps in each job, capacity utilisation in this type of industry is at most between 50 and 60%. Thus, both the demand levels investigated represent situations where demand effectively exceeds available capacity. We consider two levels for the requested lead times, namely 2 and 4 periods.

Each order consists of exactly one job with a specified structure of no-wait processing steps. The job characteristics are generated randomly upon arrival of the order. Hence, each job arriving at the system may be different. The performance of the order acceptance and capacity-loading policies might be affected by the job mix variety and the workload balance. Therefore, two levels of job mix variety and workload balance are considered. In the situation with high job mix variety the number of processing steps per job is uniformly distributed between 1 and 10, and the processing time is uniformly distributed between 1 and 49. In the situation with low job variety, the number of processing steps per job is uniformly distributed between 4 and 7, and the processing time is uniformly distributed between 20 and 30. Note that in both situations the average number of processing steps and the average processing time is the same. In generating the jobs, each processing step is allocated to a resource type. In the situation with high workload balance, the allocation probability is the same for each resource type. In the situation with low workload balance, the allocation probability is different for each resource type. On average 30%, 25%, 20%, 15% and 10% of the processing steps will be allocated to the five different resource types respectively.

Further details of the experiments can be found in Raaymakers (1999).

16.4.3 Experimental results

To compare the two policies we have measured the average realised capacity utilisation and the average replanning fraction over the runs while maintaining a minimum service level of 95%. The results are given in Table 16.4. An ANOVA showed that all main effects (four experimental factors and two policies) have a statistically significant contribution towards the value of the capacity utilisation performance measure. Below, we will discuss the relevant differences between the policies under the various scenarios.

We observe that there exists a considerable difference in performance between the makespan estimation policy and the workload policy. The differences between the performance of the two policies are especially large if β is high, whereas the differences are small for low β. In a situation with high β, many orders arrive and many opportunities exist to select the jobs that fit in well with the other jobs. With low β, most arriving orders can and will be accepted by all policies. Thus, only if β is high will the differences in selectivity between the policies show in the capacity utilisation performance measure. We further observe that the realised capacity utilisation is considerably higher if β is high. This is especially the case for scenarios that also have a high job mix variety (γ). This is explained by the fact that with high γ more opportunities exist to select jobs that fit in well, especially in combination with high β.

The results in Table 16.4 show that for high γ and high β, the makespan estimation policy results in a higher capacity utilisation than the workload policy. For the remaining twelve scenarios, the difference for the capacity utilisation performance measure between the workload policy and the makespan estimation policy is practically negligible.

Table 16.4 Simulation results: Average capacity utilisation (ρ) and average replanning fraction (rpf) for the order acceptance policies under different scenarios.

Scenario (H=high;L=low)				Policy			
β	γ	δ	ε	Makespan Estimation		Workload	
				ρ	*rpf*	ρ	*rpf*
H	H	H	H	0.61	0.18	0.58	0.30
H	H	H	L	0.61	0.16	0.59	0.28
H	H	L	H	0.58	0.16	0.55	0.30
H	H	L	L	0.58	0.13	0.56	0.30
H	L	H	H	0.59	0.13	0.59	0.12
H	L	H	L	0.59	0.14	0.58	0.13
H	L	L	H	0.53	0.15	0.53	0.14
H	L	L	L	0.53	0.16	0.53	0.16
L	H	H	H	0.59	0.17	0.58	0.30
L	H	H	L	0.59	0.14	0.58	0.27
L	H	L	H	0.56	0.17	0.56	0.29
L	H	L	L	0.55	0.13	0.55	0.27
L	L	H	H	0.59	0.15	0.59	0.16
L	L	H	L	0.59	0.14	0.59	0.16
L	L	L	H	0.54	0.15	0.53	0.17
L	L	L	L	0.53	0.13	0.53	0.15

When we consider the replanning fraction, a different picture emerges. Under the makespan estimation policy, the replanning fraction ranges from 0.13 to 0.18, whereas under the workload policy the replanning fraction ranges from 0.12 to 0.30. A closer inspection of the results shows that for scenarios with high job mix variety (γ), the workload policy consistently results in very high replanning fractions (ranging from 0.27 to 0.30) as opposed to the level of the replanning fractions (ranging from 0.12 to 0.17) in the scenarios with low γ. The makespan estimation policy, on the other hand, shows no significant difference for the replanning fraction between the different scenarios and can apparently cope very well with situations with high γ. Its resulting replanning fraction is about half of the replanning fraction of the workload policy in these high job mix variety scenarios.

16.5 REGRESSION MODELLING IN PRACTICE: THE HUMAN FACTOR

In many instances in practice we are confronted with aggregate capacity estimation questions. In some cases, estimating capacity is fairly straightforward. This is for instance the case in flow process industries, where a single resource is dominant in determining available capacity. If the level of interaction between jobs and resources and between jobs themselves is high, the estimation process becomes more complex, as has been discussed above. The structures of jobs and resources (bills-of-process) are an important indicator of this complexity. A complex bill-of-process does not however necessarily create a situation where regression modelling is the only option. In capacity loading decisions in traditional job shops, queuing models are widely used to assist the decision maker (Buzacott and Shantikumar, 1993). Additional constraints on work-in-process and intermediate inventories further trouble the aggregate picture and diminish the power of queuing models and offer increased opportunity for using regression modelling.

In industry, many people are usually involved in the operational planning, scheduling and control process if the job and process structures are complex and the interaction level is high. It is worth investigating whether a regression based approach is beneficial to estimate capacity in situations where people have a big impact on operational production control decisions. This is not only relevant in batch process industries, but also in other situations with a high level of job and resource interaction.

One such situation is the job shop under a workload control (CONWIP) policy (see, e.g., Bertrand *et al.*, 1990, and Wiendahl, 1995). It is widely known that a controlled workload may lead to increased operator productivity. The exact behaviour is, as yet, unknown. Statistical models may be built on historic data to capture the capacity of manufacturing departments under different workloads. As above, a limited insight into possible factors needs to exist. Preliminary research in estimating lead times using regression models has been documented by Ragatz and Mabert (1984) and Vig and Dooley (1991, 1993).

A second issue that is of primary importance in modelling is to model the capabilities of the human schedulers. In the experiment discussed above, the scheduling function was emulated using a simulated annealing algorithm. In most cases, algorithms do not provide the (final) schedule in business practice, but human schedulers perform this task (McKay and Buzacott, 2000). This was also the case at Pharmcom. If past data exist on job sets that have been accepted and schedules that have been completed, again regression modelling can be used to characterise the production capacity. An interesting research question is what the influence then is of the actual scheduler constructing the schedule. Would the regression model be different for different schedulers, or would experienced schedulers have comparable tactics and performance? We are not aware of any empirical research that addresses these questions.

An interesting parallel exists between the approach outlined here and the description of the way in which a scheduler makes manual estimations of run times in a textile manufacturing company as described by Dutton and Starbuck (1971). In this empirical study, they describe that the scheduler, ‘Charlie’, estimates the run time of the fabricator machine by taking into account a series of historical characteristics related to the type of product that is manufactured on the production line. This description assists in arguing that our approach is intuitively appealing to production planners working in industry.

The most challenging issue when applying regression modelling in this manner is changing reality. To build a model in industrial practice, some historical data need to be available. Insights have not yet been developed as to what extent this needs to be the case, to what extent simulation of reality can help, and how a changing reality can be modelled. The model used in the experiments above has been constructed using a fair amount of data points to calibrate the model parameters. However, a single model has been used for all situations investigated. Raaymakers (1999) shows that the model is fairly insensitive to changes in the job structure. Changes in the resource structure do affect the performance of the model. Her research results indicate that the model needs to be recalibrated if resources are added or removed from a processing department. Such a re-calibration may need to be done using data on actual performance with the new resource structure. Given the relatively long time buckets (recall that at Pharmcom they were one month in length, with about 40-60 jobs to be processed in a PD), it does not look very likely that (re)calibration of the regression model is a feasible task using actual data. Simulation of reality can however help. Using the same initial data points, different job sets can be randomly generated and schedules constructed using an algorithm. This provides us with many more data points. We then need to estimate the relation between performance of the scheduling algorithm and the performance of the human real-life scheduler. Initial results in Raaymakers (1999) indicate that there is likely to be a fairly linear relationship between the performance of the two in terms of makespan estimation quality (although the actual schedules may be very different).

All this indicates a fair amount of opportunity for using regression modelling in practical situations with human schedulers that cannot be captured with more traditional models from operational research.

16.6 CONCLUSIONS AND FURTHER RESEARCH

The question then arises as to whether this theoretically developed approach can be applied in practice. We are convinced that there are a number of factors that are beneficial for applying a regression-based approach in a practical hierarchical setting. The first one is the fact that the approach is intuitively sensible for the planners and schedulers. The planners and scheduler at the company experience the regression factors discovered as realistic and 'making sense'. The second factor is that the regression method is fairly robust. Experiments show that, as long as the physical configuration does not change, a changing variety in the job set and capabilities at the scheduling level do not have a large influence on the performance of the model. The third factor is that the model can capture any 'local behaviour' by simply looking at past performance of the departmental schedulers. This allows for a very generic application of this type of model in cases where technological constraint may be complicated and dominant.

In summary, regression modelling does provide us with a powerful and relatively straightforward tool for aggregate capacity modelling. The first completed study in the very complex environment of batch process industries scheduling has given very promising results. Further research is however needed. The use of the model in a dynamic situation of order acceptance needs to be addressed. Since the model used for estimating the capacity is the same model as the one used for accepting orders, the accepted orders tend to have a natural bias in their composition (Raaymakers, 1999). This can easily be corrected by introducing a correction factor. Although this may be sufficient from a practical point of view, it is not satisfactory from an academic point of view. Also, empirical evidence needs to be collected on the use of these models in a setting with human schedulers. We argue that we have a strong case that performance is very likely to be higher than for more traditional models like queuing.

16.7 REFERENCES

Bertrand, J.W.M., Wortmann, J.C. and Wijngaard, J., 1990, *Production Control: A Structural and Design Oriented Approach*, Amsterdam: Elsevier.

Buzacott, J.A. and Shantikumar, J.G., 1993, *Stochastic Models of Manufacturing Systems*, Englewood Cliffs: Prentice Hall.

Connor, S.J., 1986, *Process Industry Thesaurus*, Falls Church: America Production and Inventory Control Society.

Dutton, J.M. and Starbuck, W.H., 1971, Finding Charlie's run-time estimator. In: Dutton, J.M. and Starbuck, W.H. (eds.) *Computer Simulation of Human Behavior*, New York: Wiley, pp 218-242.

Fransoo, J.C. and Rutten, W.G.M.M., 1994, A typology of production control situations in process industries. *International Journal of Operations and Production Management*, 14:12, pp. 47-57.

McKay, K.M. and Buzacott, J.A., 1999, The application of computerized production control systems in job shop environments. *Computers in Industry*,

Meal, H.C., 1984, Putting production decisions where they belong. *Harvard Business Review*, 62:2, pp. 102-111.

Montgomery, D.C. and Peck, E.A., 1992, *Introduction to Linear Regression Analysis*, New York: Wiley.

Raaymakers, W.H.M., 1999, *Order Acceptance and Capacity Loading in Batch Process Industries*, PhD Thesis, Technische Universiteit Eindhoven.

Raaymakers, W.H.M. and Fransoo, J.C., 2000, Identification of aggregate resource and job set characteristics for predicting job set makespan in batch process industries. *International Journal of Production Economics*, 68:2,pp. 25-37.

Raaymakers, W.H.M. and Hoogeveen, J.A., 2000, Scheduling multipurpose batch process industries with no-wait restrictions by simulated annealing. *European Journal of Operational Research*, 126, pp. 131-151.

Raaymakers, W.H.M., Bertrand, J.W.M. and Fransoo, J.C., 2000, The performance of workload rules for order acceptance in batch chemical manufacturing. *Journal of Intelligent Manufacturing,* 11:2, pp.217-228

Ragatz, G.L. and Mabert, V.A., 1984, A simulation analysis of due date assignment rules. *Journal of Operations Management*, 5, pp. 27-39.

Schneeweiss, C.A., 1999, *Hierarchies in Distributed Decision Making*, Berlin: Springer.

Taylor, S.G., Seward, S.M. and Bolander, S.F., 1981, Why the process industries are different. *Production and Inventory Management Journal*, 22:4, pp. 15-22.

Vig, M.M. and Dooley, K.J., 1991, Dynamic rules for due date assignment. *International Journal of Production Research,* **29**, 1361-1377.

Vig, M.M. and Dooley, K.J., 1993, Mixing static and dynamic flowtime estimates for due-date assignment. *Journal of Operations Management*, **11**, pp. 67-79.

Wallace, T.F. (Ed.), 1984, *APICS Dictionary,* 5th ed., Falls Church: American Production and Inventory Control Society.

Wiendahl, H.-P., 1995, *Load-oriented Manufacturing Control*, Berlin: Springer.

CHAPTER SEVENTEEN

Engineering a Vehicle for World Class Logistics: From Paradox to Paradigm Shifts on the Rover 75[1]

Joy Batchelor

17.1 INTRODUCTION

The automotive industry presents a popular image of itself as leading the way in demonstrating the efficiency gains associated with the implementation of just-in-time 'customer-pull' production techniques. Other industries look towards the automotive industry in search of benchmarks against which they, too, can achieve the rewards of low cost, high quality and profitable operations. However, this picture is far from the truth. As management consultants KPMG report in *Europe on the Move* (1998), it is estimated that the value of unsold finished vehicle stocks within Europe are in the region of around £18 billion, with each car being stored for an average of 50 days before being sold. Furthermore, whilst a number of manufacturers aspire towards the goal of reducing customer order lead-times it is still closer to the truth to talk of industry build-to-customer order lead-times being on average 40 days or more according to research undertaken by the International Car Distribution Programme (Kiff, 1997).

Whilst theoretically, in the UK at least, automotive manufacturers have offered many millions of vehicle combinations for customers to chose from, reality has been very different. A study carried out in 1994 reported that only 64% of UK customers bought the exact specification of vehicle that they were looking for, a further 23% took an alternative specification, whilst the remaining 13% found no suitable match (Kiff, 1997). A similar picture emerges in the US where it has been reported that 11% of buyers switch to another manufacturer because their first choice vehicle is unavailable (Automotive Manufacturing and Production, 1997).

The reality of vehicle production within Europe is one of a just-in-time 'push' system in contrast to a pure just-in-time environment where resources are only pulled through the value chain in response to a direct customer order. Most of the

---

[1] This chapter was prepared prior to the sale by BMW of Rover Cars to the Phoenix Consortium and Land Rover Vehicles to Ford. Whilst the future production of the Rover 75 may deviate from the principles outlined within this chapter, a number of generic lessons can be drawn from the development of the Rover 75 in respect to overcoming the perceived trade-off between product variety and manufacturing effectiveness in a build-to-order environment.

major automotive manufacturers operate some form of fixed allocation system whereby dealers are obliged to place a set number of orders each month, often many months in advance of production, and more often than not, before they have real customers for these vehicles. Efforts have been made, through the introduction of centralised distribution centres and the manipulation of 'virtual' stocks between dealers, to improve the probability of customers finding an exact vehicle match to their requirements (Kiff, 1997; Muffatto *et al.*, 1999). When a dealer does not have a customer's first choice vehicle in stock and they cannot source a vehicle from national stocks, they then search on-line for 'virtual' orders, which are allocated to the dealer network. If an exact vehicle match cannot be found, an existing unallocated order can be amended to meet a customer's requirements. Therefore "virtual" orders, and not actual vehicles, are exchanged between dealers prior to vehicle manufacture. However, depending upon the degree of variability allowed for certain options and vehicle configurations as the vehicle draws near to its build date, the customer will often experience a delay of many weeks, if not months, before the order is built and delivered. Lean distribution appears to be a long way from realisation and the question is for how much longer will customers continue to tolerate and pay for such inefficiency?

The practice of building-to-stock thereby buffers manufacturers from fluctuations in vehicle demand. Such a system currently aids the impressive plant productivity levels recorded by some manufacturers but how many of these vehicles then sit for many months in vehicle compounds before finding a buyer who is prepared to accept a different specification to the one that they originally wanted? Regardless of how 'leanly' a vehicle has been produced, a vehicle without a buyer is waste of the worst kind.

Whilst the Japanese lean system utilised by the volume manufacturers operates effectively with stability of demand, to ensure material flow optimisation, the requirements of specialist vehicle producers such as BMW, Mercedes and Jaguar, who build few cars alike, are quite different.[2] They require a high level of agility to manage the levels of customisation required by customers. However, in the past this has meant a trade-off in delivery lead-times due to production scheduling constraints, supply chain inflexibility and high levels of product variety. Therefore both the volume and the specialist manufacturers face the same dilemma of how to build exactly what the customer wants, when they want it.

The design and development of the Rover 75 is significant for a number of reasons. The 75 was the first vehicle to be fully designed under the ownership of BMW and the vehicle was launched in a period of speculation over the future of the Rover Group after the announcement of large losses in 1998. Much rested upon the 75 re-establishing the Rover brand, not only in the UK but also around the world. The 75 was a key model to be introduced into the highly competitive compact executive market sector within Europe, alongside BMW (3-series),

[2] It should be noted that Volvo's much criticised Uddevalla plant, which closed in 1994, had moved to a build-to-order environment. In 1991 when Uddevalla started its programme to compress customer order lead-times only 20% of assembled cars were built to customer order whilst the rest were scheduled according to a central plan. By November 1992 the level of customer specified orders being built at Uddevalla had risen to 70%. In one year Volvo reduced the total lead-time to process an order from 60 days down to 30 days and was moving towards a target of 14 days (Berggren, 1995).

Mercedes (C class), Alfa Romeo (156), Audi (A4), Lexus (LS200) and others. The model is a land mark vehicle for the whole of the BMW Group because the vehicle has been specifically engineered to deliver world class logistics performance levels (Batchelor and Schmidt, 2000). Utilising a build-to-order capability, a shortest possible time to customer order lead-time of 10 working days will be achievable within Western Europe. In addition to short customer order lead-times, the Rover 75 will also offer levels of product variety akin to those of BMW, which represents a significant departure from the low product variety environment encouraged under the influence of Honda during their years of collaboration with Rover.

The research presented within this chapter is drawn from a five-year longitudinal study of Rover Group covering the period 1994-1999, which focused on the implementation of just-in-time distribution efficiency within the company. Qualitative interview data were gathered from the Rover 75 Project team between October 1996 and May 1997. In total over 40 interviews were conducted with senior members of the project management team and eight of the component core teams. For each component core team the component engineer, the buyer and the logistics representative from Rover and the supplier's project managers were interviewed. Subsequent research has been conducted on the Rover 75 Project in the final 18 months leading up to volume production and the launch of the vehicle in June 1999.

This chapter is presented in three sections. The first section presents a brief overview of the trade-off debate between variety and operational efficiency before reviewing recent trends in the provision of product variety within the auto industry, where dramatic changes have been observed over the last decade. The second section explores the relationship between variety provisioning within the Rover car brand under the influence of Honda and the move by Rover Group towards the development of a build-to-order capability. The final section addresses the apparent conflict faced by the Rover 75 Project in designing a vehicle which satisfied BMW's requirement for high levels of profitable variety, the opposite approach to that of Honda, but which retained the requirement for a highly responsive build-to-order capability. Looking at variety provisioning through the lens of Rover provides a unique perspective in which to explore the perceived trade-off relationship between high levels of product variety, and operational responsiveness and effectiveness.

17.2 VARIETY PROVISIONING AND PERFORMANCE TRADE-OFFS

High external product variety is normally associated with an increase in internal complexity thereby leading to a trade-off between the level of variety offered in the market place, the volume of production and the effectiveness of manufacturing and logistics operations (Slack *et al.*, 1998). Within the operations management literature it has been widely accepted that a dichotomy appears to exist between inventory and product variety which obstructs the implementation of just-in-time production within high variety environments (Bennett and Forrester, 1993). This dichotomy is seen to be caused by the differing flexibility requirements of JIT, which operate on *process response flexibility*, and those of high variety production, which require *product range flexibility* (Slack, 1988). Product range flexibility defines the capability of a manufacturing system to produce a wide range of variety without unduly compromising upon the overall efficiency and the costs of production. Process response flexibility increases a system's capability to produce smaller batches more frequently, thereby reducing inventory. However, this capability alone will not address the constraints which limit the variety of items that can be produced without unduly affecting manufacturing efficiency.

The application of the focused factory (Skinner, 1974) concept advocates the restriction of activities undertaken within a manufacturing operation to avoid adverse trade-offs. The traditional assembly line is still seen as working most effectively when such a system seeks to achieve classic economies of scale by reducing model mix variation in order to minimise balance loss and quality problems. Such an approach is seen to correspond to the attempts of automotive manufacturers to commonise parts, to standardise option packaging, as well as supporting the adoption of vehicle platforms (Jügens *et al.*, 1997). For example, in the mid 1990s the most productive automotive assembly facility in Europe was the Opel (GM) plant at Eisenach in Germany. The productivity of the Eisenach plant has been attributed to the reduced number of model variants of the Corsa and Astra models produced in comparison to the degree of variation found in other GM plants producing the same models (Hsieh *et al.*, 1997). Viewing product variety in this light therefore leads to a perceived trade-off between the level of variety offered to customers and the efficiency of manufacturing operations. One notable study by MacDuffie *et al.* (1996) questions the strength of this trade-off relationship between product variety and manufacturing performance, pointing instead towards the adverse effects of parts complexity upon manufacturing performance, a point we will return to later in the chapter.

Research into the management of product variety has attracted interest in recent years (Ho and Tang, 1998). On the one hand, approaches into the study of product variety management have focused upon understanding the requirements of the demand side choices provided by marketing in developing the strategy behind increased product offerings (Lancaster, 1990). On the other hand, supply-side approaches have sought to understand the implications of delivering increasing levels of variety upon the effectiveness of the total value chain (see Fisher, 1997). Product variety as a concept can be addressed in a number of ways and it is important to clarify a number of points. An important distinction to be made is between **external variety**, which represents the level of product choice offered to customers, and **internal variety,** which is the level of variation handled by manufacturing and logistics operations in satisfying the provision of external variety. External variety can be defined as 'a form of complexity arising from either a large number of product families or a large number of alternative configurations (variants) within single families' (Calderini and Cantamessa, 1998). Within this chapter it is the latter definition with which we are concerned, for this indicates the level of customisation undertaken within a single product family whilst the former indicates the level of segmentation undertaken within a particular market.

Two approaches, which seek to minimise the impact of external variety upon the effectiveness of operations, can be identified. Both seek to minimise internal variety and the uncertainty associated with such variety. Firstly, process-based approaches seek to develop sufficient flexibility within supply, production and distribution systems to accommodate high external variety at a reasonable cost, for example, the adoption of flexible production technologies such as Computer Integrated Manufacturing (Upton, 1994). Alternatively, product-based approaches can be used to develop product designs that allow for high external variety whilst minimising the level of component variation and assembly complexity. Examples of this include the adoption of product platforms, modular design concepts and component standardisation (Ulrich, 1995; Fisher *et al.*, 1997). Most attention had been focused upon the former through the vast literature on manufacturing flexibility (Slack, 1991; Upton, 1994) but increasing attention is now placed upon product design.

Ulrich (1995) contends that much of a manufacturing system's ability to manage variety resides not within the flexibility of the equipment of the factory, the prime focus of the flexibility literature, but within the architecture of the product itself. By distinguishing between modular and integral designs, Ulrich emphasises the link between the product design and the performance of the manufacturing firm with respect to the way in which product architecture influences: product performance; product variety; component standardisation; manufacturability; product change, and product development management. The interfaces between components and systems are defined by the degree to which the interface creates a 'loose coupling' or a 'tight coupling'. Modular design is defined as 'a special form of design which intentionally creates a high degree of independence or 'loose coupling' between component designs by standardising component interface specifications' (Sanchez and Mahoney, 1996). The standardisation of interfaces and the creation of independence between components and systems, through modular product architectures, enable the accommodation of variety through the interchangeability of parts (Meyer and Utterback, 1993). Alternatively, integral product architectures are tightly coupled thereby creating a high level of interdependency between parts and restricting the ease with which product variety can be accommodated. Modular design has most evidently been applied within the computer industry (Baldwin and Clark, 1997), where modularity enables companies such as Dell to configure and assemble computers quickly to individual customer specifications.

From a product design perspective, the level of component variation is also of critical importance and we can distinguish between **positive variety**, which is the level of variety demanded by customers, and **negative variety** (Galsworth, 1994), which are variations in the product design not associated with meeting customer requirements. For example, customers may demand a range of different trims and functions on a car seat, which is positive variety. However if each seat derivative has a range of different fixtures and fitting procedures, for example, then this is negative variety because it does not directly impact upon the functionality or the distinctiveness of the seat in the eyes of the customer. Design for manufacture and assembly activities seek to address such unnecessary variations in product design.

17.3 REDUCING THE LEVEL OF INTERNAL VARIATION WITHIN AUTOMOTIVE ASSEMBLY OPERATIONS THROUGH PRODUCT DESIGN

Strategic approaches towards variety provisioning within the automotive industry have shifted dramatically over the last 15 years. The mid to late 1980s witnessed an explosion in model proliferation offered by the Japanese, however by the early 1990s the Japanese approach towards mass customisation ran into significant problems (Stalk and Webber, 1993; Pine *et al.*, 1993). Up until this point the previously successful strategy of product proliferation was supported by strong demand within the Japanese market, but as demand at home fell, the previously 'hidden' inefficiencies of this strategy began to emerge. Toyota in particular, as a result of its 'fat' design, suffered from excessive product complexity and parts proliferation, as a consequence of designing products on independent vehicle platforms (Fujimoto, 1997; Cusumano and Nobeoka, 1998). Competition based upon the provision of variety had followed two distinct routes until this point in time. MacDuffie *et al.* (1993) classified external variety provisioning within the auto industry as follows:

- *Fundamental variety:* Product variety is based upon different platforms, models and body styles while minimising the use of options.
- *Peripheral variety:* Product variety is based upon the offering of many different options leading to millions of Model, Territory, Option (MTO[3]) combinations.

Up until this point the Japanese usually competed through the use of fundamental variety, offering choices of models and distinct platforms but restricting the option choices available, typically through offering three levels of trim. The Japanese also installed many items as standard features on their models which other producers only supplied as options. However, American producers traditionally used the opposite approach to that of the Japanese by having less fundamental variety but offering an enormous amount of peripheral variety through the provision of options but without altering the core design of the vehicle.

---

[3] The level of theoretical vehicle variety offered in the market place is calculated by the MTO, the model, territory or type and option. A second, more detailed measure sometimes used is that of MTOiC - model, territory, option and colour where i refers to interior trim colour and C to the exterior paint colour. MTO and MTOiC can be calculated in two different ways and this can cause confusion about the true level of product variety. For example, the theoretical MTO is derived by multiplying the model options by the number of markets thereby assuming that the bill of material is the same for all markets. However, feature listings are often restricted by market and therefore the true MTO is derived by multiplying out all of the potential combinations by each individual market and then adding these individual totals together to generate the true number of potential MTO's theoretically available. However, due to manufacturing constraints and restrictions affecting the way in which vehicles are retailed the true level of vehicle variety made available to customers is very much lower.

However, by the late 1980s and early 1990s the US producers were stepping back from these high levels of peripheral variety due to the perceived trade-off between variety and production efficiency (Fisher *et al.*, 1994). Even accounting for the fact that the Japanese had approached the provision of product variety in a fundamentally different way to the American producers, by 1993 Toyota, Nissan and Mazda had all announced significant plans to reduce both model proliferation and unnecessary parts variety. Honda too, which prided itself on its engineering originality, set itself a target of carrying over 50% of parts between the 3rd and 4th generation Accord models which were to be launched in 1993 (Muffatto, 1996). Figure 17.1 illustrates the variety reduction trends over the last fifteen years within the automotive industry.

European manufacturers also chose to actively reduce the number of individual vehicle specifications available in an attempt to improve the overall efficiency of vehicle supply. For example, in 1996 Ford Europe had the potential to make 27 million different variations of the popular small car the Fiesta. Ford's reported aim was to reduce this figure to around 10,000 variants. When work was started on stripping out complexity from the Fiesta design it was discovered that the model was offered with 132 different door trims (Sunday Times, 1997).

In the early to mid 1990s a significant change in thinking occurred over the provision of external variety through the adoption of common vehicle platforms, a move led by Fiat and VW. The introduction of vehicle platforms alters the fundamental architecture of a vehicle design so that high levels of commonality can be achieved within a range of vehicles sharing a common platform whilst retaining high levels of perceived external product variety within the market. A parallel development was the division of a vehicle design into distinct modules. Whilst this practice originated in the mid 1980s when Fiat developed the Tipo model, the true potential of developing common modules across a range of vehicles is only now beginning to be fully explored through the development of shared vehicle platforms. The popular 1980s concept of the world car has given way to the 'world platform' with production volumes of between 1 to 1.5 million vehicles per year (Financial Times Automotive World, 1999). A vehicle platform typically accounts for 60% of a vehicle's development and production costs (Wilhelm, 1997). Significant economies of scale are to be realised if a number of niche vehicles can be developed off one common platform through the reduction of both development and production costs. In the late 1990s both VW and Ford have led the industry in commercialising common platforms. Most major manufacturers are following this lead and now a number of manufacturers have, or are, on the verge of moving into modular production whereby the responsibility for the assembly of complex commodities is passed into the supply base.

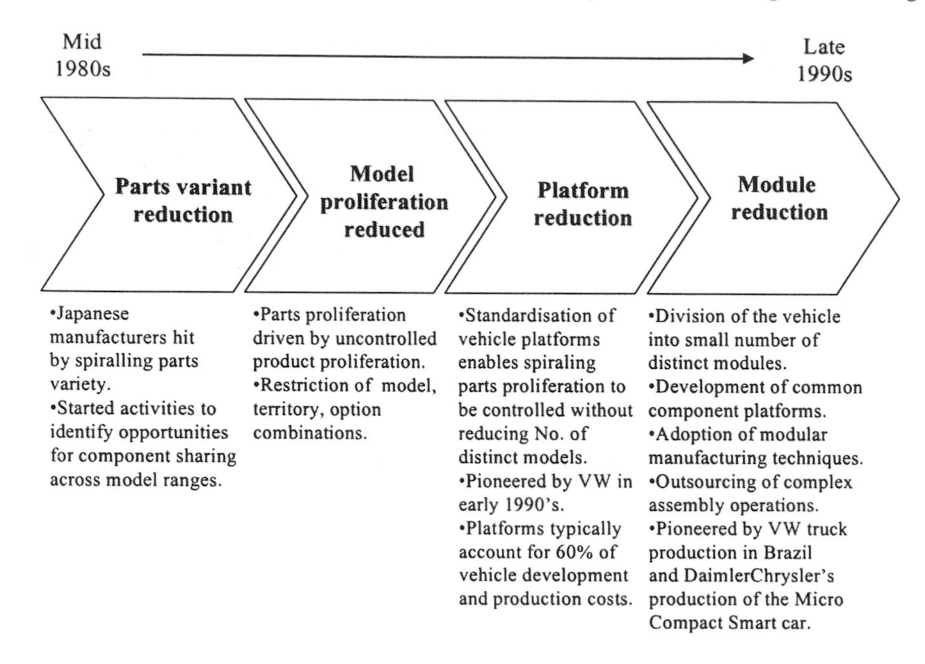

Figure 17.1 Variety reduction trends from the mid 1980s to the late 1990s.

17.4 CHANGING CUSTOMER PERCEPTIONS TOWARDS PRODUCT VARIETY AND BRAND DIFFERENTIATION

With the move towards common vehicle platforms and the increased opportunities for parts commonisation, the nature of the product variety management process has altered. Manufacturers have to take into account the impact of platform and parts commonality upon product integrity (Clark and Fujimoto, 1990). Instead of determining the level of variety to be offered on one model, consideration now needs to be given to determining variety requirements across a series of models each with distinct brand requirements. The balance between commonality and distinctiveness therefore becomes a critical factor to vehicle planners (Robertson and Ulrich, 1998). This balance will not only impact upon the physical appearance of the vehicle and its performance characteristics but will also influence customer perceptions towards distinctiveness. A senior purchasing director from VW made the following observation highlighting the issue raised by the adoption of common component platforms designed by suppliers and integrated into a range of manufacturers vehicles: 'I assume that in the future, we will have technical standardisation. If suppliers deliver to many other motor manufacturers worldwide, the feeling of "what is typical Volkswagen personality" or "what makes a Volkswagen a Volkswagen" is lost' (Automotive Engineering, 1997). Despite being a runaway success in America, in Europe the new VW Beetle is not selling in the numbers first anticipated, especially in Germany. One reason for this short fall in demand has been attributed to customer awareness of the degree of commonality between the Beetle and the highly popular VW Golf model, with which it shares a platform, and which sells for a considerably lower price than the Beetle.

Whilst companies such as Ford and VW are vigorously pursuing platform rationalisation as a means of cutting both development and purchasing costs, BMW stand against this approach on the grounds of maintaining their brand integrity. Wolfgang Ziebart, BMW's head of R&D, is highly sceptical about the adoption of common vehicle platforms. 'What we do not do is offer cars with the same concept under different brands, merely modifying the body and the interior. It would have been easier to place the Rover 75 on the platform of the BMW 3 or 5-series, instead of developing an entirely independent car. But then the character of that car would have been too similar to that of our own models, and customers would have asked what makes a BMW 3 or 5-series so special?' (Autocar, 1999a:13). BMW stand firm in their belief that independent platforms are critical to their brand integrity, however component commonisation can be applied where customer expectations are not compromised. An example of this is the sharing of common electrical architectures between vehicles where increasingly up to 20%, or more, of the costs of the vehicle can be incurred.

17.5 WHY LOOK AT THE EXPERIENCE OF ROVER GROUP AND THE DEVELOPMENT OF THE ROVER 75 MODEL?

The experience of the Rover car divisions over the last decade provides a unique insight into two different approaches towards the provision of external product variety to customers. At the beginning of the decade Rover Group, then owned by British Aerospace, worked in close collaboration with Honda on the joint development of new models. In a thirteen year period, starting in 1981, Rover and Honda collaborated on six joint vehicle projects and entered into several joint production agreements in both the UK and in Japan (see Mair, 1994a for a history of the Honda joint development projects). In a surprise move in January 1994 Rover Group was sold to BMW AG, much to the publicly voiced annoyance of Honda. Rover has therefore designed, developed and manufactured vehicles under the influence of two fundamentally different approaches towards the provision of product variety in the last decade. Honda, whilst lean in it's operations, still operates within a mass production, high vehicle specification paradigm, where a limited range of defined variety is manufactured within a 'flexible mass production' framework (Mair, 1994b). BMW, however, competes in the market place on the basis of providing customers with very high levels of peripheral variety through the provision of options. Figure 17.2 illustrates these two alternative approaches.

Under the influence of Honda, Rover moved towards the provision of fundamental variety, whilst BMW competes through the provision of high levels of peripheral variety. The step change undertaken by the Rover 75 is demonstrated through the shift away from restricting variety, as practised on the Rover 600, towards the provision of high levels of peripheral variety, more akin to that of BMW. It must be said that whilst these approaches towards the provision of product variety in the market may be viewed as being diametrically opposed, both Honda and BMW pursue these strategies successfully to the benefit of their respective brands. However, both Honda and BMW protect themselves from shifts in demand through the operation of fixed allocation schemes within their dealer networks thereby operating on a build-to-stock-basis with long order lead-times. BMW are committed to developing a more responsive sales and production system and the Rover 75 is the first vehicle within the Group designed with the capability to deliver a shortest possible order cycle of ten working days. The origins of the Rover 75 lie in the debate over the trade-off between variety, operational effectiveness and responsiveness to customer demands, a debate shaped by Honda's influence and one challenged by BMW's insistence for high levels of profitable variety.

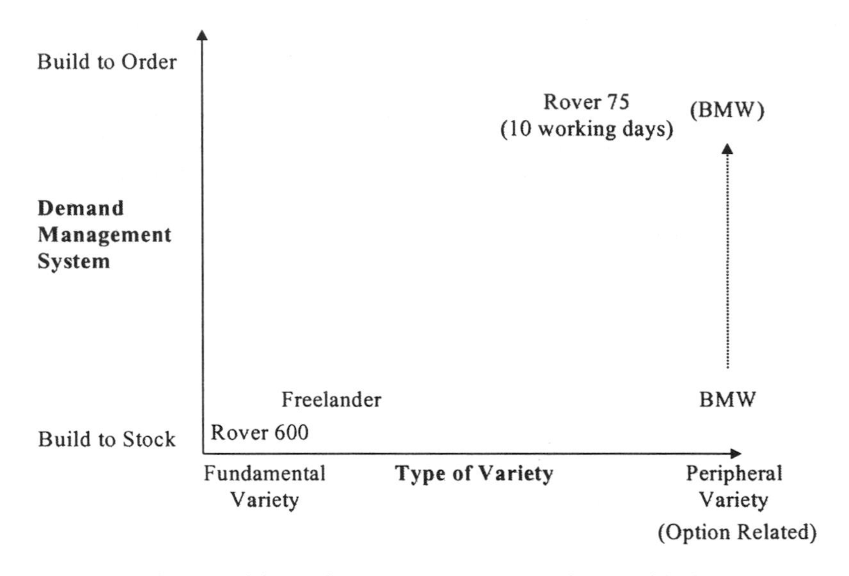

Figure 17. 2 Strategic approaches towards variety provisioning.

17.6 HONDA'S MANUFACTURING PHILOSOPHY AND THE RESTRICTION OF PRODUCT VARIETY ON THE ROVER 600

It is important to understand the manufacturing philosophy adopted by Honda because it is quite distinct to that of Toyota. Honda's approach towards reducing model variety had a significant impact upon Rover during their years of collaboration, whilst Honda's manufacturing philosophy was most influential at Rover's Cowley operations (now known as Rover Oxford) where the Rover 600 was produced. The Honda philosophy focuses upon minimising variation in the manufacturing process and this is primarily achieved through restricting the level of option-derived variety offered to customers and by producing vehicles in large identical batches. Their approach can be summed up as follows: 'Honda's strategy is not a full JIT system, because it prefers to reduce the variety of vehicles in the market place, so that orders can be satisfied from stock. However, this stock is made in a JIT manner; the amount of material needed to operate the manufacturing process is reduced, and efficiency is increased. The reduced variety allows the amount of stock needed to cover the range of choices made by the customer to be low' (Pilkington, 1996a:142).

The Rover 600, one of the models replaced by the Rover 75, was the result of the fifth joint vehicle development project undertaken with Honda. This project allowed Honda to develop its first car specifically for the European market, the Accord, which was to be assembled at Honda's new factory in Swindon. Whilst both models were visually distinct[4] they shared many common parts and a joint supply base which Honda co-ordinated through their Swindon operations.

Due to the commonality of parts supply Honda's strict scheduling system became the operational norm tying the Rover 600 production schedules to the same inflexible lead-times as those operated by Honda with the result, for example, that production schedules, based upon dealer forecasted orders, were set four months before production. Derivatives by batch were also established at this time, whilst the delivery plan for colourised parts was fixed one month before production. A Production Control Manager at Honda in Swindon explained the simplicity of their production process as follows: 'The philosophy behind it is a regular production pattern each day. So we always go in a regular pattern, which typically goes from simple to most complex, left to right hand drive and sunroof to no sunroof, its pretty simple. It means that people on the line know that they should be seeing a regular pattern of vehicles hitting them. It's the same as they use in Japan, there's no batching rules, there's no allotting rules if you like, there is a regular pattern to the day. We also aim to spread the MTO throughout the month, so we don't build batches in big lumps, right or left hand drive, high spec/low spec....'[5] Manufacture of the Rover 600 adopted the same daily pattern to production as Honda for the Accord.

Over the life of the Accord, Honda restricted the total theoretical number of MTO to only 29, with five exterior colours and one interior colour. At launch both models offered only 20 MTO combinations. This level of variety restriction was imposed to protect the integrity of the fixed production schedules thereby optimising the flow of materials through the supply base. Cowley, like Swindon, built in batches of 30 identical vehicles at the MTO level and these batches were then divided into two batches of 15 for painting (MTOC). Over the life of the Rover 600 the number of vehicle combinations made available increased gradually through the introduction of Rover sourced engines and additional exterior colour options, however the number of MTO never exceeded 80. One of the benefits attributed to batch building on the 600 was the level of manufacturing quality

---

[4] The Rover 600 had distinct external skin panels, trunk and front and rear bumpers and shared with the Accord a common roof, common underframe and the majority of mechanical parts and trim parts. Both models initially shared the same Honda-sourced engines until the Rover-sourced T series diesel engines were introduced.

[5] Interview with Honda Production Control Manager at Swindon, May 1994.

achieved, which consistently recorded better levels than the Rover 800[6]. However the level of variety restriction also proved to be a significant constraint to the 600 in the market place.

17.6.1 Just-in-time distribution efficiency: reducing time-to-customer lead-times

A number of manufacturing industries such as the automotive industry face what has been referred to as the customisation-responsiveness squeeze (McCutcheon *et al.*, 1994). This occurs where firms need to deliver differentiated products in less time than it takes to make them. This situation is expressed as the P:D ratio (Mather, 1988). Where a P:D ratio is greater than one it requires that a firm speculate upon and forecast expected demand because process lead-times (P) are greater than the required delivery time (D). Traditionally the solution for satisfying quick delivery times has been product standardisation, the use of buffering mechanisms and building-to-stock. In the early 1990s, under the banner of just-in-time distribution Efficiency (JIT/DE), Rover initiated a number of activities aimed at improving the responsiveness of the entire value chain in working towards a lean distribution customer-pull system. They declared their intention to supply vehicles to individual customer's requirements by increasing their capability to build-to-order and to do this with significantly reduced lead-times. The scope of the Programme was captured in the 14-day vehicle concept: 'Our plans for the future are very much to make the customer decide what it is that is required, to be able to give to the customer factory fresh product, built to order, to a reliable delivery promise and to very short lead-times. Within the JIT/DE Programme a target of fourteen days[7] remains our primary goal, that is, fourteen days from the placement of the order by the customer to the customer actually having the car in his or her hands.'[8]. Under the guidance of the JIT/DE Programme, later to be known as Personal Production, the following goals were set:

- the delivery of a personal car to customers;
- the offering of a guaranteed delivery promise to customers;
- to have competitive lead-times in all markets;
- factory fresh products through increasing the proportion of built-to-order vehicles.

The strategy of moving towards a customer build-to-order environment was not influenced by Honda for they themselves had no interest in the notion of JIT/DE as this ran counter to their desire to ensure manufacturing stability by buffering

---

[6] Cowley operated a mixed model line producing both the Rover 600 and the 800 models. The 600 was produced in one body style whilst the 800 was produced in 3 body styles (3 door/5 door/coupe). However, unlike the Rover 600, the Rover 800 was never produced in batches of 30 identical vehicles. Every vehicle in a 'batch' of 32 vehicles could be completely different. The co-production of these models highlighted the quality benefits gained through the consistency of producing the Rover 600 in batches of 30 identical vehicles.

[7] The 14-day absolute goal has since been superseded by the aim of developing the capability to offer customers a shortest possible order cycle of 10 working days. However the Rover 75 Project team started to design the vehicle with the 14-day goal as a target to be met by the project.

[8] Interview with the JIT/DE Programme Director, Rover Group, April 1994.

operations from unstable demand. However, the JIT/DE Programme was influenced by Honda's approach towards reducing the number of vehicle MTO. This was seen as a way of reducing the level of complexity experienced within Rover's manufacturing operations and by logistics. Studies on the sales figures of a range of the Rover models in the early 1990s showed that only a small proportion of model derivatives, available accounted for the majority of sales[9]. These research results, combined with Honda's influence on reducing the number of vehicle derivatives led to a drive to reduce the number of least popular model specifications on existing models across the company. Table 17.1 summarises the MTO reduction activities undertaken on a range of models across the car's divisions in the early 1990s. Guidelines were also established for the development of new models. For example, it was suggested that the MTO target for future models should be no higher than 300 per 100,000 vehicles per annum, that no more than two trim levels and two interior colours be made available, and that no more than eight exterior colours be considered[10].

Table 17.1 MTO reduction activities undertaken in the early 1990s by Rover Group[11]

Model[i]	No. of derivatives	No. of colours	No. of trims	No. of options	Existing permutations	Rationalised permutations	Options deleted
Metro	15	13	6	6	3,784	1,776	3
Maestro	11	12	4	7	5,172	1,644	4
Montego	16	12	4	11	32,544	19,650	5
Rover 200	7	12	4	3	562	408	1

Rover also sought to reduce levels of negative parts variety on existing models through the redesign of Rover specific components, however due to the relationship with Honda, Rover could not influence variety reduction activities in their favour on Honda-designed parts. The real benefits of reducing negative variety through product design would only be felt when an entire vehicle was designed using such principles. The first vehicle to adopt these principles was the Land Rover Freelander, launched in 1998.

Whilst Rover's attempts to reduce parts variation and the lowering of vehicle MTO did reduce the level of parts complexity faced by manufacturing and logistics, these activities had no discernible impact upon the reduction of overall total lead-times in line with the 14 day aspiration. If anything the adoption of a strict policy of MTO reduction was to the detriment of vehicle sales. An internal

[9] However it should be noted that the inflexibility in the distribution chain would have distorted true demand.

[10] Guidelines issued to the new vehicle project teams in 1995.

[11] Source: Rover Group internal documentation.

study by Rover into the restriction of variety to only 29 MTO on one model predicted a reduction in yearly sales volumes on a world-wide basis of 40%.

In summary, the vehicle concept stage of the Rover 75 originated at a time when Rover was seeking to both reduce time-to-customer lead-times and to reduce manufacturing and logistics complexity through a reduction in vehicle variety as measured by vehicle MTO counts. The experience of working with Honda promoted repeatability and consistency in manufacturing through the adoption of batch building, however this highly inflexible system placed the following constraints upon manufacturing flexibility:

- batches of 30 identical vehicles;
- assembly line loading constraints;
- fixed build programmes of 3-4 months;
- weekly order submissions.

17.7 DESIGN FOR WORLD CLASS LOGISTICS ON THE ROVER 75: 'HITTING COMPLEXITY WHERE IT HURTS'

During the pre-development stages of the Rover 75 in early 1994 it was suggested that one of the project's targets should be the restriction of product variety to 110 MTO per 100,000 units in line with the reasoning of the Personal Production ethos at the time. However, it was quickly recognised that such a restriction of variety would not support the brand proposition of the 75 and by the end of 1994 this figure had crept up to 210 MTO. There was a desire within the Project team to increase further the level of variety offered. At this stage the engineering team were in close collaboration with plant manufacturing and logistics personnel and the Project started to explore how they could increase the level of variety offered without impacting unduly upon manufacturing efficiency.

BMW's take-over of Rover Group coincided with the 75's early pre-development stages and the deciding factor in helping the Project challenge the Company stance on restricting vehicle variety was BMW's insistence that the Project increase the level of profitable variety offered on the 75. This highlighted the fundamental conflict with which the Project had been struggling: 'How could the Project increase levels of variety offered to customers without complexity crippling manufacturing and logistics effectiveness?' The pressure of BMW's profitability requirements gave legitimacy to the 75 Project openly questioning some strongly held assumptions upon the trade-off between increasing levels of product variety and increasing product and process complexity. BMW, however, gave no direction on how these conflicts might be resolved. The Project therefore sought to retain certain elements of the Personal Production approach that were beneficial to the Project, such as the world class logistics performance measures, but chose instead to reject the restraint on product variety. The Project therefore set out to achieve the following objectives:

- the level of variety offered on the Rover 75 was to be determined by margin contribution only and was not to be restricted by MTO measures;
- to engineer for a 14-day vehicle capability;

- the achievement of no manufacturing build constraints within trim and final assembly.

Statistics for the Rover 75

- Budget for the project £700 million.
- £270 million spent on the Cowley facilities: a new body-in-white facility, a state of the art paint shop, a trim and final assembly area and an integrated logistics centre.
- Production target of 2,800 models a week.
- 70% of production destined for export.

17.7.1 Minimising complexity through product design on the Rover 75

The key issue that the Project tried to address was to understand the true nature of the complexity present within a vehicle design and to control for the compounding effects which complexity creates within the manufacturing environment and the supporting supply infrastructure. The engineering approach of the Rover 75 therefore focused upon identifying and controlling for the micro effects of complexity at the individual component level rather than trying to control the overall perceived complexity of the vehicle by limiting the MTO count. The Chief Engineer of the project highlighted the misleading nature of looking towards MTO: 'I hate the term MTO ... people can understand it and you can't deny that if you lower the MTO it helps logistics... you can't deny that, but the point I am getting at is that it is the wrong thing to use because it doesn't allow you to hit complexity where it hurts'[12]. Complexity at the component level hurts manufacturing the most through causing variation in work content leading to line balancing problems, loading constraints and quality problems.

A series of multi-functional working groups were established in the early stages of the Project and two of these groups addressed product variety and complexity. The planning structure adopted on complexity and variety is illustrated in Figure 17.3. These working groups helped to establish the product engineering, manufacturing and logistics concepts adopted and activities were organised to enrol all of the Project's engineers as well as Business Unit manufacturing and logistics personnel.

The use of flow line techniques results in highly efficient production when product variety is small or non-existent. However, once variety increases production efficiency typically reduces due to the increasing variability of work content resulting in balance losses. On the Rover 75 a unique methodology to determine the build of vehicles, not at the vehicle specification level but at the feature level, has been developed. This allows for a high proportion of the vehicle build occurring within 'batches', thereby retaining build consistency, but without the restraints of building a group of identical vehicles, every vehicle can be unique.

---

[12] Interview conducted in November 1996.

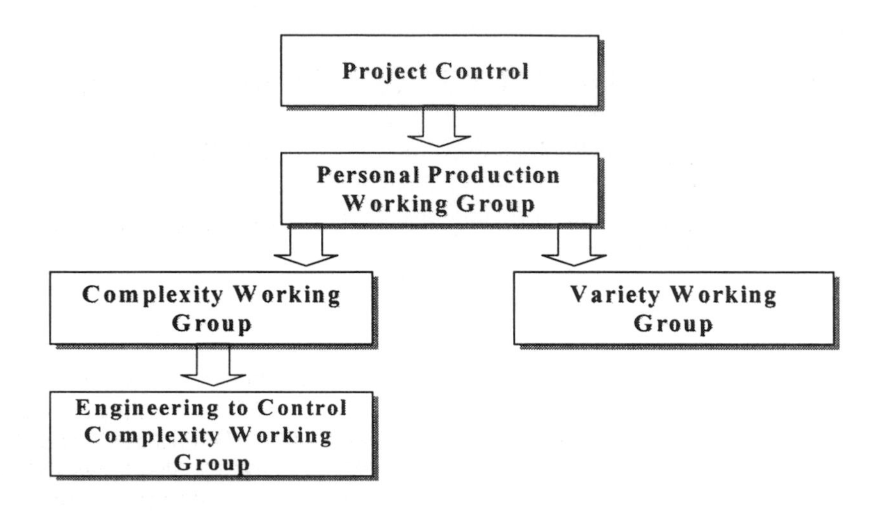

Figure 17.3 The complexity and variety planning structure adopted within the Rover 75 Project team.

Assignments are set in such a way that everyone on the line should be able to cope with a 100% worst case scenario and the design of components and systems supports the equalisation of work content aiming for no loading constraints. The worst items to fit from the point of line balancing are the sunroof, navigation packs and CD players that have a low take rate but a high level of work content, however, these items can be loaded at random under the Rover 75 production system. Table 17.2 compares and contrasts the different build methodologies employed on the Rover 75, the Rover 600 and the Rover 800. Both the Rover 600 and 800 were subject to a range of constraints which restricted the flexibility with which product variety could be handled and both models required a higher level of manning in comparison to the Rover 75.

Table 17.2 A comparison of the different production methodologies employed on the Rover 75, Rover 800 and the Rover 600 models

Methodology	R75 production system (Rover 75)	Level production (Rover 800)	Batch build (Rover 600)
	Method of production developed for the Rover 75.	Constraints imposed on sequence to equalise work content. Approach adopted on the Rover 800	Approach adopted on Rover 600 identical to the approach used at Honda in Swindon. Operated on fixed batches of 30 identical vehicles
Build constraints	None	Numerous build constraints.	Build sequence planned months in advance.
Assignment	95% of associates see work content balanced to worst model in local area. 5% assigned to meet variation of build.	Set to average model, difficulty in allocating average model.	Worst condition by area.
Consistency of build	95% of associates see the build in batches at feature level but every vehicle built can be unique.	Every Rover 800 different by design. 3 door/5 door/coupe.	100% of build in batches, MTO severely restricted due to the inability of the system to handle millions of MTO.
Headcount per shift	275	285	292

By conventional measures the Rover 600, with its very low levels of MTO variety, was conceived within Rover, outside the 75 Project, as a simple vehicle to manufacture: one body style, low parts count, built in identical batches of 30 vehicles. However, a comparison of the difference in work content between the highest and lowest standard specification models reveals that the time required to build the Rover 600 was on a par with that required for the older, and more complex, Rover 800. The difference in work content measures the impact of product/process variety upon manufacturing effectiveness moving away from an optimal process resulting in a degree of balance loss. Table 17.3 provides a comparison between the three models. Here we can see that the conventional thinking on the relationship between variety and operational effectiveness, as measured by MTOiC, is challenged. The Rover 75 which has fewer parts than either the 600 or the 800, is capable of being configured into nearly two billion combinations, yet the difference in work content between the highest and the lowest standard specification vehicles has been cut by nearly two thirds. However, the true difference between the three models is greater than the recorded difference in build times implies. The lowest and highest standard specification vehicles of the Rover 75 contain more features than either the Rover 800 or the Rover 600 and therefore there is more car to be built in less time, with fewer associates.

Table 17.3 The relationship between parts count, variety provisioning and work content

Model	Parts count	MTOiC	Ratio of MTOiC's/ part no.	Work difference[13]
Rover 600	2298	5520	2.402	60 minutes
Rover 800	3087	23926	7.751	60 minutes 4 door 90 minutes 5 door
Rover 75	2068	1918144000	927535.783	22.4 minutes

How might the reduction in the difference between build times be explained? The level of automation employed to produce all three models has remained roughly similar, although the Rover 75 uses more manual assistors to lift components; these do not automate the work. The difference cannot be attributed to manning levels as the headcount for the Rover 75 is around 5% lower than that required for the Rover 600. Furthermore the reduction in work content also cannot be attributed to the extensive use of outsourced modular assemblies.

---

[13] Difference between the highest and lowest standard specification vehicle.

Table 17. 4 Addressing the dimensions of complexity on the Rover 75

Dimensions of complexity	Impact of complexity	Actions to control complexity

Process/systems variety	Affects the number of different processes, the number of process stages, the sequence and dependency of these steps.	• Modular process designs to enable:
		  • Process independence
		  • Process postponement
		  • Process resequencing
		• Process standardisation

Product variety/product order sequence	Constraints on sequencing of the build, the procurement of parts and the flexibility of supplier operations.	• Modular product designs • Design for feature independence • Control of specific P times • Supplier flexibility
Parts variety	The number and type of parts required to produce various models.	• Design to control for parts/module complexity • Parts reduction

Whilst Design for Manufacture and Assembly principles were applied to component designs on the 75, other factors such as complexity reduction and Design for Logistics were equally addressed by the product engineers. Pugh (1990) warned of the dangers of design imbalances where design tools such as DFMA are used in isolation of a defined strategic intent, for without a guiding strategy to the design there is no way of assessing which suggestions for improvement support the overall business strategy (Whitney, 1988). In the case of the 75 the strategic intent was to develop the capability to flex vehicle supply in line with the lead-times and product specifications demanded by customers. Table 17.4 summarises three different dimensions of complexity, which product engineering sought to address. The product engineers needed to understand not only the impact of each of these dimensions of complexity upon manufacturing and logistics effectiveness but also the interdependency between each of the levels when designing components to minimise the negative impact of both product and process complexity.

Across the vehicle there are many examples of product design that have reduced not only product complexity, but the process complexity as well. One important example is the reduction in body types. The 75 and the 600 shared a common body style, a booted saloon, however the 600 had 17 different body types, whilst the 75 has reduced this to 4, right/left hand, sunroof and non-sunroof.

This reduction in product complexity has had significant knock-on effects in reducing the complexity associated with scheduling the 75. Another example is the engine mounting system. The Rover 600 experienced a number of problems, namely that the right- and left-hand engine mountings were fitted at different locations in trim and final and that the mounting brackets used different length bolts between each derivative. This meant that each body was typed by the engine derivative fitted. On the 75 there are common mounting points on the body in white, thereby avoiding the typing of bodies by engine derivative. Following the principles above, product engineering were able to align the vehicle design to support the market variety provisioning requirements with the manufacturing strategy of no build constraints.

An example of modular design: The HEVAC system

- Modular design approach adopted on the heater and air conditioning units (HEVAC) of the Rover 75.
- HEVAC systems can be configured to one of four different configurations.
- HEVAC systems designed with common piping regardless of configuration chosen.
- On the Rover 75 two associates fit the HEVAC system at one station.
- On the Rover 600 two stations were required to fit the heater (2 associates) and the air conditioning (2 associates). The heater had to be fitted before the facia but the air conditioning unit was fitted after the facia.
- On the Rover 600 loading constraints were imposed due to the different levels of work content between the two assembly stations.

17.7.2 Restructuring of the logistics processes

An investment of £10 million has been made in the development of a 240,000 ft^2 integrated logistics centre (ILC) at the Oxford facilities. EXEL Logistics provide the sequencing operations, although four suppliers late configure the headliners, front and rear bumpers, facia, and the console/gear surround in these facilities. All assemblies and components are then transported by tow truck direct to line from the ILC whilst other components such as the seats and the exhausts are supplied in sequence directly from locally based suppliers Table (17.5). The use of tow trucks for all transportation from the ILC has led to a reduction of 300 lorry movements per day around the Oxford site and the local vicinity.

Table 17.5 Materials requisitioning for the Rover 75

Material requisitioning	Components
DC sequenced message	• Parcel trays, drive shafts, cooling systems, front struts, steering wheels, door casings, carpets, heaters, and harnesses, etc.
Supplier late configuration within ILC	• Front and rear bumpers, facia, console/gear surround, and headliners.
Supplier sequenced message	• Seats, exhausts, etc.
Sequenced pallet replacement	• Engines, fuel tanks, wheel and tyre assemblies, etc.

Whilst the investment in the physical logistics infrastructure at Oxford is important, the more significant advances made on the R75 are in the sequencing of logistics processes to satisfy the shortest possible order cycle of ten working days. The logistics planning process implemented on the Rover 75 has instilled a high degree of control but not at the expense of flexibility. The significant change in logistics planning is the shift in time frames. Previously on the Rover 600 the business cycle operated on a monthly/weekly basis whereby the manufacturing schedules were fixed between three to four months before production, order submissions were made weekly and there was a weekly 'call-off' cycle. The inflexibility of the whole scheduling system, aligned with the long time frames, allowed processes to be managed in a sequential manner with each function undertaking their respective tasks before passing information onto the next. On the 75 the time frames have moved to a daily/hourly basis thereby requiring close interaction between the logistics and manufacturing functions to concurrently generate and jointly manage the process steps required to create the manufacturing schedules.

The level of co-ordination required is reflected in the co-location of all the personnel responsible for these activities into what is known as the 'hub' (Figure 17.4). Supply chain flexibility is such that up to four days before production is due to begin the specifications on most major options, such as air-conditioning, can be changed to meet customer demands. The establishment of the hub has meant that each member of staff has to not only understand their own role in the planning process but also to understand the roles and tasks of their colleagues in light of the flexibility offered by supply chain and manufacturing responsiveness.

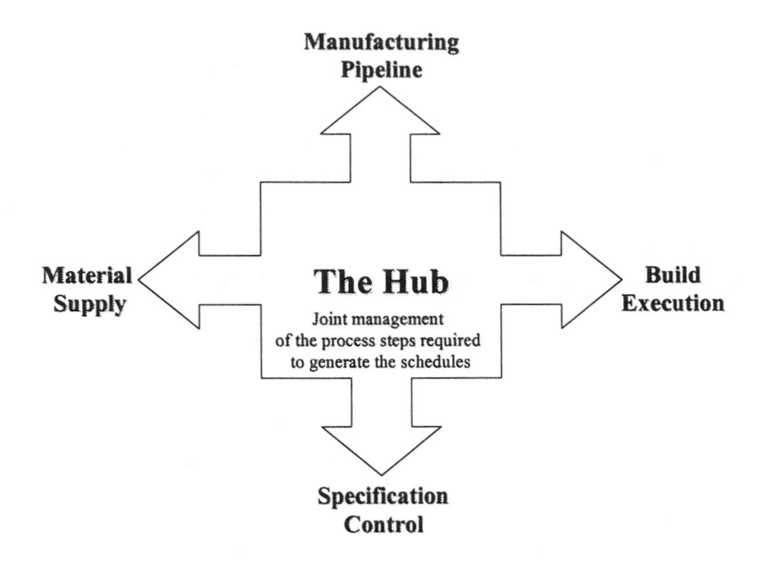

Figure 17.4 The 'hub' organisation.

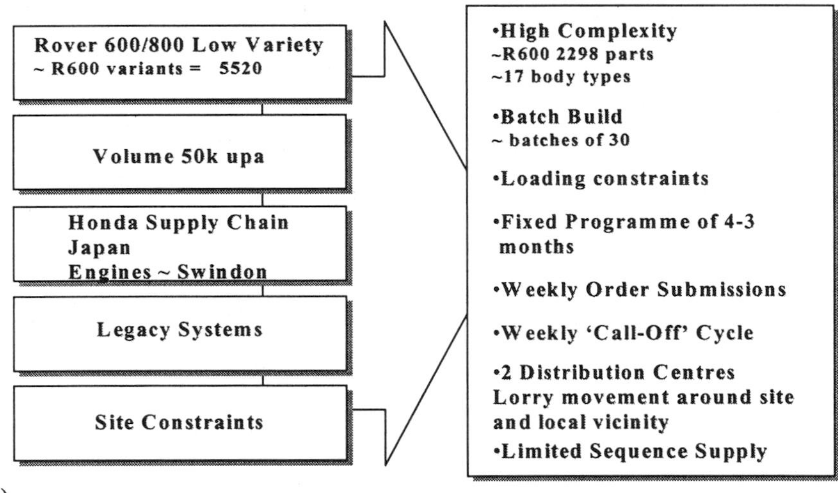

(a)

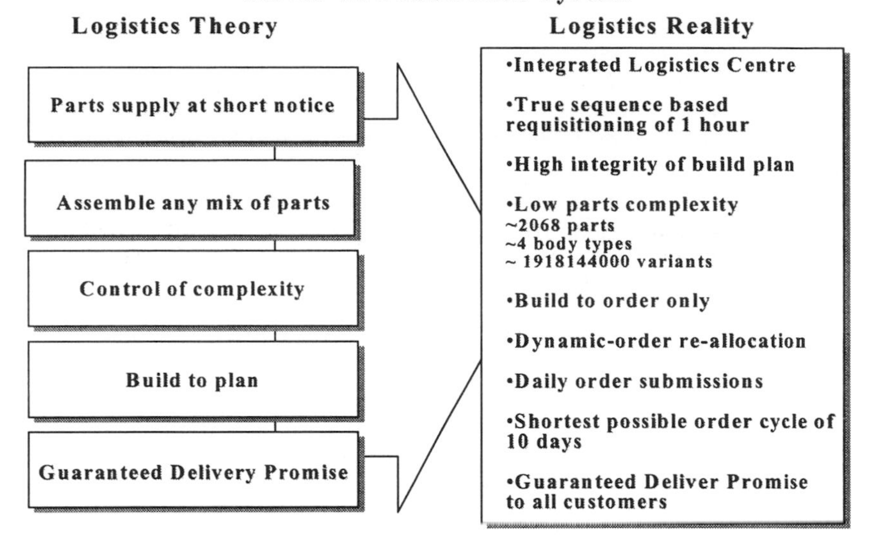

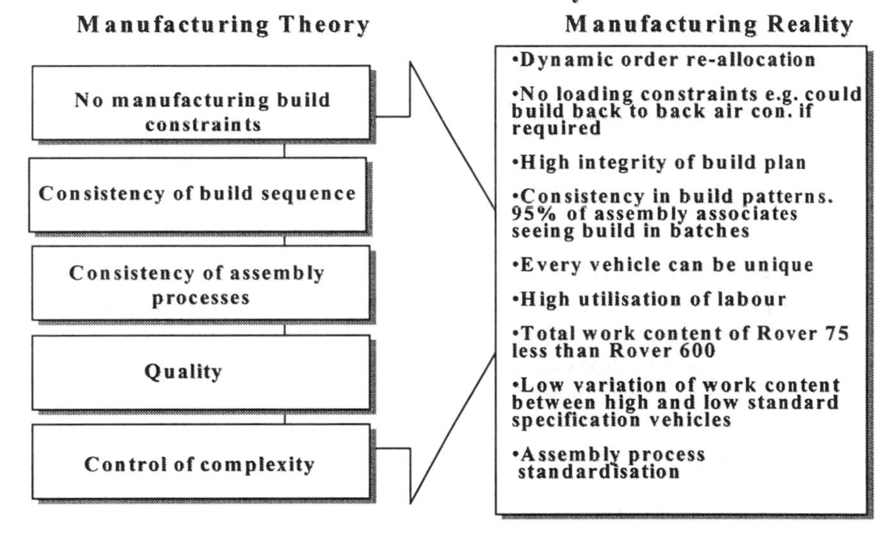

(b)

Figure 17.5 The Rover 600 production system (a) compared with the Rover 75 production system (b).

The streamlining of the order-fulfilment process for the 75 has therefore been enabled by a significant reduction in many of the manufacturing constraints associated with the late configuration of a vehicle with levels of variety akin to that of the BMW 3 and 5 series models. Ernst Baumann of BMW highlighted the benefits to Rover as follows: 'the customer will only have to decide on the vehicle's final specification a short time before production begins. This will result in greater sales potential, while avoiding large stockpiles' (Autocar, 1999b:17). Figure 17.5 summarises the main features of the Rover 600 production system, the Rover 75 production system and the outcomes achieved by manufacturing and logistics.

17.8 CONCLUSIONS

Automotive retailing is on the verge of a dramatic revolution. Within the next two to three years an increasing proportion of customers will purchase their vehicles via the Internet and this will have a profound impact upon manufacturers' order-fulfilment processes. Use of the Internet will speed up the whole buying process for vehicles already in stock, but it will also open up the possibility of creating on-line build-to-order systems directly linked into manufacturers' production schedules. The impact upon manufacturers will be twofold. Firstly, a far higher degree of their production will be exposed to fluctuating customer demand requiring that logistics and manufacturing systems are responsive and can accommodate such shifts with minimal constraints upon vehicle supply. Secondly, there will need to be a significant reduction in the time taken by the whole logistics planning cycle. Product variety will have a negative effect upon both of these factors unless vehicle designs, manufacturing operations and supply chains are engineered to accommodate the level of responsiveness required to satisfy customer demands.

The approach taken by the Rover 75 Project has resulted in the development of the combined capacity to flex vehicle supply (not total capacity) by derivative, in line with customer demand, and to have the capability to do this in the shortest possible order cycle of ten working days. These requirements were core to the vehicle concept and their integration into the early design stage although the concurrent approach adopted by product engineering, manufacturing and logistics has contributed towards the integrity of the Rover 75-production system. A careful balance between commonality and distinctiveness needs to be maintained and the experience of the 75 illustrates that significant inroads into minimising the negative impact of product and process complexity associated with product variety can be achieved through product design without undermining the integrity of the vehicle design.

17.9 ENDNOTE: THE SALE OF ROVER BY BMW

The Rover 75 was launched at a time of significant uncertainty over the future of the Rover Brand. Between February 1999 and February 2000 the Rover 75 picked up over a dozen international motoring awards, including the prestigious WhatCar? Car of the Year award for 1999 in the UK as well as being short-listed for the 1999 European Car of the Year Award.

Nine months after the launch of the Rover 75, on the 16th March 2000, the BMW board decided to sell off most of its UK based operations in light of mounting losses.

In reflecting on these developments it is worth considering the history of Volvo's Uddevalla plant, which made significant advances in implementing a build-to-order environment in the early 1990s. As a consequence of adverse market conditions, the innovations made by Volvo never had the opportunity to fulfil their potential. 'The Machine that Changed the World' (Womack *et al.*, 1990) pilloried the Uddevalla plant for engaging in Neo-craft production when compared to the productivity levels of the leading Japanese plants. However the MIT study only examined plant productivity, a very small part of the production process when compared to the overall effectiveness of the whole value chain. Yet within two years of the publication of the MIT study Volvo, at Uddevalla, was working towards a lean distribution 'customer pull' system. Volvo had controversially chosen to abandon the traditional serial flow assembly line in favour of group work production. With none of the line balancing and sequencing problems associated with serial flow lines, Uddevalla was able to accommodate a high level of building to customer order with high levels of product variety in short time-to-customer lead-times (Berggren, 1995). Adopting a different approach to that of Volvo, the Rover 75 was also designed to implement a customer pull, build-to-order, system by overcoming many of the constraints imposed by a traditional assembly line. This approach looked at the efficiency of the entire value chain and was designed to optimise the whole delivery mechanism, not just one element such as plant productivity. It is to be hoped that, after a turbulent birth, the innovative logistics systems put in place for the Rover 75 to facilitate high levels of customization will result in successful and profitable operations that meet the demands of increasingly demanding consumers.

17.10 ACKNOWLEDGEMENTS

The author wishes to thank those who participated in the research. The views expressed within this chapter are those of the author and the author assumes responsibility for any errors.

17.11 REFERENCES

Autocar, 1999a, 19 May, p. 13.

Autocar, 1999b, 17 February, pp. 17-18.

Autocar, 2000, 23 February, p. 23.

Automotive Engineering, 1997, VW integrates suppliers at Brazil, Czech Republic plants, October, p. 100.

Automotive Manufacturing and Production, 1997, Marketing changes will affect manufacturing, August, pp. 22-24.

Baldwin, C.Y. and Clark, K.B., 1997, Managing in an age of modularity, *Harvard Business Review*, September/October, pp. 84-93.

Batchelor, J. and Schmidt, S., 2000, Just-in-time - das missverständnis eines jahrzehnts, *Frankfurter Allgemeine Zeitung*, 10th January, p. 25.

Bennett, D. and Forrester, P., 1993, Product variety and just-in-time: conflict and challenge. In Pawar, K.S. (ed), *Proceedings of the 1st International Symposium on Logistics*, Nottingham, 1993, pp 67-75.

Berggren, C., 1995, The fate of the branch plants - performance versus power. In Åke Sandberg (ed.), *Enriching Production*, Avebury, Aldershot, pp. 105-126.

Calderini, M. and Cantamessa, M., 1998, Discussing the constitutive elements of product design capability. In Bartezzaghi, E. and Verganti, R. (eds), *Proceedings of the 5th International Product Development Management Conference*, 25-26 May, Como, Italy, Politecnico di Milano, pp. 171-185.

Clark, K.B. and Fujimoto, T., 1990, The power of product integrity, *Harvard Business Review*, November/December, pp. 107-119.

Cusumano, M.A. and Nobeoka, K., 1998, *Thinking Beyond Lean*, The Free Press, New York.

Financial Times Automotive World, 1999, Cutting edge platforms, pp. 30-38.

Fisher, M., 1997, What is the right supply chain for your product?, *Harvard Business Review*, March/April, pp. 105-116.

Fisher, M., Jain, A. and MacDuffie, J.P., 1994, *Strategies for Product Variety: Lessons from the auto industry*, International Motor Vehicle Programme Research Briefing Meeting, June.

Fisher, M., Ramdas, K. and Ulrich, K., 1997, Component sharing in the management of product variety, *Working Paper 96-10-03*, Department of Operations and Information Management, The Wharton School, University of Pennsylvania.

Fujimoto, T., 1997, Capability building and over adaptation - A case of 'fat design' in the Japanese auto industry, *Actes du GERPISA*, 19 (February), pp. 9-23.

Galsworth, G.D., 1994, *Smart, Simple Design*, Omneo, Essex Junction, Vermont.

Ho, T. and Tang, C.S., 1998, *Product Variety Management Research Advances*, Kluwer Academic Publishers, Boston.

Hsieh, L.H., Schmahls, T. and Seliger, G., 1997, Assembly automation in Europe - Past experience and future trends. In Shimokawa, K., Jügens, U. and Fujimoto, T. (eds.), *Transforming Automobile Assembly*, Springer, Berlin, pp. 19-37.

Jügens, U., Fujimoto, T. and Shimokawa, K., 1997, Conclusions and outlook. In Shimokawa, K., Jügens, U. and Fujimoto, T. (eds.), *Transforming Automobile Assembly*, Springer, Berlin, pp. 395-407.

Kiff, J.S., 1997, Supply and stocking systems in the UK car market, *International Journal of Physical Distribution and Logistics Management*, Vol. 27, No. 3/4, pp. 226-243.

KPMG, 1998, *Europe on the Move*, KPMG Automotive Practice, Birmingham, UK.

Lancaster, K., 1990, The economics of product variety: A survey, *Marketing Science*, Vol. 9, No. 3, pp. 198-206.

MacDuffie, J.P., Sethuranman, K. and Fisher, M.L., 1993, *Product Variety and Manufacturing Performance: Evidence from the international automotive assembly plant study*, International Motor Vehicle Programme Research Briefing Meeting, June, MIT, Massachusetts.

MacDuffie, J.P., Sethuranman, K. and Fisher, M.L., 1996, Product variety and manufacturing performance: Evidence from the international automotive assembly plant study, *Management Science*, Vol. 42, No. 3, pp. 350-369.

Mair, A., 1994a, *Honda's Global Local Corporation*, St. Martin's Press, New York.

Mair, A., 1994b, Honda's global flexifactory network, *International Journal of Operations and Production Management*, Vol. 14, No. 3, pp. 6-23.

Mather, H., 1988, *Competitive Manufacturing*, Prentice Hall, Englewood Cliffs, New Jersey.

McCutcheon, D.M., Raturi, A.S. and Meredith, J.R., 1994, The customization-responsiveness squeeze, *Sloan Management Review*, Winter, pp. 89-99.

Meyer, M.H. and Utterback, J.M., 1993, The product family and the dynamics of core capability, *Sloan Management Review*, Vol. 34, No. 3, pp. 29-41.

Muffatto, M., 1996, Reorganizing for product development: Honda's case, *International Journal of Vehicle Design*, Vol. 17, No. 2, pp. 109-124.

Muffatto, M., Roveda, M. and Verbano, C., 1999, Outbound logistics and information technology: A case study in the automotive industry. In Muffatto, M. and Pawar, K.S. (eds), *Proceedings of the 4th International Symposium on Logistics*, Florence, 1999, SGE, Padova, pp. 695-702.

Pilkington, A., 1996a, *Transforming Rover. Renewal Against the Odds 1981 to 1994*, Bristol Academic Press, Bristol.

Pilkington, A., 1996b, Learning from joint venture: the Rover-Honda relationship, *Business History*, Vol. 38, No. 1, pp. 90-115.

Pine, J.B., Victor, B. and Boynton, A.C., 1993, Making mass customization work, *Harvard Business Review*, September/October, pp. 108-118.

Pugh, S., 1990, *Total Design*, Addison-Wesley Publishing Company, Wokingham, England.

Robertson, D. and Ulrich, K., 1998, Planning for product platforms, *Sloan Management Review*, Summer, pp. 19-31.

Sanchez, R. and Mahoney, J.T., 1996, Modularity, flexibility, and knowledge management in product and organization design, *Strategic Management Journal*, Vol. 17, Winter Special Issue, pp. 63-76.

Skinner, W., 1974, The focused factory, *Harvard Business Review*, May/June, pp. 113-121.

Slack, N., 1988, Manufacturing systems flexibility - An assessment procedure, *Computer-Integrated Manufacturing Systems*, Vol. 1, No. 1, pp. 25-31.

Slack, N., 1991, *The Manufacturing Advantage*, Mercury, London.

Slack, N., Chambers, S., Harland, C., Harrison, A. and Johnston, R., 1998, *Operations Management*, Financial Times Pitman Publishing, London.

Stalk, G. and Webber, A.M., 1993, Japan's dark side of time, *Harvard Business Review*, July-August, pp. 93-102.

Sunday Times, 1997, Ford cuts Fiesta's 27m varieties, 18th January.

Ulrich, K, 1995, The role of product architecture in the manufacturing firm, *Research Policy*, Vol. 24, pp. 419-440.

Upton, D.M., 1994, The management of manufacturing flexibility, *California Management Review*, Winter, pp. 72-85.

Whitney, D.E., 1988, Manufacturing by design, *Harvard Business Review*, July-August, pp. 83-92.

Wilhelm, B., 1997, Platform and modular concepts at Volkswagen - their effects on the assembly process. In Shimokawa, K., Jügens, U. and Fujimoto, T. (eds), *Transforming Automobile Assembly*, Springer, Berlin, pp. 146-156.

Womack, J.P., Jones, D.T. and Roos D., 1990, *The Machine that Changed the World*, Rawson Associates, New York.

CHAPTER EIGHTEEN

A Socio-technical Approach to the Design of a Production Control System: Towards Controllable Production Units

Jannes Slomp and Gwenny C. Ruël

18.1 INTRODUCTION

Many companies are facing tighter market demands regarding the price, quality, variety and delivery time of their products. They will have to arrange the production organization so that these demands can be met. This often implies a restructuring of the means of production or a change in the structure and system of production. Flexible, automated means of production are purchased and thus arranged to allow an efficient and effective product flow. Many firms have also decided to apply cellular manufacturing, or team production, to be competitive in the market place. Usually production control, too, needs to be adapted. There will be growing pressure to balance sales and production and to use the means of production and the workforce as efficiently and effectively as possible. The contribution of this chapter is to explore the application of concepts from socio-technical systems design to the production control domain and demonstrate its applicability with a case study. Brown *et al.* (1988, pp. 266-267) mention the need to apply socio-technical principles in the design of a production control system. In their opinion, '*the relative failure of many "production management systems"* (here: production control systems) *can be explained, at least partially, in terms of the lack of a true socio-technical approach to the design and installation of these systems'. (p.266).* They criticize the overemphasis on the technical aspects of production control systems and argue that disappointments arise because of failure to give regard to the social aspect system. Hyer *et al.* (1999) present a case study illustrating 'a socio-technical systems approach to cell design'. Part of the cell design, as they present it, concerns the determination of production planning and activity control procedures. An important socio-technical aspect of this part of the cell design is, in their case study, the fact that cell operators were assumed responsible for material ordering, job tracking, and scheduling. This ensured a certain level of autonomy for each manufacturing cell. The decentralization of control tasks required user-friendly information systems and training of the operators to use the new simplified systems. A material council (with representatives from each cell and production planning) was made responsible for the development of information flow procedures across the cells.

Van Eijnatten and Van der Zwaan (1998) present the current Dutch socio-technical design approach to integral organizational renewal. Part of this approach is the design of a control structure, including production control. According to Van Eijnatten and Van der Zwaan, an important concept in the socio-technical design of a production control system is *the control loop in which all different control aspects merge.* Closed loops within organizational units support the autonomy of groups. This can be seen as a plea to give workers the production planning and control responsibilities needed to deal with the variances in their work, such as the absenteeism of colleagues and machine breakdowns. Although Brown *et al.* (1988), Hyer *et al.* (1999), and Van Eijnatten and Van der Zwaan (1998) stress the importance of a socio-technical systems approach for the design of a production control system, they do not explore the application of the various concepts of socio-technical systems design to the production control area. This chapter is meant to fill that gap. (For a contrasting view on socio-technical aspects of planning, scheduling and control, see Chapter 19 by Wäfler in this book).

Section 18.2 will describe the basic philosophy of socio-technical systems design. Section 18.3 gives a brief explanation of the basic elements of a production control system. Next, Section 18.4 makes a link between socio-technical principles and the (re)design of a production control system. Section 18.5 presents a redesign approach for a production control system, which offers a framework for the integration of the relevant socio-technical principles. Section 18.6 concerns a brief description of the firm for which the production control system was redesigned. The case described in Section 18.6 serves as an illustration of the various elements, or steps, in the (re)design approach. Sections 18.7 to 18.10 explain the various steps in more detail and apply them to the case. These sections also explain how the socio-technical principles are integrated in the redesign approach. Finally, Section 18.11 presents a résumé.

18.2 WHAT IS SOCIO-TECHNICAL DESIGN?

An important approach which supports the integration of human factors in industrial settings is the so-called socio-technical systems (STS) approach. The term 'socio-technical systems' originates from Trist and Bamforth (1951), at the Tavistock Institute. On the basis of an empirical case, they describe the importance of finding a 'joint optimization' between the technical and social systems of an organization, even if this leads to suboptimal conditions for the systems individually (Emery and Trist, 1972; Herbst, 1974; Cherns, 1976). The technical system concerns the technical-economical aspects and the social system refers to all of the social aspects of the functioning of an organization. Both sub-systems have an impact on the performance of a firm and must be optimized simultaneously in an organizational renewal process. An improvement in one system, without considering the effect on the other system, may deteriorate the overall performance of the organization.

The basic premise of the socio-technical systems theory is the principle of 'organizational choice' (Trist *et al.*, 1963; Hage, 1977), which means that technology does not necessarily determine the organizational arrangement of human tasks (known as the technological imperative), but that it still leaves design-space. In the socio-technical viewpoint, the organizational arrangements determine the fit between the technical and the social systems. In conformity with these thoughts, socio-technical design approaches are basically focused on the (re)design of the organizational structure.

Another essential element in STS is the notion of organizations as open systems. Open system theory states that entities and situations outside the organization can affect what happens within it, and the organization, in turn, can influence what happens in and around it (e.g. Child, 1972). Based upon the open system approach, one may define socio-technical redesign as the process of rearranging tasks and responsibilities in order to create a new 'input-transformation-output' situation. In a closed system approach, organizational redesign is only focused on the transformation process itself.

The basic starting point of the socio-technical approach is the conviction that the existence of a controllable situation is of importance for the survival of an organization. The socio-technical insight on the matter of controllability is built upon 'the law of requisite variety' of Ashby (1969), which states that variety can only be controlled by variety. The translation of this law into STS terms implies that in order to realize a controllable situation the number of control measures inside an organizational unit should be at least as large as the number of variations (from inside or outside) which affect that particular organizational unit. Or as Weick (1979) puts it: 'only variety can regulate variety'. Based upon this law, the socio-technical approach distinguishes two design strategies. In the first strategy, socio-technical (re)design attempts to reduce the number of variations. This can be done, for instance, by subdividing the organizational unit into smaller units (in terms of machines, equipment, workers, etc.), each responsible for a particular family of products. Each family unit is confronted with only a certain segment of the environment, and consequently with less variety. In the second strategy, socio-technical (re)design attempts to add (or decentralize) control tasks to an organizational unit which is facing variety. This increases the number of measures that can be taken in the organizational unit. Some socio-technical literature (especially the Dutch Socio-technical Approach - see Kuipers and Van Amelsvoort, 1990 or Van Eijnatten and Van der Zwaan, 1998) suggests a certain logical sequence of the two design strategies. First, the production structure should be redesigned so that teams, or cells, are created that are relatively independent from each other with respect to their primary tasks. This corresponds to the first design strategy of reducing the internal and environmental variety. Next, the control structure has to be redesigned in such a way that the teams, or cells, are able to deal with the variety. As will be seen in this chapter, the logical sequence of the design strategies is debatable.

The socio-technical systems approach is important for the design of a production control system because such a system involves technical and social systems. The technical system can be regarded as the set of abstract models of planning, scheduling and control as presented in Production and Operations Management literature. Also software tools and information systems can be seen as part of the technical system. Such models, tools and information systems, however, do not usually incorporate the human aspect of production control, such as the division of decision tasks and the social and psychological characteristics of the people who will be made responsible for planning and control tasks. The socio-technical systems approach advocates a strong focus on the organizational choices in the design of a production control system. These choices concern the design and division of production control tasks among employees. As will be made clear later on in this chapter, such a focus on organizational choices may have an impact on the design of the technical system. The socio-technical approach assumes that its particular focus will create the best fit between the technical and social systems of a production control system. Section 18.4 will further specify the socio-technical guidelines which may be useful in the design of a production control system.

18.3 WHAT IS A PRODUCTION CONTROL SYSTEM?

This section gives a brief overview of the elements and aspects of a production control system. Firstly, a more general introduction is given in which major production control concepts are explained. A global hierarchical structure of a production control concept is presented which encompasses the major elements in each production control concept. Secondly, a socio-technical view on production control is given by presenting the major aspects of a production control system which have to be considered in a (re)design.

A production control system is a major part of the control structure of a firm and is responsible for the planning, scheduling and control of the activities. It usually has a hierarchical character. Several arguments can be given to justify a hierarchical approach to production control problems:

- *Reduction of complexity.* Production control problems can generally be characterized by the presence of multiple, sometimes contradictory, objectives and a number of complicating and to a certain extent conflicting constraints. A hierarchical approach offers the possibility of splitting up complex interrelated production control problems into several small solvable parts.
- *Separation of short-, medium-, and long-term aspects.* Production control problems on a long-term level are generally more strategic in nature than medium-term and short-term problems and therefore demand different solution methods.
- *Improving stability and controllability.* Production control problems may arise at regular and/or irregular intervals. Without a hierarchical decision structure all the (interrelated) production control problems are affected by any disturbance. A hierarchical approach offers the possibility of solving problems on one level without the need to replan on higher levels. This improves the stability of the production control decisions and severely reduces the amount of information required.

Figure 18.1 presents a global hierarchical structure of a production control decision system. Bertrand *et al.* (1990) describe the elements of the structure as follows:

'The aggregate production planning level forms the connection with the higher levels of control in the production organization. At this level integration takes place of the various control aspects of the organization (sales, logistics, quality, finance, personnel, etc.) In a situation with standard end-items the outputs of the aggregate production planning process are the aggregate delivery plan, the capacity use plan, the capacity adjustment plan, and the aggregate inventory plan. These four plans are the driving force for short term capacity control and material control. In material co-ordination, priorities are determined for the release of work orders. These priorities are based upon detailed demand information (sales) and on the work order throughput times of the various stages in production. The detailed demand information used in material co-ordination is directed by the aggregate delivery plan (part of aggregate production planning). Actual work order releases are determined on the one hand by the priorities given by material co-ordination and on the other hand by the release possibilities from the aggregate release pattern (output of the workload control function). The aggregate release pattern is determined by the capacity use plan of aggregate production planning and by controlling the workload of the production units (or work floor). The release possibilities can be further restricted by other (finer) operational constraints of the production units and by material availability. The process that determines the releases is called work order release. The process that determines the aggregate release pattern is called workload control. Inputs for workload control and work order release are the capacity use plan of the aggregate production plan and the work order priorities from material co-ordination'. (Bertrand *et al.*, 1990, p. 56).

Production unit control (PU-control) concerns the production control decisions (who, where and when decisions) on the work floor. The functional elements presented in Figure 18.1 have to be specified in the design of a production control system.

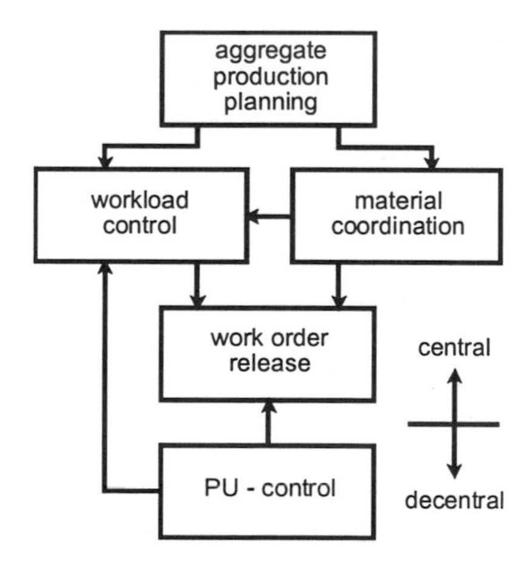

Figure 18.1 Global hierarchical structure of a production control concept (Bertrand *et al.* 1990).

Vollmann *et al.* (1992) give an overview of the main production planning and control concepts used in industry. All these concepts can be related to the hierarchical framework of Figure 18.1. The Material Requirements Planning (MRP) concept is a job-oriented concept that translates the overall plans for production into the detailed individual steps necessary to accomplish the plans. The elements of Figure 18.1 can be seen as separate functions in the MRP concept. The Just-in-Time (JIT) concept is oriented towards a smooth flow of materials through the firm in such a way that customer demands can be met without controlling the progress on the shop floor. In terms of Figure 18.1, JIT streamlines the execution on the shop floor and, by doing so, simplifies the PU-control (e.g. KANBAN-system). Workload control and material co-ordination are integrated into the aggregate production plans. The basic principle of the production control concept of Optimized Production Technology (OPT) is that bottleneck operations are of critical scheduling concern. OPT calculates different batch sizes throughout the plant, depending on whether a work center is a bottleneck. In terms of Figure 18.1, OPT attempts to combine the material co-ordination and workload control function by means of a bottleneck approach. Hierarchical Production Planning (HPP) is a planning concept which starts from the information of an aggregate capacity analysis. Disaggregation of this information provides the required information on lower levels of the production control hierarchy. In terms of Figure 18.1, HPP performs the workload control and material co-ordination function at various levels of a production control hierarchy, applying different levels of aggregation. Bitran *et al.* (1982), who developed the concept of HPP, stress the need to match product aggregations in HPP to decision-making levels in the organization. Disaggregation should follow organizational lines. This statement comes close to a more socio-technical definition of a production control system as is presented in the remainder of this section. Larsen and Alting (1993) indicate that the various production planning and control philosophies are usable in different, sometimes overlapping, industries. MRP is meant for complex multi-level batch manufacturing and assembly, JIT, as a production control mechanism, can be used in high volume production and assembly of simple products. OPT is developed for multi-level batch and process flow manufacturing and assembly of high volume and high complexity products. HPP is basically developed for a low volume, high variety process industry. Larsen and Alting (1993) also present the concept of Distributed Production Planning (DPP). This concept proceeds from the idea that information technology enables the move from a centralized concept, such as MRP and OPT, to a decentralized system in which each production unit has to control its own materials flow. Within the DPP approach, concepts like MRP, OPT and JIT may support the production control in the various production units. As will be seen later in this chapter, the socio-technical design approach to production planning, scheduling and control can be seen as a particular elaboration of the DPP approach.

From a socio-technical viewpoint, a production control system can be regarded as consisting of three sub-systems: i) a decision hierarchy, ii) an organization hierarchy and iii) an information system and decision support tools (see Figure 18.2). The decision hierarchy of a production control system concerns the arrangement of decision tasks. Each level of the decision hierarchy reduces the complexity but also limits the decision space of lower levels. Figure 18.1 can be seen as a global structure of the production control decision hierarchy. The decision tasks, described in the decision hierarchy, have to be fulfilled by one or more levels of the organization hierarchy of a firm. Each level of the organization hierarchy reduces the scope of responsibility for subsequent lower levels. The decision hierarchy and organization hierarchy determine the framework in which the information system and the decision support tools should fit appropriately. A well-working information system is essential for a successful implementation of a decision and organization hierarchy. To enable the decision tasks at the various organizational levels, decision support tools may be needed. The three aspects of a production control system are strongly interrelated. The need for computerized decision support tools, for instance, depends on the chosen decision hierarchy. Furthermore, the qualification and experience of the people responsible for certain decision tasks may determine the desired abilities of the decision support tools. The decision hierarchy, the information system and decision support tools can be seen as the technical system in the socio-technical system approach. The design of the organization hierarchy encompasses, in socio-technical philosophy, the aspects which determine the fit between the technical and the social systems.

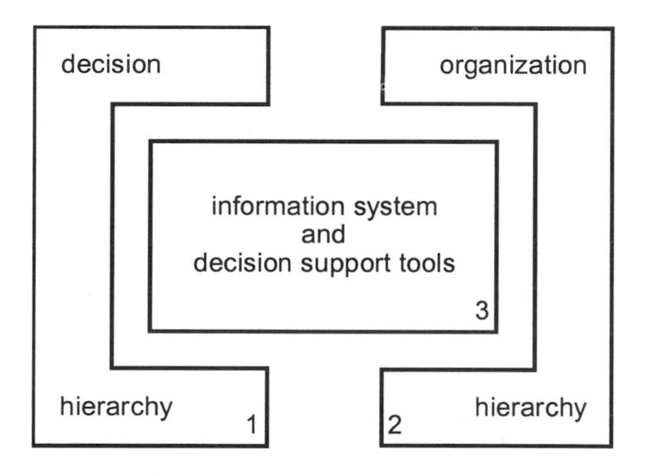

Figure 18.2 Aspects of a production control system.

18.4 SOCIO-TECHNICAL DESIGN PRINCIPLES AND GUIDELINES

A key term which may cover the original meaning of the socio-technical idea is 'self-organization'. The open system approach and the focus on gaining a controllable situation points in the direction of the development of self-organization. Self-organization has to be realized through the design of a production structure as well as the design of a control structure. Self-organization has to be seen as a means to realize the objectives of an organization and its workers. Each situation requires its own degree of self-organization. The more complex tasks are, and the more variety there is to deal with, the higher the required level of self-organization. Self-organization, furthermore, refers to the result of the design process as well as to the design process itself (see e.g. Cherns, 1987).

Table 18.1. Socio-technical design principles

Compatibility	The way in which design is done should be compatible with the design's objective.
Minimal critical specification	No more should be specified than is absolutely essential. What is essential should be specified.
Variance control	Variances should not be exported across unit, departmental, or other organizational boundaries.
Boundary location	Boundaries should not be drawn so as to impede the sharing of information, knowledge, and learning.
Information flow	Information for action should be directed first to those whose task it is to act.
Power and authority	Those who need equipment, materials, or other resources to carry out their responsibilities should have access to them and authority to command them.
The multifunctional principle	If the environmental demands vary, it then becomes more adaptive and less wasteful for each element to possess more than one function.
Support congruence	Systems of social support (systems of selection, training, conflict resolution, work measurement, etc.) should be designed so as to reinforce the behaviors which the organization structure is designed to elicit.
Transitional organization	The design team and its process should be seen as a vehicle of transition.
Incompletion	Design is a reiterative process. The closure of options opens new ones. At the end we are back at the beginning.

A well-known list of socio-technical design principles, which supports the idea of self-organization, is given by Cherns (1976, 1987). Table 18.1 gives a brief summary of the principles. Several authors have used the list of Cherns as a starting point for analysis. Huber and Brown (1991) use Cherns' list to recover human resource issues in cellular manufacturing. Hyer *et al.* (1999) apply the principles of Cherns in the design of manufacturing cells.

The socio-technical principles can be applied in the design process of a production control system, as will be shown in the remaining part of this chapter. The sequence of the principles in the list, however, is somewhat random. Some of the principles refer to the (re)design process, some principles concern the characteristics of an ideal design, and, finally, some principles refer to the environment of the ultimate design. The next three subsections reorder the principles. Subsection 18.4.1 transforms the principles into procedural guidelines, i.e., guidelines which are helpful in the process of gaining a planning, scheduling and control system. Subsection 18.4.2 presents guidelines by which the ultimate design of a production control system can be evaluated. Subsection 18.4.3 contains the guidelines for obtaining a good fit between the design of a production control system and its environment. Subsection 18.4.4 summarizes all of the guidelines.

18.4.1 Procedural guidelines

As mentioned before, the application of socio-technical principles supports the idea of self-organization. An important question is *how* to gain an organization, or a production control system, which is based upon the idea of self-organization. Three design principles mentioned by Cherns (1987) may be helpful in answering this question: compatibility, minimal critical specification, and transitional organization.

Compatibility. Cherns argues that the process of design must be compatible with its objectives, saying: 'If the objective of design is a system capable of self-modification, of adapting to change, and of making the most use of the creative capacities of the individual, then a constructively participative organization is needed. A necessary condition for this to occur is that people are given the opportunity to participate in the design of the jobs they are to perform.'

Application of this principle on the subject of this chapter means that the planning and control system should be developed and implemented in participation with the people who will have a task in the system. Generally speaking, participation will improve the quality and the acceptability of a new planning, scheduling and control system. The last statement, however, is relative. Research in the area of Management Information Systems has shown that user participation in the development of a system is less important for highly structured and well-defined systems. Participation in the development is critical when information required to design the system can only be obtained from the users, or if the system causes significant changes to the jobs of employees (Ives and Olson, 1984). Barki and Huff (1990) have studied the necessity of participation in the implementation of Decision Support Systems. Their survey showed a strong correlation between user participation and the success of the support system (i.e., user satisfaction and system use). However, it may be clear that participation has to be regulated in order to avoid a hotchpotch of opinions.

Minimal critical specification. As Cherns states, this principle has two aspects, a negative and a positive one. The negative aspect simply states that no more should be specified than is absolutely essential; the positive aspect requires that we identify what is essential. The two aspects can be applied in the process of gaining a production control system. The negative aspect demands that design problems should not be formulated too tightly. A problem of a limited availability of cutting tools, which may cause serious scheduling problems, for instance, may be solved by purchasing extra cutting tools, or by implementing an intelligent planning or scheduling procedure. If these two possibilities are present, then the designer of a planning and scheduling system (or the design group) should not exclude one. The positive aspect of the principle of minimal critical specification requires that the designer identifies the essential, or critical, design problems or constraints. If, for instance, the firm is not willing to invest in new machines in order to simplify the production control, then this fact has to be accepted as an important design constraint. These examples also illustrate some links between the design of a production system and the design of a production control system.

Transitional organization. Cherns (1987) states that 'the design team and its process need to be seen as a vehicle of transition'. It is not unusual in design practice to distinguish a design and an implementation problem. The designer (or design team) is responsible for the first problem, the client (or firm) has to deal with the second one. Cherns stresses the need to close the gap between design and implementation. As a consequence of the principle of 'transitional organization', one should give the designer (or design team, including the participating employees) the responsibility for the design as well as for the implementation of the new production control system. The division between design and implementation is also criticized by Ackoff (1979). A solution for a practical problem which in the event is not implemented, is not a solution. Design and implementation are two connected activities. Implementation will, almost inevitably, ask for redesign, and redesign asks for implementation.

According to the socio-technical systems theory, the design principles mentioned above have to be used in the process of designing and implementing a production control system. Not using the principles may be the reason for a failure.

18.4.2 Design guidelines

A production control system can be described by its decision hierarchy, organization hierarchy, the information system and the decision support tools. Socio-technical design principles may be helpful in taking those design decisions that optimize the self-organization on all levels of the production control hierarchy. The following principles seem to be applicable: minimal critical specification, variance control, boundary location, the principle concerning the information flow, power and authority, the multifunctional principle, and the incompletion principle.

Minimal critical specification. As mentioned above, this principle has a negative and a positive aspect. Both aspects can be applied in the design of a production control system. The negative aspect states that on each hierarchical level no more should be specified than is absolutely essential. In this way flexibility is left for subsequent levels of the hierarchy. The positive aspect states that each hierarchical level should give the necessary directives which enables an optimal functioning of a manufacturing system. In a bad design of the decision-making hierarchy it would be possible to frustrate the decision-making tasks at certain levels by inaccurate solutions derived at previous levels. The principle of minimal critical specification is, to a certain extent, a plea for decentralization of production control tasks. Several authors support this plea. Child (1984) stresses the need to increase human abilities and to decentralize decision-making in order to react appropriately to disturbances of the production process and changes in the market. Furthermore, decentralization offers the possibility to activate hidden human abilities. This is likely to improve the quality of labor (Kuipers and Van Amelsvoort, 1990). Another aspect of the principle of minimal critical specification concerns the question of what has to be established for each level of the production control structure. Cherns (1987) distinguishes objectives and methods, and discusses the overall need of establishing methods: 'While it may be necessary to be quite precise about what has to be done, it is rarely necessary to be precise about how it is done'. This statement may depress the sometimes irresistible challenge of designing methods which optimize a certain objective.

Variance control. Cherns states that variances should not be exported across unit, departmental, or other organizational boundaries. This means that each organizational level in a production control system should be able to cope with the variances that may arise at that level. In other words, decision-making tasks (levels) should reflect the variances that may arise at the organizational level.

Boundary location. This principle says that boundaries should not be drawn so as to impede the sharing of information, knowledge, and learning. The principle contributes to the considerations with respect to the assignment of decision tasks and responsibilities to the organizational levels. Principally, all levels should contribute to the overall objectives of the production control system. Because of the complexity of the planning and scheduling functions, however, each of the levels of the decision and organization hierarchy may have its own objectives. It is required that these objectives are tuned to one another. A system of co-ordination (= sharing of information, knowledge, and learning) is needed to avoid sub-optimizations.

Information flow. This principle states that 'information systems should be designed to provide information in the first place to the point where action on the basis of it will be needed' (Cherns, 1976). This principle has to be seen as a design and evaluation criterion for the information system. The required design of the information system depends on the division of tasks and responsibilities. In a situation of centralized responsibility the information system should be able to collect and transfer detailed information, such as order status, actual level of capacity and loading, actual scheduling and the availability of material. A detailed data recording system as well as a control system for short-term instructions will probably have to be installed in case of centralized responsibilities. In case of more distributed responsibilities, where production control tasks are performed more locally within extensive margins, it may be possible to take advantage of the personal know-how of the workers at each organizational level. The appropriate support for these workers may be specific decision support tools, which enables them to keep control over their local sub-area.

Power and authority. This principle states that 'those who need equipment, materials, or other resources to carry out their responsibilities should have access to them and authority to command them' (Cherns, 1987). This means that in a production control system, people cannot be made responsible for taking good decisions if they do not have the means and/or authority to take and execute those decisions. This principle may seem a matter of course and of common sense. However, in many practical cases, people are given responsibility for a high delivery performance without having sufficient decision support tools and information to plan and control the process.

The multifunctional principle. This principle states that multifunctionality of employees will make the organization more adaptive and efficient with respect to a varying environmental demand (Cherns, 1987). The principle of multi-functionality is also of importance for the design of a production control system. Decision support tools, which can only be used by one employee, for instance the foreman of a work group, are not usable if the employee is absent. In order to serve the multifunctional principle, it can be argued to construct the decision-making hierarchy in such a way that the complexity of each decision task is as small as possible. It will then be easy to instruct more than one person taking the decisions. Complex decision tasks can be simplified by implementing intelligent decision support tools.

Incompletion. Cherns (1987) criticizes the myth of stability, which so easily accompanies the designer (or design team). Cherns says: 'Although the myth of stability is essential to enable us to cope with the demands of change, we all know that the present period of transition is not between the past and a future stable state but really between one period of transition to another. The stability myth is reassuring but dangerous if it leaves us unprepared to review and revise'. Following this warning, the changeability of the production control system should be helpful in dealing with a changing situation. The changeability, for instance, can be expressed by a modular structure or by the simplicity of the system.

The design principles mentioned above are to be seen as guidelines and evaluation criteria for the designer (design group) of a production control system.

18.4.3 Environmental guidelines

A production control system has to perform in an organizational environment. The organizational environment needs to support the proper functioning of the production control system. One socio-technical principle refers to the organizational environment: the principle of 'support congruence'.

Support congruence. 'Systems of social support should be designed so as to reinforce the behaviours which the organization structure is designed to elicit' (Cherns, 1976). The introduction of a new production control system can urge the need for a training program, or even the hiring of new employees. Also the payment system may be changed because of the new production control system. One may think of the impact of more control responsibility on the salary of workers. Important, furthermore, is the congruence of work measurement systems with the responsibilities required in the production control system. It would, for instance, be incorrect to measure only the efficiency of operators, if they have also the responsibility of processing the orders in time. The lack of support congruence can be the reason for the failure of a production control system.

18.4.4 Summary of the guidelines

Previous sections have presented several guidelines for the socio-technical redesign of a production control system. Table 18.2 summarizes the guidelines. These guidelines are independent from the particular production control concept (i.e., MRP, JIT, OPT, HPP). The guidelines do not give suggestions about how and

Table 18.2 Guidelines for the redesign of a production control system

Procedural Guideline 1 *(Compatibility)*	Users of the production control system should participate in the design.
Procedural Guideline 2 *(Minimal critical specification)*	Only essential constraints should be specified.
Procedural Guideline 3 *(Transitional organization)*	The designer is also responsible for the implementation of the production control system.
Design Guideline 1 *(Minimal critical specification)*	Only essential decisions should be taken at each level of the production control hierarchy. These decisions concern merely objectives instead of procedures for lower levels.
Design Guideline 2 *(Variance control)*	Decision-making tasks (levels) should reflect the variances that may arise at the organizational level.
Design Guideline 3 *(Boundary location)*	Each level in the decision hierarchy may have its own objectives. Co-ordination between levels (e.g. by sharing of information, knowledge, and learning) may be required to avoid sub-optimizations.
Design Guideline 4 *(Information flow)*	Information systems should be designed to provide information in the first place to the point where action on the basis of it will be needed.
Design Guideline 5 *(Power and authority)*	People can only be made responsible for decision tasks if they have the means (decision support tools, information) to deal with the decision problems.
Design Guideline 6 *(The multifunctional principle)*	More than one employee should be able to deal with each decision task in the production control hierarchy. This can be easily realized if the decision complexity at each decision level is low. Decision support tools may be helpful in making decision problems less complex.
Design Guideline 7 *(Incompletion)*	The production control system must be easy to redesign. This can be realized by means of modularity and/or a basic simplicity of the decision problems.
Environmental Guideline 1 *(Support congruence)*	Training programs, salary systems, work measurement systems, etc. need to be congruent with the design of the production control system.

when they can be integrated in a systematic (re)design approach. The remainder of this chapter contains an illustration of how to integrate the guidelines into a systematic (re)design approach. Section 18.5 will briefly present the steps of the proposed (re)design approach. Section 18.6 describes a real-life case situation for which a production control system has been redesigned. Sections 18.7 to 18.10 illustrate the integration of socio-technical principles into the steps of the (re)-design approach.

18.5 APPROACH FOR THE DESIGN OF A PRODUCTION CONTROL SYSTEM

The production control system of a manufacturing company can be compared to the nervous system of the human body: all positions within a company are connected to the co-ordinating role of the production control system. A proper design of the system is therefore crucial to an effective and efficient operations management. This section presents a design-oriented approach to production control. The basics of the approach are derived from the work of Bertrand *et al.* 1990). Figure 18.3 gives a schematic overview of the approach. Firstly, operations are defined based on the company's means of production. These operations can be regarded as its smallest, independently controllable units. Secondly, a distinction is made between items (output of operations) requiring central control and items to be controlled in a decentralized way. Centrally controlled goods are called 'Goods Flow Controlled Items' (GFC items). To a certain extent, these GFC items determine the limits of the production units (PUs) that are controlled centrally but comprise operations that can/should be controlled in a decentralized way. The PUs are defined in such a way that they are relatively independent of each other in the short-term and responsible for the production of a specific set of part-products/end products. A PU can be regarded as a set of operations. Thirdly, the operational characteristics of the PUs need to be carefully defined, first in qualitative terms. When releasing work for the PUs, central production control will have to take these characteristics into account.

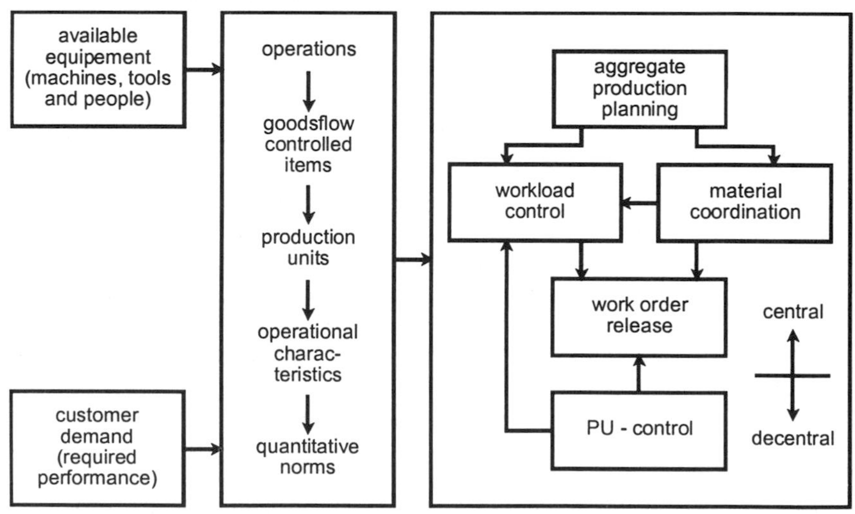

Figure 18.3 Approach to the design of a production control system.

On the basis of the market situation (what is important for the customer) and the operational characteristics, the firm has to decide on a logistic structure (what to make on stock, and what to make on order) and has to define the order lead times of the PUs to be reserved. Simultaneously, norms for the operational characteristics need to be set. The order lead times to be reserved can be seen as important quantitative information to be used in the central production control functions. Subsequently questions regarding aggregate production planning, workload control, material co-ordination, work order release and PU-control have to be answered. Section 18.3 gives a brief explanation of these production control functions. As mentioned in Section 18.3, the socio-technical approach can be seen as an elaboration of the Distributed Production Planning concept. This means that the relation between each PU and the central production control may be based on its own specific concept (which may have the character of MRP, JIT, OPT, HPP, or such).

Sections 18.6 to 18.10 will explain the approach in more detail, illustrating how socio-technical principles may guide the designer of a production control system. This is done on the basis of a real-life case study in which the production control system has been redesigned using the elements of Figure 18.3. At the time of the redesign, the socio-technical principles were not explicitly integrated into the approach. Therefore, the case study only serves as a tool to show the possible integration of socio-technical principles; it does not prove the usefulness of the principles. Section 18.6 presents the particular case. Section 18.7 gives an explanation of how the firm has defined its production units (PUs). Each PU has its own specific operational characteristics that production control (the central control) should take into account. The definition and content of these characteristics will be addressed in Section 18.8. Section 18.9 concerns the order lead times of the PUs and the choice of a logistic structure. Next, in Section 18.10, the contents of the elements of a production control system (PU-control, work order release, material co-ordination, workload control and aggregate production planning) are discussed.

18.6 CASE STUDY

The firm presented in this case study concerns a small company of about 60 direct employees, in the north of The Netherlands. Since 1915, the firm has produced a large variety of perforated sheet metal which it supplies as half-products or end-products to nearly all branches of industry. Perforating is an industrial process in which numerous holes of random shape and size and in various patterns are made fast and efficiently in plate material. The company has its own tool manufacture where perforating equipment is manufactured and serviced. Other activities besides perforating include rolling, cutting, setting and rounding corners. Figure 18.4 gives a schematic representation of the primary production process.

The firm manufactures standard perforated and special purpose perforated plates. The perforations of the standard perforated plates are listed in a catalogue for customers to choose from. They are made with existing equipment. The standard perforated plates (with current perforation and merely rolled) are supplied to wholesalers. However, in view of the huge variety of simple standard perforated plates, no stock is kept. Furthermore, the company supplies more complex standard perforated plates to various customers. These plates have standard hole patterns and undergo one or a few additional modeling processes besides perforating and rolling. For instance, check plates, filters, balcony rails and ceiling boards. The special purpose perforated plates refer to custom-made perforating tools. The catalogue does not include this sort of perforation so production requires special equipment. They make up around 6% of the total number of orders per year. Naturally many customers place repeat orders for their special purpose perforated plates.

The firm is facing tightening market demands. There is a call for extra processes other than perforating as well as a greater variety of perforations. Moreover, there is a growing need for short and reliable delivery times together with competitive price levels for perforated plates. To meet the demands the firm has introduced tighter production control. The approach used to redesign the current production control system is schematically presented in Figure 18.3. The next section will illustrate the approach in more detail, showing the relation with socio-technical guidelines.

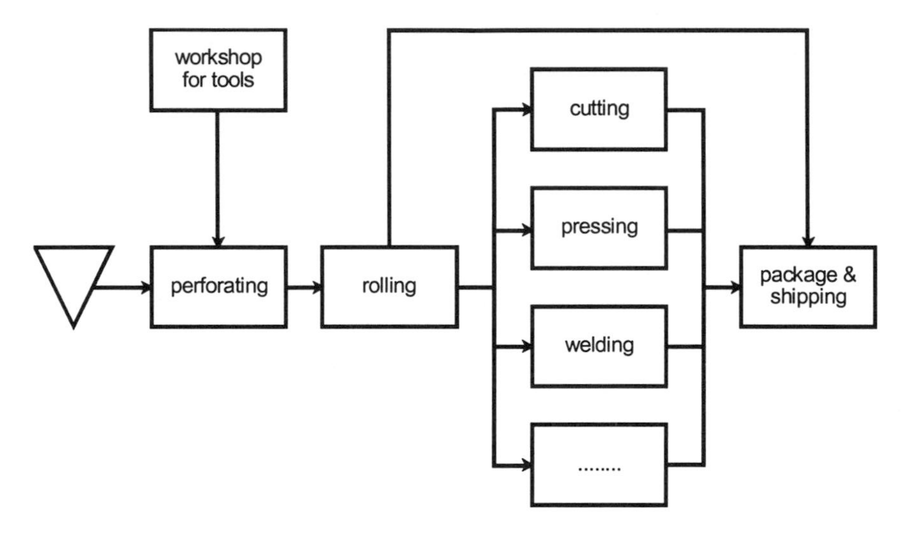

Figure 18.4 The primary production process.

18.7 OPERATIONS, GFC-ITEMS, AND PRODUCTION UNITS

As can be derived from Figure 18.3, a proper definition of production units is at the basis of good production control. To this end, first operations and GFC items have to be carefully identified, and then the limits of the production units in which several operations are combined.

18.7.1 Operations

In the production process, an operation is the smallest unit that can be controlled independently. In practice an operation comprises a group of processes that have little autonomy with respect to the moment of release. The operation processes are carried out together in a certain, logical sequence. Within the firm, the perforating process is an example of such an operation. It consists of clamping a roll of plate material, adjusting tools, making a hole pattern and cutting to size. Also rolling can be included in the perforating process. Almost all perforated plates need to be rolled. This is done immediately after the perforating process. Defining operations can be best done participatively with the employees of the firm *(socio-technical procedural guideline of compatibility)*. They know best the activities to be performed on the shop floor and they are able to describe the operations in some detail. The following operations were distinguished within the firm:

- perforating and rolling
- (circular) cutting
- eccentric pressing
- rounding corners
- squaring
- stretching (straightening; smoothing)
- welding
- manufacturing tools
- packing and shipping.

Distributions, i.e. shipping the end-products to customers, is done by an external transport company working on call. This means that the distribution operation is quite an easy, administrative task (the firm merely has to phone the haulage company) so it is not included here any further.

18.7.2 GFC items

The next step is to identify those stages in the goods flow requiring central control. At this point the so-called GFC items are defined; these involve the materials, part-products, and end-products that are generated during the goods flow and preferably need feedback coupling to and from a central production control department. Centralized production control decisions are needed to deal with deviations from scheduled times and quantities. These deviations may have an impact on the whole goods flow. As few GFC items as possible should be defined *(socio-technical design guideline of minimal critical specification)*. Three elements are of major importance in selecting GFC items: fluctuations/uncertainty in demand and supply of items, the product structure and capacity bottlenecks (Bertrand, 1990).

Fluctuations/uncertainty can be found in the demand as well as in production. In practice, uncertainty in demand occurs with the end-products (customer demand) and with the products/parts that have to be transformed into end-products on customer order. The latter are kept in the so-called CODP (customer order de-coupling point) and are produced to stock and/or to forecast. In view of an unstable demand, the end-products and the CODP products/parts are preferably GFC items. In the specific situation of our case the end-products and the raw material, which is purchased on stock, are GFC items. Information on (the availability of) these items in a centralized production control function is important for realizing and maintaining an efficient and effective production system. Production uncertainty may also be the reason for defining GFC items. The uncertainty may be related to the risk of machine breakdown, variations in the required production time and capacity, the varying availability of production capacity, etc. As for the particular firm presented in this section, uncertainty is found especially at the perforating machines and some of the finishing operations. Therefore, products which have undergone a perforating (including rolling) operation and products which have undergone a finishing operation are considered GFC items. The central production control will need information on them for a smooth running of follow-up activities (such as the extra start-up of a series).

Product structure is the second element that plays a role in defining GFC items. It is particularly important for complex assembly situations that require the timely availability of all parts for (sub-) assemblies. These parts make up GFC items. Central control is advisable specifically for parts that are used for various (part-) products (i.e., parts with a large 'commonality'). The firm in our case study manufactures a single product. Still, two GFC items can be distinguished. At the start of production two elements must be available at the same time: plates/rolls (i.e., the 'raw materials') and perforating tools. Consequently, they can be regarded as GFC items (this had already been concluded for the plates/rolls).

The third element used for defining GFC items relates to the importance of the production means and whether or not it constitutes a capacity bottleneck. The occupation of such a process has a considerable impact on the productivity of the entire company. The incoming and outgoing products/parts of a bottleneck are GFC items; central control of the bottleneck is required in view of its impact on the entire goods flow. At the particular firm the perforating machines are considered the bottlenecks of the company.

Table 18.3 represents the GFC items distinguished in the particular case. They are the starting point for the definition of production units (PUs). A change in product structure or the purchase of new, and better, equipment may reduce the number of GFC items. The designer of a production control system should take care of the *socio-technical procedural guideline of minimal critical specification.* New finishing equipment, for instance, may lead to avoiding the fifth GFC item (i.e. plates after finishing operation).

Table 18.3 GFC items

GFC items
raw material
tools for perforating
perforated and rolled plates
plates after finishing operation
final products

18.7.3 Production units

As mentioned before, a PU is relatively independent and responsible for the production of a specific set of part-products/end-products. In principle the GFC items represent the incoming and outgoing flows of a PU and must be planned by a central control department. By doing this, the PUs are buffered from variances which they cannot control *(socio-technical design guideline of variance control).* In the case presented here, four PUs are identified: i) the tool manufacture, ii) perforating and rolling, iii) finishing processes, and iv) packing. Each PU is responsible for one or more operations. If the finishing-PU comprises many machines and people, the final processing department, if necessary, can be split up into smaller PUs (using group technology). This, however, is not necessary for the firm involved here.

18.8 OPERATIONAL CHARACTERISTICS OF THE PRODUCTION UNITS

To facilitate the release of orders from a centralized production control function to the independent production units, the characteristics of the PUs should be considered. In general four operational characteristics can be distinguished (Bertrand *et al.* 1990): i) batch quantity constraints; ii) sequence constraints; iii) workload constraints; and iv) capacity constraints.

Batch quantity constraints. The set-up times during perforation are considerable so it is advisable to produce large batches in the perforating and rolling PU. The production-order size is determined by customer-order size. This means that the firm's sales department and customers must agree on which batch size is minimally acceptable. In case of very large customer orders, the production control department may decide to split some orders. The batch quantity constraints of other PUs are less severe than with perforating. Given the dependence of the perforating/rolling PU it is not necessary to include batch quantity constraints in the set of operational characteristics of these production units.

Sequence constraints. As for the perforating process, careful consideration of the order sequence may result in a significant reduction of changeover times. Time is saved only if the roll of plate material has to be replaced and tools can remain in place (or vice versa). This could be arranged at central control level. However, there is more local knowledge on the shop floor, for instance on the standing time of moulds, so decisions on sequence are preferably taken at workshop level *(socio-technical design guideline of minimal critical specification).* Here, 'preferable order sequences' can be more effectively balanced against 'realizing internal delivery times'. This means that central production control should allow the perforating/rolling PU some play concerning the time of starting an order. This can be realized by offering the production unit a significant amount of Work in Process to choose from. The other PUs have no or hardly any sequence constraints.

Workload constraints. The accepted orders determine the utilization, or occupation, to be realized (workers and machines). From a financial point of view, a maximum utilization of each source of capacity would naturally be best. However, as Figure 18.5 demonstrates, this would have negative consequences for the lead times to be realized. With a high degree of utilization, long waiting times will occur. The choice of the maximal workload to be assigned to people and machines should therefore allow for reasonable lead times. The capacity of the workers in the perforating/rolling PU can be easily adjusted to the number of perforating machines in use and does not have to be included as one of the operational characteristics. In the final processing PU, human capacity is the determining factor for the production to be maximally realized (with an acceptable lead time). Thus, the number of people present in the final processing PU, together with their 'occupation', can be viewed as operational characteristics. The utilization of the machines in the final processing PU is generally low so they form no operational characteristics to be reckoned with by central production control. In tool manufacture the production to be realized depends on the number of people. Work pressure here is determined by the number of new perforating devices (tools) to be made and the number of tools to be serviced. For the packing department, the number of workers, again, determines how much work can be done. The work volume varies considerably.

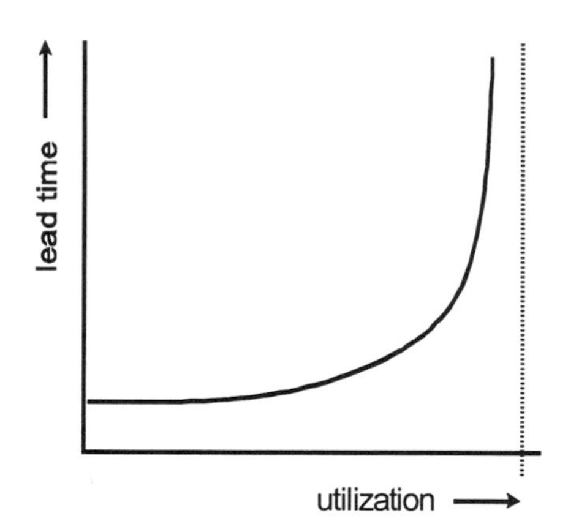

Figure 18.5 Relation between utilization and lead time.

Capacity constraints. The capacity constraints of the perforating/rolling PU are due to the number of machine hours available. Extension is only possible through overtime work. If the need for extra capacity continues, the switch to a three- or four-shift system may be considered. Capacity extension of the final processing PU can be realized by hiring temporary workers. After all, the available capacity here is determined by the number of man-hours available and training periods are generally brief. The other PUs, too, have capacity constraints (in this case the number of workers). Capacity extension of the tool manufacture, however, will only be necessary in case of a substantial increase in orders. The capacity need in the tool manufacture can be balanced by repair jobs. If necessary, packing can be done with workers from the final processing PU or with temporary workers.

The operational characteristics of the various PUs are summarized in Table 18.4. These characteristics have to be dealt with in a central production control function. By doing this, autonomy of the PUs can be facilitated by central control (*socio-technical design guidelines of variance control and boundary location*). Furthermore, the characteristics play a major role in balancing sales and production. The sales department will have to take account of the operational characteristics when making arrangements with customers. The specific interpretation of these characteristics, on the other hand, such as the necessary capacity and workforce, will depend on market characteristics. Tuning of sales and production is based on the definition of the norm lead times of orders by the various PUs. This will be discussed in the next section.

Table 18.4 Operational characteristics of the production units

Production unit	Operational characteristics
Tool manufacturing	• Number of employees • Workload per employee (number of new and repair tasks)
Perforating and rolling	• Minimal batch sizes of orders • Minimal amount of Work In Process (to enable the PU to optimize the sequence of production orders) • Number of perforating machines and their capacity • The capability for overtime work • The capability to go to another shift system
Finishing processes	• Number of employees and their capacity (in hours per day) • The capability to make use of temporary employees
Packing	• Number of employees and their capacity (in no. of orders per day) • The capability to extend the capacity with workers from the final processing PU or with temporary workers.

18.9 LEAD TIME FOR NORMS ORDERS

The definition of the lead time norms of orders within the various PUs is of crucial importance when tuning sales and production. These lead time norms depend on market characteristics (what is important for customers) and the logistical structure (what is made-to-order, what is made-to-stock) of the firm. The lead time norms should take into account the operational characteristics of each PU, and can generally be calculated using the following formula:

$$LTR_{ij} = p_{ij} + w_j n_{ij} \tag{18.1}$$

LTR_{ij} equals the lead time to be reserved for order i in PU j, p_{ij} is the total processing time (including set up times) required for order i in PU j, n_{ij} is the number of processing steps (or operations) of order i in PU j, w_j is the waiting time per processing step in PU j. In view of the sequence and batch quantity constraints it is important for the perforating/rolling PU to have the option of combining orders and lining them up. Consequently, ample waiting time, w_j, must be reserved. The final processing PU can do with a shorter waiting time, w_j. For the tool manufacture (particularly the manufacturing of new tools/perforating equipment) and for the packing department fixed lead times can be chosen independent of specific orders. It is important that the total lead times are acceptable to the market. If total lead times are generally not acceptable for customers, the firm may choose to perform several PU operations on forecast. This may decrease the number of operations (n_{ij}) to be performed on order and, therefore, the lead time. Another structural possibility to reduce lead time norms concerns changes, or investments, on the shop floor. If, for instance, the firm enlarges the capacity of the perforating unit through investments, then it may be acceptable to reduce the reserved waiting time (w_j) for this unit. Keeping this option open in the design of a production control system is part of *the socio-technical procedural guideline of minimal critical specification.*

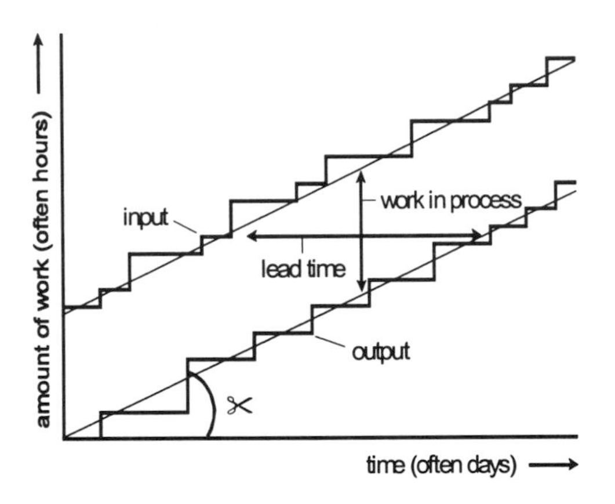

Figure 18.6 Lead time diagram.

Actual lead times of orders are strongly related to the work-in-process (WIP) of a production unit. This can be illustrated by a so-called lead time diagram (Figure 18.6). The diagonal lines in the diagram reflect the cumulative 'input' and 'output' of the PU; the gradual change in input and output lines indicates that a number of production hours (required for an order) has come in or is ready for the next PU. The horizontal spacing between the diagonal lines reflects the mean order lead time. The vertical spacing is the mean volume of work-in-process in the PU expressed in production hours. The mean productivity of a PU (in production hours per time unit) is equal to tangent α or the quotient of the mean volume of work in execution and the mean lead time:

$$P = WIP / LT \qquad (18.2)$$

where P is the mean productivity, WIP is the mean volume of work-in-process, and LT is the mean lead time. The mean productivity is also equal to the sum of the effective working hours per day of the machines (or workers) in a PU. In conformity with the formula above, it therefore applies that a certain degree of effective productivity and a mean lead time require a mean volume of work-in-process. More work on the shop floor will result in a longer mean lead time unless the productivity of the PU can be increased. In turn, a decline in the work-in-process will be accompanied by a shorter lead time. There is, however, always a minimal lead time (see Figure 18.5). A decline in work-in-process, therefore, may also lead to a worse productivity (idle machines and/or workers). Bechte (1994) describes the principles of load-oriented manufacturing control.

The relation between productivity, work-in-process and lead times is very basic and has to be seen as a fundamental law (also known as Little's Law in queueing theory) for production control. By controlling the workload of a PU, central production control reduces the variations with which PU-control has to deal. This is especially important when variations in the demand are not caused by the PU itself (which is usually the case) or if the PU does not have the means to deal with the variations. Workload control can be seen as an interpretation of *the socio-technical design guideline of variance control.*

18.10 PRODUCTION CONTROL FUNCTIONS

Having defined the production units, the operational characteristics and agreements on lead time, the production control functions can be identified. The following functions can be distinguished (see also Figure 18.3):

- aggregate production planning
- material co-ordination
- workload control
- work order release
- PU control.

In this particular case, PU control is done within the PUs, work order release is the responsibility of the manufacturing department, material co-ordination is partly performed by the sales department and partly by the centralized planning department, workload control is the responsibility of the centralized planning department and, finally, aggregate production planning is the task of the management team of the firm. This assignment of responsibilities is in conformity with *the socio-technical design guideline of variance control.* The specific decision tasks will be explained in the remainder of this section.

Aggregate production planning. Aggregate production planning can be regarded as a long-term planning for the goods to be delivered (aggregate delivery plan) and the capacity to be used (aggregate capacity use plan) and adjusted over time (aggregate capacity adjustment plan). This planning specifies the goods to be made on stock and on customer order. It also describes the agreed minimal and maximal stock levels (aggregate inventory plan). The firm in the case study has opted to order raw materials only on stock; the entire production process is based on customer order. Specifying the aggregate production plans, involves further identification of the operational characteristics and the corresponding norm lead times. By doing so, aggregate production planning specifies the objectives of the decision levels of the remaining production control hierarchy. It serves *the socio-technical design guideline of minimal critical specification.*

Material co-ordination and workload control. Material co-ordination is partly performed by the sales department and partly by the central planning department. The sales department submits quotations together with specifications of the dates of delivery. The sales department has to inspect the delivery times for feasibility. In the particular case described in this chapter, the planning department suggests earliest delivery weeks for new orders, depending on the overall workload of the factory. If necessary, the sales department may request earlier delivery weeks for particular orders. This can be realized through an adjustment of the productivity of the PUs (see previous paragraph). All this requires co-ordination between the sales, the planning and the manufacturing departments (*socio-technical design guideline of boundary location*). Having obtained a final confirmation of an order it may be necessary to check once more the feasibility of the delivery date mentioned in the quotation, especially when there is quite some time between quotation and order confirmation. Meanwhile capacity can be reserved if there is a fair chance of an order being accepted. This also requires co-ordination between the sales and planning departments. The planning department makes the planning of confirmed and expected orders. This planning is based on the norm lead times calculating back from the date of delivery. For each PU an order has an internal delivery time that should be met as closely as possible. By calculating back from the date of delivery (instead of planning an order as early as possible) the capacity available for any short-term orders is extended as long as possible (*socio-technical design guideline of minimal critical specification*). Next to this 'medium- term' planning, the planning department provide a short-term planning using confirmed orders in which the various PUs are loaded maximally to their capacity limit. In case the PUs are under-loaded for the coming weeks, it must be possible to move orders forward. This demands an availability of raw materials for, at least, fast running orders. If it is not possible to load PUs under their capacity limit for some periods in the future, then some measures have to be taken, for example, allowing extra productivity (overtime work, temporary workers, or working in more shifts) and/or subcontracting orders. This requires co-ordination between the planning and the manufacturing departments (*socio-technical design guideline of boundary location*). Material co-ordination and workload control are closely connected activities and are performed by a planning manager who is also a member of the management team of the firm.

Work order release. The work order release function should release orders in conformity with material management and workload control. However, there should also be alertness as to the actual status on the shop floor (e.g. ill people and/or machine breakdown.). Decisions on the necessary measurements can be taken in regular meetings usually held weekly. Examples of measurements are the hiring of temporary workers and the decision to have overtime work for one or more days. A planning officer who works in the manufacturing department performs work order release. The meetings are attended by the planning officer, the operation manager, the foremen of the PUs, and the shipping manager.

PU control. The PU foremen are responsible for the detailed scheduling of orders so that the internal delivery times can be met (which are derived from the norm lead times). In the case study, the scheduling of the perforating machines is particularly complex because it has to pay heed to both internal delivery times and sequence relations (the possibility to save on set-up time).

The specification of the production control tasks of the management team, the sales department, the planning department, the manufacturing department and the PUs, as described here, is done in conformity with *the socio-technical design guidelines of minimal critical specification, variance control, and boundary location*. Only critical elements are specified at higher decision levels. The decision levels are located at organizational levels where appropriate measures can bc taken. Arrangements are taken to ensure effective co-ordination between the levels of the production control. After defining the tasks of each level in the production control, the information system has been adapted in conformity with *the socio-technical design guideline of information flow*. Information should be provided in the first place to the point where action on the basis of it will be needed. Decision support tools may be needed for certain functions in the production control. This refers to *the socio-technical design guideline of power and authority*, which indicates that decision support tools may be needed to be able to make good decisions, *the multifunctional principle*, which suggests that decision support tools may simplify problems so that they can be solved by more than one employee, and *the principle of incompletion*, which asks for simplicity so as not to frustrate the necessity to redesign, if needed.

The success of the production control system, furthermore, depends on aspects described in the *procedural guidelines as compatibility*, which asks for participation of all people involved in the production control, and *the guideline concerning the transitional organization*, which gives the designer of the production control system a role in the implementation of the system. It may be clear *that the environmental guideline of support congruence* forms another element which may support or frustrate the success of the production control system.

The case study presented here concerns a firm for which a production control system has been (re)designed. At the moment of design, the socio-technical guidelines were not expressed explicitly. The guidelines were followed implicitly, partly, and sometimes simply not. Also, the performance of the new production control system has not yet been evaluated. The case study, therefore, only serves as a tool to show the possible integration of socio-technical guidelines. It does not prove the usefulness of the guidelines.

18.11 RESUME

This chapter has described socio-technical guidelines for the design of a production control system. These guidelines fit well in the systematic approach to the design of a production control system described in Section 18.5. This is illustrated by means of a case study. Important elements of the approach are the determination of operations, goods flow controlled items, production units, the specification of operational characteristics of the production units, and the concept of workload control. The stepwise approach gives employees the opportunity to participate in the design of a production control system. Furthermore, its bottom-up philosophy supports the idea of autonomous, controllable production units responsible for the major part of the planning, scheduling and control of their tasks. The integration of socio-technical guidelines into the approach is helpful in the design and allocation of production control tasks and responsibilities over the various departments/people of the firm.

An interesting aspect of the approach is the fact that production units are defined on the basis of getting autonomy with respect to production control. The Dutch socio-technical literature (Van Eijnatten and Van der Zwaan, 1998) suggests first designing the production structure, by means of production flow analysis or other clustering methods, and then the control structure. This chapter has stressed the need to focus on autonomy with respect to production control. This can be seen as an implicit plea for, at least, a simultaneous design of the production structure and the control structure.

18.12 REFERENCES

Ackoff, R.L., 1979, Resurrecting the Future of Operational Research, *Journal of the Operational Research Society,* Vol. 30, No. 3, pp. 189-199.

Ashby, W.R., 1969, Self-regulation and Requisite Variety, in: Emery, F.E. (ed.), *Systems Thinking: Selected readings*, Penguin Books, Harmondsworth, pp. 105-124.

Barki, H. and Huff, S.L., 1990, Implementing Decision Support Systems: Correlates of User Satisfaction and System Usage, *INFOR,* Vol. 28, No. 2, pp.89-101.

Bechte, W., 1994, Load-oriented Manufacturing Control Just-In-Time Production for Job Shops, *Production, Planning & Control*, Vol. 5, No. 3, pp. 292-307.

Bertrand, J.W.M., Wortmann, J.C. and Wijngaard, J., 1990, *Production Control, A Structural and Design Oriented Approach,* Elsevier, Amsterdam.

Bitran, G.D., Haas, R.A. and Hax, A.C., 1982, Hierarchical Production Planning: A Two-Stage System, *Operations Research*, Vol. 30, No. 2, pp. 232-251.

Brown, J., Harhen, J. and Shivnan, J., 1988, *Production Management Systems – A CIM Perspective*, Addison-Wesley Publishing Company, Wokingham.

Cherns, A., 1976, The Principles of Socio-technical Design, *Human Relations,* Vol. 29, No. 8, pp. 783-792.

Cherns, A., 1987, Principles of Socio-technical Design Revisited, *Human Relations*, Vol. 40, No. 3, pp. 153-162.

Child, J., 1972, Organizational Structure, Environment and Performance: The Role of Strategic Choice, *Sociology,* Vol. 6, pp. 1-22.

Child, J., 1984, New Technology and Developments in Management Organization, *OMEGA, International Journal of Management Science,* Vol.12, No.3, pp. 211-223.

Eijnatten, F.M. van, and Zwaan, A. van der, 1998, The Dutch IOR Approach to Organizational Design: An Alternative to Business Process Re-engineering?, *Human Relations*, Vol.51, No.3, pp. 289-318.

Emery, F.E. and Trist, E.L., 1972. *Towards a Social Ecology.* Plenum Press, London.

Hage, J., 1977, Choosing Constraints and Constraining Choice, in Warner, M. (ed.), *Organizational Choice and Constraint*, Saxon House, Hants.

Herbst, P.G., 1974. *Socio-technical Design,* Tavistock Publications, London.

Huber, V. and Brown, K., 1991, Human Resource Issues in Cellular Manufacturing, *Journal of Operations Management*, Vol.10, No.1, pp. 138-159.

Hyer, N.L., Brown, K.A. and Zimmerman, S., 1999, A Socio-technical Systems Approach to Cell Design: Case Study and Analysis, *Journal of Operations Management*, Vol.17, pp. 179-203.

Ives, B., and Olson, M., 1984, User Involvement and MIS Success: a Review of Research, *Management Science,* Vol. 30, No. 5, pp. 586-603.

Kuipers, H. and Amelsfoort, P. van, 1990, *Slagvaardig Organiseren – Inleiding in de sociotechniek als integrale ontwerpleer*, Managementreeks, Kluwer Bedrijfswetenschappen (in Dutch), Deventer.

Larsen, N.E. and Alting, L., 1993, Criteria for Selecting a Production Control Philosophy, *Production Planning & Control*, Vol.4, No.1, pp. 54-68.

Trist, E.L. and Bamforth, K.W., 1951, Some Social and Psychological Consequences of the Longwall Method of Coal-Getting, *Human Relations*, Vol. 4, No.1, pp. 3-38.

Trist, E.L., Higgin, G.W., Murray, H. and Pollock, A.B., 1963, *Organizational Choice*, Tavistock Publications, London.

Weick, K.E., 1979. *The Social Psychology of Organizations*, Addison-Wesley, Massachusetts.

CHAPTER NINETEEN

Planning and Scheduling in Secondary Work Systems

Toni Wäfler

19.1 INTRODUCTION

In a world of increasingly turbulent markets and growing technical complexity of products and production processes it is a core problem of work systems to cope with uncertainty. Many different activities are performed within work systems to deal with this problem. Temporal planning and scheduling are part of these activities, meant to cope with uncertainties that arise in matching dynamic production demands with potentially unstable production resources. This sets up some peculiarities of temporal planning and scheduling (e.g. McKay, 1987, 1992; Weth von der and Strohschneider, 1993; Schüpbach, 1994; Wiers, 1997):

- Information to be processed may be incomplete, ambiguous, dynamic and of stochastic nature.
- Information flow follows feed forward as well as feed back and formal as well as informal structures.
- Decisions to be taken may be highly interrelated not only in content but also time wise. This causes dynamic complexity in planning and scheduling: today's solutions entail tomorrow's problems.
- Goals to be followed may not be independent - even if set clearly: minimising lead times as well as machine idle time are contradictory.
- Information processing and decision-making may be distributed among many different (human and non-human) actors.
- Result oriented performance measurement and even more process-oriented evaluation of planning and scheduling practices may be constrained due to temporal delays between actions and effects as well as unclear mutual relations.

These characteristics indicate that temporal planning and scheduling cannot be considered as tasks that can be isolated and allocated to a clearly defined agent, be it a human scheduler or a sophisticated technology. They must rather be considered as a process that takes place within a complex system consisting of humans, organisational structures and technology that are highly interrelated. Optimising one part of the system e.g. the planning and scheduling technology or the human scheduler's task might lead to a sub-optimal performance of the system as a whole. What is needed instead is a joint optimisation of people, organisation and technology, as the sociotechnical systems approach demands (cf. Emery, 1959; Alioth, 1980; Ulich, 1998; with a special focus on planning and scheduling cf. Strohm, 1996; and Slomp and Ruël, see Chapter 18 in this book).

The independent variable of sociotechnical systems design is the task. It is understood as a subject to be jointly performed by the social and the technical sub-system. The technical sub-system consists of a work system's technical devices, production materials, technical conditions of the production processes, and spatial circumstances. The social sub-system comprises all the work system's members with their individual and collective abilities and needs, as well as the formal and informal relations between individuals and work groups. The task is the point of articulation between the two sub-systems (see Figure 19.1) linking jobs in the technical sub-system with role behaviour in the social sub-system (Blumberg, 1988). Consequently the task is the core of sociotechnical systems as well as the focus of sociotechnical system design.

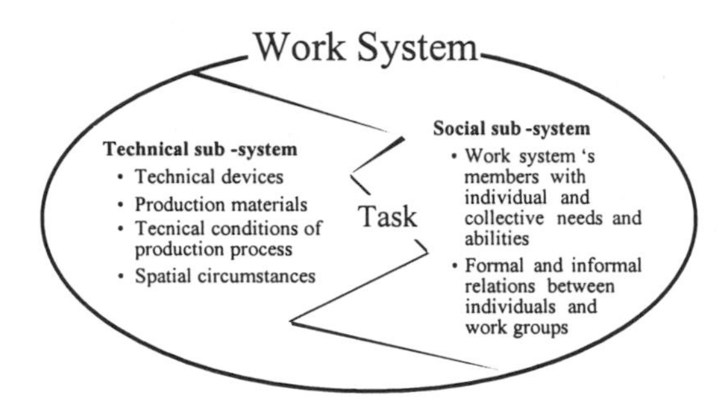

Figure 19.1 The technical and the social sub-system of a work system. They are interrelated by the work system's task, that is to be performed jointly by the two sub-systems (Grote, 1993, transl. TW).

Current methods of sociotechnical system analysis and design distinguish between different organisational levels (see Figure 19.2). For each level design criteria are derived. Designing tasks according to such criteria aims at an integral optimisation of social and technical sub-systems, by providing opportunities for local regulation of variances and disturbances on the organisational level as well as for the design of individual work tasks that promote motivation and qualification (Ulich, 1998).

Even though temporal planning and scheduling is a process that takes place in a sociotechnical system, a direct adoption of the 'classical' sociotechnical design criteria is not possible. It is the purpose of this chapter to elaborate the deficiencies in the 'classical' sociotechnical concepts. It will be argued that the sociotechnical approach focuses too much on separation of organisational units, whereas planning and scheduling processes aim at co-ordinating and integrating the activities performed within different organisational units. To overcome this deficiency the concept of *secondary work systems* will be introduced. It is the function of secondary work systems to integrate the primary work systems (cf. the formal units) of an organisation. Furthermore, some reflections regarding design parameters of secondary work systems will be made. In the following sections the sociotechnical approach will be received with reference to planning and scheduling processes. Conceptual differences will be elaborated in Section 19.2. In Section 19.3 sources of uncertainties as well as strategies to deal with uncertainties will be presented. It will be argued that an attempt to minimise uncertainty can itself cause homemade uncertainty. Hence, a strategy of coping with uncertainty will be presented. In compliance with that strategy, design requirements for organisation, individual task design and human-computer interaction will be elaborated (see Section 19.4). Finally, a case study will be presented that has been carried out in a Swiss SME (see Section 19.5). It demonstrates that the company, although it has an up-to-date organisation and has implemented ERP software of the latest generation, still shows many deficiencies regarding design implications as presented in this chapter.

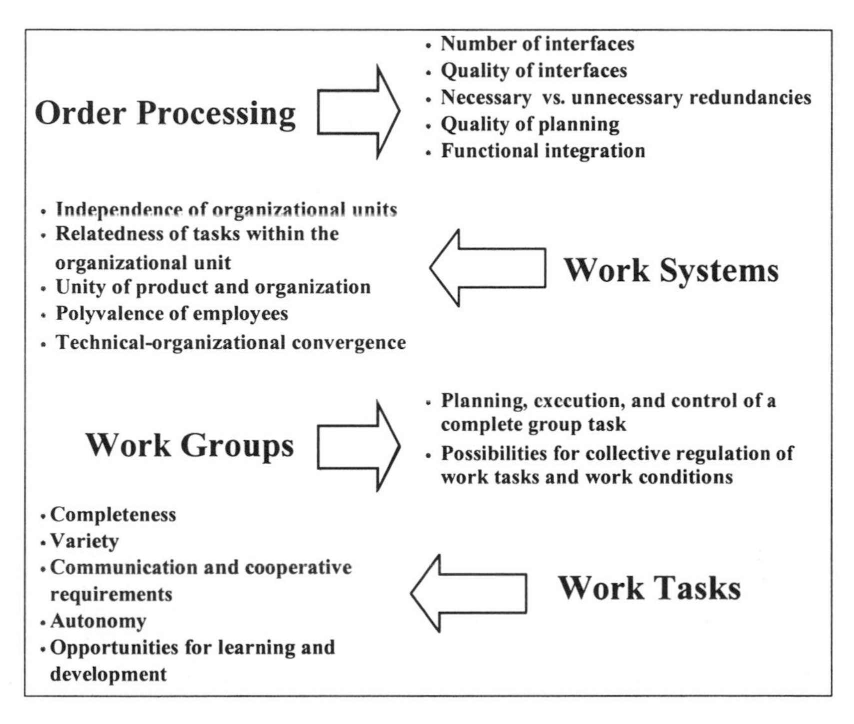

Figure 19.2 'Classical' criteria for analysis and design of sociotechnical systems from four points of view: order processing (business process), work systems (organisational unit), work groups (organisational sub-units), and individual work tasks (Strohm, 1998).

19.2 SOCIOTECHNICAL APPROACH TO PLANNING/SCHEDULING

Temporal planning and scheduling - as seen above - might be considered as a task performed within a sociotechnical system. But the direct application of sociotechnical design criteria on planning and scheduling tasks turns out to be rather difficult. The following two examples might serve as illustrations for these difficulties. The sociotechnical approach aims at designing complete work tasks for individuals, i.e. tasks including aspects of preparation, planning, execution, controlling and maintenance/repair. Due to the fact that complete tasks include both planning and executing aspects, it may be impossible to assign complete tasks to the human scheduler and the human operator at the same time. On the one hand, the human operator's task will be less complete, if more planning is allocated to the human scheduler. On the other hand, the human scheduler's task will by definition never be complete, because he never executes what he plans - he only executes the planning itself. The second example focuses on the organisational level. On that level the sociotechnical approach aims at allocating tasks to organisational units in a way that minimises mutual dependencies. However, the more independent the scheduling department is from the production unit, the more dependent the production unit is on the scheduling department, and vice versa.

A closer look at the sociotechnical concept might help for a better understanding of the differences between 'classical' sociotechnical systems such as production units and 'temporal planning and scheduling' sociotechnical systems. In the sociotechnical systems approach, organisational units are named 'primary work systems' (Trist, 1981; Strohm, 1996). This term is meant to mark the lowest of three levels from which work organisations might be considered. The two higher levels are called 'whole organisation systems', and 'macrosocial systems' (Trist, 1981). Whereas the former refers to the whole work organisation, the latter refers to the organisation's environment. For the internal structuring of the organisation it is only the concept of the 'primary work system' that remains. This exclusivity could lead to the conclusion that all types of organisational units need to be designed according to the same criteria, which - as seen above - results in contradictory design goals (e.g. independence of the production unit and the scheduling department at the same time). In order to avoid such contradictions as much as possible, it may be useful to distinguish between different types of work systems.

A criterion for this differentiation might be the sociotechnical concept of work tasks. We can distinguish between primary and secondary tasks of a work system. The primary task is the task that the work system has originally been created to fulfil (Rice, 1958). The secondary work task comprises all tasks that must additionally be performed in order to be able to fulfil the primary task. Both the production unit and the scheduling department have a primary and a secondary task (Strohm, 1996). Whereas the primary task of a production unit is the manufacturing of certain workpieces, its secondary task includes securing the system's preservation (e.g. maintenance, training etc.) as well as regulation (e.g. input control, internal co-ordination). The primary task of a scheduling department might be the elaboration of 'good' schedules. Its secondary task includes training of its people, maintenance of its tools, planning and co-ordination of its own resources, etc. But there is one major difference between the two systems. The primary task of the production unit stands on its own while the primary task of the scheduling department (i.e. co-ordinating both the order flow through the production unit and the boundary regulation between production units) relates to the production unit's secondary task.

Why is this difference important? It is true that all sub-units of an organisation are interrelated and mutually dependent but from a design point of view some can be considered to be more capable of demarcation than others. This might be especially true for a work system that has a primary task on its own (e.g. a production unit) whereas a work system that has a primary task which consists of (parts of) the secondary task of other work systems (e.g. the scheduling department) cannot easily be demarcated. Consequently the two types of work system can be named 'primary work systems' and 'secondary work systems'. Conceptually they differ as follows:

- Whole organisation systems consist of several primary and secondary work systems.
- Primary work systems have a primary work task that is not directly related to the secondary work task of other work systems.
- Secondary work systems have a primary work task that directly supports/takes over (part of) the secondary work task of other work systems.
- There are no secondary work systems without primary work systems.
- Primary and secondary work systems might partly or fully overlap. Full overlapping occurs when a primary work system performs its secondary work tasks on its own. In these cases distinguishing between primary and secondary work systems makes sense only from an analytical point of view.
- A secondary work system can overlap with several primary work systems at the same time.
- A whole organisation system is more than the sum of its primary work systems. It is constituted by primary work systems that are interconnected by secondary work systems.

These differences must be considered in system design. The more an organisational unit shows demarcation from others, the more it is useful to design it according to criteria mentioned in Figure 19.2.

Temporal planning and scheduling takes place in secondary work systems. These systems consist of all people and technology that explicitly or implicitly influence the matching of dynamic production demands with unstable production resources. On the technical side these are planning and scheduling tools as well as plans, schedules and other relevant information sources (e.g. empty KANBAN boxes, large material buffers etc.). The social system consists of all people that provide or process information relevant to temporal planning and scheduling. These might be planners and schedulers as well as shop floor supervisors and even operators that on site take *ad hoc* decisions on whether that job or another will be carried out next. Within that system, artificial and/or human agents take decisions. The system as a whole is a network of formal and informal information flow and decision-making (see Figure 19.3). It is a distributed human-machine (or computer) system that amplifies cognitive performance or - in that sense – is a distributed joint cognitive system (cf. Hollnagel and Cacciabue, 1999).

The next diagram shows four primary work systems (a planning department, a scheduling department, and two production units). These primary work systems are interrelated by a network of people who are members of the primary work systems and at the same time are constituents of the overlapping secondary work system that controls the planning and scheduling process. This secondary work system consists of all humans and all technological devices that take part in the matching of dynamic production demands with unstable production resources. Information flow and decision-making takes place in technical (full lines) as well as in social (dashed lines) networks, following formal as well as informal channels.

The secondary work system overlaps and penetrates the whole organisation. It comprises far more than the formal planning and scheduling departments.

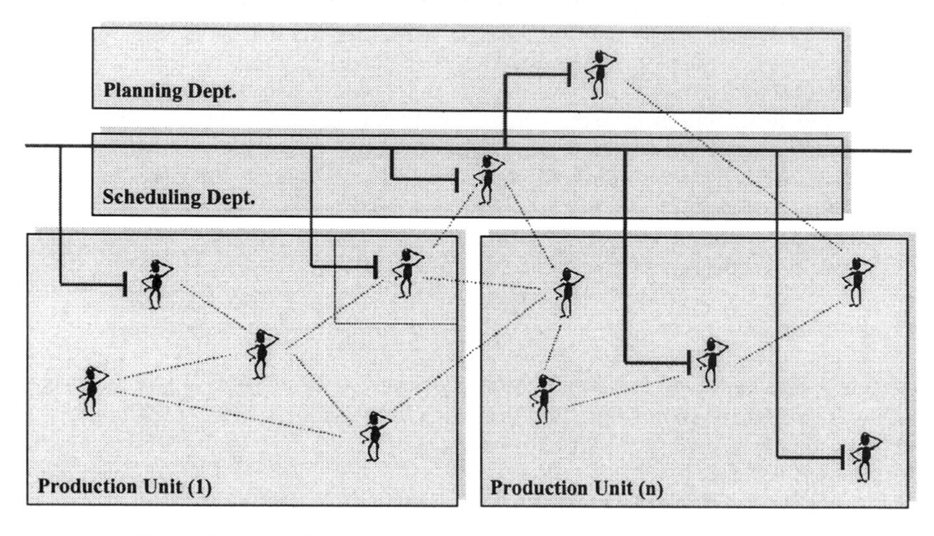

Figure 19.3 Technical and social information flows and decision-making.

In accordance with the overlapping character of secondary work systems, a sociotechnical planning and scheduling system is not limited to the formally defined boundaries of planning or scheduling departments. McKay (1987, p.133f) supports this point of view describing real planning and scheduling processes as following:

- 'The informal scheduling system is the "realtime" scheduling system that deals with the subjective and dynamic issues of scheduling.
- In an open job shop environment, the world is ever changing and when a change occurs, its impact is normally greater than individual components involved.
- The scheduling hierarchy is not straightforward and contains many interdepartmental communications, which are affected by the departments, the individuals, the current world, and the work involved.
- Scheduling duties are not clear; overlaps occur and organisational positions do not necessarily reflect duties, responsibility, and authority.
- Scheduling decision points are not isolated as various levels of scheduling are distributed throughout the organisation.'

A sociotechnical planning and scheduling system may in its technical as well as in its social constitution penetrate a large part of the whole organisation system, i.e. at least the whole logistics chain. One human might easily belong to a primary work system like a production unit and at the same time perform activities within the planning and scheduling process. This analytic dissolution of formal organisational boundaries has two advantages. On the one hand, it helps to avoid contradictions in system design. The design question is no more how to make production units and scheduling departments independent of each other, but how to interrelate primary work systems by the means of overlapping and penetrating secondary work systems. On the other hand, it provides a framework for a holistic view not reduced to e.g. algorithms, cognitive aspects, isolated scheduling tasks or the question of allocation of function between organisational units. Such selective approaches are criticised by Hollnagel and Cacciabue (1999, p.2) when they state: 'Until recently there has however been much less concern for theories and models that enable us to maintain sight of the whole.' Wiers (1997, p.62) identified a similar deficiency: 'However, in many companies, foremen and shop floor operators have not been regarded explicitly as part of the production planning and control system.'

19.2.1 Intermediate conclusion

Temporal planning and scheduling is a process that takes place in a secondary work system. Design of planning and scheduling systems has to take into explicit consideration the overlapping and penetrating character of secondary work systems.

19.3 CAUSING UNCERTAINTY BY MINIMISING UNCERTAINTY

To handle uncertainties that arise in matching dynamic production demands with potentially unstable production resources is the function of temporal planning and scheduling. These uncertainties can be caused by an organisation's environment (i.e. by its input-output relations) as well as by its processes (i.e. transformation process). Both input-output relations and transformation processes can be dynamic and unpredictable and consequently cause variances and disturbances. Individual customer demands, unreliability of suppliers to deliver on time, or quality variances in raw materials are examples of variances and disturbances that originate in an organisation's environment. Internally caused variances and disturbances can result from an increasing complexity of products and a consequently decreasing process control as well as from machine breakdowns or absence of personnel. Susman (1976) distinguishes four fields of uncertainty by combining these two types of uncertainties and by dichotomising the degree of uncertainty into 'high' and 'low' (see Figure 19.4). Two principally different strategies to handle uncertainty can be distinguished (Grote, 1997). If an organisation is confronted with high uncertainty (quadrant D) it can try to get to quadrant A by minimising the uncertainty or it can try to master quadrant D by coping with the uncertainty.

Uncertainty originating in the environment		Uncertainty originating in transformation processes: low	high
	low	A	B
	high	C	D

Figure 19.4 Situational fields with reference to the degree of uncertainty (Susman, 1976).

19.3.1 Strategy 1: Minimising uncertainty

The strategy of minimising uncertainty is an attempt to prospectively recognise uncertain occurrences and compensate for them by the means of planning. Every eventuality should be predicted and the measures to handle it should be prepared. For that purpose a centralised planning unit is put between the production system and its environment. This apparatus tries to create the conditions needed for a smooth working of the production systems. To reach this the planning unit needs to have as much information as possible. Valid due dates must be known, availability of materials and resources must be checked, etc. Most of this information is uncertain. It can change frequently. The planning unit needs to bring order into this mess. It tries to do that by measures such as reserving material for certain orders, waiting with the release of an order until the final due date is absolutely clear, etc. All the planning unit's action serve the final goal to make itself - i.e. the planning unit - the performer in control. As a consequence the production units that it tries to optimise may become reactive. The stability the planning unit must create may reduce the production units' abilities and responsibilities to develop self-dynamics, which would produce additional uncertainty for the planning unit. If this happens, the production units cannot prepare for eventualities anymore. They don't even know them. To tackle them is not their task.

Although this strategy is meant to minimise uncertainty, it paradoxically also causes uncertainty. It is uncertainty that originates within the organisation but out of the influence of the organisational unit that is confronted with its effects. Input-output relations between organisational units that cause mutual dependencies create this type of uncertainty. The units face variances and disturbances but are only confronted with their symptoms. Their causes are out of each unit's scope of influence. This third type of uncertainty is homemade. It does not originate in dynamic environments or complex processes that can be mastered to a greater or a lesser extent. It arises from the structure the organisation gives to itself. It comes from an inadequate allocation of tasks to organisational units, due to which variances and disturbances have consequences in units other than those they originate from. On the one hand, the creator of such variances and disturbances is not (directly) confronted with its effects. Consequently his interest to prevent them is (at least) reduced. On the other hand, the one who is confronted with the variances and disturbances has little influence on their causes. Not only shop floor units but also planning and scheduling departments can be confronted with such home-made uncertainty. Wiers (1997) describes one typical form of the planners' uncertainty. He calls it 'execution uncertainty'. It arises when the shop floor, due to a plan's inadequacy, has to deviate from the plan it gets from the planning unit. This deviation is necessary because abstract and formal regulations and automated procedures must always be adapted to concrete circumstances (cf. Volpert, 1994). This is a quite difficult and may be a subversive process, because it often means to illicitly straighten out unsuitable procedures in order to meet actual requirements of the situation (ibid.). It is necessary to be able to turn (more or less inadequate) plans into reality. However, the more the shop floor deviates from the original plan, the higher is the execution uncertainty the planning department is confronted with, because it does not know anymore whether and how the original plan has been adapted. In the planning unit only the plans are known, but not what really happens on the shop floor. But these plans, which consequently contain unreliable information, build the basis for further planning.

The lack of possibilities for local regulation leads to variances and disturbances that spread over the organisation uncontrolled, causing after-effects. Volpert (1994) as well as Wiers (1997) state that it is an inadequate allocation of autonomy that causes these uncertainties. The goal must therefore be to create organisational structures that competently cope with given uncertainty without causing home-made uncertainty. This leads to the second strategy, which is fundamentally different from the one described so far.

19.3.2 Strategy 2: Coping with uncertainty

In contrast to the strategy of minimising uncertainty the strategy of coping with uncertainty proceeds on the assumption that one will never succeed at recognising all relevant occurrences in advance. Consequently, in the second strategy preventing uncertainty is not tried but attempts are made to create structures that cope competently with uncertainty. Variances and disturbances should be tackled at their point of origin. Their effects should be kept and handled locally. Due to the fact that many variances and disturbances occur on the shop floor, opportunities for flexible reaction as well as mechanisms for local regulation must be given there. These opportunities do not only enhance an organisation's flexibility they also are seen as preconditions for individual and organisational learning (Frei *et al.*, 1993).

The strategy of competently coping with uncertainty leads to a different division of tasks between planning and producing organisational units. The shop floor and with it the direct productive workers on the shop floor must (partly) become the actor in control. Foremen and operators on the shop floor are not only allowed but even expected to develop self-dynamics. To make that possible for them is the remaining task of the centralised planning unit. Enabling the shop floor to recognise eventualities, to predict their possible effects, and to prepare for them comprises this (new) task. For that purpose it can, for example, be necessary to communicate uncertain information to the shop floor, if possible with specification of likelihood.

Table 19.1 The two strategies for dealing with uncertainty by comparison

Strategy 1: Minimising uncertainty	Strategy 2: Coping with uncertainty
- complex, centralised planning and scheduling systems - reduction of scope of operative action by means of regulation and automation - disturbances seen as preventable symptoms of inefficient system design	- planning as a resource of situated action - extended scope of operative action - disturbances seen as opportunities for learning and system development

19.3.3 Intermediate conclusion

Dealing with uncertainty is the function of temporal planning and scheduling, facing a joint cognitive system problem of a corresponding secondary work system. There are two fundamentally different strategies to do so (see Table 19.1). Although it is meant to minimise uncertainty, strategy 1 may cause new, home-made uncertainty. Strategy 2 is considered to be superior as it aims at coping with uncertainty competently. In the following sections design requirements that allow for strategy 2 will be elaborated.

19.4 LEVELS IN SOCIOTECHNICAL WORK SYSTEMS

Designing a temporal planning and scheduling system as a secondary work system requires the meaningful allocation of functions to its different actors, be it humans and/or technical devices. Thus, three sub-problems have to be considered: design of the organisation, design of individual tasks, and design of human-computer function allocation. These three design problems, that are not totally independent of each other, are the subject of the following sections. Design requirements that allow for coping with uncertainty competently (strategy 2, Table 19.1) are elaborated for each level.

19.4.1 Organising planning and scheduling

With reference to the organisation of temporal planning and scheduling processes, many authors criticise the classical Tayloristic structure that causes a separation of thinking and doing (cf. e.g. McKay, 1987, 1992; Schüpbach, 1994; Strohm, 1996; Grote, 1997; Yeatts and Hyten, 1998). It leads to a hierarchical, centralised control-oriented allocation of planning and scheduling tasks where the centralised planning and scheduling departments are promoted to be actors in control. Fitzgerald and Eijnatten (1998) identified five core assumptions that lie behind such thinking:

- *Empiricism*: Everything must be recognisable through the five senses, extreme empiricists are virtually blind to the rich but subtle aspects of reality.
- *Reductionism*: Assuming, that the whole is identical with the sum of its parts. This leads to a fragmentation of considerations.
- *Determinism:* Assuming that one cause has always one determined effect. This leads to thinking in endless, one-dimensional chains of cause-effect relations.
- *Conservatism:* Believing that the prevailing order of equilibrium not only can but even should be sustained indefinitely.
- *Interventionism:* Believing that the world - if left to itself - is destined to disintegrate into maximum disorder (chaos). Every opportunity to rise to new and higher levels of order must be taken advantage of.

Assumptions like these lead to a hierarchical design of planning and scheduling systems, which is characterised by specialisation, aggregation and top-down constraints (McKay, 1992). *Specialisation* refers to the fact that responsibilities are clearly assigned to the different levels. Each level collects the information it needs in order to carry out its tasks. Part of this information is not accessible for other levels. The reductionism that is inherent in specialisation '... artificially limits a level in what can be detected and what responses can be generated. This limitation then impacts a level's ability to plan and avoid problems before they occur' (ibid, p.25). Each level must take its decisions using *aggregated information* and abstract models because it never can dispose of all information from lower levels. Such aggregation increases '... threshold levels for detection, control, and analysis ... since it is not possible to observe individual trends and events within the grouped information. It is always hoped, that the information lost is unimportant for the type of decision being made' (ibid, p.25). Finally, each level has a domain over which it has control. It can take decisions about this domain, which *constrain* lower levels. The assumption that levels can sequentially be constrained top-down implies that no lower level would take better decisions. Schüpbach (1994, 1998) describes the effects of such hierarchical, control-oriented structures of planning and scheduling as counterproductive. The attempt to minimise complexity by minimising a system's openness as well as by deterministic top-down instructing leads to an organisation that is characterised by a high degree of division of labour, by a minimal degree of scope of action on lower levels, and by a minimum of co-operative relations. The rigidity that results from such work structuring entails few possibilities for flexible reacting, which in turn amplifies the effects of variances and disturbances. This amplification reduces the reliability of hierarchical planning and control. As a result, high material and time buffers are required, causing long lead times and high stocks. This leads to the fact that often only about ten percent of lead-time is production time; the remaining ninety percent is idle time (Schüpbach, 1994).

Suggestions on how to overcome the organisational deficiencies described as related to the hierarchical, control-oriented approach mainly concern the assignment of autonomy and control, the organisation of the information flow and the simultaneity of decision-making.

19.4.1.1 Distributed autonomy and control

Grote (1997) distinguishes between autonomy and control. On the one hand, autonomy refers to the decision latitude regarding goal setting and the definition of rules and procedures. Control On the other hand, concerns abilities and possibilities to influence situations in order to reach certain goals. Competence-oriented system design requires opportunities for (collective and individual) self-regulation, which incorporates both autonomy and control. The assignment of autonomy without adequate control is to be avoided, because autonomy does not incorporate the possibility to influence the situation. The (formal) competence of a planner to prescribe goals and procedures without concrete influence on the situation might easily lead to unrealistic prescriptions that impede the executor's opportunities of control. Hence autonomy must always be assigned together with appropriate control in order to increase an organisation's competence in coping with uncertainty. Promoting collective and individual self-regulation not only faces the human need of self-determination as an anthropological constancy, it substantially influences a system's capability to locally regulate variances and disturbances. Therefore resistance against decentralised self-regulation does not base itself on a rational reflection of its necessity. It rather arises from the myth of centralised plannability and steerabilty, from value systems incompatible with the idea of self-determination, and from power struggles (cf. Grote, 1997).

With reference to planning and scheduling processes, it is argued that decisions on the assignment of autonomy and control must be taken under explicit consideration of contingencies of the production process. There must be a fit between regulation requirements determined by the amount of uncertainty, and regulation opportunity determined by the amount of autonomy and control (Grote *et al.,* 1999). Distributed autonomy and control enhance the potential for human recovery in terms of the human's abilities to detect and correct possible system failures (McKay, 1992; Wiers, 1997). It optimises the openness and complexity of the system by making possible (individual and collective) processes of coping, learning, and development (Schüpbach, 1994). The higher the uncertainty to be coped with, the higher the need for self-regulation.

19.4.1.2 Multidirectional information flow

In the classical, hierarchical approach to planning and scheduling decisions are taken in specialised planning and scheduling departments and executed in the production units. Consequently formal information flow is organised in a vertical manner. Whereas decisions are communicated top-down, bottom-up information flow mainly concerns feedback on actual states in the production units. As a result, higher levels constrain lower levels using aggregated constructs or models of lower levels (McKay, 1992). However, an autonomy-oriented approach requires a different organisation of information flow regarding both, direction as well as content of information flow (McKay, 1992; Schüpbach, 1994; Grote, 1997).

Direction of information flow should not be restricted to organisational verticality. Information must not only flow top-down and bottom-up. What is required as well is a lateral information flow between units on the same level, be it directly or indirectly between productive units. As long as information flows vertically only, there is no possibility for units on the same level to laterally co-ordinate their actions. Consequently, disturbances in one production unit must always be communicated to the centralised scheduling department, which has to take adequate decisions. The decisions then have to be communicated again down to all concerned units. Independently of decision quality this organisation of information flow gives official channels with long information distances and many interfaces. This takes time and involves the danger of information loss at the interfaces. Moreover, it might also lead to a refusal of responsibility on lower levels, which in turn consider neither lateral information-giving (in terms of active inputs) nor lateral information-seeking (in terms of active out-takes) as their responsibility (Hoc, 1988).

Content of information flow restricted to bottom-up telling and top-down constraining may reduce decision quality. Higher decision-making levels are highly dependent on the aggregated information they get from the bottom. They may not be able to consider the more detailed and specific information that is available on-site. That detailed information only affects the decision-making process when lower levels are allowed to constrain higher levels as well (McKay, 1992).

19.4.1.3 Simultaneous decision-making

Decision-making in the classical, hierarchical approach to planning and scheduling is organised sequentially. The attempt to centrally control the process not only leads to the situation that higher level decisions constrain lower level decisions (see above), it also requires that higher level decisions always precede lower level decisions. Whereas the former (i.e. constraining) reduces the lower level's decision latitude, the latter (i.e. preceding) reduces the speed of lower level decision-making. Both have a large impact on the lower level's possibilities to act in a situated fashion (Suchman, 1987). In order to overcome that deficiency, co-resident control (McKay, 1992) or simultaneous decision-making (Schüpbach, 1994) is required. Decisions should be taken continuously throughout the organisation. On all levels people should be actively seeking problem solutions or - even better - possibilities to prevent problems. This is possible only when decision-making is simultaneous rather then sequential. Co-ordinating this simultaneity rather than controlling the manufacturing process remains the (more demanding) task of centralised planning and scheduling units. This comprises

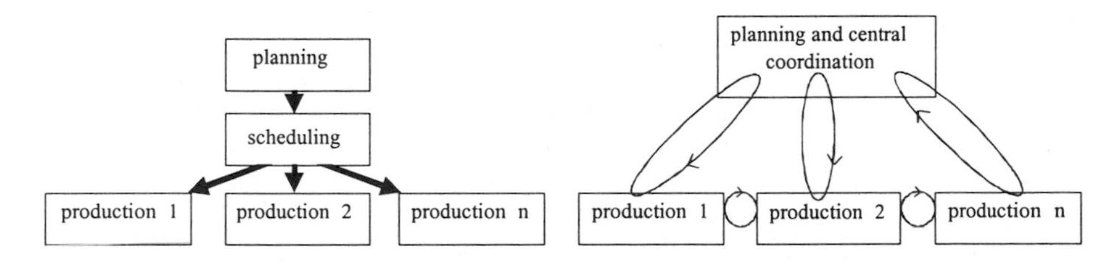

Figure 19.5 Main differences between the classical, hierarchical approach to planning and scheduling (on the left) and the autonomy-oriented approach (on the right) regarding information flow between organisational units (blocks).

responsibilities such as: ensuring an adequate information-base for the various decision-makers, providing them with appropriate decision support and making available adequate communication channels and decision bodies.

19.4.2 Intermediate conclusion

Regarding the organisation of temporal planning and scheduling processes, three aspects have to be considered carefully (see Figure 19.5): autonomy and control must be distributed throughout the manufacturing process; the information flow needs to be multidirectional in both senses, vertical as well as horizontal; and decisions should be taken simultaneously at meaningful decision points.

19.4.3 Individual planning and scheduling tasks

The nature of the work task assigned to an individual has a large impact on work behaviour (Emery, 1959). Motivation is not dependent on an individual's personality only. It depends to a high degree on the worker's task, which can be boring or challenging. Complete work tasks are a precondition for task orientation, a state of interest and commitment concerning the work task. Providing individuals with complete work tasks promotes intrinsic motivation as well as competencies and know-how. Complete work tasks contain opportunities for (Ulich, 1998):

- autonomous setting of personal sub-goals embedded in the organisation's super-goals;
- preparing actions in terms of carrying out planning of goal attainment;
- choosing means and relevant interactions to reach goals;
- executing the planned actions including feedback of progression to make possible action correction; and
- feedback of results including the opportunity to compare the attained results with the original goals.

The opportunity to plan one's own actions regarding both aspects, temporal as well as procedural planning, is a significant aspect of task completeness. People who plan their actions reach higher performance with reduced workload and emotional load than people that just act reactively (Hacker, 1998). To be able to act on the basis of self-planning has a higher effect on performance than intelligence (Battmann, 1984) or memory capacity (Wiesner, 1995). Of course, people who act on the basis of self-planning do not always find better ways. But as obvious as this insight is, the fact is that people who do not have the opportunity to act on the basis of self-planning will show work-alienated rather than task-oriented behaviour.

People who act on the basis of self-planning show a significant share of preparing and preventing actions (Hacker, 1998). The basic cognitive processes comprise goal-orientation, the availability of process schemes, and an inventory of signals that enable the individual to comprehend current process states and to predict possible process developments. Furthermore, such aspects as motivation, capability, and opportunity to actively influence the process are preconditions of acting on the basis of self-planning (ibid). Consequently such acting is possible only when the situation is (at least partly) controllable for the individual. It just makes no sense for a human to plan in a situation that is not controllable. Controllability requires comprehensibility and predictability of the situation as well as opportunities to influence the situation (Brehmer, 1993; Hacker *et al.*, 1994, Grote *et al.*, 1995). This relation of controllability and acting on the basis of self-planning has to be taken into explicit consideration in the design of sociotechnical planning and scheduling systems. Both the planners and schedulers as well as the executors need to be provided with means that render the situation controllable. Consequently, planners and schedulers that have no means to really control the situation may have low performance, i.e. may for example produce plans and schedules that are far from optimal. On the other hand, executors on the shop floor may have low performance too if the situation is not controllable for them, even when they get 'good' plans.

However, it is in the nature of sociotechnical planning and scheduling systems that total completeness of individual tasks is not possible. On the one hand, the planners and schedulers take over (part of) the temporal planning functions from the executors. On the other hand, the planners and schedulers lack executing functions. So planning and scheduling as well as executing remain incomplete work tasks. Deficiencies occur in autonomous goal setting and in choosing the means, as well as in feedback on progression and results. A closer look at the way humans plan is needed to find design parameters that are appropriate to compensate for reduced task completeness that lies in the nature of labour-dividing work structures.

19.4.3.1 Hierarchical versus opportunistic planning

In traditional theories of action planning it is assumed that people plan in a hierarchical manner (Miller *et al.*, 1960; Schank and Abelson, 1977; Hacker, 1998). It all starts from a main goal that is divided into several layers of sub-goals. The lowest layer contains sub-goals that are reachable by performing simple routines. The sequential execution of these routines is considered to be the observable part of acting. The mental structure of action regulation that lies behind it, i.e. the working through the goal hierarchy, is understood as a hierarchical-sequential process. After the execution of each routine people test whether they have reached the sub-goal. If the test is positive they turn towards the next routine. If it is negative, they have to perform the routine again, to choose an alternative routine, or to dissolve the sub-goal into lower level sub-goals. A goal of a higher level is attained when all its sub-goals are attained.

This conception of humans as totally rational information-processing beings is criticised by some authors. Planning is considered to be an opportunistic (Hayes-Roth and Hayes-Roth, 1979) or a situated (Suchman, 1987) process rather than a hierarchical one. The main criticism concerns the fact that hierarchic planning is possible only if the main top goal is set clearly as well as non-contradictorily, and if the different levels of sub-goals can be completely set before the start of any action. It is argued that in real-life planning this is seldom possible. Planning is mostly a multidirectional process rather than a top-down one and it turns out to be an incremental process rather than a complete one. Moreover, goals can often be contradictory, which indicates rather a heterarchical than a hierarchical goal structure. Only if these characteristics of real-life planning are taken into consideration can situated acting in terms of reactive as well as proactive flexible adaptation on the situation be possible. Whereas hierarchical planning requires complete planning ahead, opportunistic/situated planning provides the planner with various opportunities for plan development. Triggered by new ideas, observations or realisations the planner can, at each point in the planning process, revise previously taken decisions. The subsequent decisions follow up on selected opportunities. Interim decisions can lead to subsequent decisions at arbitrary points in the planning space at higher or lower levels of abstraction, specifying actions to be taken at earlier or later points in time (Hayes-Roth and Hayes-Roth, 1979).

In order to make possible opportunistic planning and situated acting on the different levels of a work organisation, the functions and related plans must be carefully reflected. Volpert (1994) addresses the former when he states that detailed plans are required in extreme cases only. The actor does not need to plan ahead for actions he has performed several times before. There is another mode of action for such cases of trained, experience-guided action: intuitive-improvising action. Suchman (1987) refers to the latter. For her - wherever plans are needed - their function is not to serve as detailed specifications, but rather to orient or position people in a way that will allow them, through local interactions, to exploit some contingencies of their environment and to avoid others. This is a fundamentally different function of plans. It is comparable to plans in terms of city maps in contrast to plans in terms of detailed route-descriptions. A city walker who gets a detailed description on how to get from point A to point B is lost whenever he meets with an obstacle, even though the description provides him with the optimised way. Confronted with the obstacle he has no other choice than to go back the very same way and ask for a new plan. However, if he is equipped with a reliable city map and if he is well-trained in reading it, he might not always find the most efficient way, but he will find his way whenever obstacles emerge. Plans, like city maps, are neither detailed orders nor instruments of controlling but rather an enabler for situated acting.

19.4.3.2 Limitations of planning for others

Planning for others is limited by the fact that opportunistic planning and situated acting is possible on site only (see above). Planning for others can be considered from the point of view of the planner as well as from the point of view of the 'other'. In his conceptual framework Resch (1988) focuses on the planner. He distinguishes between an actual and a referential field of action. The actual field is the field in which a person actually acts. For the planner this is the field of planning. Within this field he executes 'only' the planning, he never executes the plan. The referential field is the field to which the planner's actions refer, the production process on the shop floor. The product of the planner's actions consists of (more or less) fixed sequences of actions that have to be executed in his referential field by somebody else. Resch (1988) describes seven phases of action for planners (which of course are not meant to be always followed in this sequential way):

- *Orientation in the actual field*: Procurement of information about the planner's planning opportunities as well as the possible planning procedures (e.g. How could I plan?).
- *Planning in the actual field*: Definition of planning procedures and products (e.g. How and what will I plan?).
- *Orientation about the referential field*: Elaboration of a mental representation of the world to be planned (e.g. What is the executor's situation?).
- *Planning for the referential field*: Mentally going through acting possibilities for the world to be planned. Selection of one possibility (e.g. What must the executor do to reach optimal performance?).
- *Execution of planning in actual field:* Producing the plan (e.g. How do I symbolically represent the result of the last phase's mental work?).
- *Feedback from the actual field*: Check whether the planning process went as planned, with reference to efficiency of the planning process, completeness of the planning result, and the like (e.g. Did I plan well?).
- *Feedback from the referential field*: Check whether the planned actions and results have been realised (e.g. Has my plan been executed properly?).

This detailed description shows clearly that the planner as a mental worker acts in a material as well as in a symbolic world. In his material world he executes the planning itself. His actions get materialised in the form of plans. In contrast to physical work, the materialisation of the planner's work (the plan) is *not* directly related to its object. Consequently for the planner there always results a gap between the symbolic and the physical object of his work. This gap - that is inherent in the nature of a planner's/scheduler's task - limits planning for others. When the planner's/scheduler's symbolic object of work guides actions of people in the production department, its quality depends on the distribution of knowledge about the production department, on the share of constant conditions in the production department that do not have to be updated permanently, and on the encoding state of the information related to the production department. Resch (1988) summarises: 'The knowledge of the real object remains a precondition for planning for others. Otherwise planning is not realistic and the product of mental work will not satisfy its practical value' (p.29, transl. TW).

From the point of view of the 'other', planning for others is limited too. Dahme (1997) distinguishes between two major phases of acting. In the first phase actions are constituted, in the second phase actions are executed. The constituting phase requires:

- the definition of the goal of an action;
- the ability and the opportunity to assess the action's goal, circumstances, and expected results; and
- the ability and the opportunity to decide on the execution of the action.

Only when an action is constituted, can it be executed. The executing phase requires:

- the ability and the opportunity to execute the action;
- the object of the action; and
- the means to execute the action.

As seen above, it is questionable whether these two phases - that again represent autonomy and control (Grote, 1997) - can be separated and assigned to different people, because a high degree of mutual knowledge is required.

However, the psychological effects that a separation of the constituting and the executing phase on the executors also results on limits of planning for others. Humans that only execute what somebody else determines perceive themselves to be controlled externally. In such a situation they locate the locus of control of whatever happens with and around them out of their own sphere of influence. It is clear that such attribution leads to low motivation, due to experience of low self-efficacy (Bandura, 1982). With reference to planning processes humans that have no opportunities to participate in the constituting phase are reduced to an object of other people's decisions. Consequently they react passively to these decisions, and follow none of their own goals, nor are they aware of the meaning of their actions. They lose decision-making abilities and show general passivity regarding their work task (Dahme, 1997). Those people that are able to react to external control try to prevent themselves becoming an object. To regain subjectivity they have to set their own goals (Dahme, 1997). These goals might or might not be work-related, but in any case they must be different from the goals set by the planner. Otherwise they are not appropriate for the regaining of subjectivity. Such self-set goals can spread from setting priorities other than the ones prescribed by the planner, as far as to consciously sabotaging the externally-set prescriptions. The meta-goal of such behaviour is not to sabotage the company but to regain subjectivity. Strohschneider (1993, p. 41, transl. TW) states in this context: 'It can be more important for the human to control the whole situation (including all personal and social aspects) than to reach concrete, pertinent goals'. However, whether people that are reduced to the execution phase of acting are or are not able to react, in both cases their behaviour is not suitable to support the planner's goals.

As a conclusion it can be said that an allocation of planning and executing to different people may be problematic due to several reasons. On the one hand, the planner always plans in a theoretical (symbolic) world. Consequently he might not have an adequate information base of the physical world he makes plans for and he might be tempted to optimise his own work, which is the planning. On the other hand, planning for others may also be limited from the point of view of the executor. The simple fact that other people plan for him excludes him from the decision loop. This may cause passivity because it provokes a feeling of being controlled externally. Hence he may lose motivation and commitment regarding both the plans and the work itself.

19.4.3.3 Consequences for the assignment of planning competencies

The reflections on human behaviour and planning made so far make clear that planning and executing aspects of a task cannot be fully separated and assigned to different people. Such a separation would not only have effects on the objective controllability of the situation, but also on the motivation and performance of the individuals, who are reduced to planning or to executing functions. In order to prevent such deficiencies an adequate balance of autonomy and control is required on an individual level too (as much as on an organisational level). It is the only way to make it possible for individuals to act in a plan-based fashion as well as a situated fashion. Consequently, as many planning competencies as possible should be allocated to the executor in order to make possible opportunistic planning and situated acting on site as well as to enable task orientation. The demanding task of co-ordinating and integrating such distributed autonomy remains for the planner. However, in labour-dividing work structures the executor will never reach total autonomy and the planner will always lack control opportunities. Hence, the traditional sociotechnical call for the design of complete tasks by always assigning planning together with executing functions has problems.

The question arises of whether there is another possibility to compensate for the lack of completeness of individual tasks. Mutual knowledge might help by substituting the assignment of complete tasks; individual tasks of executors should be designed in a way that incorporates as many planning aspects as possible, together with transparency over other people's planning and plans. Mutual transparency of planning and plans can be reached by explicitly assigning tasks to individuals that are related, i.e. that incorporate the necessity of joint problem-solving and therefore set communication requirements (Dunckel *et al.*, 1993). On the other hand, making planning and plans more explicit might also enable mutual transparency. In their concept of opportunistic planning Hayes-Roth and Hayes-Roth (1979) assume the existence of a kind of a mental blackboard with different zones (plan, plan-abstractions, knowledge-base, executive, meta-plan). Within these zones decisions can be taken and changed permanently at arbitrary points of time. A task design that allows for a kind of shared blackboard might promote transparency of the planning process and therefore could be an appropriate compensation for incomplete tasks (see also McKay's (1992) concept of shared work space). In that sense mutual knowledge regarding each other's actual situation, goals, and intentions is supposed to compensate for the cracks in the planning process that arise at the interfaces between and among planners and executors. This conclusion is supported by Zölch (1997) who found in her analysis of shop floor scheduling the following hindrances to action co-ordination:

- insufficient knowledge about work process;
- insufficient opportunities for mutual information update;
- insufficient correspondence of documented and actual state;
- insufficient correspondence of goals;
- hindrances in communication;
- actively created hindrances (e.g. withholding of information);
- lack of resources (e.g. no redundant machines).

It is only the last of the identified hindrances that is not directly related to mutual knowledge. All others can only be compensated for by appropriate task design that incorporates communication requirements and transparency of workflow. In addition, adequate (organisational as well as technical) channels of information flow must be provided.

19.4.4 Intermediate conclusion

The assignment of complete individual work tasks is a precondition for motivation as well as to develop and utilise people's know-how. Whereas the executors' tasks usually lack planning opportunities, the planners' tasks normally comprise execution of planning but no execution of plans. In order to allow possibilities for opportunistic planning and situated acting as well as to compensate for limitations in planning for other people, two aspects must be considered carefully. Adequate autonomy and opportunities for control must be allocated to the executors as well as to the planners. Mutual knowledge must be guaranteed in order to overcome problems in co-ordination.

19.4.5 Human-computer interaction in planning and scheduling

Temporal planning and scheduling as an individual or as a multi-person process is subject to automation. A lot of effort is put into the design of sophisticated information processing technology that is meant to either support or to replace the human planner or scheduler. The implementation of systems of both types, however, mostly does not lead to expected results (e.g. Hafen, 1999). As a consequence, many systems are not really used by the human planners and schedulers (Wiers, 1997). Support systems fail to support co-ordination of jobs because they do not provide users with complete information, with clarity of temporal requirements (e.g. due dates, priorities), or with actual representations of process states (Zölch, 1997). Systems that aim to replace the human are criticised more fundamentally. Resch (1988) questions the automation of mental work in principle. He argues that mental work in a referential field of action will always produce unique results. 'Mental work comprises such activities that can be rationalised and such that cannot. But it cannot in its essence, i.e. activity planning for others, be rationalised, but can be replaced by the means of formalisation and standardisation. However, such replacement changes the quality of mental work. How far an enterprise can accept this loss of quality without higher frictional loss than productivity gain is a question that can be clarified empirically only' (Resch, 1988, p. 104, transl. TW). He reasons further that it is quite possible to partialise a referential field of action. But this does not lead to higher productivity of mental work, because the referential outcome will also be changed. The efficiency of mental work is not increased by this strategy, only its task is reduced. It is questionable whether this would produce more advantages or more disturbances in terms of problems in information flow. Sanderson (1989) has described the deficiencies of the two 'classical' approaches to scheduling automation, analytic/algorithmic control systems and expert systems. Her criticisms mainly concern the simplifying character of models and the incompleteness of knowledge bases and inference engines. Sanderson argues that in either of the two approaches the presence of a human monitor is crucial to successful performance. For three reasons she pleads for hybrid or human-computer interactive systems. i) Humans are needed to troubleshoot because no control algorithm or expert system can handle all system abnormalities or changes in production requirements. ii) If the human is becoming too remote from the process he might not have an up-to-date mental picture of the current process state and even lose his long-term mental model of system functioning and structure. iii) Human-computer systems often outperform human or computer individually.

Improved computer technology has led to automation approaches that follow the idea of artificial intelligence. Scheduling problems might be distributed to virtual agents. Each of these agents supports a clearly defined task and has its own specific problem-solving strategies. Agents have information about the problem area they are 'responsible' for (e.g. a group of equipment) and they can request services from other agents. The agents use simulation to determine the future consequences of their decisions. Events from the ongoing process are forwarded in real time to the agents. The agents compare the actual state of the process with the planning data, taking appropriate actions if necessary (Luethi *et al.*, 1996). Like the 'classical' approaches to automation, the agent-based approach also considers technology as an enabler of dynamic process control. It is based on (quantitative) models of the production processes. The quality of these models influences efficiency and effectiveness of the automated control system. In the process of generating and comparing different scheduling solutions by computer simulation, only those variables and interactions represented by the models are considered. Effects of variables that are not represented by the models (e.g. the human operators' motivation) are neglected. Because of this fundamental problem, even the more sophisticated technology of artificial intelligence does not guarantee technical control over highly dynamic production processes (Bossink, 1992; Fischer, 1993; Wiers, 1997; Vernon and MacCarthy, 1998). Hence Sanderson's reasoning for hybrid systems has not yet lost its relevance.

19.4.5.1 The necessity of complementary function allocation

When designing human-computer interactive systems the degree of automation must be determined, which includes the allocation of functions between human and technical systems. This has important consequences for the specification of technical requirements, for the design of the jobs, and for the efficiency, quality, and safety of the automated processes. The allocation of functions will significantly affect flexibility, not only because computers are still inflexible by comparison, but also because functions can be allocated in a way that renders it very difficult or even impossible for the human to use his or her flexibility. Two important issues in this context are the opportunities for the development and maintenance of practical skills provided by a given allocation of functions and the shift generally required in more automated systems from practical to theoretical systems knowledge. How both of these issues are dealt with in terms of technical and job design, as well as the skills required of the human, will strongly influence the degree to which human, potential can be employed.

There are several strategies concerning decisions on function allocation (see Table 19.2). Besides overriding economic considerations, a still frequently adopted strategy is the leftover principle, which allocates whatever functions cannot be automated to the human. These functions do not necessarily form a meaningful job, they might even be impossible to carry out, as Bainbridge (1982) has argued convincingly. Her criticism also applies to a third approach to function allocation, he comparison principle (Bailey, 1989). Based on lists comparing characteristics of humans and machines, functions are allocated to the one who supposedly can perform that function better (Fitts, 1951). With increasing technical possibilities, these lists change since more functions can be performed better by machines, leaving monitoring and dealing with deviations and disturbances to the human. Jordan (1963) pointed out more than thirty years ago that humans and machines are fundamentally different and therefore cannot be quantitatively compared and assigned functions. Instead, they are to be seen as complementing each other in joint task fulfilment. Function allocation should allow for the support and development of human strengths and compensation of human weaknesses by the technical system in order to secure quality and efficiency of processes as well as humane working conditions. Complementary design takes into explicit consideration that human and technical systems - based on the strengths and weaknesses of both - can achieve through their interaction a new quality possible neither to human nor technical systems alone. There have been some attempts to operationalise criteria for complementary design, but to date neither a fixed set of criteria, comparable for instance to software usability criteria, nor widely accepted methods for their measurement exist (see also Older *et al.*, 1997).

Table 19.2 Criteria for the decision on allocation of functions to humans and technology respectively, as well as the implicit assumption that lay behind such criteria (Wäfler *et al.*, 1999, transl. TW)

Criterion for the decision on the degree of automation	**Implicit assumptions about**	
	Humans	**Technology**
Cost efficiency: Costs of automation and operation should be minimised.	Humans as well as technologies are primarily considered to be *cost factors.*	
Leftover: The human is excluded from the process as much as possible. Whatever is technically feasible should be automated.	The human is considered to be a *disturbance and risk factor.*	Technology is considered to be an *effectiveness and safety factor.*
Comparison of performance: Technology should replace the human wherever it is more powerful.	Humans as well as technologies are considered to be *competing factors* that can replace one another.	
Situated flexibility: Humans and technologies should be able to develop their different strengths in dependency of process requirements.	Humans as well as technologies are considered to be *complementary valuable resources* that can compensate for each other's deficiencies and further each other's strengths.	

19.4.5.2 The KOMPASS approach to complementary function allocation

KOMPASS is a method that supports complementary systems design in industrial settings (Grote *et al.*, 1995; Grote *et al.*, 2000). It provides guidelines for an integral approach to systems analysis (Grote *et al.* 1999), as well as to participatory systems design (Wäfler *et al.*, 1999). In contrast to other approaches based on a similar theoretical background (Dunckel *et al.*, 1993; Clegg *et al.*, 1995) KOMPASS provides measurable criteria for the quality of the human-machine/computer interaction, which are directly relevant for technical design specifications. The KOMPASS criteria are based on the objective of furthering human control over technical systems and automated processes. As suggested by control theory (Brehmer, 1993; Hacker *et al.*, 1994), three different prerequisites for control are distinguished: comprehension, predictability, and influence, which are operationalised by the criteria of process transparency, dynamic coupling, decision authority, and flexibility (see Figure 19.6).

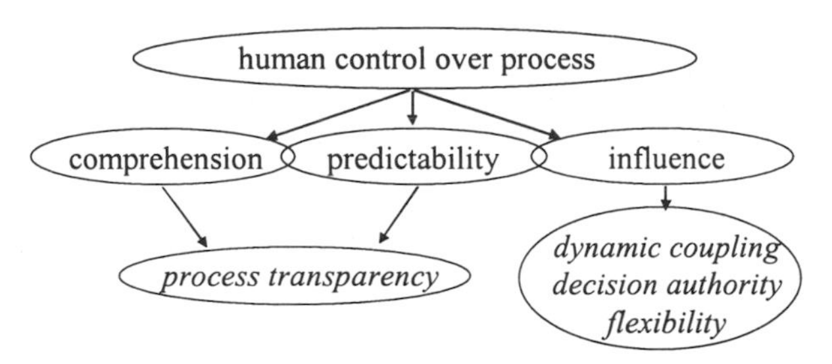

Figure 19.6 KOMPASS-criteria for analysis and design of human-computer function allocation (adapted from Corbett, 1985; Sheridan, 1987; Clegg *et al.*, 1989; Kraiss, 1989).

Process transparency. A crucial prerequisite for fulfilling supervisory control tasks is the transparency of processes for the human operator, which permit mental models adequate to the tasks to be performed by the human to be formed and maintained (e.g. Wilson and Rutherford, 1989). This requires active involvement in the processes and direct process feedback. To determine the amount of process transparency provided in a particular human-computer system, the opportunities for gaining an understanding of the general nature and temporal structure of the process and the required process interventions are analysed. In addition, types and extent of available process feedback are determined and evaluated on the basis of a sufficient amount of direct feedback and a clear distinction between information feedback stemming from process simulation versus the real process. For scheduling tasks, process transparency refers to the transparency of the scheduling process itself. In order to enable the human to make meaningful interpretation of automatically generated scheduling solutions, it may be important that they understand *how* these solutions are derived. If they don't understand the underlying models and algorithms they might not be able to judge the solutions.

Dynamic coupling. This criterion describes the degree of control the human operator has with respect to the coupling between him or her and the technical system. Tight coupling as well as total decoupling of the human are to be avoided. Instead, technical options should be provided that allow for a dynamic coupling in the sense of the concept of loose coupling (Weick, 1976; Orton and Weick, 1990). Coupling is determined with respect to time, place, work procedures, and required cognitive effort. The latter refers to the level of information processing required by the human operator in regard to the technical system, independent of his or her physical proximity to the technical system. For all these four kinds of coupling, technical options provided to the operator for varying the degree of coupling are determined. With reference to scheduling tasks an example for high dynamic could be the option to use different job scheduling procedures. This is possible only if the human-computer interaction is designed in such a way that provides the human with a variety of procedures as well as with the possibility to choose among them.

Decision authority. The distribution of decision authority in a human-computer system determines to what extent the human operator and the technical system control the actual processes (e.g. Sheridan, 1987). Decision authority is analysed with respect to controlling both the access to information from the production process as well as the process itself. The distribution of decision authority can vary between fully manual, manual with automatic support or constraints, automatic with manual influence (even veto) to fully automatic. Scheduling tasks must be automated in a way providing the human with access to the information needed as well as with possibilities to influence the scheduling process. Otherwise they will be excluded from the decision loop. While the design goal for all other criteria is ‘the more, the better’, the design goal for decision authority is an 'internal' fit between the authority regarding information access and process control and an 'external' fit between overall decision authority and responsibility assigned to the human operator for process efficiency and safety.

Flexibility. Human-computer systems fulfil the criterion of flexibility if it is possible to switch between different levels of decision authority for a given function. The way decisions are made concerning the use of such flexible function allocation can be considered as a higher order decision authority. If this authority lies with the human operator, the human-computer system is adaptable; if it lies with the technical system, the system is adaptive. In most cases, adaptable flexibility is the design goal because it allows the human to freely assign functions to the technical system or to him/herself based on considerations of over- or underload. Such flexibility is also a basis for dynamic coupling. Adaptable automated scheduling tasks provide the human with the possibility of choosing between (semi-)automated and manual scheduling procedures. This might be useful to keep the human in the decision loop and make use of the advantages of highly sophisticated computer technology.

19.4.6 Intermediate conclusion

In the design of the technical part of the planning and scheduling system, in terms of a secondary work system, two aspects must be considered carefully. On the one hand, technology supports efficient job co-ordination only if it provides the user with complete and actual information. On the other hand, human-computer interaction must be designed following a complementary approach in order to make possible human control over automated planning and scheduling processes. Consequently the human-computer function allocation must guarantee process transparency, dynamic coupling, human decision authority, and flexibility.

19.5 A CASE STUDY

19.5.1 Method

An exploratory case study has been chosen to illustrate the nature of information that is processed in a sociotechnical planning and scheduling system as well as the yet unknown relations between causes and effects. Wiers (1997) tried to quantitatively assess scheduling systems. He found that such an approach does not really lead to useful results in terms of the derivation of hints towards a 'better' design. He reasons that a classical, quantitative assessment must fail because of its inherent necessity of oversimplification of complex real-world processes as well as its causality-oriented theoretical framework that leads to a measurement of input-output relations and leaves the actual scheduling process hidden. In their research agenda for scheduling, MacCarthy and Wilson (2001) also deplore the poor effects of the classical approaches:

> 'There has been a vast amount of academic research on theoretical and algorithmic aspects of production scheduling in the last forty years. However, there is a very wide gap between theory and practice. In fact, the classical theory has had almost no impact in actual scheduling practice. ... In reality, it is the human scheduler who must address the multiplicity of factors influencing real-time schedules' (p.2).

Hollnagel and Cacciabue (1999) convincingly plead for field studies instead of laboratory experiments: 'In research concerning human-machine systems, the real world is the world of work' (p.5).

To be a fruitful method a case study must incorporate a battery of instruments that are to be used systematically in order to ensure a scientific basis, without being too rigid by forcing the research subject to adapt to the theoretical concepts (Robson, 1993). Hence in this study semi-structured interviews, whole-shift observations, and written examination tools have been employed to investigate the company's planning and scheduling system. A total of 22 people from different hierarchical levels and with different functions (planners, schedulers, shop floor supervisors, foremen, and workers) have been investigated.

19.5.2 The company

The company produces high-end plumbing articles such as kitchens and bathrooms for private households as well as for commercial use (e.g. canteen kitchens, hospitals, hotel businesses). Its products are sold world-wide. Independent traders buy directly from the company and resell to the plumbers. The company employs about 500 people. Around 300 of them work in the production unit (250 directly productive, 50 indirectly productive). The database of the company's ERP-system includes around 15,000 customer order positions and 4,000 articles that are purchased from about 250 suppliers. The scope of production is fairly wide (casting, pressing, milling, drilling, grinding, polishing, surface treating, varnishing, assembling). Additionally, purchased articles are mainly accessories and plastic parts. The production unit is divided into a manufacturing part (250 employees, 6 sub-units) and an assembling part (50 employees, 5 sub-units). A stock buffer separates the manufacturing from the assembling. The sub-units of the manufacturing unit work sequentially, i.e. most work-pieces pass through all sub-units. Order processing is planned and scheduled centrally by an MRPII-like system based on forecasts. The assembling unit is organised in independent assembly lines; each assembles final products. If pre-assembling is necessary it is separated from the final assembling by the same interim stock that separates it from the manufacturing unit. The assembly lines produce for customer orders mainly. The orders are communicated directly to the assembly lines. Minimum lead-time of assembly is two days. Between the interim stock and the assembly lines a KANBAN-system is implemented. The lines are allowed to assemble on a small on-site stock, which they have to control themselves, in order to give them the possibility of smoothing out variances in customer orders.

19.5.3 Results

The following results of the case study are presented. Firstly, the company's secondary work system for temporal planning and scheduling is described. Then its characteristics on the three levels of organisation, individual planning-task, and human-computer interaction are presented, illustrated by statements picked up during the interviews and observations from planners, schedulers, and shop floor representatives. These statements are - of course - translated from Swiss-German to English by the author.

19.5.3.1 The secondary work system of temporal planning and scheduling

The general information flow as well as the decision-making process in the company's temporal planning and scheduling system is presented in Figure 19.7. The main task of the planners is to adjust the ERP-software's parameters in order to reach certain goals (e.g. low stock, service level). Their main information sources are the sales people and the historical data. The ERP-software makes job proposals to the schedulers, whose task it is to check the proposals' plausibility, to check the readiness to fulfil the proposed job (e.g. availability of material and production resources), and to release it (including make-or-buy decisions). The manufacturing units are controlled centrally by the scheduler, in a top-down cascaded way via supervisors and foremen down to the workers. The assembly units get their jobs directly from the ERP-software. The foremen do the scheduling for their work groups. However, assembling is highly standardised in a way that no critical decision-making is required by the foremen.

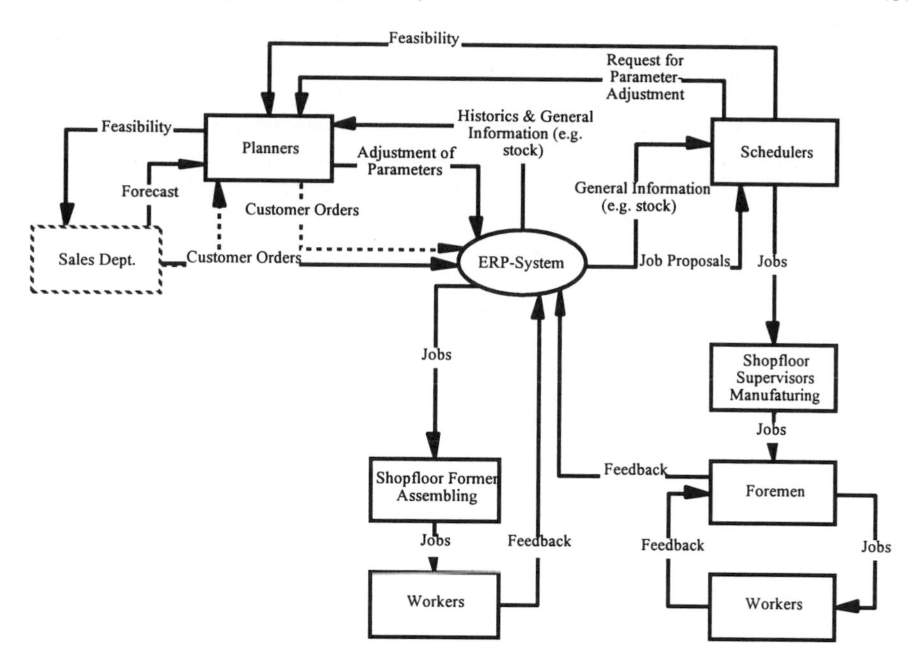

Figure 19.7 The company's general system of information flow and decision-making in temporal planning and scheduling.

19.5.3.2 Organisational aspects

Temporal planning and scheduling is performed with the close co-operation of planners and schedulers who are located in the same office. Whereas planners care more about long- and middle-term planning, schedulers concentrate on the scheduling. Short-term planning is performed jointly. Multidirectionality of information flow as well as mutual constraints are guaranteed by permanent communication and joint decision-making regarding short-term planning. By checking the plausibility of automatically generated order proposals the schedulers implicitly check the usefulness of the system's parameters that are fixed by the planners. The opportunity to request alterations to parameters and the readiness of the planners to do so in mutual agreement, as well as the inclusion of the schedulers in feasibility-based decisions regarding special orders, allows for bottom-up constraints. The schedulers' rather broad scope of decision-making allows for simultaneous decision-making.

The interaction between the schedulers and the shop floor is quite different. The schedulers strongly constrain the manufacturing units. They try to optimise the schedules for those resources they consider to be the bottlenecks (which are located in the milling-drilling unit, the second unit in the process). For these resources only the jobs of the actual day are released with detailed prescriptions regarding sequencing. Although the ERP-system provides the scheduler with order proposals for several weeks, the schedulers hold them back, even if they are plausible. They do this in order to keep flexibility, as they never know whether rush orders will make their schedules obsolete. But the statements they made in the interviews indicate also a need for central control (see Table 19.3).

Table 19.3 Schedulers' statements regarding centralised control

Statements of schedulers
- 'We *release* orders *when we see* that they run out of work.' - '... to have control that they really do the work according to the dates *we have set.*' - '*I think* there will be a problem with a certain machine because *I know* what might come next.'

As a consequence, lateral co-ordination between the units does not take place. On the one hand, the schedulers criticise this situation. On the other hand, the shop floor supervisors complain that they have not the information base to actively co-ordinate with their colleagues (see Table 19.4).

Table 19.4 Statements regarding lateral co-ordination on the shop floor

Statements of schedulers	Statements of shop floor supervisors
- '*I must* do the co-ordination between the units.' - '*I must* call and ask them to produce and deliver certain parts immediately because the next unit urgently needs them.' - 'Although they sit next door, *he calls me* and asks me to look after parts he needs from his colleague.'	- 'The due dates on the schedules *correspond poorly* with the due dates of the next unit.' - 'I *only* get information *from the scheduler.*' - 'Where the next operations will be performed is *not really clear* from the documents.' - '*I don't know* about the others' problems.'

The centralised decision-making leads clearly to the situation that everybody delegates responsibility to the schedulers. If a problem occurs, they are expected to work out solutions. Bottom-up information flows operate as formally prescribed, i.e. the shop floor (mostly) gives feedback on job completion through the ERP-system. But this information does not nearly cover the information need of the schedulers. They are permanently gathering information by means of walking through the factory and making intensive use of their telephone. In order to get an up-dated mental representation of the current state of production there are many hints they pay attention to. Remaining up-dated is a rather hard job, as can be illustrated by an incident that occurred during the observation of a scheduler. Confronted with a large rush order to be delivered as soon as possible he was asked by a planner to determine a realistic delivery date that could be communicated to the customer. The scheduler set a date five weeks ahead. Afterwards he told the observer that the actual production would take only one week: 'But I can only rush up to three orders through the organisation at once. Otherwise I would be permanently phoning and have no time to do anything else.'

Top-down information flow from the schedulers to the shop floor is restricted to the communication of decisions. Consequently, the shop floor's comprehension of goals and working of the scheduling departments is low (see Table 19.5).

Table 19.5 Shop floor supervisors' statements regarding comprehension of goals and working of the scheduling department

Statements of shop floor supervisors
- 'I have *no idea* of the procedures in the planning and scheduling departments, I could *only guess* from statements they make on how they might work.'
- 'You *have to ask* if you want to know why certain decisions have been taken. If you know the circumstances you often agree with their decisions.'
- 'I have *never been up there.*'

The interaction between the schedulers and the assembly units is slightly different. As mentioned above, the assembly units get the orders directly out of the ERP-system. They schedule their orders themselves. However, there is not really a need to schedule the orders. Priorities are pre-determined by delivery dates. The whole order is completely assembled in one unit. Hence (almost) no co-ordination with other units is needed. Set-up times are short and consequently make co-ordination of jobs necessary. Components are available on site, controlled by a KANBAN-system. If there are any disturbances the assembly units' foremen delegate the problem to the schedulers or to the head of the interim stock, who himself chases for jobs in the manufacturing units or asks the schedulers for help.

In the relationship between the scheduling departments and the shop floor, autonomy is centralised, there is no lateral information flow (e.g. there is no direct mutual co-ordination between production and assembly units), constraints only go top-down, and simultaneous decision-making has not been observed.

19.5.3.3 Individual planning tasks

The execution of planning has been observed with planners and schedulers as well as with some shop floor supervisors. All of them take a lot of planning-related decisions intuitively based on their experience (see Table 19.6). At the same time none of them really has the opportunity to assess the quality of decisions (see Table 19.7).

However, planners seem to plan most systematically. By the means of spread-sheets which they have programmed themselves they try to compute optimal parameter adjustments on the basis of historical data and forecasts. Due to the nature of their task immediate feedback is poor. Conclusions regarding the quality of decisions can only indirectly be drawn from the observation of trends in the development of stocks, service levels, and the like.

Table 19.6 Statements regarding the intuitive nature of planning and scheduling

Statements
- 'To take good decisions one must have a *good nose.*' (planner) - 'Apart from the automatically generated job-proposals there is neither decision-support nor are there decision-rules. Decisions are mostly made *in good feeling.*' (scheduler) - 'How much material you order from the stock is a *matter of personal judgement.*' (shop floor supervisor) - 'We always *fly blind.*' (head of interim stock)

Table 19.7 Statements regarding feedback on the quality of planning decisions

Statements
- 'The quality of single decisions *cannot be seen.*' (planner) - 'It comes to light *later* if you - what happens seldom only - hit and miss.' (scheduler) - 'As long as nobody complains ... .' (shop floor supervisor) - 'You get a *'Thank you'* when it works.' (shop floor supervisor) - 'If I don't know what I don't know, *I cannot ask.*' (shop floor supervisor)

Table 19.8 Statements regarding top-down constraining

Statements of schedulers	Statements of shop floor supervisors
- 'If there is an overload on a certain group of equipment then *I say* where to produce. Maybe in arrangement with the shop floor supervisor, but what has to be done *comes from me.* ... because *I know* what to do better.' - 'The problems on the shop floor are probably manifold. But I don't think they have problems with scheduling. Their problems are related to the *production technology.*'	- 'With regard to the order sequencing and the assignment of jobs to machines we have to decide *according to the schedulers' decisions.*' - 'All I can do is *wait and see* whether everything was right.' - 'We *lack information* from above.' - 'The due dates on the documents are *not reliable.*' - 'We *can't work* according to the due dates on the documents.'

Sandwiched between planning and executing, the schedulers have to plan opportunistically because it is they who (try to) match the plan with reality. Scheduling is a multidimensional problem:

> 'We don't want to produce too early. But we always want to have enough stock of every product. We don't want to produce anything we don't need afterwards. We want to have it when we need it. But we also want to ensure that there is always something to do on the shop floor. We don't want to force them to work like crazy in Spring and to have nothing to do in autumn. So ... there are many different things we have to consider. It can also be that we release orders too early, when we think that they have little work in a group of machines. That can be a factor too.'

To reduce the scheduling problem's complexity and to keep in (imaginary) control, the schedulers are forced to simplify the problem by reducing the variables that are considered, and - of course - by constraining the executing units on the shop floor. Simplification could be reached by non-consideration of capacities: 'When an order is released, lead time can take up to six weeks. You never know the workload in five weeks, hence ... we don't just look at capacities. If a capacity problem occurs we look for other solutions: activate other machines, try to get it produced externally.' Constraining as far as scheduling goes is clearly seen as task of the schedulers (see Table 19.8).

The shop floor supervisors react in different ways to the fact that scheduling is clearly assigned to the schedulers. Some of them are just reactive, refusing any responsibility (see Table 19.7). They rely heavily on the instructions they get directly from the schedulers. They do not (even try to) schedule according to the information they get from the documents, because they have the experience that this information is not reliable, and that - in case of lack of clarity - it is better to ask the schedulers, because they are supposed to have an overview. On the other side, some (but not many) of the supervisors also try to be more active. As an example, the following describes one of them in more detail. The casting unit is the first in the manufacturing process, followed by the milling-drilling unit with the bottleneck machines. As the schedulers schedule these bottlenecks on a daily basis (see above), the assignment of jobs to these machines might change daily. However, casting is a process that takes several days, setting-up the furnaces is time-consuming. Hence, jobs cannot be rescheduled frequently. Consequently, the casting unit often produces the 'wrong pieces', i.e. not the pieces that would be needed actually on the bottleneck machines of the subsequent unit. The casting unit's supervisor described the situation as follows:

> 'This is a problem for me. I cannot sort the orders according to due dates, because there are no due dates for casting, only for the whole manufacturing process. I have information on the latest start only. When I add it to the allowed production time, I can guess when I should have completed an order. Now - latest start is never right, I cannot rely on it. So I have complained and I have asked the schedulers to give me reliable information, to enable a meaningful scheduling of the casting unit. As a provisional solution they provide me with the job-list of the bottleneck machines of the subsequent milling-drilling unit. They just give me this list of usually about twenty jobs and fill in - with a pencil - priorities for eight jobs. To be able to work according to these priorities - which actually are priorities for the milling-drilling machines, that are not in my unit - I had to help myself, by introducing my own table, in which I list all the sub-operations of casting and sequence them according to the prescribed priorities. That's how we work. And everything that comes in between or that is not priorised by the schedulers we just work off ... hoping to do it the right way.'

The casting unit's supervisor (still) tries to actively optimise the scheduling for his unit and to adapt to the need of the subsequent milling-drilling unit. But he lacks the information he would need to do so. The schedulers do their best to support him. But as long as they see it as their task to control the whole manufacturing process instead of making possible local regulation, they will never succeed in handling all the relations. Consequently, they will not be able to provide the casting unit's supervisor with the information he requests. It might be a detail, but nevertheless surprising, that the casting unit's supervisor is well known among the schedulers and planners to be a notorious grumbler.

All in all, expected symptoms of centralised scheduling could be observed. Planners and schedulers are (at least partly) overloaded when attempting to control the whole manufacturing process. Hence they mostly decide reactively. Proactive deciding could be observed - if at all - in attempts to gain time by introducing buffers. The shop floor supervisors On the other hand, show reactive behaviour too. Most of them do not even see it as their task to schedule but only to execute prescribed schedules. Those who try to act in a situated fashion lack the information required to do so.

19.5.3.4 Human-computer interaction

On the level of human-computer interaction it must be stated that the implemented ERP-system is mainly used as an information system. There are no algorithms for planning and scheduling. In fact, the system does make proposals on jobs to be released. This process is under human control, due to the planners' competence to freely set the corresponding parameters. The control is restricted not by the technology but by the fact that the interrelations of the parameters and the variables are not really clear to anybody and the system does not explicitly help to make it clearer. However, planners as well as schedulers say that the system provides them with a better information base than they had before. Anyway, the schedulers have pointed out two main problems regarding information presentation. First, it is quite difficult to identify the real-world relations in the system's abstract representations. It happens that the same raw material or the same component is used in different products. When scheduling according to the ID-number-based proposals of the system, one must have a deep knowledge of the products in order to be able to realise that different IDs might refer to the same or to similar work pieces, that are better produced together. This causes problems for the schedulers: 'It is an unpleasant situation for me to release two identical jobs within a short time, or to slow down a running job in order to let a similar job catch up. Sometimes I let the first job go through and delay the second one in order to prevent bad blood.' The second deficiency refers to the clarity of the information presented. One has often to navigate through many windows in order to gather the information required to take a certain decision. Therefore the schedulers use a lot of additional instruments where they collect information in an arranged way (e.g. tables). They even still use the planning-board they planned with before they had the system. On that board they stick all the jobs after having printed the corresponding documents. At a glance they can see what jobs are released, and where they are in the manufacturing process. The schedulers update this board after every round they make through the factory.

The shop floor supervisors use the system both to feedback states of jobs (e.g. in progression, completed) and also as a source of information. The type of information they look for in the system mainly concerns the availability of material as well as job states in preceding units. But this information is not really used to plan and schedule proactively, rather it is used to check whether any disturbances in the schedulers' plans are to be expected. If that is the case, the next job in the list is manufactured and the schedulers are informed. However, that kind of active information seeking has been observed infrequently. Many of the shop floor supervisors when asked, did not really know how to find the relevant information in the system.

To summarise, it can be stated that the ERP-system provides the planners and supervisors with information and does not restrict human control over planning and scheduling processes. However, information presentation needs improvement. On the shop floor level the system's information potential is not really used. Here, the system does not explicitly promote transparency regarding the scheduling process, local regulation of variances and disturbances, and direct lateral co-ordination between production units. It represents instead a clearly structured data-base.

19.6 CONCLUSIONS

The findings of the case study show an absence of autonomy-oriented design of temporal planning and scheduling systems understood as secondary work systems. The organisation of the planning and scheduling process, the assignment of planning tasks to individuals as well as the human-computer interaction mainly follow the goal of putting single persons into control positions. Distribution of autonomy and control, considered to increase an organisation's competence in coping with uncertainty, are still not envisioned. What is required are concepts for organisational as well as for technical support of planning and scheduling processes that locally empower situated acting. The secondary work system for planning and scheduling should be considered as a network of people's brains that has to be supported in its capability to flexibly adapt to dynamic situations. Organisation as well as technical support of information flow and decision-making should aim at interconnecting people's creativity. Instead, organisation kills creativity by centralising autonomy and control, whilst technical support - by the means of clearly structured abstract databases - does not really interconnect people, but terminals. Integrated technical and organisational design solutions which aim at realising autonomy-oriented planning and scheduling systems still remain to be achieved.

19.7 REFERENCES

Alioth, A. (1980). *Entwicklung und Einführung alternativer Arbeitsformen.* (*Development and Implementation of Alternative Ways of Work.*) Bern: Huber.

Bailey, R.W. (1989). *Human Performance Engineering* (2nd Edn). London: Prentice-Hall International.

Bainbridge, L. (1982). Ironies of automation. In G. Johannsen and J.E. Rijnsdorp (Eds.), *Analysis, Design and Evaluation of Man-machine Systems*. Oxford: Pergamon Press, pp. 129-135.

Bandura, A. (1982). Self-efficacy mechanism in human agency. *American*

Psychologist, **37**, pp. 122-147.

Battmann, W. (1984). Regulation und Fehlregulation im Verhalten IX: Entlastung und Belastung durch Planung. (Regulation and mis-regulation in behaviour IX: Relieving and loading through planning.) *Psychologische Beiträge*, **26**, pp. 672-691.

Blumberg, M. (1988). Towards a new theory of job design. In W. Karwowski, H.R. Parsaei and M.R. Wilhelm (Eds), *Ergonomics of Hybrid Automated Systems I.* Amsterdam: Elsevier, pp. 53-59.

Bossink, G. (1992). *Planning and Scheduling for Flexible Discrete Parts Manufacturing.* Twente: Universiteit.

Brehmer, B. (1993). Cognitive aspects of safety. In B. Wilpert and T. Qvale (Eds), *Reliability and Safety in Hazardous Work Systems.* Hove: Lawrence Erlbaum, pp. 23-42.

Clegg, C., Ravden, S., Corbett, M. and Johnson, G. (1989). Allocating functions in computer integrated manufacturing: a review and a new method. *Behaviour and Information Technology*, **8**, pp. 175-190.

Clegg, C., Older, M. and Waterson, P. (1995). A tool for allocating tasks between humans, and between humans and machines. Internal Report, Institute of Work Psychology, University of Sheffield.

Corbett, J.M. (1985). Prospective work design of a human-centred CNC lathe. *Behaviour and Information Technology*, **4**, pp. 201-214.

Dahme, C. (1997). *Systemanalyse Menschlichen Handelns.* (*Systems Analysis of Human Behaviour.*) Opladen: Westdeutscher Verlag.

Dunckel, H., Volpert, W., Zölch, M., Kreutner, U., Pleiss, C. and Hennes, K. (1993). *Leitfaden zur Kontrastiven Aufgabenanalyse und -gestaltung bei Büro- und Verwaltungstätigkeiten. Das KABA-Verfahren.* (*Guidelines for Contrastive Analysis and Design of Office and Administrative Work. The KABA-method.*) Zürich: Verlag der Fachvereine; Stuttgart: Teubner.

Emery, F. (1959). Characteristics of socio-technical systems. *Document No. 527.* London: Tavistock.

Fischer, K. (1993). *Verteiltes Kooperatives Planen in einer flexiblen Fertigungsumgebung.* (*Distributed Co-operative Planning in a Flexible Production Environment.*) Sankt Augustin: infix.

Fitts, P.M. (Ed.) (1951). *Human Engineering for an Effective Air-navigation and Traffic-control System.* Washington, DC: NRC.

Fitzgerald, L.A. and Eijnatten, F.M. (1998). Letting go for control: The art of managing in the chaordic enterprise. *International Journal of Business Transformation*, **1**(4), pp. 261-270.

Frei, F., Hugentobler, M., Schurman, S., Duell, W. and Alioth, A. (1993). *Work Design for the Competent Organization.* Westport, Conn.: Greenwood Press.

Grote, G. (1993). *Schneller, besser, anders kommunizieren.* (*Communicate Faster, Better, Differently.*) Zürich: vdf Hochschulverlag.

Grote, G. (1997). *Autonomie und Kontrolle. Zur Gestaltung automatisierter und risikoreicher Systeme.* (*Autonomy and Control. The design of automated, risky systems.*) Zürich: vdf Hochschulverlag.

Grote, G., Weik, S., Wäfler, T. and Zölch, M. (1995). Criteria for the complementary allocation of functions in automated work systems and their use in simultaneous engineering projects. *International Journal of Industrial Ergonomics*, **16**, pp. 367-382.

Grote, G., Wäfler, T., Ryser, C., Weik, S., Zölch, M. and Windischer, A. (1999). *Wie sich Mensch und Technik sinnvoll ergänzen. Die _Analyse automatisierter Produktionssysteme mit KOMPASS.* (*How Humans and Technology Complement. The analysis of automated work systems with KOMPASS.*) Zürich: vdf: Hochschulverlag.

Grote, G., Ryser, C., Wäfler, T., Windischer, A. and Weik, S. (2000). KOMPASS: A method for complementary function allocation in automated work systems. In John McCarthy, Liam Bannon and Enda Fallon (Eds), *The Role of Function Allocation in Design.* Special Issue of the International Journal of Human Computer Studies, pp. 267-287.

Hacker, W. (1998). *Allgemeine Arbeitspsychologie.* (*General Work Psychology.*) Bern: Huber.

Hacker, W., Heisig, B., Hinton, J., Teske-El Kodwa, S. and Wiesner, B. (1994). *Planende Handlungsvorbereitung* (*Forschungsberichte Band 3*). (*Planning in Action Preparation.*) Dresden: Institut für Allgemeine Psychologie und Methodenlehre der Technischen Universität.

Hafen, U. (1999). Ansätze zur nachhaltigen Reorganisation in der Unternehmenslogistik von KMU. (Approaches to Sustainable Reorganisation in Logistics of SMEs.) Dissertation ETH Zürich.

Hayes-Roth, B. and Hayes-Roth, F. (1979). A cognitive model of planning. *Cognitive Science,* **3** (4), pp. 295-310.

Hoc, J.-M. (1988). *Cognitive Psychology of Planning*. London: Academic Press.

Hollnagel, E. and Cacciabue, P.C. (1999). Cognition, technology and work: An introduction. *Cognition Technology and Work,* **1**(1), pp. 1-6.

Jordan, N. (1963). Allocation of functions between man and machines in automated systems. *Journal of Applied Psychology*, **47**, pp. 161-165.

Kraiss, K.-F. (1989). Autoritäts- und Aufgabenteilung Mensch-Rechner in Leitwarten. (Allocation of authority and tasks to humans and computers.) In Gottlieb Daimler- und Karl Benz-Stiftung (Hrsg.), *2. Internationales Kolloquium Leitwarten*. Köln: Verlag TÜV Rheinland, S. 55-67.

Luethi, H.-J., Ulrich, H. and Duerig, W. (1996). Innovation by simulation using the example of an automated work-cell. *Central European Journal for Operations Research and Economics*, **4**, pp. 135-154.

MacCarthy, B. and Wilson J. (2001). Human Performance in Industrial Scheduling: A framework for understanding. *Journal of Human Factors and Ergonomics in Manufacturing*. In press.

McKay, K.N. (1987). Conceptual Framework for Job Shop Scheduling. Master Thesis. Waterloo, Canada: University of Waterloo.

McKay, K.N. (1992). Production Planning and Scheduling: A Model for Manufacturing Decisions Requirement Judgement. PhD Thesis. Waterloo, Canada: University of Waterloo.

Miller, G.A., Galanter, E. and Pribram, K.H. (1960*). Plans and the Structure of Behavior.* London: Holt, Rinehart and Winston.

Older, M.T., Waterson, P.E. and Clegg, C. (1997). A critical assessment of task allocation methods and their applicability. *Ergonomics*, **40** (2), pp. 151-171.

Orton, J.D. and Weick, K.E. (1990). Loosely coupled systems: A reconceptualization. *Academy of Management Review*, **15**, pp. 203-223.

Resch, M. (1988). *Die Handlungsregulation geistiger Arbeit. (Action Regulation in Mental Work.)* Bern: Verlag Hans Huber.

Rice, A.K. (1958). *Productivity and Social Organization: The Ahmedabad experiment*. London: Tavistock.

Robson, C. (1993). *Real World Research*. Oxford: Blackwell Publishers.

Sanderson, P.M. (1989). The human planning and scheduling role in advanced manufacturing systems: An emerging human factors domain. *Human Factors*, **31**(6), pp. 635-666.

Schank, R.C. and Abelson, R. (1977). *Scripts, Plans, Goals, and Understanding*. Hillsdale, New Jersey: Lawrence Erlbaum Associates.

Schüpbach, H. (1994). *Prozessregulation in rechnerunterstützten Fertigungssystemen*. (*Process Regulation in Computer-aided Production Systems*.) Zürich: vdf Hochschulverlag, pp. 153-171.

Schüpbach, H. (1998). From central planning and control to self-regulation on the shop floor. In Eric Scherer (Ed.), *Shop Floor Control – A systems perspective*. Berlin: Springer.

Sheridan, T.B. (1987). Supervisory control. In G. Salvendy (Ed.), *Handbook of Human Factors*. New York: Wiley, pp. 1243-1268.

Strohm, O. (1996). *Produktionsplanung und -steuerung im Industrieunternehmen aus arbeitspsychologischer Sicht*. (*Production Planning and Scheduling from a Work Psychological Perspective*.) Zürich: vdf Hochschulverlag.

Strohm, O. (1998). Integrated design according to the concept of people, technology and organization. In P. Vink, E.A.P. Koningsveld and S. Dhondt (Eds), *Human Factors in Organizational Design and Management VI*. Amsterdam: Elsevier.

Strohm, O. and Ullich, E. (Eds.) (1997). *Unternehmen arbeitspsychologisch bewerten*. (*Work Psychological Assessment of Organisations*.) Zürich: vdf Hochschulverlag.

Strohschneider, S. (1993). Die Aufrechterhaltung der Handlungsfähigkeit. (Maintaining the ability to act.) In Stefan Strohschneider and Rüdiger Weth von der (Hrsg.). *Ja, mach nur einen Plan. Planen und Fehlschläge - Ursachen, Beispiele, Lösungen*. (*Yes, Just Make a Plan. Planning and Failures – Reasons, examples, solutions*.) Bern: Verlag Hans Huber. pp. 36-50.

Suchman, L.A. (1987). *Plans and Situated Action. The problem of human-machine communication*. Cambridge: Cambridge University Press.

Susman, G.I. (1976). *Autonomy at Work: A Sociotechnical Analysis of Participative Management*. New York: Praeger.

Trist. E.L. (1981). *The Evolution of Socio-technical Systems. Issues in the quality of working life, No.2*. Ontario: Ministry of Labor.

Ulich, E. (1998). *Arbeitspsychologie, 4. Aufl.*. (*Work Psychology, 4th edn*) Zürich: Verlag der Fachvereine; Stuttgart: Poeschel.

Vernon, C. and MacCarthy, B. (1998). *The Effectiveness of Decision Support for the Planning and Scheduling Function - a critical appraisal of contemporary practice in UK manufacturing*. Paper presented at INFORMS/CORS, April 26-29, Montreal, Canada.

Volpert, W. (1994). *Wider die Maschinenmodelle des Handelns. Aufsätze zur Handlungsregulationstheorie. (Against the Machine Model of Acting. Essays on action regulation theory.)* Lengerich: Pabst.

Wäfler, T., Windischer, A., Ryser, C., Weik, S. and Grote, G. (1999). *Wie sich Mensch und Technik sinnvoll ergänzen. Die Gestaltung automatisierter Produktionssysteme mit KOMPASS. (How Humans and Technology Complement. The design of automated work systems with KOMPASS.)* Zürich: vdf Hochschulverlag.

Weick, K.E. (1976). Educational organizations as loosely coupled systems. *Administrative Science Quarterly*, **21**, 1-19.

Weth von der, R. and Strohschneider, S. (1993). Planungsprozesse aus psychologischer Sicht. (Planning processes from a psychological perspective.) In Stefan Strohschneider und Rüdiger Weth von der (Hrsg.), *Ja, mach nur einen Plan. Planen und Fehlschläge - Ursachen, Beispiele, Lösungen.* (*Yes, Just Make a Plan. Planning and failures – reasons, examples, solutions.*) Bern: Verlag Hans Huber. pp. 12-35.

Wilson, J.R. and Rutherford, A. (1989). Mental models: Theory and application in human factors. *Human Factors,* 31, pp. 617-634.

Wiers, V. (1997). *Human-Computer Interaction in Production Scheduling. Analysis and design of decision support systems in production scheduling tasks.* Eindhoven: Technisch Universiteit Eindhoven.

Wiesner, B. (1995). *Diagnostik individueller Planungsprozesse.* (*Diagnostic of Individual Planning Processes.*) Regensburg: Roderer.

Yeatts, D.E. and Hyten, C. (1998). *High-performing Self-managed Work Teams. A comparison of theory and practice.* Thousand Oaks (CA): Sage Publications.

Zölch M. (1997). Aktivitäten der Handlungsverschränkung. (Activities of action co-ordination.) Dissertation: Universität Potsdam.

PART V

Defining the Future Research Domain

CHAPTER TWENTY

Influencing Industrial Practice in Planning, Scheduling and Control

Bart MacCarthy and John Wilson

20. 1 INTRODUCTION

The factors that motivated this book have been discussed in chapter one - the importance of planning and scheduling processes to successful manufacturing enterprises, the importance of adopting a holistic view that incorporates technical, organisational and human dimensions and the need to understand the human contribution. The human role is, in the broadest sense, to manage these processes. The corollary is that best use must be made of people's knowledge, abilities and skills. The studies reported in the preceding chapters show that there is much potential for performance improvement across industrial sectors. It is a rich domain with both extensive research needs and research opportunities that can have a significant impact on industrial practice.

The study of planning and scheduling has the ultimate goal of improving industrial practice. The studies reported here provide a unique platform to develop a reliable body of knowledge that can support systems design and re-engineering. In this chapter we firstly synthesise generic issues of most significance to industry and then outline briefly a research agenda that highlights the most promising areas for future study in terms of industrial relevance. Finally, we discuss how a concerted effort is needed to support World Class planning, scheduling and control processes in the extended enterprise.

20.2 IMPROVING PERFORMANCE

Over the last two decades there have been numerous manufacturing 'revolutions', accompanied by clarion calls for universal adoption of some new paradigm e.g. – Manufacturing Resources Planning (MRPII), Just-In-time (JIT), Optimised Production Technology/Theory of Constraints (OPT/TOC), Flexible Manufacturing Systems (FMS), Total Quality Management (TQM), Lean Manufacturing, Agility, Time-based Competition (TBC), Quick Response Manufacturing (QR/QRM) and Business Process Re-Engineering (BPR). Undoubtedly many of these philosophies have had a significant impact but equally there have been many casualties in their adoption and many failures to achieve desired or anticipated levels of performance improvement. There is now greater realisation across business and industry of the limitations of simplistic 'magic bullet' or 'one size fits all' solutions, characteristic of some of the new manufacturing paradigms. It is clear that effective solutions must address the complexity of environments. A strong theme throughout this book has been the importance of understanding the context in which planning, scheduling and control processes are managed. However, there are generic issues that can impact on performance improvement. Here we identify some of the key issues that should be addressed where performance improvement is desired.

20.2.1 Understanding planning, scheduling and control processes

It is essential to realise that these processes have strongly organisational and social components that are key to achieving formally assigned tasks, responsibilities and goals. To design, re-engineer and support these processes within an organisation the centrality of the human contribution must be acknowledged and the roles held within these processes need to be understood. Equally, the event-driven nature of manufacturing environments, particularly as we move closer to manufacturing operations, must be acknowledged.

20.2.1.1 Roles, responsibilities, authority, autonomy

Across manufacturing sectors, a very wide range of formal positions, duties, roles and responsibilities (given and assumed) are held by personnel involved in planning, scheduling and control processes. They may incorporate aspects of many business functions e.g. sales, planning, engineering, production management, logistics, materials management or supervisory management. Some responsibilities may be very widely distributed e.g. requirements planning. It is evident that the roles and responsibilities held can affect how the process is carried out and its overall effectiveness.

When schedulers are studied in detail it is evident that *schedule facilitation* and *schedule implementation* are equally, if not more dominant in practice than *schedule generation*. Facilitation and implementation activities are concerned with ensuring that the prevailing conditions (e.g. manufacturing resources, materials and personnel, quality issues) enable a desired schedule to be carried out. If a desired schedule cannot be carried out then acceptable alternatives must be found and a strategy put in place to complete work that has not been carried out. Schedulers may bear the ultimate responsibility for schedule implementation. Failure to achieve a desired schedule, for whatever reason, may be very visible. The burden placed on schedulers by these *facilitation and implementation roles* is often very significant – the need to be proactive, to problem-solve, to negotiate, to be ‘in control’, to exploit opportunities that arise.

Deeper analysis of the work of personnel within the planning and scheduling spectrum indicates a range of informational, decisional, interpersonal and networking roles. Thus such individuals or teams can become the *nerve centre of operations*. These ‘making it happen’ roles rely on strongly social and organisational networks. When viewed in this wider context of the realities of people’s jobs we see that such people occupy positions that balance responsiveness, efficiency and stability in manufacturing business. *Schedule managers* and *manufacturing controllers* may be more appropriate designations for some of these roles. The skills and training needed and the selection processes for these roles need careful consideration.

20.2.1.2 Event-driven environments

The reality of many manufacturing environments is that they are dynamic and event-driven and must be managed accordingly. Major sources of disturbances include materials and supply uncertainty, customer-related order uncertainty and production or process uncertainty. Materials and supply uncertainty impacts predominantly at the scheduler level whilst customer-related order uncertainty has more impact at the planning level. Process and production disturbances may be specific to particular environments. Quality issues are major factors that impact on schedules, typically emanating from materials problems and production processes.

Visibility and position in the internal supply chain affects the likely importance and success of event-handling strategies. A key issue in handling unplanned events at the planning level is the need to balance the maintenance of internal company relationships with external customer relationships. The nature and frequency of events a scheduler faces are linked to the responsibilities held and have a strong bearing on the breadth of knowledge they require, the degree of environmental monitoring necessary, the number and type of information sources that need to be sampled and the span of informal communication networks employed. The likelihood of unplanned events also affects the degree of house-keeping undertaken, the scheduling strategies employed, expediting and urgency decisions made, protection policies used and the level of proactivity required. It is important to acknowledge that such unplanned events may be unavoidable and must be managed, and that planners and schedulers need to be trained in their management. Robust schedules and scheduling strategies that give flexibility within time windows may facilitate effective management of unplanned events.

Many planners and schedulers act as *information hubs* for gathering, filtering and disseminating information relevant to real-time operations. As such they are subject to many interruptions and experience many kinds of interactions with other personnel. Such events may not only be viewed positively but as essential for successful operations.

20.2.1.3 Authority, responsibility and autonomy

The levels of responsibility and authority assigned to planners and schedulers need to be carefully considered, in particular at the scheduler level. Assigning formal responsibility for carrying out tasks without ensuring that the individual carries the appropriate level of authority results in a difficult working environment and poor performance is likely. Merely assuming that responsibilities will be matched by the requisite level of authority will not in itself ensure that this is the case.

The work of planners and schedulers has a direct impact on others in the organisation, including production personnel, logistics and procurement personnel, departmental managers, internal customers etc. This raises non-formalised but very real expectations of colleagues, peers and other staff that have to be considered. In order to balance these often competing and difficult expectations, planners and schedulers need strong social and organisational networks. Their perceived status within such networks is important. Again these issues are important in selection, support and training.

It is important that schedulers have sufficient latitude to manage disturbances and to influence the performance measures with which they are targeted. Businesses that dictate prescriptively and precisely what the shop floor should make and when it should be made may not only stifle peoples' initiative but also prevent shop floor personnel from being sufficiently flexible to manage events or balance competing goals such as satisfying urgent orders and achieving efficiency.

Related to this is the issue of autonomy. In general, devolving responsibility for scheduling close to execution is desirable for many reasons. It avoids reliance on the 'in-the-head' knowledge of one person and it enables jobs to be designed that are complete with a degree of autonomy. Effort and complexity can also be reduced for the scheduler by devolving some responsibilities to those who are closer to the problem. However, the degree of latitude requires careful consideration.

20.2.2 Helping planners and schedulers

Accurate and timely information is essential in carrying out planning and scheduling roles successfully. However, as the studies in this book show, a limited technical view of the domain fails to come to terms with reality. Information is critical but IT in itself is not the solution. The entire process must be designed, including IS/IT, and a human-centred perspective is essential. Here we limit our discussion to a re-evaluation of support needs for planners and schedulers and identify problems associated with high levels of compensatory activity.

20.2.2.1 Reinterpretation of support needs - teamwork and training

The view that computer-based scheduling systems can replace human schedulers tends to be based on the premise that schedule generation is the dominant issue. The automation of routine tasks may be useful but does not address the core issues. Even in this area computer-based scheduling has had patchy results. It is apparent from the discussion above that planning and scheduling personnel tend to have far wider remits than purely schedule generation. A technical IT-dominated view fails to address the social and organisational support required by planning managers, planners, schedule managers and manufacturing controllers. A more comprehensive view of support is required that ensures compatibility and linkages between the roles, tasks, expected goals and organisational structures.

Planning and scheduling processes may well involve a team of people. Even where effort is concentrated on one person, interaction with others will be necessary - thus a scheduler on one shift may have to hand over formally to a scheduler on a following shift or may require detailed interaction with a materials controller or a scheduler in another department. Consideration needs to be given as to how to form effective teams and how to support them to achieve shared goals. Roles within a team, along with expected behaviour and performance, need to be clarified. Organisational support may include regular, proactive meetings related to the state of the schedule and the shop floor and regular reviews on performance over a previous period. Organisational learning can be encouraged by providing a channel for communication between planner and scheduler. Where organisational factors hinder communication or promote conflict, then communication channels need to be encouraged.

Environments where the planners and schedulers are encouraged to interact and have a greater understanding of each others constraints, will result in the generation of more realistic plans. When people are in close proximity, information and ideas are exchanged easily. However, when people are geographically separated a mechanism to support information flow needs to be considered explicitly. Geographical dispersion needs to be compensated for using procedures to encourage communication and appropriate technologies need to be put in place to facilitate it. This is especially important in co-ordinating planning and scheduling roles as they are often geographically separated. It is essential that procedures and IT provide a means of communication, negotiation and collaboration. It is also important that existing communication channels, informal as well as formal, are taken into account when reviewing changes to organisational location and structure. Only an understanding of how people *really* work can provide the necessary insight to avoid jeopardising the effectiveness of a system through change.

An important aspect of organisational support is effective selection and training of people for these roles. This is essential for both individuals and teams. By appreciating the scope of these roles their components become more transparent and understandable to other personnel within a business and appropriate people can be selected to fit the requirements and expectations. Effective training will enhance performance. For instance, the social and organisational networks that are essential for effective performance do not come ready-made. This may be particularly important in larger businesses. Identifying the need to develop such networks may be an important aspect of training. A scheduler that is likely to have a strong materials focus may need to be trained in the skills and contacts needed to be successful e.g. developing an understanding of the supply chain operations and developing a network of contacts in the supply departments through job rotation. It may also require that such a scheduler is trained in negotiation and bargaining skills.

20.2.2.2 Reduce compensation activities within planning, scheduling and control processes

The importance of accurate and timely information has already been noted. However, computer-based support systems often fail to provide it and furthermore they often lack the appropriate functionality to support planners and schedulers. The problems associated with legacy systems, particularly in larger organisations, can be a very significant. Planners and schedulers may expend large amounts of time on cleansing data, transferring data between systems and checking data accuracy against numerous sources. Clearly redesign of information systems needs considerable attention in these environments. The latest breed of ERP systems, if properly engineered for the environment, may offer some hope in these respects.

This is an example of a more general issue. The context and environment determine to a large extent the human activity that takes place – redesign the environment and the required human contribution may change. The human ability to adapt in order to manage within the environment is unquestioned but when and how much adaptation should be necessary? We need to distinguish between the level of *human effort* involved in schedule generation, facilitation and event/disturbance management and the *human compensation* required to handle and interpret raw information, to account for data inaccuracy, to make up for poor systems functionality or lack of integration, to work around lack of co-operation between departments, to manage in the face of persistent quality problems or persistent late materiel delivery problems. A considerable level of such non-value-added, compensatory activity is apparent in some environments. Failure to address the core problems means that planners and schedulers are deflected from their main focus to the detriment of the business.

The production planning and control system can be audited by examining the informal systems in place. Informal systems should exist only when they are complementary to the formal system and not contradictory or compensatory. Reviewing the informal system can provide a means of performing a health check of the production planning and control system. Excessive sampling of information points and sampling of more information points than is necessary can indicate deficiencies in the formal information process or can indicate that there is a control issue in the organisation generating additional work. For example, a scheduler may be forced to sample information and make decisions, neither of which are part of his role, due to organisational deficiencies or the ineptitude of others. Duplication of effort, particularly in larger organisations, should also be explored. It may indicate ineffectual formal systems that require review and overhaul.

20.2.3 Operations strategy

Business policy and operations strategy influence the environment in which planning, scheduling and control processes take place and how they are performed. Here we discuss briefly two important issues – responsiveness policy and performance measurement.

20.2.3.1 The level of responsiveness

Essentially this is a strategic issue – how responsive should a business be to market demands or specific customers within a market and how can a business as a whole potentially realise the desired level of responsiveness? Lack of agreement across an organisation on the appropriate level of responsiveness is potentially a source of organisational conflict. The consequences of a responsiveness policy need to be understood. Relying on the planning and scheduling process to provide the flexibility the business needs without acknowledging the realities and limitations of the business may lead to stress on the personnel involved and ultimately to poor business performance. A highly responsive organisation may need to be designed quite differently than a less responsive one e.g. by the maintenance of inventory buffers. Businesses need to generate consensus on this strategic issue and need to have in place appropriate mechanisms to support it across the critical business interfaces – between sales/marketing and planning, between planning and scheduling, and between scheduling and production/shop floor management.

Timely communication across these interfaces is key to successful dynamic operations with front office personnel understanding real production constraints and planners and schedulers taking on board the need for quick response when necessary. Power and authority relationships may impact on what actually happens in practice. In some companies front office functions that deal directly with customers may have more authority than those closer to the operational side of the business. This can also manifest itself in front office personnel's attitudes towards the shop floor and its information requirements. For the effectiveness of the business as a whole the front office functions need to respond as quickly as possible to the requests for information from planning, scheduling or production. In some businesses the reverse situation occurs and production personnel must be encouraged not to place unnecessary restriction on production resources.

20.2.3.2 Performance measurement

This is a critical, sensitive and difficult area for planning, scheduling and control processes and is identified as an important area for research. Here we note some general issues.

Performance measures need to be designed holistically, ensuring that they are in line with, and represent, business objectives, that financial and non-financial measures are well integrated and that they support continuous improvement. These ideas are not new to those familiar with current thinking on performance measurement. Yet many businesses have limited appreciation of the impact of poorly designed measures. Such measures drive behaviour contrary to management desires and expectations and are ultimately detrimental to the business performance. Merely identifying specific production planning and control measures such as due date compliance or throughput volume and using them in isolation as a measure of individual, team or process performance is likely to prove ineffective.

Personnel themselves need to be involved in designing holistic measures that truly reflect performance. Trade-offs need to be agreed between key business objectives up front at an appropriate managerial level. Expectations, both implicit and explicit, that the production scheduler is expected to achieve need to be understood and defined. The linkages between formal and informal performance measures, and task and role behaviour and performance need to be understood by planners, schedulers, the production control department as a whole and the personnel who interact with them. Performance appraisal of individuals and teams may need to be divorced from formal performance measures of processes.

20.3 A RESEARCH AGENDA

A dominant underlying theme throughout this book has been the centrality of planning, scheduling and control processes to successful manufacturing operations. They are needed to manage complex environments, handle hard and soft constraints, manage dynamic events, and deliver good solutions efficiently, consistent with business goals. Organisational support must facilitate communication, teamwork and skill development. Information and decision systems must be capable of providing accurate and timely information that facilitates rescheduling, updating and flexible human interaction.

Here we discuss the way forward for an integrated research effort that will have industrial relevance in the future. A concentrated multidisciplinary research effort is needed and there is wide scope for a number of disciplines to contribute. Addressing the gaps in knowledge is a priority. The research domain extends to supply networks, distribution channels and logistic systems. International comparisons can be particularly valuable.

20.3.1 Research goals

The competitive environments in which firms operate were noted in chapter 1 and highlighted in a number of the case studies. Five research goals can be identified that should provide a backdrop for research to underpin future planning, scheduling and control processes:

1. Flexible planning, scheduling and control processes are needed where human management and decision-making and computer-based information and decision support systems are integrated within supportive organisational structures. The philosophy should not be to replace the skills of people but to better understand their strengths and weaknesses and utilise their abilities to best effect.
2. Context-sensitive frameworks need to be developed for the optimal design and organisation of planning and scheduling functions with integrated human involvement in modern manufacturing environments.
3. Performance measurement and benchmarking approaches are needed for planning and scheduling processes that facilitate enterprise goals.
4. Approaches are needed for selection, job design and training of planners, schedulers and associated personnel.
5. Approaches are needed that can improve practice in planning and scheduling.

20.3.2 A framework for research

A research framework to underpin this field, with a structured and very detailed set of research questions has been described by MacCarthy *et al.* (2001). The framework provides a basis for considering the research questions that are central to developing our understanding of the human contribution to planning and scheduling in the context of contemporary manufacturing practice. We do not repeat the discussion here but very briefly note the approach, which focuses on three areas:

1. understanding the *context* within which planning, scheduling and control processes take place;
2. understanding the *processes* themselves, specifically how the activities are carried out and what influences them;

3. understanding *performance*, specifically how well they are carried out and what influences performance.

The number and complexity of research issues within this framework have been evident throughout the book. A concerted research effort can lay the basis to improve practice across industrial sectors. Research design in this domain requires care if data are to be of value in supporting strong and robust conclusions. Industrial field studies designed to observe, to analyse and to understand actual practice and performance are central to the goals above. However, as was evident in a number of studies, conducting this type of research is difficult and raises fundamental methodological issues.

20.3.3 Conducting field studies with industry

Field study investigations that are not geared to address the central questions of relevance to contemporary manufacturing environments are unlikely to add to the state of useful knowledge. Field research needs to be driven by research questions and supported by valid methodologies. Research methods need to be practical, robust and supportable. We note some of the difficulties.

Planning, scheduling and control are spatially and temporally distributed activities that are necessarily difficult and time-consuming to study. This kind of research may be disruptive and inconvenient to some degree for the organisation participating in the study. Issues of commercial confidentiality may have to be addressed, as planning and scheduling are at the heart of an organisation's real operational systems. There may be difficulties in identifying and selecting the appropriate staff for study or in obtaining a representative selection across a large organisation with complex planning and scheduling systems. For comparative purposes it may be beneficial to compare planning and scheduling activities across related facilities in different parts of an organisation, but this is not always feasible. Pilot studies in which research methods can be practised, refined and used with confidence are always valuable. Crawford *et al.* (1999) discuss these issue in greater depth and describe a range of methods for observation, probing and analysis that have been tested in the field.

20.3.4 Management science and other disciplines

Although we have emphasised field-based empirical research in order to make strong advances in the body of knowledge, there are developments in other domains that are of relevance to planning, scheduling and control. Management Science and Operational Research in particular have long concentrated on the modelling and algorithmic aspects of generating sequences and schedules. There are some areas within these more theoretical fields that may be of relevance to the process view of planning, scheduling and control advocated here. Complexity theory, distributed decision-making, multi-level and multi-criteria decision-making are noted in MacCarthy *et al.* (2001) as having potential to planning and scheduling in practice. In computer science, both constraint-based and case-based reasoning approaches are of relevance, as are some aspects of modern control theory.

20.4 WORLD CLASS PLANNING, SCHEDULING AND CONTROL PROCESSES

The term 'World Class Manufacturing' came into general usage with the work of Schonberger (1986, 1996), although some of the ideas can be traced back earlier. The concept has found a ready audience across manufacturing sectors over the last decade but with disagreements over definition, interpretation and implementation. Naïve interpretations and simplistic comparisons are not useful and can be damaging. We lack strong empirical evidence from which any kind of general theory can be advanced. Not surprisingly perhaps, there have been difficulties in 'operationalising' the concept for specific business processes. Notwithstanding this lack of clarity, theory and practice suggest that certain concepts are particularly relevant in the world class paradigm: an emphasis on competencies, teamwork and empowerment; on practices rather than slavish adherence to performance metrics; on continuous improvement and business 'excellence'; on achieving customer-focus with goals emanating from within the business.

There is strong interest in the concept across industrial sectors. Can we articulate the world class concept in the context of planning, scheduling and control? Can we create benchmarks for world class planning and scheduling? Can we draw a routemap to guide us towards world class performance in planning, scheduling and control and identify the practices that are capable of sustaining it?

The importance of responsiveness has been noted in Section 20.3 above. It is becoming a core competence in many sectors. Achieving the requisite responsiveness level in the extended enterprise is likely to be an important component of World-Class environments. The central role of people in these processes is also likely to be emphasised. A major research effort is needed to focus on the characteristics of responsive performance: the human contribution, roles and responsibilities, support needs and responsive teams; performance measurement and performance monitoring; and world class practices. The critical interfaces and linkages within manufacturing and the initiatives and enablers that can accelerate beneficial change need to be identified.

It can be argued that the world class concept is essentially a 'state-of-corporate-mind' and refers to a journey rather than a destination. This book provides an ideal starting point to begin the journey with respect to the critical processes of planning, scheduling and control.

20.5 REFERENCES

Crawford, S., MacCarthy, B.L., Vernon, C. and Wilson, J.R. (1999). Investigating the work of industrial schedulers through field study. *Cognition, Technology and Work*, 1, 63-77.

MacCarthy, B.L., Wilson, J.R. and Crawford, S. (2001). Human performance in industrial scheduling: a framework for understanding, *International Journal of Human Factors and Ergonomics in Manufacturing.* In press.

Schonberger, Richard J. (1986). *World Class Manufacturing: the Lessons of Simplicity Applied.* Free Press, New York Collier; Macmillan, London.

Schonberger, Richard J. (1996). *World Class Manufacturing: the Next Decade: Building Power, Strength, and Value*, Free Press, New York.

Key Word Index

This key word index has been compiled by the editors. The page numbers refer to the first page of the relevant chapter.

Advanced Planning and Scheduling (APS) 15

Business process re-engineering (BPR) 451

Classifications and taxonomies 3, 15, 45, 165, 201, 311, 339
Communication 67, 83, 135, 217, 231, 245, 281, 311, 339, 411, 451
Complexity 3, 15, 45, 67, 135, 179, 201, 217, 245, 311, 339, 355, 383, 383, 451
Costs 3, 15, 45, 135, 179, 201, 217, 231, 281, 311, 355, 411

Data and information issues in planning and scheduling
accessibility 15, 311
accuracy, checking, reliability, verification 15, 45, 67, 83, 135, 217, 311, 451
dynamic, real time 45, 67, 135, 165, 201, 231, 311, 451
flow and feedback 45, 135, 179, 311, 339, 383, 411
gathering, searching for, recording 15, 67, 135, 201, 217, 231, 311, 451
managing, processing, using 15, 45, 67, 83, 135, 165, 231, 245, 281
object-oriented 45, 231, 245
systems 3, 15, 45, 83, 135, 165, 179, 231, 245, 311, 383
types of data and information used 15, 45, 67, 83, 135, 165, 217, 231, 245, 281, 311, 339, 383, 411
Decision-making
aiding 15, 45, 165, 245
automated 15, 245
behaviour 15, 83, 135, 245
centralised, de-centralised, distributed, devolved 3, 45, 67, 83, 135, 165, 179, 245, 339, 383, 411
factors affecting 15, 67, 135, 165, 179, 383
human 3, 15, 45, 67, 135, 165, 179, 245, 281, 383
methods to study 15, 135
types of, nature of 15, 45, 67, 105, 135, 165, 179, 245, 281, 383
Decision Support Systems (DSS)
description of, design of 15, 45, 135, 165, 179, 201, 217, 231, 245, 281, 383
knowledge-based systems 15, 165, 201
knowledge acquisition, elicitation 15, 45, 135, 201, 217
Demand and demand management 3, 15, 45, 135, 201, 245, 281, 311, 383
order fulfillment processes 3, 15, 67, 83, 135, 165, 179, 201, 231, 339, 355, 383
Disturbances, events, exceptions, interruptions
general 15, 67, 83, 105, 135, 165, 179, 201, 231, 245, 311, 339, 411, 451
recovery from, managing, trouble-shooting 15, 45, 83, 135, 165, 201, 217, 411, 451

Enterprise Resource Systems (ERP) 3, 15, 45, 201, 231, 411, 451
Expertise, knowledge and skills 3, 15, 45, 83, 135, 165, 179, 217, 217, 231, 245, 281, 311, 383

Forecasting, prediction 83, 135, 165, 217, 231, 245, 281, 311, 339, 355

Future research in planning and scheduling 15, 45, 83, 105, 165, 217, 311, 339

Human Factors
general 3, 15, 179, 217, 245, 311, 339, 383
of planning and scheduling 3, 15, 45, 67, 135, 179, 201, 217, 231, 245, 411, 451
Human-computer interaction (HCI) 217, 231, 245, 281
displays and presentation 15, 135, 201,
Gantt charts 15, 201, 231, 245
general 15, 135, 165, 179, 245, 281, 411
methods, models, needs 245, 281

Industries
aerospace 3
automotive 15, 45, 311, 355
brewing 339
chemicals 339
clothing and textiles 15, 135
electronics 15, 311
engineering 15, 67, 83, 179, 231, 383
general, operations 3, 15, 45
oil 339
paper 339
pharmaceuticals 339
plumbing components (furniture) 411
printing 15, 245
process 45, 201, 281, 339
steel, metals 15, 165, 217, 339
Inventory, stock
buffers, work in progress 3, 45, 67, 83, 105, 135, 179, 201, 217, 231, 311, 339, 355, 383, 411, 451
general 45, 105, 135, 231, 281, 355, 383, 411

Logistics 3, 15, 83, 217, 355

Materials 3, 15, 45, 67, 83, 135, 165, 217, 231, 311, 355
Materials Requirements Planning (MRP, MRPI and II) 3, 15, 45, 83, 105, 135, 179, 231, 281, 311, 383, 411, 451
Models and modelling
general 15, 135, 165, 179, 201, 217, 231, 245, 339, 383, 411
mental 15, 83, 135, 245, 411
of planners and planning 15, 179, 339, 411
of schedulers and scheduling 15, 45, 83, 165, 245, 311, 411

Operators 3, 15, 67, 135, 165, 179, 201, 231, 281, 311, 383
Optimality and optimization 3, 15, 45, 135, 165, 179, 201, 217, 245, 281, 311, 339, 355, 383, 411

Performance
criteria, goals 15, 165, 201, 217, 245, 311, 451
factors affecting 3, 15, 45, 67, 83, 105, 135, 179, 245, 311, 355, 383, 451
human 3, 15, 411
improving 3, 15, 105, 217, 245, 451
in planning 3, 105, 411
in scheduling 15, 67, 83, 105, 217, 231, 311, 339
measuring 3, 15, 45, 83, 105, 135, 165, 179, 217, 231, 245, 311, 339, 451
World Class 355, 451
Planners
authority, autonomy, responsibilities, roles 45, 135, 179, 281, 339, 411, 451
behaviour 15, 105, 135, 179, 339, 411
tasks done by 15, 135, 165, 179, 281, 311, 339, 411
training and selection 15, 451
Planning
capacity 15, 83, 165, 201, 281, 311, 339, 383
defining 15, 105, 179, 311
factors affecting 15, 45, 105, 411
hierarchical, levels of, feedback to 15, 45, 105, 135, 281, 339, 383, 411
human contribution to 67, 135, 217, 281, 339, 355, 411
methods of, process of, approaches to 67, 83, 105, 135, 165, 179, 201, 217, 281, 339, 383, 411
support for 15, 45, 217, 281, 339, 383, 451
systems, tools 135, 231, 281
Product design, product mix and variety 3, 15, 45, 67, 105, 135, 179, 201, 217, 311, 339, 355, 383

Production management
production control 45, 83, 135, 165, 245, 281, 383
production managers 3, 135, 217, 311
Productivity, throughput 3, 135, 217, 311, 339, 355, 383

Quality related issues 3, 15, 45, 83, 135, 201, 217, 245, 281, 311, 355

Research methodology and methods
analysis methods 15, 83, 105, 135, 311, 339
case studies and case study approach 15, 45, 67, 83, 135, 165, 179, 201, 217, 231, 245, 311, 339, 383, 411, 451
data gathering and observation 67, 83, 135, 217, 231, 245, 311, 355, 451
emergent design 135
empirical data 3, 15, 83, 105, 201, 451
methodological issues 15, 45, 83, 105, 135, 311, 451
qualitative approaches/methods 45, 83, 135, 311, 355, 451
questionnaires, surveys, interviews 15, 355
statistical, regression and simulation techniques 3, 15, 339
the study of planners and schedulers 3, 15, 67, 83, 135, 451
Resources 3, 15, 45, 67, 105, 201, 281, 311, 339, 355, 411

Schedulers
attributes, capabilities, competencies 45, 83, 135
authority, autonomy, responsibilities, roles 3, 15, 45, 83, 105, 135, 179, 201, 245, 339, 355, 383, 411, 451
behaviour 15, 45, 83, 135, 245, 311, 411
tasks done by 15, 45, 67, 83, 105, 165, 179, 201, 231, 339, 411
training and selection 15, 83, 451
Scheduling
about the study of 3, 15, 45, 67, 83, 201, 217, 245, 311, 451
algorithms 15, 45, 135, 165, 179, 201, 217, 231, 245, 311, 339, 411
approaches to, methods of, techniques and rules for 3, 15, 67, 83, 105, 135, 165, 179, 201, 217, 231, 245, 311, 339, 355, 411
on defining, meaning of, place of 3, 15, 83, 105, 165, 217, 245, 311, 339, 411
factors affecting, constraints on 15, 67, 83, 105, 135, 165, 179, 201, 217, 245, 311, 339, 355, 411
human contribution to 15, 45, 67, 83, 105, 135, 165, 179, 201, 217, 245, 339, 411
rescheduling 105, 135, 179, 217, 231, 311, 411
sequencing 45, 67, 83, 105, 135, 179, 201, 217, 231, 245, 281, 339, 355
support for 3, 45, 67, 83, 105, 135, 165, 201, 217, 231, 245, 383, 411, 451
systems and tools 3, 15, 45, 67, 83, 105, 135, 165, 179, 217, 231, 245
theory 3, 15, 45, 67, 105, 135, 201, 217, 231, 245, 311, 411
Simulation 217, 231, 339, 411
Socio-technical Systems
general 45, 165, 201, 245, 383, 411
in planning, scheduling and control 383, 411
principles and theory 383, 411
Supervisors 15, 105, 179, 281
Supply chains 3, 15, 165, 355
suppliers 3, 15, 45, 355

Teams, teamwork and work organisation 3, 15, 83, 105, 105, 179, 355, 383, 411, 451
Types of manufacturing and types of manufacturing systems
agile manufacturing 451
assembly lines, flow lines 355
assembly processes, systems 15, 45, 83, 105, 281, 355, 383, 411
batch manufacturing 45
cellular manufacturing 45, 83, 383
Computer-Numerical Control (CNC) 67,
Flexible Manufacturing Systems (FMS) 179, 311, 451
flow shop 3, 105
job shop 3, 45, 105, 339
Just in Time (JIT) and Lean Manufacturing 3, 67, 355, 383, 451

Make to Order, Build to Order 67, 135, 231, 355, 383
Make to Stock 67, 231, 383
manufacturing technologies 3, 15, 201
Optimised Production Technology (OPT) 45, 383, 451
project manufacturing, Engineer to Order 231
responsive manufacturing 135, 355, 451